Communications
in Computer and Information Science   1952

Series Editors

Gang Li, *School of Information Technology, Deakin University, Burwood, VIC, Australia*
Joaquim Filipe, *Polytechnic Institute of Setúbal, Setúbal, Portugal*
Zhiwei Xu, *Chinese Academy of Sciences, Beijing, China*

Rationale

The CCIS series is devoted to the publication of proceedings of computer science conferences. Its aim is to efficiently disseminate original research results in informatics in printed and electronic form. While the focus is on publication of peer-reviewed full papers presenting mature work, inclusion of reviewed short papers reporting on work in progress is welcome, too. Besides globally relevant meetings with internationally representative program committees guaranteeing a strict peer-reviewing and paper selection process, conferences run by societies or of high regional or national relevance are also considered for publication.

Topics

The topical scope of CCIS spans the entire spectrum of informatics ranging from foundational topics in the theory of computing to information and communications science and technology and a broad variety of interdisciplinary application fields.

Information for Volume Editors and Authors

Publication in CCIS is free of charge. No royalties are paid, however, we offer registered conference participants temporary free access to the online version of the conference proceedings on SpringerLink (http://link.springer.com) by means of an http referrer from the conference website and/or a number of complimentary printed copies, as specified in the official acceptance email of the event.

CCIS proceedings can be published in time for distribution at conferences or as post-proceedings, and delivered in the form of printed books and/or electronically as USBs and/or e-content licenses for accessing proceedings at SpringerLink. Furthermore, CCIS proceedings are included in the CCIS electronic book series hosted in the SpringerLink digital library at http://link.springer.com/bookseries/7899. Conferences publishing in CCIS are allowed to use Online Conference Service (OCS) for managing the whole proceedings lifecycle (from submission and reviewing to preparing for publication) free of charge.

Publication process

The language of publication is exclusively English. Authors publishing in CCIS have to sign the Springer CCIS copyright transfer form, however, they are free to use their material published in CCIS for substantially changed, more elaborate subsequent publications elsewhere. For the preparation of the camera-ready papers/files, authors have to strictly adhere to the Springer CCIS Authors' Instructions and are strongly encouraged to use the CCIS LaTeX style files or templates.

Abstracting/Indexing

CCIS is abstracted/indexed in DBLP, Google Scholar, EI-Compendex, Mathematical Reviews, SCImago, Scopus. CCIS volumes are also submitted for the inclusion in ISI Proceedings.

How to start

To start the evaluation of your proposal for inclusion in the CCIS series, please send an e-mail to ccis@springer.com.

Manish Gupta · Shikha Agrawal ·
Kamlesh Gupta · Jitendra Agrawal ·
Korhan Cengis
Editors

Machine Intelligence and Smart Systems

Third International Conference, MISS 2023
Bhopal, India, January 24–25, 2023
Revised Selected Papers, Part II

 Springer

Editors
Manish Gupta
Amity University Madhya Pradesh
Gwalior, Madhya Pradesh, India

Shikha Agrawal
University Institute of Technology RGPV
Bhopal, India

Kamlesh Gupta
Rustamji Institute of Technology
Gwalior, India

Jitendra Agrawal
School of Information Technology RGPV
Bhopal, India

Korhan Cengis
Trakya University
Edirne Merkez, Türkiye

ISSN 1865-0929   ISSN 1865-0937 (electronic)
Communications in Computer and Information Science
ISBN 978-3-031-69114-0   ISBN 978-3-031-69115-7 (eBook)
https://doi.org/10.1007/978-3-031-69115-7

© The Editor(s) (if applicable) and The Author(s), under exclusive license
to Springer Nature Switzerland AG 2025

This work is subject to copyright. All rights are solely and exclusively licensed by the Publisher, whether the whole or part of the material is concerned, specifically the rights of translation, reprinting, reuse of illustrations, recitation, broadcasting, reproduction on microfilms or in any other physical way, and transmission or information storage and retrieval, electronic adaptation, computer software, or by similar or dissimilar methodology now known or hereafter developed.
The use of general descriptive names, registered names, trademarks, service marks, etc. in this publication does not imply, even in the absence of a specific statement, that such names are exempt from the relevant protective laws and regulations and therefore free for general use.
The publisher, the authors and the editors are safe to assume that the advice and information in this book are believed to be true and accurate at the date of publication. Neither the publisher nor the authors or the editors give a warranty, expressed or implied, with respect to the material contained herein or for any errors or omissions that may have been made. The publisher remains neutral with regard to jurisdictional claims in published maps and institutional affiliations.

This Springer imprint is published by the registered company Springer Nature Switzerland AG
The registered company address is: Gewerbestrasse 11, 6330 Cham, Switzerland

If disposing of this product, please recycle the paper.

Preface

Nowadays, when the world is becoming so ambiguous that it cannot be comprehended by a single individual, information is growing at a tremendous rate, and software systems are becoming uncontrollable, this has inspired computer scientists to design alternative intelligent systems in which control, pre-programming, and centralization are replaced by autonomy, emergence, and distributed functioning. The field of research focused on developing such systems and applying them to solve a wide variety of problems is termed 'machine intelligence'. Machine intelligence is a methodology involving computing that provides a system with an ability to learn and/or to deal with new situations, such that the system is perceived to possess one or more attributes of reason, such as generalization, discovery, association, and abstraction. Since its origin, the number of its successful applications has grown rapidly, and the use of these machine intelligence algorithms has increased. Machine intelligence (MI) techniques are ideal for such applications as tools of 'knowledge discovery from data' or in short 'data to knowledge' for complex and often apparently intractable systems. There is a need to expose academicians and researchers to machine intelligence (MI) and its multidisciplinary applications for better utilization of these techniques and their future development for developing smart systems.

The papers in this proceedings not only deal with MI techniques and their applications, they also try to cover several novel applications of combining MI techniques and utilizing the hybrid forms in different practical areas such as engineering systems used in agriculture, military and civilian applications, manufacturing, biomedical and healthcare systems, as well as education. Equally important, this proceedings aimed to demonstrate successful case studies, identify challenges, and bridge the gap between theory and practice in applying machine intelligence to solving all kinds of real-world problems. Since machine intelligence is a truly interdisciplinary field, scientists, engineers, academicians, technology developers, researchers, students, and government officials will find this text useful in handling their complicated real-world issues by using machine intelligence methodologies and assisting in furthering their own research efforts in this field. Moreover, by bringing together representatives of academia and industry, this proceedings is also a means for identifying new research problems and disseminating the results of research and practice. The main goal of this proceedings is to provide scientific researchers and engineers with a vehicle where innovative technologies for developing smart systems through machine-intelligent techniques are discussed.

The 61 revised papers presented were carefully reviewed and selected from a total of 203 submissions. Double-blind review procedures were used and an average of 2 to 4 reviews were conducted for each paper.

Manish Gupta
Shikha Agrawal
Jitendra Agrawal
Kamlesh Gupta
Korhan Cengis

Organization

General Chair

Korhan Cengiz   Istinye University, Turkey

Conference Chair

Sanjeev Sharma   RGPV Bhopal, India

Conference Co-chair

Pawan Agrawal   SRIIT College, India

Publication Chair

Jitendra Agrawal   RGPV Bhopal, India

Session Chair

Sanjay Agrawal   NITTTR Bhopal, India

Program Chairs

Kamlesh K. Gupta   Rustamji Institute of Technology, India
Shikha Agrawal   RGPV Bhopal, India
Manish Gupta   Amity University, Madhya Pradesh, India

Advisory Committee

Rajkumar Buyya   University of Melbourne, Australia
Sheng-Lung Peng   National Taipei University of Business, Taiwan

Korhan Cengiz	Trakya University, Turkey
Vivek Tiwari	IIIT Naya Raipur, India
Dac-Nhuong Le	Haiphong University, Vietnam
Yogesh Kumar Meena	Swansea University, UK
Arindam Gjendra Mahapatra	Yamaguchi University, Japan
B. Shameedha Begaum	NIT Trichy, India
Rupesh Dewang	MNNIT Allahabad, India
Oscar Castillo	Tijuana Institute of Technology, Mexico
Yu-Chuan Li	Taipei Medical University, Taiwan
Mohammad-R.	Ferdowsi University of Mashhad, Iran
Lokesh Chauhan	NIT Hamirpur, India
Noor Zaman	King Faisal University, Saudi Arabia
Chakchai So-In	Khon Kaen University, Thailand
Milan Tuba	Singidunum University, Serbia
Dhananjay Singh	Hankuk University of Foreign Studies, South Korea
Luis M. Bergasa	University of Alcalá, Spain
Bernard Goossens	University of Perpignan, France
Yao Zhao	Beijing Jiaotong University, China
Ernesto Damiani	Khalifa University, United Arab Emirates

Technical Program Committee

Sandeep Raghuwanshi	Samrat Ashok Technological Institute, India
Bhanu Pratap Soni	Fiji National University, Fiji
Ankit Kumar Sharma	University of Engineering and Management, Jaipur, India
Revathi S.	National Institute of Technology Puducherry, India
Pronaya Bhattacharya	Nirma University, India
Mohit Tiwari	Bharati Vidyapeeth's College of Engineering, India
Elammaran Jayamani	Swinburne University of Technology Sarawak Campus, Malaysia
Pankaj Sharma	IPS College of Technology and Management, India
Sandhi Kranthi Reddy	Vignan Institute of Technology and Science Hyderabad (JNTU Hyderabad), India
Niraj Gupta	Dr. A.P.J. Abdul Kalam Technical University, India
Pratap M. S.	Reva University, India

Varsha Sharma	RGPV Bhopal, India
Pavithra M.	Jansons Institute of Technology, India
Rajwant Singh Rao	Guru Ghasidas Vishwavidyalaya Bilaspur, India
G. Saravanan	Erode Sengunthar Engineering College (Autonomous), India
Harish B. G.	Visvesvaraya Technological University, India
Prajna S.	SJCE Mysuru, India
Subhadip Sarkar	Seacom Skills University, India
Shashidhar R.	JSS Science and Technology University, India
Pulavar Sona Mathialagan	Kurichi Nagapattinam Dist., India
Shikha Agrawal	RGPV Bhopal, India
Chandra Sekhar Kolli	GITAM, India
Ashish Sharma	GLA University, India
Raghunandan K. R.	NMAM Institute of Technology, India
Umesh K. K.	J S S Science & Technology University, India
Shilpa Prashant Kodgire	Maharashtra Institute of Technology, India
Binny S.	Kristu Jyoti College of Management and Technology, India
Sapna Katiyar	Impledge Technologies, India
Prateesh Kumar S.	Anna University, India
Abhishek Dixit	Rajiv Gandhi Proudyogiki Vishwavidyalaya, India
S. Srinivasulu Raju	JNTU Kakinada, India
P. Selvaraj	Sri Venkateswara Engineering College Tirupati, India
Sangram Ray	National Institute of Technology Sikkim, India
M. R. Ramesh	Indira Gandhi Centre for Atomic Research, India
Sushma S. J.	GSSS Institute of Engineering and Technology for Women, India
Maithri C.	Visvesvaraya Technological University, India
P. Gunasundari	Bhaktavatsalam Memorial College for Women, India
T. Manonmani	Mepco Schlenk Engineering College, India
Kapil Sharma	Amity University, Madhya Pradesh, India
S. Jayachitra	PSNA College of Engineering and Technology, India
Rituz Ranvijay Kumar	Alpha Classes, India
Ajesh F.	Sree Buddha College of Engineering, India
Pushpa D.	Maharaja Institute of Technology, Mysore, India
Devidas	NMAM Institute of Technology, India
Karuna Nidhi Pandagre	RGPV Bhopal, India
Navnit Kumar Shukla	M J P Rohilkhand University, India
Swati Sah	Jain University, India

T. Ananth Kumar	IFET College of Engineering, India
S. Sivaranjani	Bon Secours College for Women, India
Karuna Valisammagari	JNTUA, India
M. Manimaran	Sri Krishna Adithya College of Arts and Science, India
Kavyashree M. K.	JSS Science and Technology University, India
Sushmitha Vadone V.	Kuvempu University, India
Anjana Goen	Rustamji Institute of Technology, India
Munish Kumar	Maharaja Ranjit Singh Punjab Technical University, India
Vikas Verma	Kanpur Institute of Technology, India
Srinidhi N. N.	Manipal Institute of Technology Bengaluru, MAHE, India
Charu Jain	Manav Rachna University, India
Sumathi Pawar	NMAMIT, India
Sarika Chaudhary	Amity University Haryana, India
Krishan Kumar	Manav Rachna International Institute of Research and Studies, India
Rashmi Soni	New Horizon College of Engineering, India
Raghav Prasad Parouha	Indira Gandhi National Tribal University, India
Pooja Tiwari	Indira Gandhi National Tribal University, India
Harish Kumar Shakya	Amity University, Gwalior, India
Vikas Thada	Anand International College of Engineering, India
Shweta Mongia	Manav Rachna International Institute of Research and Studies, India
Asif Khan	Integral University, India
Juhi Singh	Amity University, Haryana, India
Arun Kumar Singh	Amity University, Haryana, India
Munish Sabharwal	Galgotias University, India
Nivedita Tiwari	Maharshi Dayanand College, India
Shyam Sunder Gupta	Amity University, Gwalior, India
Dasari Anantha Reddy	Koneru Lakshmaiah Education Foundation (KL University), India
Davinder Paul Singh	Shri Mata Vaishno Devi University, India
Abhishek Vishnoi	Kanpur Institute of Technology, India
Ayesha Taranum	VVCE, India
Krishnaraj Rao N. S.	NMAMIT, India
Sachin Saxena	ABES, AKTU University, India
Gulivindala Suresh	Aditya Engineering College, India
Harshdeep Kaur	Khalsa College for Women, Panjab University, India
Prashant Kumar	Woxsen University, India
Santhosh Krishna B. V.	New Horizon College of Engineering, India

Khajamannanuddin	Sumathi Reddy Institute of Technology for Women, India
Jaimy James Poovely	KTU, India
Ripal Ranpara	Atmiya University, India
Suresh Kaswan	Sharda University, Uzbekistan
Mukesh Choubisa	Indrashil University, India
V. Chandrasekar	Jain University, India
Shamik Tiwari	University of Petroleum and Energy Studies, India
Shiv Kumar Verma	Galgotias University, India
Priyanka Makkar	Amity University, Haryana, India
Sherin Zafar	Jamia Hamdard, India
Ramesh Cheripelli	G Narayanamma Institute of Technology and Science, India
Deepali Virmani	Ggsipu, India
Manoj Eknath Patil	SSBT's College of Engineering and Technology Jalgaon, India
Abhishek Rawat	IITRAM Ahmedabad, India
Prasad Lokulwar	G H Raisoni College of Engineering, India
Nidhi Singh	Gautam Buddha University, India
Kiran Dhanaji Kale	Presidency University Bangalore, India
N. A. Natraj	Symbiosis International, India
Sapna Juneja	KIET Group of Institutions Ghaziabad, India
Nidhi Saxena	Malwa Institute of Science and Management, India
Madhu B.	Dr. Ambedkar Institute of Technology, India
Rajesh G.	New Horizon College of Engineering, India
Saroja B.	Siddartha Institute of Science and Technology, India
Savita Sindhu	Manav Rachna International Institute of Research and Studies, India
Kaptan Singh	Truba Group of Institutes, India
Nancy Arya	Shree Guru Gobind Singh Tricentenary University, India
Taskeen Zaidi	Jain University, India
Ankita Vaish	Banaras Hindu University Varanasi, India
Pooja Chaturvedi	Nirma University, India
Meenu Vijarania	K R Mangalam University, India
Prabhat Kumar	Banaras Hindu University, India
Neeraj Gupta	Amity University, Haryana, India
Rashmi Gupta	Amity University, Haryana, India
Mrityunjay Singh	Indian Institute of Information Technology Una, India
Santosh Kumar Satapathy	PDEU, India

Punit Gupta	University College Dublin, Ireland
Naganna Chetty	A J Institute of Engineering and Technology, India
Manoj Kumar Patra	National Institute of Technology, Rourkela, India
Amit Kumar	Jaypee University of Information Technology, India
Yuvika Gupta	Uttaranchal University, India
Bodhi Chakraborty	ITM University, India
Jyotir Moy Chatterjee	Lord Buddha Education Foundation, India
Naman Garg	Indian Institute of Information Technology, Una, India
Deepak Batham	Madhav Institute of Technology & Science, India
Lakshmi Simhan	Georgia Southern University, USA
Hardeep Singh	IK Gujral Punjab Technical University, India
Gaganpreet Kaur	Chitkara University, India
Sunil Joshi	Samrat Ashok Technological Institute, India
Indrajit De	Institute of Engineering and Management Kolkata, India
Pariza Kamboj	Sarvajanik College of Engineering & Technology, India
Om Prakash	Hemvati Nandan Bahuguna Garhwal University, India
Saurabh Bilgaiyan	KIIT, India
Ashima Gambhir	Amity University Gurgaon, India
Lim Tion Hoo	Universiti Teknologi Brunei, Brunei
Kavi Kumar Khedo	University of Technology, Mauritius
Punit Kanuga	Walmart, India
Kavin K. S.	AB Technologies, India
Neelima K.	Sree Vidyanikethan Engineering College, India
Aayush Shrivastava	Sagar Institute of Science, Technology and Engineering, India
K. Sujatha	Dr. MGR Educational and Research Institute, India
Sumitra Singar	Bhartiya Skill Development University, India
Joshua Abolarinwa	NUST, Namibia
Pawan Kumar Verma	MIT Art, Design & Technology University, India
Ajay Kushwaha	CSVTU, India
Mohammad Husain	Islamic University of Madinah, Saudi Arabia
Anirudh Agarwal	LNM Institute of Information Technology, India
T. Sukumar	Anna University, India
A. Senthil Kumar	Sanskrithi School of Engineering, India
Khadim Moin Siddiqui	Shri Ramswaroop Memorial College of Engineering and Management, India
Chetan Dhule	G H Raisoni College of Engineering, India

Amrit Pal	Vellore Institute of Technology, Chennai, India
Sajjan Kumar	Aditya Engineering College, India
Ankit Gupta	Chandigarh College of Engineering and Technology, India
Bhavana Narain	MSIT-Mats School of Information and Technology, India
Dariusz Jakóbczak	Koszalin University of Technology, Poland
Archana R. Raut	G H Raisoni College of Engineering, India
V. Ananthaswamy	Madura College (Affiliated to Madurai Kamaraj University), India
Panem Charanarur	National Forensic Sciences University, India
Veeraswamy Ammisetty	Koneru Lakshmaiah Educational Foundation, India
Priyamwada Sharma	RGPV Bhopal, India
Rana Mukherji	ICFAI University Jaipur, India
Anshul Gupta	S P Jain School of Global Management, UAE
Nihar Ranjan Roy	Sharda University, India
A. Kanaka Durga	Stanley College of Engineering and Technology for Women, Osmania University, India
Mukesh Soni	Chandigarh University, India
Diwakar Agarwal	GLA University, India
Devendra Bhavskar	JK Lakshmipat University, India
Aliseri Govardhan	Jawaharlal Nehru Technological University Hyderabad, India
Diana Moses	Methodist College of Engineering and Technology, India
Saumendra Kumar Mohapatra	SRM University Sikkim, India
Saravanakumar Kandasamy	Vellore Institute of Technology Vellore, India
Suresh Penagaluru	SV College of Engineering, India
Kasa Sudheer	Sri Venkateswara College of Engineering, India
Abhishek Tiwari	MGCPS, Lucknow, India
Abhishek Choubey	Sreenidhi Institute of Science and Technology, India
Md Habibur Rahman	Islamic University, Bangladesh
Sabyasachi Pramanik	Haldia Institute of Technology, India
Virendra Kumar Shrivastava	Alliance University, India
Garima Singh	Amity University Noida, India
Shivangi Agarwal	Ramrao Adik Institute of Technology, DY Patil University, India
Kiran Singh	MGCPS, Lucknow, India
Shashi Kant Gupta	ITM University, India
Monica	MITS Gwalior, India
Tanya Singh	Amity University Jharkhand, India

Deepak Motwani	Amity University Gwalior, India
Sonu Mittal	Guru Gobind Singh Inderprastha University, India
Roopsandeep Bammidi	Aditya Institute of Technology and Management, India
Shilpi Tomar	Samrat Ashok Technological Institute, India
Suresh Vishnudas Limkar	AISSMS Institute of Information Technology, India
Sharad Sharma	MMEC, MMDU, Mullana, India
Jaideep Kumar	AKTU, India
Kiran Deep Singh	Chitkara University, India
Gnanajeyaraman Rajaram	Saveetha School of Engineering, India
B. Dhiyanesh	Dr. N.G.P. Institute of Technology, India
Aniket K. Shahade	Shri Sant Gajanan Maharaj College of Engineering, India
Varun Gupta	KIET Group of Institutions, India
Hari Kishan Kondaveeti	VIT-AP University, India
Priyanka V. Deshmukh	Sant Gajanan Maharaj College of Engineering, India
Gowrishankar S.	Dr. Ambedkar Institute of Technology, India
Naveen Kumar	Larsen & Toubro Infotech Ltd. India
Sonam Mittal	B K Birla Institute of Engineering & Technology, India
Nagasundara K. B.	JSSATE, India
Piyush Sharma	Rajasthan Technical University, India
Rashi Agarwal	HBTU, India
Deepak Sharma	Aryabhatta College, University of Delhi, India
Sagar Damodar Padiya	Shri Sant Gajanan Maharaj College of Engineering, India
Ata Jahangir Moshayedi	Jiangxi University of Science and Technology, China
A. Velayudham	Jansons Institute of Technology, India
Swati Chowdhuri	Institute of Engineering & Management, Kolkata, India
Mirnalinee T. T.	SSN College of Engineering, India
P. Chinnasamy	MLR Institute of Technology, India
Deepjyoti Das	Tripura University, India
Aashi Singh Bhadouria	RGPV Bhopal, India
Navneet Kaur	Guru Nanak Dev Engineering College, India
Gurjot Kaur Walia	IKG PTU, India
Mijanur Rahaman Seikh	Kazi Nazrul University, India
Aniket Avinash Muley	Swami Ramanand Teerth Marathwada University, India
Kauser Ahmed P.	VIT Vellore, India

Sradhara Rinkal Mansukhbhai   L J University, India
Seema Maitrey   KIET Group of Institutions, India
Richa Sharma   JK Lakshmipat University, India
Ashish Kumar Mourya   Greater Noida Institute of Technology, India
Anil Kumar   Starex University, India
Nahid Fatima   Prince Sultan University, Saudi Arabia
K. Prabu   Annai College of Arts & Science, India
Abid Hussain   Career Point University, India
Chahat Jain   Guru Nanak Dev Engineering College, India
Nitish Jain   Broadway Public School, India
Mohammad Aasim Khan   Integral University, India
Srinivas Aluvala   SR University, India
Priti Maheshwary   Rabindranath Tagore University, India
Rajeev Tiwari   UPES, India
Ankitha K.   Canara Engineering College, India
Krishna Kumar Joshi   Sinhgad Institute of Technology Lonavala, India

Contents – Part II

Contents – Part I

Theme 2: Smart Systems

Design and Implementation
of a High-Performance Solar-Based Wireless
Sensor Network

Gopal M. Dandime[(⊠)] and Manish D. Sawale

Oriental University, Indore, India
gopaldandime@gmail.com

Abstract. In this study, the research aimed to address the growing global energy consumption and related environmental issues by exploring ways to improve the efficiency of solar PV cells, a crucial form of renewable energy. We focused on the nonlinear characteristics of solar PV cells, which result in low efficiency and highlighted the importance of using MPPT (Maximum Power Point Tracking) to extract maximum power. To enhance the traditional MPPT controller performance, we proposed incorporating a PID controller and using optimization techniques to predict optimal tuning parameters. Simulation results showed that our proposed solar energy harvesting system achieved an efficiency of 97.3% using the Mutated Fire-fly Algorithm (MFA) for tuning the PID controller, revealing superior performance over comparative techniques. Overall, this research highlights the potential for improving the efficiency of solar PV cells through the integration of advanced controllers and optimization techniques.

Keywords: Solar Energy Harvesting · Maximum Power Point Tracking (MPPT) · Mutated Fire-Fly Algorithm (MFA) · Wireless Sensor Networks

1 Introduction

With advancement of energy harvesting technology in recent years, solar light, mechanical vibration, heat, and existing forms of energy may now transform into useable electrical energy, allowing sensor nodes to autonomously absorb energy from their surroundings and have self-energizing properties. It is now one of the most tremendously influential technologies in the twenty-first century [1], with extensive development opportunities in numerous domains. Wireless Sensor Networks have largely employed in medical and health care, environmental monitoring, military, and various existing domains due to the extensive improvement of microelectronics and wireless communication technologies [2]. In industry and science, WSNs are frequently utilised [3]. However, the mutual interference of each communication technology, such as Wi-Fi, ZigBee, and Bluetooth, greatly limits future growth of WSNs [4]. The network has limited lifetime owing to energy constraints of sensor nodes powered by batteries, although other applications are intended to run endlessly. As a result, the question of how to increase working duration of wireless sensor networks has piqued academics' interest. Energy-saving solutions have

© The Author(s), under exclusive license to Springer Nature Switzerland AG 2025
M. Gupta et al. (Eds.): MISS 2023, CCIS 1952, pp. 3–24, 2025.
https://doi.org/10.1007/978-3-031-69115-7_1

been developed by certain studies [5, 6], however these approaches simply extend the restricted operating period [7]. Wireless charging technology has been found in recent research to successfully extend network lifetime [8].

Sensor node normally consists of radio transceiver, microcontroller unit, and interface with embedded or separate sensors, as well as energy provision, which commonly provided through battery or super-capacitor, which correctly controlled by a specialised system [9]. Environmental energy harvesting, which uses sun, wind, vibration, and radio-frequency (RF) signals as energy provision, may also be used to provide partially or entirely the energy [10]. The main issue with WSNs design is sensor node battery energy with limitations and can be depleted in a short amount of time depending upon duty cycle of application. Utilizing renewable energy sources namely solar photovoltaic energy [11] is novel notion for extending the life of sensor networks. Solar energy is big, clean, and abundant energy source. Solar energy may currently be harnessed in two ways: solar photovoltaic and solar thermal [12]. Solar photovoltaic device operates by turning solar energy into electrical energy immediately [13]. Weixing Song et al. "However, utilising such energy storage units that must be charged or replaced on a regular basis, it is tough to supply sustainable power for electronic devices. Self-powered system used in harvesting environmental energy and transform into sustainable electricity to overcome this problem [14]." o maximize the efficiency of large PV module arrays, MPPT is commonly employed in conjunction with a power converter (such as a dc-dc converter or inverter). By adjusting the PV arrays, MPPT ensures that the system always captures the maximum amount of generated power [15]. Afef Badis et al. state that many researchers are currently working to enhance the performance of PV systems through the development of new or the improvement of existing MPPT techniques. One such technique, known as the Perturb and Observe method, which is widely utilized due to its simplicity, involves adjusting the duty cycle of the DC-DC converter to achieve the MPP and generate an output voltage reference V_{ref} [16].

2 Literature Review

Himanshu Sharma et al. by "proposed an effective approach for utilizing ambient solar energy to overcome the limited energy and battery difficulties in Wireless Sensor Networks (WSNs), which are the main building blocks of IoT infrastructure in smart parks, smart cities, and smart buildings. Their method involved using PWM and MPPT to create a solar energy harvesting system that had an efficiency of 87% with PWM control and 96% with MPPT control, as shown in their simulation results [17] 2018".

Afef Badis et al. [18] 2017, for Photo Voltaic system, had proposed PID controller design in controlling DC-DC boost converter. GA (Genetic Algorithm) utilizes to tune controller's three parameters. While fast-changing solar irradiation was experienced, both PID MPPT controller optimised through GA and PID controller tuned using PP (Pole Placement) were tested. At last, Perturb and Observe MPPT approach was used to produce reference of output voltage (Vref) to get optimal system performance. Using MSE criterion, percentage overshoot criterion, and rising time criterion, outcomes of simulation could be distinguished.

Srisailam Sreedhar and Devadi Jagadeesh [19] 2016, had proposed all PV systems two fundamental drawbacks: PV power generation efficiency is relatively poor, and PV system output power is nonlinear, depending on meteorological factors namely ambient temperature and solar irradiation. So, for PV systems, MPPT approaches namely CVT (constant voltage tracking) approach, P&O (perturbation and observation) approach, INC (incremental conductance) approach, curve-fitting approach, look-up table approach, and so on have been presented.

B. Ashok Kumar et al. [20] 2015, for photovoltaic (PV) system, had proposed employing PID (proportional integral and derivative) controller to create an optimal controller based MPPT control. The boost converter's duty cycle was varied using the proposed PID controller. Here, a PV system is put to the test in a variety of environments. When SIMULINK findings are compared to traditional approaches like P&O and incremental conductance approach, the outcomes demonstrate power extracted for PID MPPT controller was at its maximum.

Kashif Ishaque et al. [21] 2012, had proposed modified PSO algorithm for an enhanced MPPT approach for photovoltaic (PV) systems. Because the technique was simple and could be computed quickly, it could be implemented using low-cost microcontroller. To test proposed method's efficacy, MATLAB simulations are performed under extremely difficult conditions, such as irradiance variations, load changes, and partial shading of PV arrays. Simulation and experimental studies have shown that proposed technique outperforms HC regarding tracking speed and steady-state oscillations.

3   Problem Statement

The Wireless Sensor Network (WSN) spatially distributed collection of sensor nodes to monitor conditions of physical or environmental, namely sound, temperature, pressure, and so on and cooperatively pass its data through network towards the primary location. WSN nodes have a significant design limitation for battery energy to get restricted; they only operate for few days depending on operation duty cycle. Generally, solar energy harvester circuit's efficiency is critical. If solar energy harvester system's capability is inadequate, battery will not be properly recharged, and wireless sensor network's lifetime will be reduced. By employing ambient solar PV energy, efficient solution of solar energy harvesting in restricted battery energy problem of WSN nodes is required. Using PWM (Pulse Width Modulation), to attain best results, many models for solar energy harvester systems were constructed, and iterative simulations during solar-powered DC-DC converters. MPPT were done in MATLAB/SIMULINK. For instance, PWM and MPPT are insufficient for effectively harvesting solar energy. Conventional P&O and INC are the most widely utilized MPPT techniques because to their simplicity of use and excellent performance under continual irradiation. Oscillations around MPP are major drawback since they become lost and track MPP in erroneous direction under quickly varying atmospheric circumstances. To resolve, following proposed methodology intended to employ for the benefit of efficient solar energy harvesting.

4   Research Objective

- The main objective of the research is to enhance the performance of solar energy harvesting systems beyond traditional PWM and MPPT controllers, to address the issue of limited battery energy in WSNs.

- To achieve this objective, the research incorporates established optimization techniques such as PSO and FA to tune the PID controller for MPPT
- The research further improves performance by mutating velocity updating strategy in the top performing FA approach.

5 Proposed Methodology

In this study, the research aimed to enhance the efficiency of a solar energy harvesting system by investigating and improving the methods used for Maximum Power Point Tracking (MPPT). The previous research in this field has primarily used the Perturb and Observe (P&O) and Pulse Width Modulation (PWM) methods for solar energy harvesting system design. However, simulation results from these methods have shown room for improvement. In order to optimize the performance, it is crucial to identify appropriate gain and tuning parameters for voltage regulation in PID control. However, manually predicting these parameters can be a time-consuming and complex process. Therefore, the proposed methodology of this study is to incorporate optimization techniques such as PSO, FA (Firefly Algorithm) and a proposed Mutated Firefly Algorithm (MFA) to identify optimal tuning parameters and simplify the process. The overall research flow is illustrated in Fig. 1.

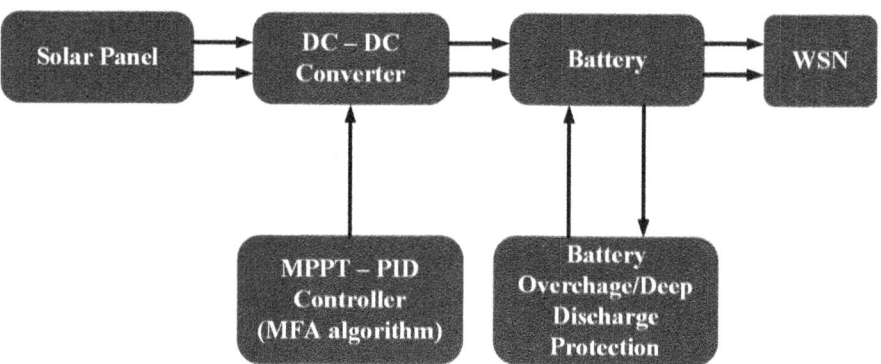

Fig. 1. Overall Research Flow of Solar Energy Harvesting System'

Power supply of DC, 3.6 V, Tektronix, Inc., Beaverton, OR, USA, is received by node of WSN from solar energy-harvesting system. Solar panels are utilized to harvest this voltage from ambient sunlight. Light energy is converted directly into DC electrical energy by solar panel. To charge battery, DC-DC converter adjusts voltage of DC. WSN node is powered by a rechargeable battery. The sensor measurement unit in WSN node measures necessary physical quantity (for example, light, temperature, humidity and pressure). This sensed data is processed by a microcontroller in the computation unit. Using transmitter unit, measured or sensed data can be wirelessly communicated within data packets from neighbouring network node. Through cluster head nodes, Data is transferred from end nodes to gateway node of USB. At last, in smart city, user monitors and controls process of application remotely, such as forest monitoring, temperature monitoring, traffic management, industrial boiler plant control, battlefield monitoring applications, volcano monitoring, glacier monitoring and air conditioner cooling system control.

6 Modelling of a Solar PV Panel

Semiconductor device is photovoltaic cell (familiarly termed as solar cell). It transforms light energy into electrical energy [8]. The electron-hole pair (EHP) is produced when light energy's photon (hv > Eg) strikes solar cell. This freshly generated EHP adds an electric current termed as light generated current that can be indicated as (OI_L). Solar cell's ideal theoretical current-voltage (I-V) equation:

$$Solar\ Cell\ current\ (I) = OI_L - I_O \left[\exp\left(\frac{qV}{kT}\right) - 1 \right] \tag{1}$$

here, OI_L = Light generated current through solar cell, I = solar cell's total output current, q = electron charge (1.6×10^{-19} C), T = Temperature of Solar cell (300 K), I_o = Reverse Saturation current due to recombination, k = Boltzmann's constant (1.38×10^{-23} J/K), V = open circuit voltage of photovoltaic cell. Analysing Fig. 2a, symbol of solar cell can be represented. In Fig. 2b, solar cell equivalent circuit model is illustrated. Through equation of Shockley, and parallel resistances as well as two series, it contains OI_L (light generated current source), D (diode) modelled. A MATLAB Simulink model for a solar panel is shown in Fig. 2c. Figure 2b shows the characteristic current equation (Kirchhoff's current law) for this similar circuit.

$$Output\ Current\ of\ Equivalent\ Cell\ Model\ (OI) = OI_L - OI_D - OI_P \tag{2}$$

here, OI_L = Light generated current, OI_D = diode current, and OI_p = current in parallel resistance.

$$Diode\ Current\ (OI_D) = I_o \left[\exp\left(\frac{V + IR_s}{nV_T}\right) - 1 \right] \tag{3}$$

The Eq. (3) I = solar cell output current (1 for ideal, 2 for practical diode), V = open circuit voltage of solar cell, R_s = series resistance, "I_o = Reverse Saturation current due to recombination, V_T = Thermal voltage (kT/q), T = Temperature of Solar cell (300 K), n = diode ideality factor, k = Boltzmann's constant (1.38×10^{-23} J/K) is used to represent the current in parallel resistance" (Himanshu Sharma, 2018).

$$Current\ in\ parallel\ resis\tan ce(OI_p) = \frac{V + IR_s}{R_p} \tag{4}$$

Now, by plugging OI_D and OI_p values into current Eq. (4), complete IV equation of single solar cell's equivalent circuit can be received whereas entire output current and voltage parameters can be provided:

$$Solar\ Cell\ Current\ (I) = OI_L - I_0 \left[\exp\left(\frac{q(V + IR_s)}{nkT}\right) \right] - \left(\frac{V + IR_s}{R_p}\right) \tag{5}$$

The equation q = electron charge ($1.6 \times 10–19$ C), R_p = Parallel Resistance, IL, k, Io, R_s, q, V, T, I, n explained earlier is used to define the efficiency (η) of a solar cell in Eq. (5).

$$Solar\ Cell\ Efficiency\ (\eta) = \frac{V_{oc} \cdot I_{sc} \cdot FF}{P_{in}} \qquad (6)$$

In Eq. (6), the Open Circuit Voltage (V_{oc}), Short Circuit Current (I_{sc}), and Fill Factor (FF) are represented, where "P_{in} = incident optical power. The Solar cell's Fill Factor (FF) can be explained by Eq. (7)" (Himanshu Sharma, 2018).

$$Fill\ Factor(FF) = \frac{P_{max}}{P_{dc}} = \frac{I_m \cdot V_m}{I_{sc} \cdot V_{oc}} \qquad (7)$$

where, I_m denotes maximum current whereas V_m denotes solar cell's maximum voltage. "Solar cells come in a various types, including c-Si (monocrystalline silicon solar cells), a-Si (amorphous silicon solar cells), multi-Si (polycrystalline solar cells), and TFSC (thin-film solar cells)" (Himanshu Sharma, 2018). Solar cells made of a-Si efficiency, on the other hand, have a higher efficiency than all others, nearly 18%.

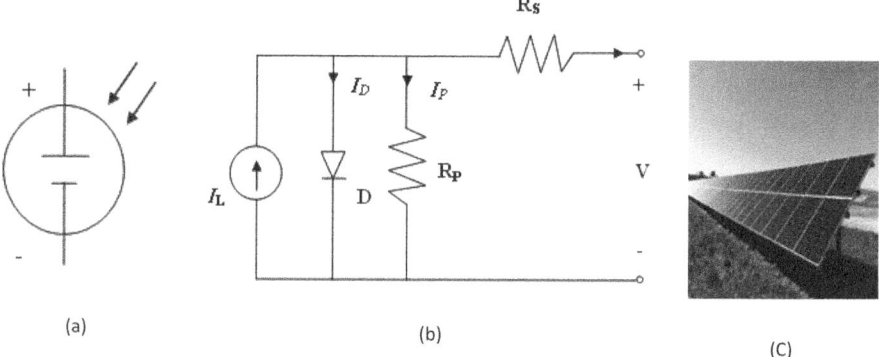

(a)   (b)   (C)

Fig. 2. Modelling of Solar cell (a) Symbol; (b) Equivalent circuit of Solar Cell; (c) Solar Panel

6.1 Impact of Solar Radiation (G)

The efficiency of solar cells (η) is directly correlated with variations in solar radiation. As solar radiation increases, the efficiency of the solar cell (η) also increases and vice versa. "Fig. 4a illustrates the current-voltage (I-V) characteristics of a commercial 10 W solar panel (Dow Chemical DPS 10–1000) at different levels of irradiance. The panel, which has a module area of approximately 0.13 m2 and measures 546 mm × 232 mm, shows that as the irradiance level rises, the current in the solar panel increases. For a solar irradiance of 1000W/ m^2, the solar cell current reaches its maximum of 6.2 A" (Himanshu Sharma, 2018) Fig. 3 depicts Power-Voltage characteristics of Solar Panel at

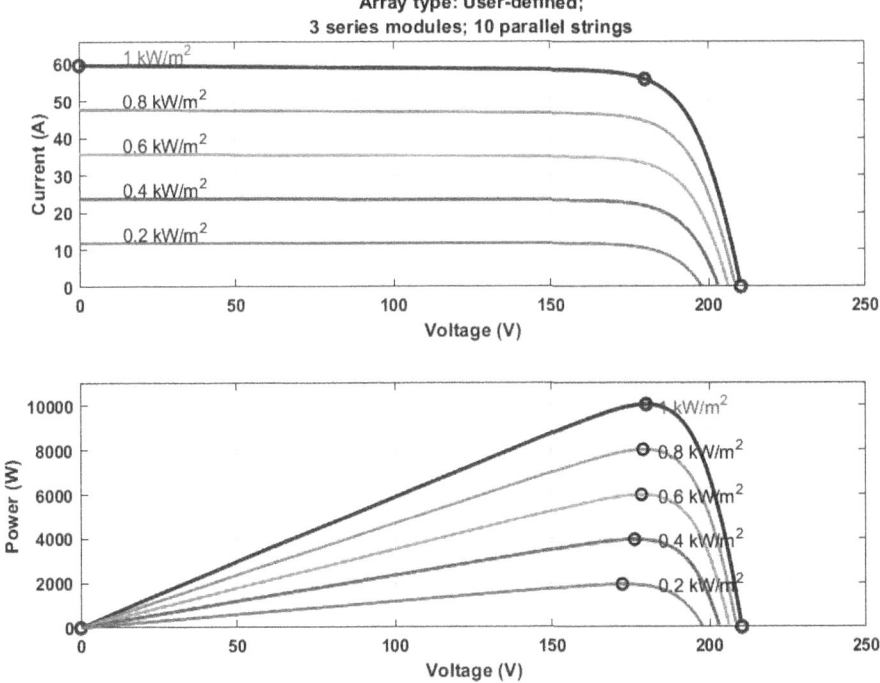

Fig. 3. Solar Panel characteristics with variations in Irradiance

various radiation levels. For the highest solar irradiance, 1000 W/ m^2, harvested power will be maximum (9.8 W).

"The I-V and P-V characteristics of the 10 W solar module DPS-10–1000 of Dow Chemical Company (Midland, Michigan, United States) are simulated using the MATLAB/Simulink simulation software, as shown in Fig. 3" (Himanshu Sharma, 2018).

6.2 Effect of Temperature (T)

From Fig. 4, as solar panel temperature rises, output current falls and vice versa. Therefore, temperature changes have an inverse relationship with output current. Similarly, as temperature rises, output power drops and vice versa in Fig. 5b. Therefore, output power will be inversely proportional to changes in temperature.

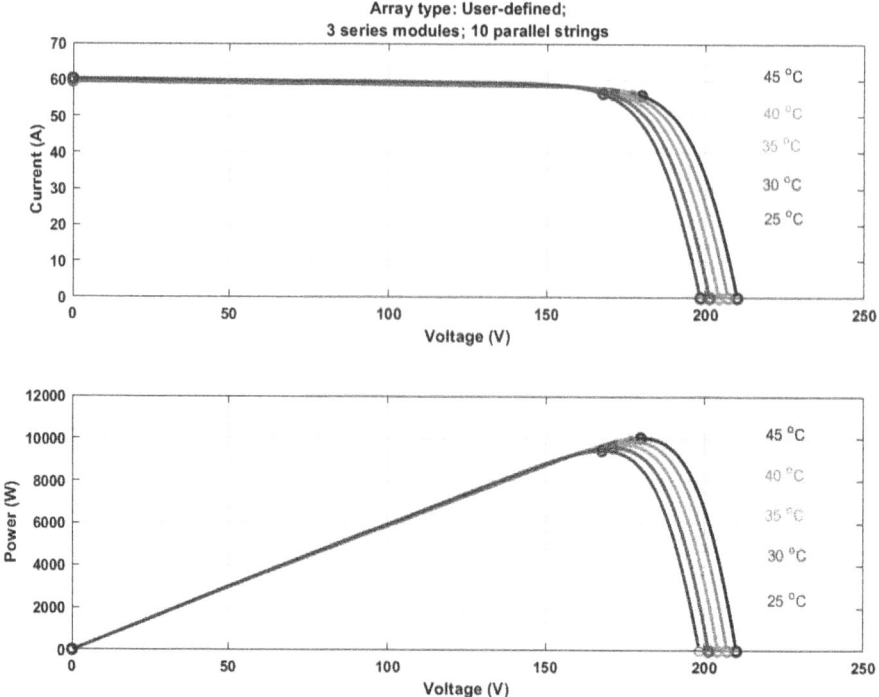

Fig. 4. Solar Panel characteristics with variations in temperature

7 Modelling of DC-DC Converter

According to [15], in the design of photovoltaic systems, there are three types of DC-DC converters: Buck-Boost Converter, Buck Converter and Boost Converter. It has utilised DC-DC Buck converter since it has higher efficiency than converters of Buck-Boost and Boost. DC-DC buck converter will be converter of power electronics with output voltage that will never be higher than input voltage. As illustrated in Fig. 5, Buck converter is made up of capacitor (C), V_{dc} (DC voltage source), L (inductor), MOSFET (switch), and D (diode). Input voltage V_s occurs across load resistor while "MOSFET switch (S) can be closed at time t_1. Output voltage (V_o) has smaller amplitude than input voltage V_o" (Himanshu Sharma, 2018). By altering time period t_1, Duty Cycle (D_c) changes from 0 to 1. $D_c = V_o/V_{in}$ is Duty Cycle of the Buck converter. Buck converter's average output voltage is provided:

$$V_o = \frac{1}{T} \int_o^{t_1} v_o dt = \frac{t_1}{T} V_{in} = f \cdot t_1 \cdot V_{in} = V_{in} \cdot D_c \qquad (8)$$

Here, input voltage referred by V_{in}, output voltage denoted by V_0, duty cycle referred by D_c, t_1 = MOSFET switch ON time duration, frequency of operation denoted by f, T = Total Time period.

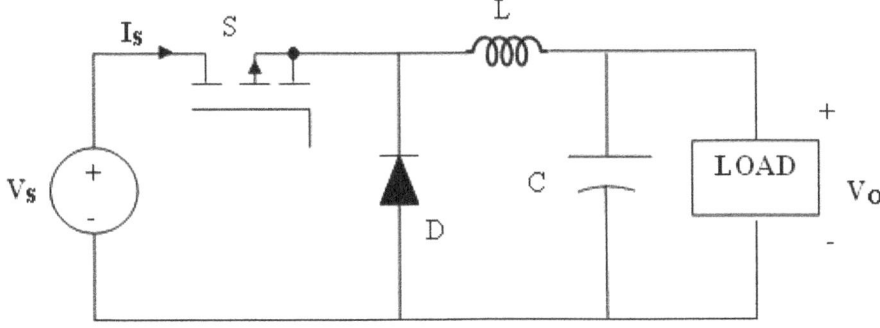

Buck Converter

Fig. 5. Circuit Model of DC-DC Buck Converter

At output, load current average can be defined:

$$I_o = I_L = V_o/R = D_c \cdot V_{in}/R \tag{9}$$

here, f = chopping frequency, $D_c = t_1/T$ denotes duty cycle, T = chopping period.

8   Losses in Power for DC-DC Buck Converter

In DC-DC buck converter, power dissipation is divided into three sections: losses of MOSFET conduction, losses of inductor conduction, and losses of MOSFET switching.

"In all types of DC-DC converters, the inductor is the primary source of power consumption" (Himanshu Sharma, 2018). Although losses from MOSFET switching and diode conduction are present, they are considered insignificant in comparison to inductor losses, and are typically disregarded in practice. The power consumption loss of an inductor can be calculated using:

$$P_L = I_{L(rms)}^2 \times R_{L(dc)} \tag{10}$$

here, $R_{L(dc)}$ = DC resistance of Inductor, $I_{L(rms)}$ = Inductor RMS current, P_L = Power loss in Inductor (mW).

8.1   Firefly Algorithm

Firefly Algorithm mimics is based on social behaviour of fireflies soaring across summer sky in tropics. Fireflies use bioluminescence in a variety of flashing patterns to communicate, seek for food, and locate mates. Various metaheuristic algorithms may be built by mimicking nature. Some of flashing properties of fireflies were idealised in this research to create firefly-inspired algorithm. Only three rules observed for the sake of simplicity:

1) As all fireflies are unisex, they will be attracted to each other regardless of their gender.

2) The brightness of fireflies has an impact on their attraction. One of two flashing fireflies will move toward the other when one of them is brighter. The brightness of the firefly is related to their attractiveness, which diminishes as the distance between them grows. If there are no brighter fireflies in the area, specific firefly will travel about at random.

3) Brightness of a firefly is connected to cost function's analytical form in some way. The brightness is directly proportional to the value of the cost function in the maximization problem. Remaining types of brightness established similar fashion how genetic algorithms define fitness function.

Three criteria stated above are summarised by pseudo-code displayed in Fig. 1 that contains basic phases of the FA.

It is worth noting that FAs and BFA (bacterial foraging algorithm) [17] have some conceptual similarities. There are, nevertheless, some significant variances. Initially, in BFA, bacteria's attraction is determined in part by their fitness and in part by their distance. In FA, on the other hand, the attraction is determined by the cost function which monotonically decays as the distance between fireflies. Secondly, individuals of FA have greater versatility in terms of varied attraction and changeable visibility: this typically leads to increased mobility as well as permits search space to get explored effectively. Finally, Firefly Algorithm contains two important limit cases, and the algorithm can be fine-tuned to combine benefits of both limit cases in a more efficient search space exploration. The flow chart of MFA is exposed in Fig. 6.

8.2 Distance, Attractiveness, and Limit Cases

In FA, formulation of attraction as well as variation of light intensity becomes two prominent challenges. For sake of simplicity, firefly's attractiveness can be governed with brightness and can be connected with encoded objective function.

As the distance from the source increases, the changes in light intensity and attractiveness should decrease in a monotonic manner as the light intensity and attraction decrease. In most cases, the combined effect of the inverse square law and absorption can be modelled using the Gaussian form below.

$$I(r) = I_0 e^{-\gamma r^2} \tag{11}$$

where coefficient of light absorption γ examined to be constant. It can be defined firefly's attractiveness β by multiplying attractiveness by intensity of light seen through adjacent fireflies.

$$\beta(r) = \beta_0 e^{-\gamma r^2} \tag{12}$$

Here β_0 denotes attractiveness in $r = 0$. Characteristic distance $\Gamma = 1/\sqrt{\gamma}$ over can be formulated in Eq. (12) which attractiveness varies primarily from β_0 to $\beta_0 e^{-1}$.

At x_i and x_j, distance between any two fireflies i and j individually, explained as Cartesian distance $r_{ij} = \|x_i - x_j\|$.

Attraction of one firefly movement i to several other brighter (attractive) firefly j can be explained.

$$\Delta x_i = \beta_0 e^{-\gamma r^2}\left(x_j^t - x_i^t\right) + \alpha \varepsilon_i, x_i^{t+1} = x_i^t + \Delta x_i \tag{13}$$

Initial term is a result of attraction. Randomization is the second term, and the randomization parameter is α. In this case, from Gaussian distribution, ε_i will be vector of random numbers drawn. That will be worth noting using a Levy distribution, this second term is enhanced even more. Step size can be determined at random.

$$L(S) = A_S^{1-q}, A = q\Gamma(q)\sin(\pi q/2)/\pi, \tag{14}$$

here Gamma function denoted by $\Gamma(q)$, whereas q refers to distribution exponent. During present case, it has been used q = 3/2.

To create solution using (13) from an implementation standpoint, it should replace the final word with α L (s). It may mostly use fixed value of α = 0.01. q = 1.5 can be utilised for entire simulations. It may utilise conventional transformation approach and remaining common approaches on producing pseudo-random numbers to obtain a decent, random step size L(s) through Eq. (14). Because (14) is only applicable for big steps, it sets the size of minimum step to 0.1, whereas simulations are performed using a conventional Gaussian distribution N(0, 0.1). It has been constructed symmetrical step sizes due to symmetry, L(s) accepts values of positive and negative.

It is clear from Eq. (13) when is small or large, there are two limit instances. The attractiveness and brightness of the firefly remain consistent as it γ approaches zero, allowing firefly to be seen through remaining fireflies; it will be unique occurrence. When γ is extremely vast, on the other hand, attractiveness (and hence brightness) drops considerably, and entire fireflies become short-sighted. This implies about entire fireflies travel in nearly random manner, similar to how a random search strategy works. Firefly Algorithm, in general, corresponds to the condition in between these two limit instances, allowing these parameters to be fine-tuned in order to FA outperforms random search and PSO. In reality, FA can locate all of local optima as well as global optima with efficient manner. Later in implementation, this benefit will be demonstrated in greater detail. FA also has the benefit of allowing various fireflies to function independently. As a result, FA lends itself easily to parallel implementation. Fireflies beat GAs and PSO because they aggregate closer to every optimum (without jumping around while regarding GAs). In parallel implementation, interactions between distinct sub regions will be low.

8.3 The Influence of Internal Factors on the Firefly Algorithm's Convergence Behaviour

User must define a number of internal parameters in Firefly Algorithm, this may have a variable impact on convergence behaviour.

For starters, practically entire stochastic population-based algorithms require selection of initial locations or solutions. To guarantee that the optimal design is insensitive to the initial population, the initial solution/guess should be insensitive as widely as feasible.

In the current study, two strategies were used: (a) to distribute initial solutions across entire design space in a fairly uniform manner, thereby without biasing random regions of search space; (b) to create every firefly away from existing fireflies as possible, allowing for more efficient exploration of search space.

For each initial population, 100 optimization runs were performed. Results of optimization were found to be nearly independent of initial guess. Statistical measurements namely the cost function mean value whereas related standard deviation support. This method could be significantly more versatile than depending on a few optimization runs alone.

On influence of population size and attractiveness, extensive sensitivity experiments were conducted.

For most applications, population size of n $= 10^{-25}$ was found to be enough, however slightly larger population size might be employed in highly difficult situations. Almost all issues should be solved with a population of 50 fireflies. Higher numbers are not advised because they will greatly lengthen the computation time.

Furthermore, 100 optimization runs performed for every combination of internal parameters in order to get statistically meaningful findings. The best solutions were discovered to be unaffected by the original parameter settings. The worst solutions and standard deviation, on other hand, are slightly dependent on initial conditions: due to metaheuristic algorithms' random nature. Performing multiple optimization runs helps to reduce impact of dependencies as much as possible. γ parameter, describes variation of attractiveness, has an impact on FA algorithm's convergence speed and overall behaviour. In theory, γ might be range within interval $(0,\infty)$.

Thus, the value is based upon γ characteristic length of system to be optimized: this is often varied from 0.01 to 100 in most situations. Preliminary investigation showed easiest and most efficient technique to be set $\gamma = 1/\sqrt{L}$ here typical design variables length denoted by L. Initial attractiveness value has been discovered not to influence significant results of optimization, fixed value $\beta_0 = 1$ had been used.

Although fundamental FA formulation is relatively efficient, when the optimum design is approached, oscillatory behavior might arise. Reduce randomization parameter α using geometric progression reduction strategy equivalent to simulated annealing cooling schedule to improve the solution quality. Specifically,

$$\alpha = \alpha_0 \theta^t \tag{15}$$

here reduction factor of randomization denoted by $0 < \theta < 1$. This strategy had discussed in this study through decreasing α from 0.5 to 0.01.

Because scales can range widely for various issues and even several variables within similar issue, each design variable's real scale should α ideally be related to randomization parameter. In such situation, it's typically good idea to replace α with α^{sk}, with scaling parameters S_k $(k = 1,...,d)$ in d dimensions being specified by problem of interest's actual scales.

In associate with movement of all Fireflies to their better solution the velocity computation process is also employed for the performance enhancement. This mechanism used to move (evolve) particle position to search for optimal solutions. Figure 6 shows flow chart of mutated firefly algorithm, where velocity computation strategy incorporated with the fireflies 'existing updating strategy for performance enhancement.

8.4 Velocity Update Equation

$$v_{id}^{t+1} = v_{id}^t + c_1 r_1 (p_{id}^t - x_{id}^t) + c_2 r_2 (p_{gd}^t - x_{id}^t) \tag{16}$$

The position can be updated by position update Eq. (17).

8.5 Position Update Equation

$$x_{id}^{t+1} = x_{id}^t + v_{id}^{t+1} \tag{17}$$

here particle index is denoted as i = 1, 2,…, S whereas dimension is referred by d = 1, 2,…,D. S represents swarm size, whereas c1 and c2 denote constants, sometimes known as simply acceleration coefficients or cognitive and social scaling parameters. In range [0, 1], random numbers produced within uniform distribution represented as r1, r2. This can be seen in Eqs. (16) and (17) that every dimension of each particle can be updated without relying on others. Only link between problem space dimensions can be demonstrated through objective function, for example, gbest and pbest are locations of best positions discovered so far. Basic PSO algorithm version is represented in Eqs. (16) and (17). Algorithmic method for MFA procedure has been provided:

Pseudo-Code

begin

Objective function f(**x**), **x**=(x₁, …., xₐ)T

General initial fireflies population xᵢ (i=1,2, …,n)

Light intensity lᵢ at xᵢ is determined through f (xi)

Define coefficient of light absorption γ

while *(t<MaxGeneration)*

for i=1: *n all n fireflies*

　　for j=1: *i all n fireflies*

　　　if (lⱼ>lᵢ)

　　　L'evy flights are utilized to move firefly I towards firefly j in the d-dimension.

　　　end if

Attractiveness changes with distance r via exp [-γr²]

Updated light intensity and new solutions should be considered.

end for j

end for i

Determine the current best firefly by ranking them.

end while

Visualization of post-processed findings

End

Create and configure a D-dimensional swarm, as well as the S and velocity vectors accompanying;

for *t= 1 to maximum bound on number of iterations* ***do***

  for i=1 to S **do**

    for d=1 to D **do**

      Apply velocity update equation 1;

      Apply position update equation 2;

    end

    Calculate updated position fitness;

    If required, pbest and gbest's historical data should be updated.;

  end

  Terminate if gbest satisfies the problem's conditions;

end

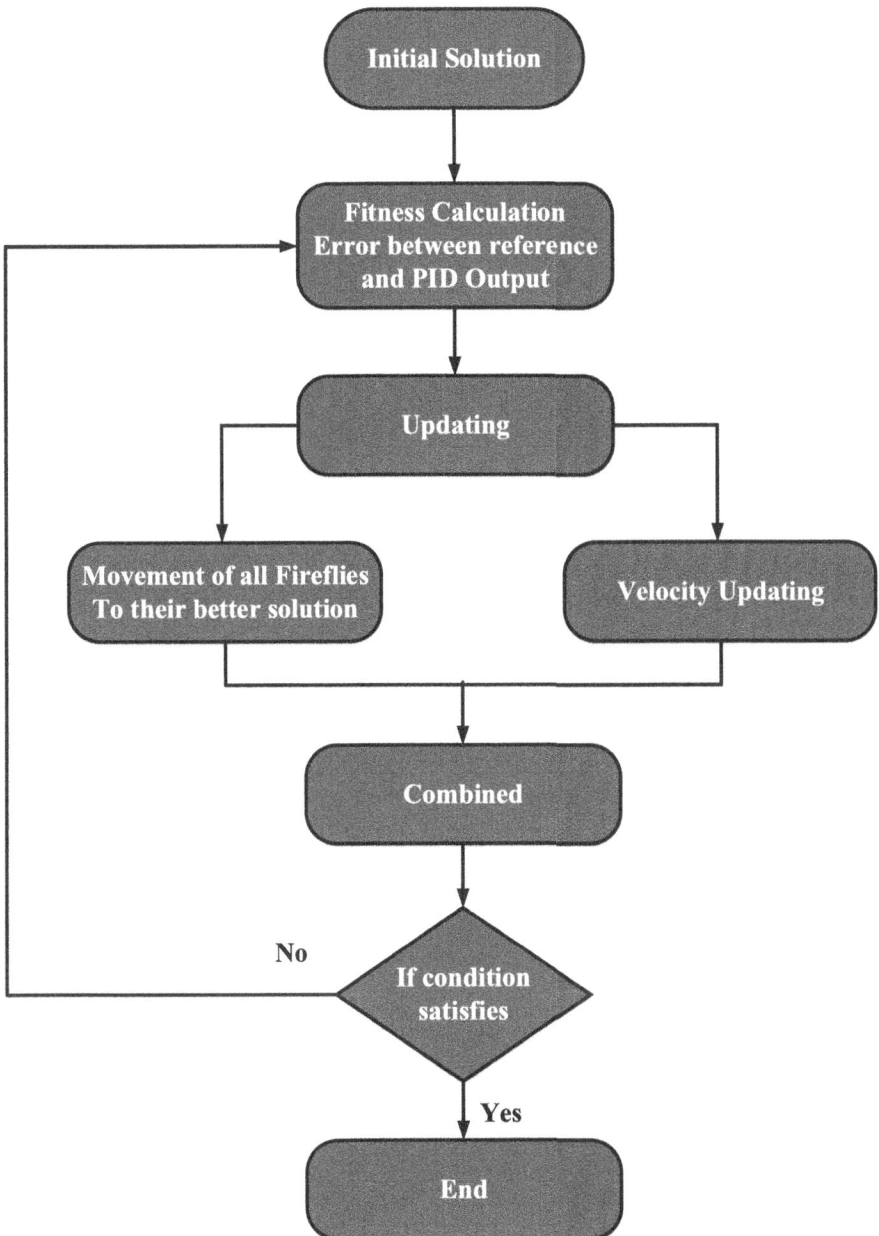

Fig. 6. Flow chart of Mutated Firefly Algorithm

8.6 3-Level Bridge Inverter

The Fig. 7 indicates S3L inverter (soft-switching three-level inverter) output; this is a high-efficiency power electronic inverter that may be used with three-phase drives, as a grid-tie inverter for solar systems, and as a power supply.

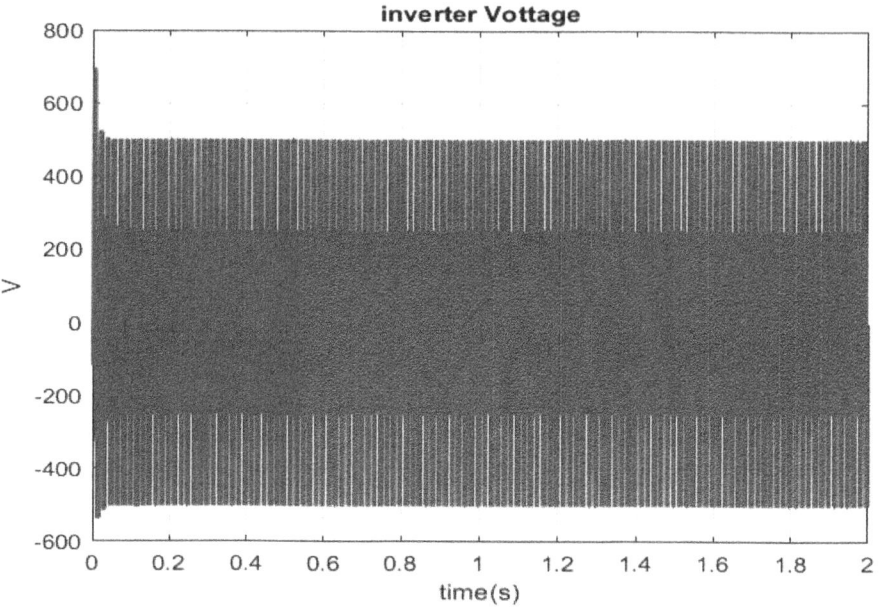

Fig. 7. Inverter output

9   Results and Discussion

Table 1 shows the parameters for simulating a solar energy harvesting system. "MATLAB Simulink 2018 was used to simulate a solar-powered Boost converter with MPPT control for battery charging in a WSN node. The simulation assumes 1000 W/cm^2 of solar irradiance incident on a solar panel at a constant temperature of 25 °C. The solar panel, which has an efficiency of 15%, captures 15 mW/cm^2 of solar energy. When the solar panel is exposed to full irradiance, the simulated output voltage is 3 W, 6 V, and 500 mA. The electrical energy from the solar cell is then directed to a DC-DC boost converter, which increases the output voltage. The rechargeable battery is charged using the output voltage from the Boost converter, which then powers the WSN node. In this illustration, the WSN load is represented as an output with a 100 Ω DC load resistance. Table 1 lists the various simulation parameters, including duty cycle, WSN load model, power losses, irradiance, DC-DC converter type, temperature, battery voltage and type, and solar panel power, current, and voltage" (Himanshu Sharma, 2018).

Table 1. Simulation Parameters.

Parameters	Value	Parameters	Value
Irradiance (W/m^2)	1000 Watts/m^2	Capacitors (C)	100 uF
Maximum Solar Panel output voltage (Vm)	6 V	Initial duty cycle	0.5
Temperature (T)	25 °C	Inductor (L)	200 uH
Maximum Power from Solar Cell (Pm)	3 watts	Switching Voltage Loss (V_{sw})	0.2 V
DC-DC Converter	Boost Convertor	MOSFET Switching Frequency (f)	5 kHz
Maximum Solar Panel output current (Im)	500 mA	MOSFET Switching Power Losses (P_{sw})	0.5 mW
Battery Voltage	3.6 V	Inductor conduction Power Loss (P_L)	50 mW
Rechargeable Battery Type	NiCd	WSN Load Model	10-Ω resistor

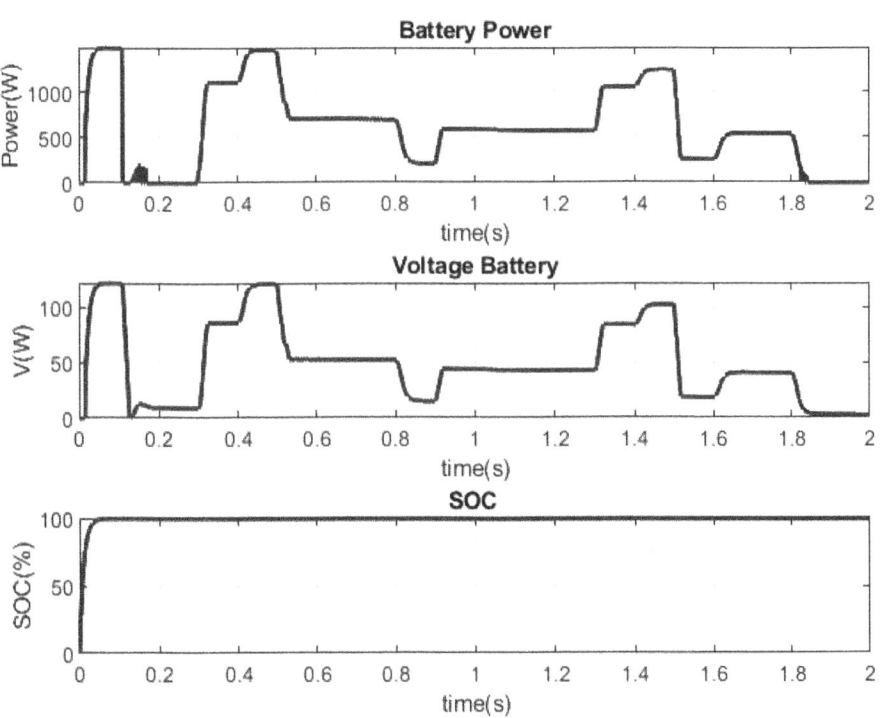

Fig. 8. MFA-MPPT controlled solar energy harvesting Battery Power, voltage and SOC during charging

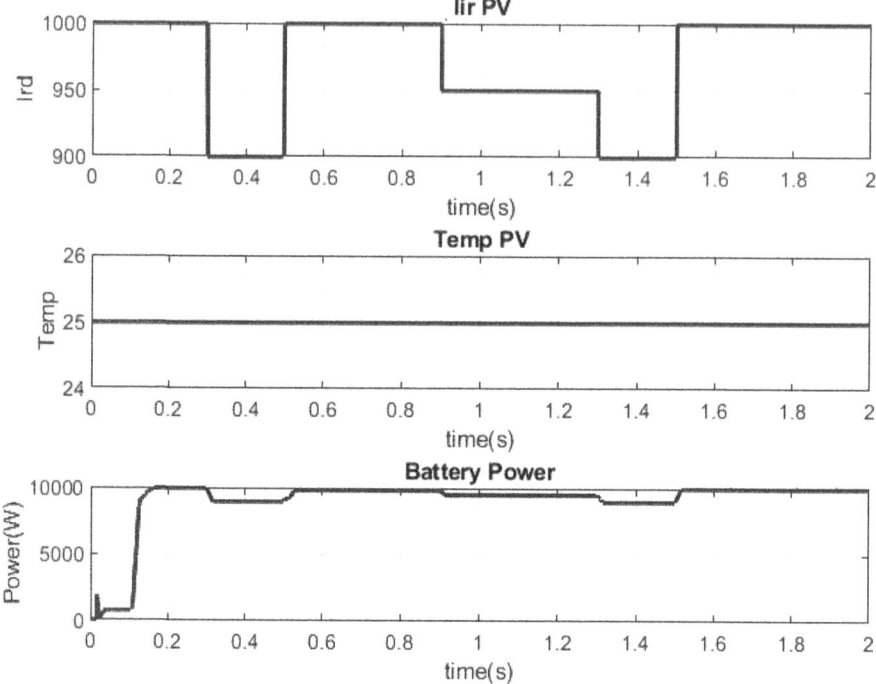

Fig. 9. Power output for a solar energy system

Figures 8 show the outcomes of simulations utilizing battery SoC (State of Charge), battery voltage, and battery power for MPPT-controlled solar energy harvesting systems. The SoC Estimator is utilized to calculate a battery's SoC during charging conditions. The input current of the battery when it is being charged by a Solar PV array is utilized to calculate the SoC of the battery. SOC refers to the battery's current capacity to nominal capacity ratio. The nominal capacity, or maximum amount of charge contained within the battery, is specified by the manufacturer. Figure 9, illustrates battery power with respect to solar irradiation and temperature for a given time. Subsequently, the Fig. 10 exhibits the techniques wise comparison for an output power. It is obvious from the results that proposed having minimum spikes and it's almost closest to the actual output. This is possible because of incorporating the mutated strategy in associates with traditional FA for performance enhancement. The proposed mutated strategy identifies the optimal tuning parameters for PID controller to regulate the output power over FA, PSO and P&O existing techniques.

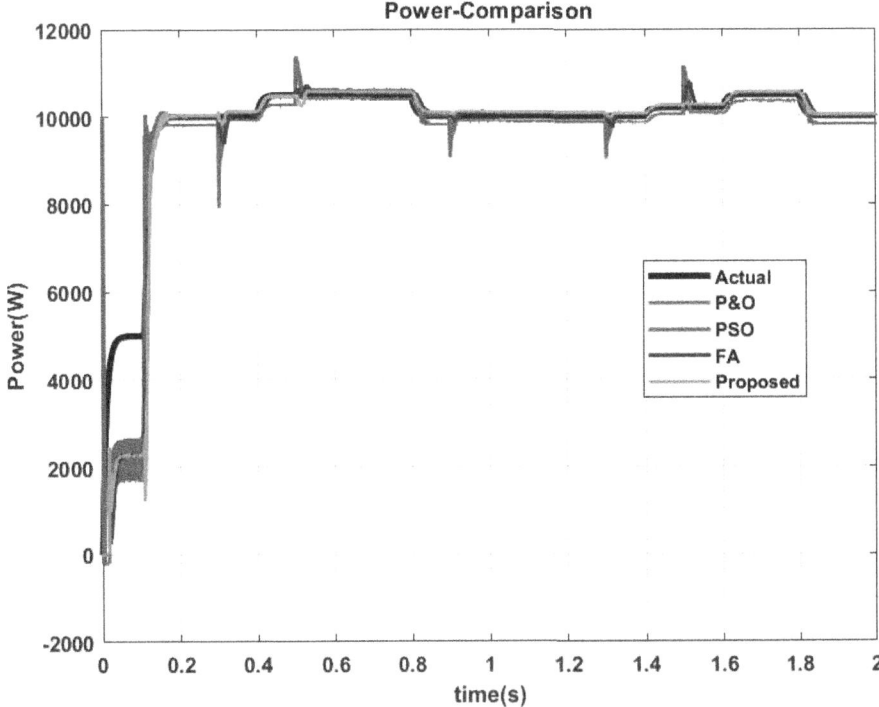

Fig. 10. Techniques wise output power comparison for solar energy harvesting system

The Fig. 11 solar energy harvesting system for actual vs output voltage and the Fig. 12 shows the solar energy harvester system efficiency comparison for traditional and proposed technique. The efficiency of energy harvester system can be computed for methods of PWM control and MPPT control [11]. The proposed MFA associates PID controller achieves the solar energy harvester system efficiency as 97.3%, whereas FA associates PID controller and PSO associates PID controller achieves 96.83% and 96.52% respectively. Traditional techniques like P&O associates PID controller and PWM controller attains 96.06% and 87.76% respectively.

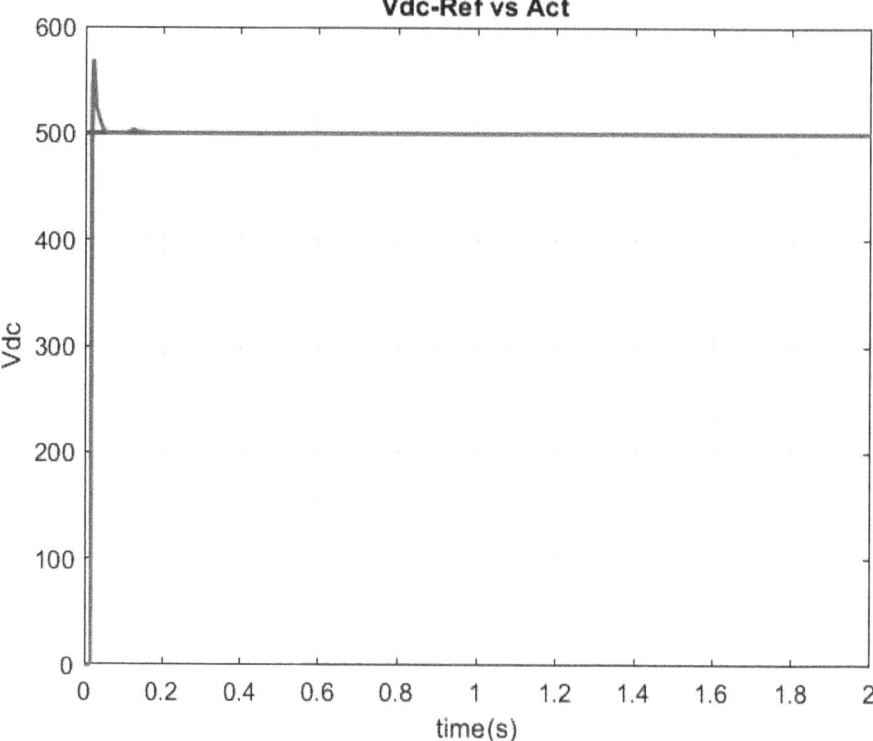

Fig. 11. Solar energy harvesting system for actual vs output voltage

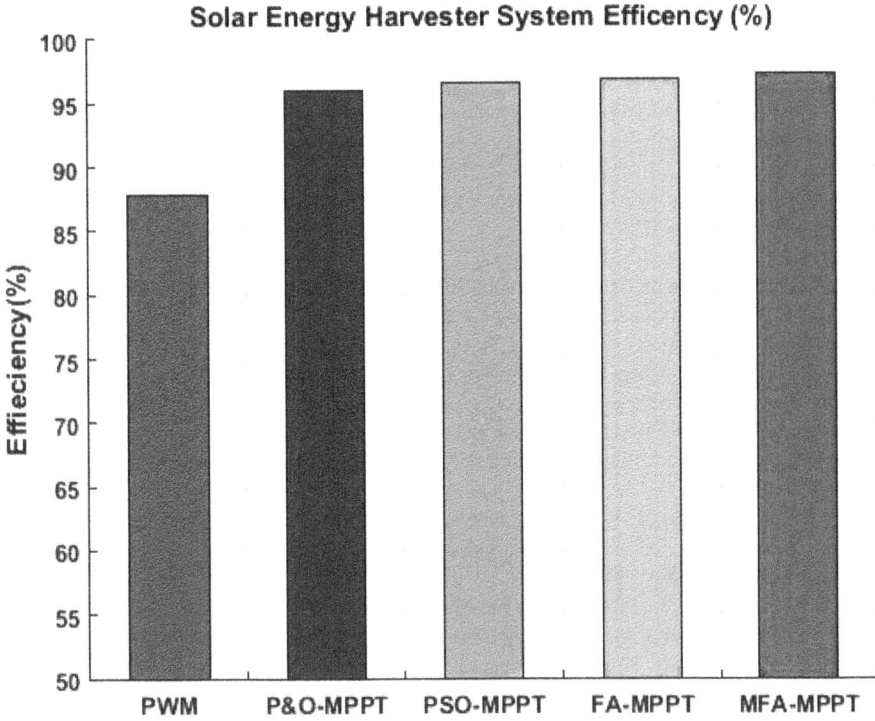

Fig. 12. Solar energy harvesting system efficiency

10 Conclusion

In conclusion, this study aimed to extract maximum power from solar PV cells by incorporating a PID controller and optimization techniques such as FA and MFA. The proposed PID controller for MFA achieved a solar energy harvesting system efficiency of 97.3%, showing superior performance over traditional approaches and other comparative techniques like FA-MPPT, PSO-MPPT, P&O-MPPT, and PWM. The mutation strategy in particular was found to improve efficiency by identifying the maximum power point of solar energy harvesting systems. The results demonstrate the effectiveness of the proposed MFA-associated PID controller in improving solar energy harvesting system efficiency. Future research should consider other optimization techniques to further improve this efficiency.

References

1. Yukun, Y., Zhilong, Y., Guan, W.: Clustering routing algorithm of self-energized wireless sensor networks based on solar energy harvesting. J. China Univ. Posts Telecommun. **22**(4), 66–73 (2015)
2. Huang, Z., Niu, Q., Xiao, S., Li, T.: Energy harvesting algorithm considering max flow problem in wireless sensor networks. Comput. Commun. **150**, 626–633 (2020)

3. Wan, J., Chen, B., Wang, S., Xia, M., Li, D., Liu, C.: Fog computing for energy-aware load balancing and scheduling in smart factory. IEEE Trans. Ind. Inf. **14**(10), 4548–4556 (2018)
4. Huynh, V.V., Nguyen, H.S., Hoc, L.T.T., Nguyen, T.S., Voznak, M.: Optimization issues for data rate in energy harvesting relay-enabled cognitive sensor networks. Comput. Netw. **157**, 29–40 (2019)
5. Chen, M., Miao, Y., Jian, X., Wang, X., Humar, I.: Cognitive-LPWAN: towards intelligent wireless services in hybrid low power wide area networks. IEEE Trans. Green Commun. Network. **3**(2), 409–417 (2018)
6. Chen, M., Hao, Y., Lai, C.F., Wu, D., Li, Y., Hwang, K.: Opportunistic task scheduling over co-located clouds in mobile environment. IEEE Trans. Serv. Comput. **11**(3), 549–561 (2016)
7. Zhao, C., et al.: Maximizing lifetime of a wireless sensor network via joint optimizing sink placement and sensor-to-sink routing. Appl. Math. Model. **49**, 319–337 (2017)
8. Zhao, C., Zhang, H., Chen, F., Chen, S., Wu, C., Wang, T.: Spatiotemporal charging scheduling in wireless rechargeable sensor networks. Comput. Commun. **152**, 155–170 (2020)
9. Guimarães, D.A., Frigieri, E.P., Sakai, L.J.: Influence of node mobility, recharge, and path loss on the optimized lifetime of wireless rechargeable sensor networks. Ad Hoc Netw. **97**, 102025 (2020)
10. Prauzek, M., Konecny, J., Borova, M., Janosova, K., Hlavica, J., Musilek, P.: Energy harvesting sources, storage devices and system topologies for environmental wireless sensor networks: a review. Sensors **18**(8), 2446 (2018)
11. Sharma, H., Haque, A., Jaffery, Z.A.: Maximization of wireless sensor network lifetime using solar energy harvesting for smart agriculture monitoring. Ad Hoc Netw. **94**, 1–14 (2019)
12. Bhalla, V., Tyagi, H.: Solar energy harvesting by cobalt oxide nanoparticles, a nanofluid absorption based system. Sustain. Energy Technol. Assess. **24**, 45–54 (2017)
13. Di Piazza, M.C., Pucci, M., Vitale, G.: Intelligent power conversion system management for photovoltaic generation. Sustain. Energy Technol. Assess. **2**, 19–30 (2013)
14. Song, W., et al.: A highly elastic self-charging power system for simultaneously harvesting solar and mechanical energy. Nano Energy **65**, 103997 (2019)
15. Seng, L.Y., Lalchand, G., Lin, G.M.S.: Economical, environmental and technical analysis of building integrated photovoltaic systems in Malaysia. Energy Policy **36**(6), 2130–2142 (2008)
16. Badis, A., Mansouri, M.N., Sakly, A.: PSO and GA-based maximum power point tracking for partially shaded photovoltaic systems. In: 7th International Renewable Energy Congress (IREC), pp.1–6 (2016)
17. Sharma, H., Haque, A., Jaffery, Z.A.: Modeling and optimisation of a solar energy harvesting system for wireless sensor network nodes. J. Sens. Actu. Netw. **7**(40), 1–19 (2018)
18. Badis, A., Mansouri, M.N., Boujmil, M.H.: A genetic algorithm optimized MPPT controller for a PV system with DC-DC boost converter. In: International Conference on Engineering & MIS (ICEMIS), pp.1–6 (2017)
19. Sreedhar, S., Jagadeesh, D.: A review on optimization algorithms for MPPT in solar PV system under partially shaded conditions. IOSR J. Electr. Electron. Eng., 23–32 (2016)
20. Kumar, B.A., Venkatesh, M.S., Muralikrishna,G.M.: Optimization of photovoltaic power using PID MPPT controller based on incremental conductance algorithm. In: Power Electronics and Renewable Energy Systems, vol. 326, pp. 803–809 (2015)
21. Ishaque, K., Salam, Z., Amjad, M., Mekhilef, S.: An improved particle swarm optimization (pso)–based MPPT for PV with reduced steady-state oscillation. IEEE Trans. Power Electron. **27**(8), 3627–3638 (2012)

IoT Security: Challenges, IDS Evolution and AI Defensive Schemes: A Review

Neeraj Kumar[✉] and Sanjeev Sharma

SoIT UTD RGPV, Bhopal, Madhya Pradesh, India
neerajkumardob95@gmail.com, sanjeev@rgpv.ac.in

Abstract. In the current scenario Internet of Things (IoT) is playing a vital role in life's betterment of mankind. The word "smart" is being utilized in almost every aspect of modern day's life. From smart cities to smart grids, smart buildings to e-healthcare systems, smart cars to smart agriculture etc. the world is now becoming Internet of everything. According to research by Forrester, businesses will lead the surge in IoT adoption in 2019, with 85% of companies implementing or planning IoT deployments this year. Security and privacy has always been a crucial concern for any kind of computer communications and networks. When it comes to IoT the effects would even more hazardous since it involves small devices communication with sensing abilities without any human intervention. There are so many security threats in IoT environment which motivates the researchers to focus on and try to resolve in order to achieve greater results and new possibilities. The IoT based systems can be exploited by its inherent security vulnerabilities. Because IoT environment usually operates in a heterogeneous manner and it deals with interoperability issues also, it is very tedious task to design specific security mechanism. The resource constraint nature of the devices (Things) imposes additional overhead in order to develop an efficient solution for security related problems. We can consider security issues as a hurdle in widespread adoption of IoT devices. One vital security mechanism is Intrusion Detection System (IDS).

Keywords: IDS · IoT · Threats · AI

1 Introduction

The internet of things (IoT) is a paradigm in which numerous devices are connected online, including sensors, microprocessors, etc [1]. With the constant use of the internet in our daily lives, cyber attacks are constantly rising in the modern world. As a result, one of the main problems facing users of information and communication technology is the security of data transmitted via networks (ICT) in coming years. In the year 1999, Kevin Aston introduced IoT and since then it has become an integral part of our day to day life.It has been recognized as one of the most important and significant factors in the Information and communication technology (ICT) in coming years. According to Somnath Paul and T. V. Sarath [2], the IoT is affecting multiple areas of human life, including how individuals drive, make transactions, and acquire power for their houses and so on. From the small body sensors to the current CC services, the IoT encompasses it all.

© The Author(s), under exclusive license to Springer Nature Switzerland AG 2025
M. Gupta et al. (Eds.): MISS 2023, CCIS 1952, pp. 25–55, 2025.
https://doi.org/10.1007/978-3-031-69115-7_2

Through IoT various equipment's, networks and individuals can be connected together which results in a complex distributed system. Moreover, it improves the human lives by enabling reliable and effective M2M (machine to Machine) and M2H (Machine to Human) communication [3]. Smart homes are one of the most prevalent application of IoT where a number of IoT devices which include, lamps, cameras and door locks and so on are placed and linked to automate daily operations and monitor the status of the homes from distant locations. In addition to this, IoT is prominently used in Industrial areas as well which sues Wireless sensors to offer a real time updates on the status of a manufacturing line [4, 5]. Another application of the IoT is the Healthcare and medical area, where the patient's health is monitored and analyzed by connecting various monitoring machines together [6, 7]. Due to the widespread use of IoT devices in our day to day life, it has been estimated that by the end of 2030, there will be around 50 billion IoT devices including everything i.e. from smartphones to kitchen appliances [8]. Despite these benefits, a considerable impediment stands in the way of the IoT's widespread implementation in research and industrial applications. Security, trust,reliability, accessibility and portability are just a few of the issues that hinder the performance of IoTs and hence, needs to be addressed [9]. Since, IoT devices are interconnected via the global internet with authentic and vulnerable protocols, it becomes susceptible to various security threats [10]. Intruders might take advantage of these flaws and implant different anomalies which in return hinders the decision ability of the system and hence can cause disastrous consequences for people's lives, property and economics [11, 12]. Over the years, continuously evolving cyberattacks pose substantial problems to IoT systems. Aside from the many techniques utilized by the IoT industry, the diverse and distributed nature of IoT applications enhances the complexity of IoT networks, making them more prone to security risk. In order to address the IoT security issues, experts initially implemented various traditional security measures such as, Encryption, security of network and applications, authentication, access control etc. Unfortunately, these security technology deployments have proven insufficient and have required improvement to meet the diverse contextual demands of their respective contexts. Nonetheless, deploying security precautions for significant security risks has proven out to be productive, despite new attack Methods and Tactics (M&T) [13], frequently thwarting them. The first and most critical step in any network security system is to figure out how to detect invader activity because when intruder is found, only then it can be mitigated. Because of this, development in the field of IoT to make it more efficient and secure continuous to be the hot research area. In a typical IoT network, the intrusion detection system serves as the guardian to servers that enable protection against intrusions and attacks. Figure 1, depicts the IoT network in which IDS is implemented to detect attacks in the network. A large number of IoT servers and devices are susceptible to attacks as they are directly connected to the public internet thorough remote sensing feature. Intruders will exploit the flaws to get access to IoT devices. Therefore, in order to recognize and then secure IoT servers from intruders, an IDS is essential. Intrusion Detection System (IDS) is one of the strong mechanisms for tracking the IoT systems and performs best at the network layer of the IoT system. The researchers in [14], defined IDS as the system in which the process of monitoring activities is done automatically in a system and analyzes the malicious content. On the hand, the authors in [15], described the intrusion Prevention system

(IPS) which not only detects the intrusion or anomalies in the network but also takes the necessary measures to mitigate them. Therefore, it can be said that every IPS is an IDS while the reverse is not true. Moreover, a typical Intrusion prevention system (IPS) can also sandbox the anomalous traffic, that is something the regular firewalls and Intrusion detection systems are not able to do. Some of the commonly utilized Intrusion detection and prevention systems are, check point IPS, KFSensor, NIPS (Network intrusion prevention system), AlienVault Unified Security management, NGIPS (Next-generation Intrusion Prevention System), Suricate, Snare and so on.

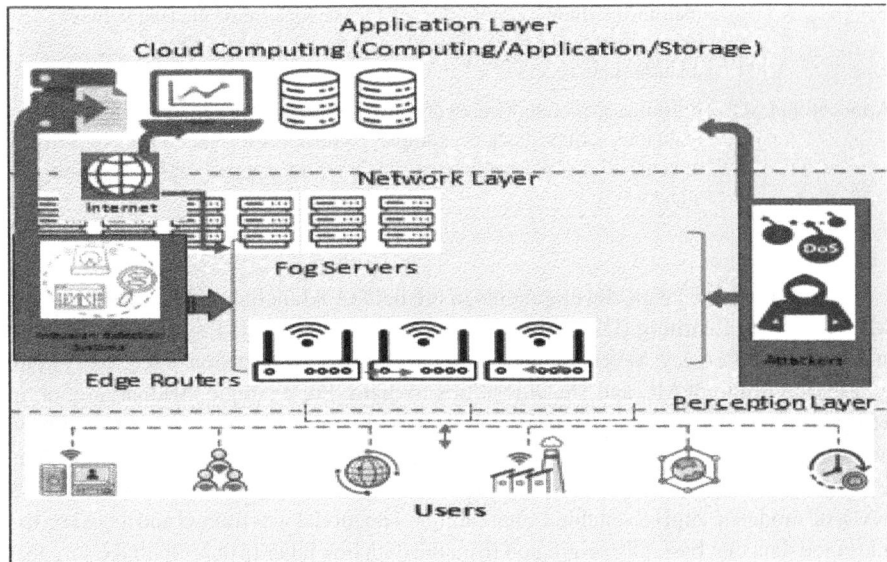

Fig. 1. IoT framework with IDS deployment

Generally, a typical IoT structure comprises of three layers those are, application layer, network layer and perception layer. The functioning of all these layers is interlinked where, the perception layer is responsible for collecting information from the surrounding areas with the help of different sensors. This information is then transmitted over to the network layer, which comprises of various fog servers and edge routers that are sued for processing and routing information. This layer is responsible for creating and maintaining connections between smart devices, servers and network devices. Furthermore, this layer is also responsible for processing and transmitting the sensor data to the next layer. This layer is most vulnerable for attacks as its connected to the public internet. The intruders or hackers try to alter the network resources to disrupt the functioning of the network layer. Finally, in the application layer of IoT, users interact and use the specified applications as per their needs. It comprises of various application which include, smart homes, cities and healthcare sectors. Different types of attacks can be launched by the attackers in the network and application layer. Among all the attacks, the DOS attack on the network layer is prominent. Some of the commonly occurred threats that occur at each layer mentioned in Table 1, along with their brief description.

Table 1. Attacks that occur on different layers

Layer	Security threats	Description
Perception layer	side channel, malicious code injection, false data injection, replay,Node capture, cryptanalysis, eavesdropping and interference, sleep deprivation etc	The perception layer's safety concerns are focused on generating pooled data from IoT devices and eliminating perception equipment
Network layer	DoS, sinkhole, wormhole, man-in-the-middle, routing information,spoofing, Sybil and unauthorized access	In this layer, the security concerns are focused on altering network resource availability
Application layer	Phishing, malicious virus/worm, malicious scripts attack is example of some of the commonly occurred attacks	Security issue lie with the software attacks as its main objective is to provide support services to users

Over the last few years, developments in the field of AI including Machine Learning (ML) and Deep Learning (DL) have been implemented in the IoT systems in order to enhance their efficiency. Several authors proposed IoT based Intrusion detection system, by utilizing different ML and DL algorithms in them. For example, Sridharan et al. in [16], developed various autoencoders for identifying anomalies and later on also used the RF (Random Forest) as the classifier for categorizing attacks in an IoT network like, IP cameras and smart lamps controlled via Wi-Fi. Moreover, they also analyzed their WADAC model on ZigBee-enabled smart lamps. The model was trained and tested on the encrypted data that basically originated from the data link layer in order to make sure that the privacy and confidentiality of the user is maintained. The system was able to detect anomalous traffic with an accuracy of 97.4%. Similarly, the experts in [17], proposed a 3 layers IoT framework in which various types of IoT devices were categorized in the first layer. In the second layer, several anomalies were detected and finally they were classified into 4 defined categories of DDOS, Man in the Middle/spoofing, reconnaissance and replay in the third layer. The suggested model achieved good results in terms of F1. Also, a wireless based IDS was developed by authors in [18] in which they utilized DL based techniques for detecting intrusions. They trained their model using feedforward NN on the NSL-KDD database. The detailed information about the datasets is provided in the latter sections of this paper. No doubt that such system was able to generate good results in detecting anomalies in the IoT environment. However, there were some issues in the traditional IDS that degraded their performance. For Instance, it was yet unclear that which database, ML or DL algorithm must be used to generate effective and beneficial system for detecting intrusions in IoT. Secondly, although being a significant component for the success of 'online' IDSs [19], the total time taken for developing and evaluating IoT based IDS was not taken into account while assessing some IDSs approaches. Keeping this in mind, a brief overview of review about the various IoT based IDS is reviewed in this paper, where our main focus is on below mentioned points;

- Intrusion Detection systems
- Classification techniques of IDS in IoT
- Comparison of various signature-based IDS techniques
- IoT based IDS datasets
- ML based IDS Techniques along with their comparison
- DL based IDS Techniques along with their comparison
- Section III discusses interesting Findings. Lastly, Section IV discusses the Conclusion.

1.1 Intrusion Detection System (IDS) and It's Classification in IoT

An Intrusion Detection System (IDS) can be defined as the proactive technology that automatically detects and classifies various intrusions, attacks, and breaches in security protocols at the network, host, or hybrid infrastructure layer. On the basis of the predefined configurable detection levels, a typical IDS examines the traffic on the network so that it can identify and detect a breach. Early identification of an intrusion will prevent it and remove it from the system before it causes any harm to the confidential data. Basically, an IDS presumes that the intrusion behavior is different from the behavior of an authorized user, as a result, the intrusions in IDS are quantifiedin terms of its features [20]. An IDS can be deployed in a number of fashions, but the essential functionality remains the same i.e. detecting harmful hacking attempts and preventing such attacks from inflicting further damage to the company.On the basis of how responsive an IDS is, it can be classified into two categories, those are, Active IDS and Passive IDS. An active Intrusion Detection System is a type of IDS that has the ability to automatically detect and block malware attacks without any human intervention. On the other hand, a passive Intrusion Detection System simply monitors the traffic on the network and notifies the user when something malicious is found. In this section, some of the recently proposed IoT Based IDS are reviewed. The following parameters were used to characterize each study, IDS installation or data collection approaches, detection or Data analysis methodand validation strategy.Our main focus in this paper, will be on the data collection techniques and Data analysis techniques that will be defined one by one in the next section.

1.2 IDS Deployment Strategies

Depending on the deployment position of an Intrusion detection system, it can be divided into three types (see Fig. 1), one is Host Based Intrusion Detection System (HIDS) and the other is Network based Intrusion Detection Systems (NIDS) and the third is the hybrid (NIDS and HIDS).Whenever, an IDS is deployed on individuals' workstations, it is referred as HIDS while as when IDS, is installed on a network segment, it is referred as the NIDS. However, HIDS undergoes through a variety of disadvantages that make them unsuitable for study. While as, the NIDS identifies the fraudulent or invasive activity in the host networks. The significance of the NIDS keeps on increasing continuously with the increase in the variability, size and value of the network data [21]. On the other hand, in case of hybrid IDS, both NIDS and HIDS along with the Wireless IDS (WIDS) for detecting intrusion in the dynamic and heterogenous IoT network. Nevertheless, the

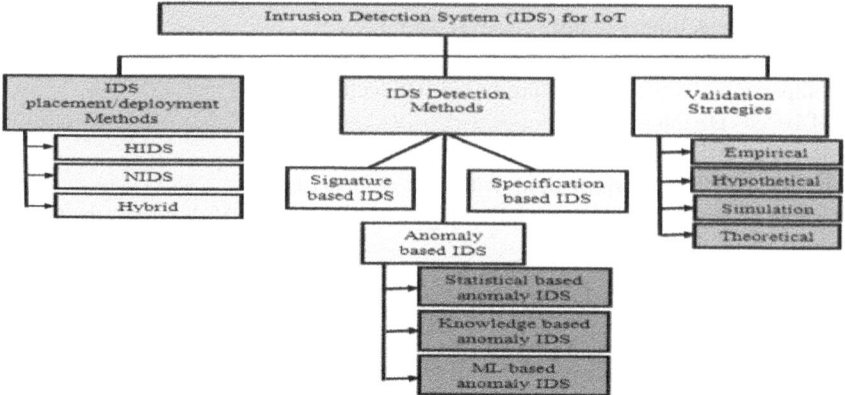

Fig. 2. Classification of IDS in IoT

concept of detecting intrusions on the basis of the gathering low level host data is not new and several researchers have worked in this domain.

Garfinkel & Rosenblum in [22], proposed a framework in which they tried to maintainthe visibility of a host-based IDS while moving the IDS outside of the host for increased intrusion resilienceby utilizing the virtual Machines (VM). The authors were able to separate the IDS from the monitored host while still having an exceptional view into hosts state. In addition to this, the authors used the VMM to entirely mediate the communications between the host software and underlying hardware. A complete analysis of the suggested approach was provided that included Livewire as well. The authors also showcased the Livewire in action by developing a set of basic intrusion detection policies and utilized them to identify intrusions in real time. Similarly, the authors in [23], carried out an experiment in which they utilized the Leipzig Intrusion detection Database (LID-DS) that was released in the year 2018. Moreover, the authors of this paper also suggested a mechanism for detecting anomaly in the IoT frameworks in which number of steps like, pre-processing, vector to image processing, training and testing of the model was accomplished to enhance the overall performance of the system. Also, a Siamese Convolutional Neural network (Siamese-CNN) was developed during the training and testing stages by utilizing the Few-shot learning method. This proved out to be effective for small sample size data in order to determine the attack type, the Siamese-CNN model assessed the similarity score for every cyberattack transformed into an image. Finally, the effectiveness of the vanilla-CNN and Siamese-CNN were compared to prove the supremacy of the suggested model. The results demonstrated that the value for recall in Siamese-CNN was improved by 6% than Vanilla-CNN.

1.3 IoT Intrusion Detection System Methods

As discussed earlier, that any unlawful action or behavior which affects the IoT system is referred as intrusion. It can also be defined as an assault which compromises the privacy, authenticity or access to information in any way. A simplest and easy example of an intrusion is when the authentic and authorized user is not able to use its computer services. The primary goal of every IDS is to detect unauthorized computer usage and harmful traffic on the network, that is impossible to do with typical firewalls. As a reason, computer systems are strongly protected from hostile operations which threaten computer systems' accessibility, authenticity, or privacy. The data analysis techniques of IoT based IDS is basically divided into three types of signature-based IDS, Anomaly based IDS and Specification based IDS. Out of the three, the anomaly-based IDS is very prominent and is further divided into three types, those are; statistical based IDS, Knowledge based IDS and Machine-Learning based IDS (see Fig. 1). Each category is described individually in the upcoming sections along with the work done by authors in that field. The difference between the signature-based IDS, anomaly-based IDS and specification-based IDS is mentioned in the tabular form and is shown in Table 2. The major difference between these three techniques is that Anomaly based IDS is able to identify zero-day attacks which is not the case in signature-based IDS that can only identify and detect previously known intrusions. Nevertheless, since anomalies might still represent new normal behaviors instead of true invasions, AIDS can lead to high false-positive rate. In case of specification-basedIDS, some rules and threshold values are specified by the system administrators in advance. Here, the IDS monitor the present condition of the network and also identifies the rules and thresholds established by administrators. An attack is identified by the specification-basedIDS when the threshold value if crossed or the rules are broken.

1.3.1 Signature Based IDS (SIDS)

Signature based intrusion detection system (SIDS), sometimes also called as the knowledge-basedIDS or Misuse Detection is a technique in which a signature identifier for known attacks is determined in order to detect them in future [24]. It identifies the malicious content in the IoT network every time a match is found with the specific signature. Generally, the accuracy in detecting intrusions is very high in SIDS especially for previously detected attacks. Figure 2 depicts how SIDS techniques function conceptually. The fundamental concept is to create an intrusion database schema, analyze current operations to known signatures, and alert the user if a similarity is detected. A rule like "if: antecedent -then: consequent" might result in "if (source IP address=destination IP address) then mark as an assault". Many researchers have proposed different approaches to enhance the performance of Signature based intrusion detection systems. Some of them are defined in [25–33].

The authors in [25], proposed an IDS model that can protect the IoT network from various attacks including external and internal. To do so, the authors of this paper used the signature-based method for detecting intrusions which includes centralized and distributed IDS modules. Furthermore, they used a Cooja Simulator for detecting DOS attacks in IoT devices. The RPL system that was extensively used for routing in low-power networks as well as in IoT networks was also employed. They also developed two types of attacks in specific, one if Hello flooding and the second one is version number alteration. After conducting experiments, it was found that such attacks can have an influence on the reachability and energy usage of certain IoT systems. Similarly, the authors in [26], proposed a signature-based intrusion detection system. SIDS struggles with the large number of signatures that need to be stored in its dataset. In addition to this, the

Table 2. Comparison for IDS methodologies

Detection method	Advantages	Disadvantages
Anomaly based intrusion detection system (AIDS)	• It has the capacity to detect ongoing assaults • It can be used to create an intrusion signature	• Since AIDS cannot process data communications, an attack may go undiscovered and be dangerous • A lot of false positive alarms occur • It's challenging to design a standard configuration for a computer system that is very dynamic • Unclassified alerts are another example • Some introductory training is required
Signature based intrusion detection system (SIDS)	• Very effective at identifying intruders with the fewest false alarms (FA) • Quickly detects intruders • Very good at spotting known attacks • Design is straightforward	• New signatures must be added on a regular basis • Because they are renowned for spotting known attacks, they are unable to recognise even tiny variations from earlier attacks • The zero-day assault went unnoticed • Incapable of seeing multiple-staged assaults • The intelligence of the assaults is not understood
Specification based IDS methods	• As soon as the specified threshold value is reached or any rule is broken, it detects intrusions	• If the intrusion data is significantly altered by the intruder, it cannot detect attacks. High false alarm rate takes a lot of time

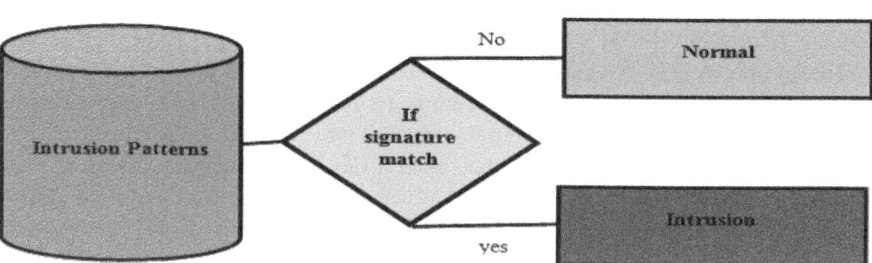

Fig. 3. Working mechanism of SIDS approaches

authors also analyzed that traditional signature-based IDS suggested regular updations in its dataset to avoid dataset size problems. However, all these methods were unable to explain how to deal with the new signatures and obsolete signatures. The authors of this paper hence, described a suggested module in which they combined the parallel processing with small datasets and an updating agent for managing the signatures uploaded frequently. The suggested model was able to detect intrusions or attacks in both the networks i.e. host and network IDS. In [27], the researchers developed a signature-based intrusion detection system that can detect intrusions or unlawful activities on android smart phone and also takes the picture of the intruder and stores it in the log for users. The main purpose of this paper was to enhance the knowledge of users about highly prevalence of attacks on Android smart phones as well as create a mitigating mechanism for make the functioning safe. In order to accomplish the given task, the authors designed an object oriented analysis and design method (OOADM). Furthermore, to identify the intrusions effectively and efficiently in the system, the suggested method was utilized so that the system can be improved by using the real intrusions traits and procedures. Signature detection was also used to detect assaults by focusing on particular sequences, harmful effort, and the attacker's address. The authors in [28], observed that it is extremely important to develop the efficient and reliable intrusion detection system, because of the continuously increasing size and speed of the IoT network. For this, a number of experts utilized the signature- based IDS. However, the major issues faced in such detection systems is keeping up with the massive amounts of the data traffic when each and every packet must be compared with every signature present in the database. Due to this whenever an IDS is overwhelmed by traffic, all it can do is discard packets, which means this could miss possible assaults. To overcome this challenge, the authors of this paper proposed a dynamic Multi-layered SIDA along with the mobile agents. These agents are responsible for detecting the possible attacks with high accuracy through creating and using a dynamic and automatic small and effective dataset, while also providing a means to upgrade such limited signature datasets at periodic intervals utilizing Mobile Agents. As the number of people using the internet grows, so does the likelihood of an assault. On the basis of the nature of assault, signatures can be found in various portions of data packet. The primary function of an IDS is to assist computer networks in preparing for and responding to cyberattacks.IDS have become an important part of assuring network systems security and implant a security policy by scanning the incoming traffic for the already defined signatures. For this, a number of researchers used Snort in their methods that is one of the commonly used SIDS, as it is lightweight and open source software. Using this benefit of Snort, the authors in [29], developed and proposed a SIDS by utilizing Snort and WinPcap. The authors in [30], proposed a method for improving the matching process by utilizing the ML based clustering techniques. The suggested scheme generated a decision tree by using the given set of signatures, each mandating number of constraints that must be satisfied by the inputs values for activating it. Later on, this was used, to discover harmful events with the as minimum repetitive comparisons as possible. Furthermore, the authors in [31], also used the Snort and proposed a data parallelism method for detecting intrusion via Signatures. The accuracy rate of the of detection was enhanced while the total time taken to analyze and drop packets were decreased. Moreover in [32], studied and analyzed that due to the fast development in ICT has altered the whole scenario of network security. There is not a single computer system that is completely safe from the attacks as demonstrated by the DOS attacks on prominent internet services. They also analyzed that conventional methods of network security, such as firewalls and data encryption were no longer appropriate or useful. Even some of the newly proposed IDS were found to be ineffective for new attacks. Keeping this in mind, the authors of this paper proposed a new Signature based IDS that makes use of multithreading. The concept of multithreading was used to manage the traffic on the network. The suggested techniques full advantage of the parallel processing on intercepted data packets. In addition to this, the researchers also proposed an approach that would aid in the update of the IDS database. Although majority of the researchers suggested that the SIDS are best suitable for dealing with the attacks on the application layer, nut few experts also stated that all attacks were not detected. Moreover, several security vulnerabilities were found to be correlated to software development, as well as design or implementation issues. As a result, the authors in [33], developed a new signature-based web application layer for detecting attacks. They analyzed the logs, web servers and dataset logs by using the genetic algorithm. The main motive of this study was to propose a system that can detect mutated signatures. Cross-Site Scripting, SQL injection and remote file are the example of few attacks that were detected in this approach. The abovementioned signature-based intrusion detection system techniques are summarized and recorded in tabular form, given in Table 3.

Therefore, it was observed from the above works that the Signature based IDS were efficient and effective as long as database were not updated and it mainly raised three concerns. To begin with, malware's unpredictable behavior makes it simple to deceive signature-based systems. This technique fails to resemblance test since, not any signature is matched with the signature in the IDS repository, allowing the intruder access system easily. Secondly, it takes longer time to process and analyze the vast volumes of data as the size of signature repository increases. Finally, and perhaps most crucially, SIDS has trouble identifying zero-day attacks because the signature isn't saved in the database [34]. The application of AIDS methods could be a possible answer to this problem. Instead of characterizing what's really abnormal, AIDS operates by distinguishing among appropriate and undesirable conduct, as discussed in the next section.

1.3.2   Anomaly Based Intrusion Detection Systems (AIDS)

In Anomaly based IDS, the network traffic is monitored so that the patterns can be identified which differ from the usual traffic, because of this, it is also referred as behavior-based intrusion detection system. This form of identification takes a sampling of networking data and uses statistical techniques to look for anomalies; if a benchmark is surpassed, it will notify the administration. Anomaly-based detection has the ability to explore new anomalies, however, it required a lot more computing power to monitor behavior patterns constantly. In addition to this, any minor deviation from the normal traffic might raise an alert which in return leadsto high false positives rates. On the basis of the training approach used, anomaly-based intrusion detection systems can eb divided into three categories, those are; statistical, knowledge based and ML based system, as shown in Fig. 1 [35]. One of the major advantages of the Anomaly based IDS is that it

Table 3. Different Signature IDS Techniques

References	Technique Used	Achievements	Drawback
[25]	• Signature based IDS based on RPL for routing	• Effectively attacks detected DOS	• It cannot detect attacks that were not present in the signature database
[26]	• Combined parallel with small datasets and updating agent	• Detected intrusions in host and network based IDS	• Regular updations in the Database make it complex
[27]	• Object oriented analysis and design method (OOADM) for intrusion detection using real time logs and patterns	• Detected intrusions effectively with high accuracy on android phone	• SIDS method is not effective for detecting intrusions when dealing with large datasets
[28]	• Dynamic multi-layered SIDA with mobile agents	• Upgraded datasets at regular intervals by using the mobile agents	• High processing time
[29]	• Snort and WinPcap	• Lightweight mechanism for detecting intrusions	• Complex consuming and time
[30]	• Machine learning clustering techniques	• Enhanced matching process in Signature based IDS	• ML algorithms didn't perform well with large datasets
[31]	• IDS using Snort	• Improved detection accuracy, drop packets were decreased	NA
[32]	• Multithreading based IDS	• Good accuracy in detecting intrusions	• High processing time
[33]	• GeneticAlgorithmfor analyzing signatures, logs and patterns	• Detected mutated signatures effectively	NA

can detect and identify zero-day attacks, as it doesn't depend on signature datasets to identify aberrant user behavior. There are number of other benefits of using AIDS over SIDS. They can, for starters, uncover internal hostile activity. It triggers an alarm when an attacker begins performing activities in a hacked account which are identifiable in standard user behavior. Furthermore, because the structure is implemented from personalized profile, it is difficult for a malicious user to recognize what is regular user activity without triggering an alert.

A. Statistical Anomaly Based IDS

In this type of detection system, the behavior of the individual or system is analyzed by frequently verifying the values for different factors including user's login activities, resource overflow flags and time durations. In this strategy, it's necessary to seek exact threshold limit and reduce the false alarm rate. The harmful behavior of the intruder is differentiated from regular network traffic in the Statistical Anomaly based IDS (SABIDS) by employing statistical properties like mean and variance of regular activities and statistical measures that assess the divergence of actions from acceptable behavior. An aberrant behavior is scored using a scoring system, and once the resultant score reaches a predefined threshold level, an alert is produced. SABIDS' major benefit is that it requires no prior understanding of security vulnerabilities and may effectively detect threats. In this regard, the researchers in [36], issued various guidelines for statistically based IDS by keeping situations and application regions in mind. They studied the importance of IDS as a result of security risks to computer systems. They analyzed that with the launch of innovative faults in computer networks, various detections approaches have been employed. As mentioned earlier that statistical based IDS utilize statistical features to assess if the seen behavior differs significantly from the anticipated behavioror not. Statistically based anomaly detection has piqued researchers' attention because it offers a complete foundation for building a promising technique. In [37], the researchers reviewed a range of Data mining strategies in order to detect the anomalies in the IoT systems and to gain a better grasp of what's out there and how interested scholars can work in this area in the coming time. Similarly, Ding et al. in [38], suggested a non-cooperative differential strategy which employed statistical approaches so that all endpoints in an IoT paradigm are enabled in order to determine the best amount of network resources to invest the information security based on the game's present state. The interactions between selfish and mischievous nodes were modelled as a differentiated game in this study. The findings revealed how harmful activity can be detected with a significant level of likelihood and precision, as well as with excellent quality and reduced resource usage. The experts in [39], utilized statistical based approach such as Chi-square distribution, Gaussian mixture distribution, and Principal component analysis to examine an anomaly-based IDS for detecting anomalies in hardware profiles. No doubt that the AIDS can detect new anomalies effectively, however, they are more susceptible to the false alarms. This type of techniques was mostly used in HIDS for detecting malicious content such as, infiltration. The characteristics were represented by a vast variety of dimensions, and the computer gets sluggish when dealing with such a large amount of data. to depict the supremacy of the suggested technique, the authors of this paper, compared the outcomes achieved in PCA, Chi- square distribution and cluster with Gaussian mixture distribution.

B. Knowledge Based Anomaly IDS

Knowledge based detection system is sometimes also referred to as cognition based or expert based system. This technique uses a classification method for analyzing the data and classifying it as per the previously defined rules, classes and features attained from the training datasets. In addition to this, the categorization rules, attributes and methods were also implied. In this technique, the two standard intrusion detection techniques (Signature and anomaly) are utilized. Knowledge-based detection makes use of the information obtained about the attacks, which can be utilized to identify any threats or system flaws. However, if somehow the system doesn't have a specific information about a particular attack, it won't be able to recognize it, so it needs to know a lot about attacks. The fundamental benefit of such systems is that they often create fewer false alarms and more precise outcomes. For this type of technology, the knowledge database must be updated on a frequent basis [40]. This method utilizes periodic knowledge upgrades in systems that employ it to identify new threats and potential vulnerabilities.

The authors in [41], alleged that Hidden Nave Bayes (HNB) method may be used to solve intrusion detection problems with high networking streaming data rates, complexity, and strongly correlated characteristics. HNB is a data mining paradigm which lowers the assumption of independence assumption in Nave Bayes technique. The results demonstrated that the HNB approach outperforms the classical NB method as per their accuracy, error rate, misclassification cost. In terms of predicting accuracy, the proposed model outperformed other leading traditional models like SVM. Moreover, the suggested methodology considerably enhanced the accuracy of identifying denial- of-service (DoS) assaults, according to the findings.

C. ML Based Anomaly IDS

On the other hand, Machine Learning based Anomaly detection systems are used more frequently and commonly in IoT-IDS.ML techniques are now widely used to solve a wide range of real-world challenges. Machine learning, as described in [42], allows machines to understand without having to be explicitly programmed. Its purpose is to enable a machine to learn from history or current and apply that knowledge to make upcoming decisions and predictions. Despite the fact that "training" is critical in machine learning, it is not the goal. Machine learning's main goal is to develop a system that can automatically and reliably discover meaningful patterns in data [43]. These approaches use an explicitly or implicitly framework to categorize the behaviors being investigated. Genetic Algorithms, Fuzzy Logic, Neural Networks, Bayesian Networks, and Outlier Detection [44] are some of the categories. The brief overview of these ML algorithms are given in Table 4.

However, the majority of the current IDSs, developed intrusion detection system using traditional ML techniques [45]. These IDS model was established using a variety of machine learning approaches. Because of the exponential growth and volume of data, majority of the traditional ML algorithms which required well crafted feature

Table 4. Various Machine learning based anomaly detection techniques

Technique	Pros/cons
Fuzzy Logic	• Generalized logic is preferred over accurate reasoning • promising against port scans and probes, in particular • There is a significant amount of resource usage required
Genetic Algorithm	• often uses characteristics such as Selection, Crossover, and Mutation. –ability to develop classification rules and determines optimal parameters
Neural Network	• Has the capacity to recognize potential unknown trends based on minimal, noisy, and incomplete information
Bayesian Network	• Ability to integrate both previous information and expertise into probabilistic correlations among the variables of interest

engineering demand substantial investigation in order to retrieve important and crucial features from mammoth size unstructured data generated by IoT devices. As a result, the conventional ML based solutions for detecting intrusion in IoT networks face various issues. To overcome the limitation of ML approaches, Deep learning-based algorithms are gaining popularity in IDS. DL accelerates the extraction of meaningful information from swift and actual data streams in order to forecast the potential of the IoT sector. DL methods are considered to be more dependable than conventional ML algorithms as, it can readily obtain information which in return gives superior accuracy results [46]. Because of these advantages of DL methods, a significant number of researchers have employed DL algorithms for anomaly and malware detection. Since, ML and DL-based IDS techniques are increasingly used everywhere, therefore, we will be discussing, few of the recently evolved ML and DL-based methodologies for detecting attacks. But before discussing the work done in this domain, we will discuss some foR the commonly utilized datasets in the IDS.

i. IoT Based Intrusion Detection System Datasets

The evaluation datasets are critical for validating any IDS technique since they enable us to examine the suggested technique's potential to perceive intruding behavior. However, because of the data privacy and confidentiality concerns, the network evaluation databases are not readily available in commercial products. IoT Botnet, NSL-KDD, and ADFA-LD are a few of the well-known and often used publicly accessible datasets.

- **IoT botnet:** The Bot-IoT database is yet another publicly available dataset that comprises of both typical IoT network activity and a range of intrusions. This Database depicts a real IoT network that aim in determining different kind of attacks which include, DDoS, DoS, OS and Service Scan, Keylogging, and Data Exfiltration and so on.
- **NSL-KDD:** The KDD cup99 database gave rise to the publicly accessible NSL-KDD database. The NSL- KDD training data has 125,973 entries, whereas the testing data contains 22,544 entries. The NSL-KDD database is big enough to allow for the usage

of the dataset without arbitrary sampling. The NSL KDD database contains 22 training intrusion attacks and 42 characteristics (i.e., features). In this database, a relationship by itself has 21 characteristics, but a relationship type within a host has 19 features [9].

- **CAIDA:** This dataset was obtained in 2007 and contains internet traffic logs from Distributed Denial-of- Service (DDoS) threats (Hick et al., 2007). This form of denial-of-service threat tries to disrupt a target computer's or network's normal activity by flooding the victim with data packet, blocking normal traffic from entering its authorized recipient machine.

ii. Machine Learning-Based IDS Techniques

Liu et al. [48] For IoT networks, a proposed artificial immune IDS was built. They claimed the system could automatically pick up new assaults and adapt to the IoT environment. The technology is built on a signature- based model and machine learning. The machine learning strategy used is modelled after the workings of synthetic immune systems. The system's goal is to make the IoT network more secure; as a result, it is a network IDS. Self-adaptation to different situations and self-learning of new attacks are this system's two key characteristics.

Kasinathan et al. [49] a 6LoWPAN-based IDS for IoT networks was developed to detect Denial of Service (DoS) attacks. A DoS protection manager, a Suricata IDS, and an IDS probe make up their suggested DoS detection architecture. The analysis of the IP-based WSNs' weaknesses served as the foundation for the construction of this system. The Suricata IDS runs on a host computer, therefore its benefit is that it can solve the issue of power consumption, conserving power resources in WSNs.

Kasinathan et al. [50] for detecting DoS attacks in IoT networks have proposed an enhanced IDS based on 6LoWPAN. The Security Incident and Event Management System (SIMEMS) and Frequency Agility Manager (FAM) are the system's two main new components, and they are both dependent on the DoS detection architecture depicted in [2]. (SIEM). Together, these elements form a monitoring system that can keep a watch on large networks.

Jun and Chi [51] IDS with complex event processing (CEP) technology has been proposed. The capacity to detect intricate patterns through real-time data processing is a benefit of CEP technology. An event filtering unit, an event database unit, a CEP unit, and an action engine unit make up the event-processing IDS architecture. In order to detect intrusions, the system relies on an event-processing model that applies the rule model methodology. The primary characteristics of this system are its ability to identify intrusions into an IoT system with high performance and real-time operation.

Zhou et al. [52] proposed a multi-model driven IDS and an industrial anomaly based on a Hidden Markov Model to separate attacks from real flaws. The proposed system performs well in terms of high precision and good real-time capabilities, according to experimental findings. Attacks may be quickly identified while reducing the false positive rate. The system's advantage is that it is not relied on attack signatures; as a result, the suggested IDS can identify new or unidentified types of attacks as well.

Pan et al. [53] offered an automated method to create a hybrid IDS for electric power systems that learns temporal state-based specifications to distinguish between disturbances, regular control activities, and cyberattacks.

Surendar and Umamakeswari [54] presented an IDS with a constraint-based specification for 6LoWPAN- based IoT networks. While identifying sinkhole assaults, our system maintains efficiency in terms of QoS metrics. The technology removes malicious nodes from the network and reconstructs it without them. This IDS uses the protocol model approach and is a specification-based IDS that relies on behavioural rules. This system's primary characteristics are its ability to recognise sinkhole attacks, maintain QoS, and isolate hostile nodes.

Ambusaidi et al. in [55] suggested a mutual information-based IDS that uses a feature selection algorithm to choose the best features for classification. Three benchmark data sets (KDD Cup 99, NSL-KDD, and Kyoto 2006 +) were used to assess their methodology.

Bostani and Sheikhan [56] For the detection of various RPL attacks, a hybrid IDS leveraging 6LoWPAN has been developed for IoT networks. This system relies on specification-based intrusion detection modules in the router nodes that act as IDS agents and an anomaly-based intrusion detection module in the root node that acts as the primary IDS. The key advantages of this system are its applicability to large-scale networks and the decrease in communication messages caused by the absence of additional control messages or monitor nodes in the IDS design. When collector and selective attack were launched simultaneously, experimental results showed that the proposed approach acquired a true positive rate of 76.19% and a false positive rate of 5.92%.

Kabir et al. in [57] showed a statistical method for detecting intrusions. Both static and incremental data can be used with the proposed IDS technique. The suggested IDS refers to this as the optimum allocation based least square support vector machine and leverages the sampling concept (OA-LS-SVM). The KDD 99 dataset is used to discuss and validate the suggested methodology. The results of the studies demonstrated the effectiveness of the suggested strategy for intrusion detection of static and incremental datasets for different types of attacks (Table 5).

iii. DL Based Anomaly IDS Techniques

The researchers in [75], aimed to use the DL technology in their work for detecting anomalies and intrusions in IoT networks. The suggested model effectiveness was analyzed by comparing its performance with the traditional ML based IDS approaches and distributed attack detection was also compared with the centralized system. The simulating results showcased that the suggested distributed attack detection system outperforms the centralized DL- based detection solutions. Furthermore, it was also analyzed that the deep learning model are more effective in detecting intrusion when compared with its shallow counterparts. Similarly, the authors in [76], observed that the non-scalable nature of AIDS which may result in misinterpretation of huge data generated by IoT devices. In order to eliminate scalability issue, the authors of this paper developed a AIDS, that was based on vector convolutional deep learning (VCDL). The suggested model can be scalable only in one condition which is if the incoming traffic can be dispersed to the fog layers nodes for preprocessing.This is well reflected in the VCDL method, that

Table 5. Comparison of various ML based IDS techniques

References	Work done	Feature extraction/Classifier	Outcome
[48]	Proposed a ML based IDS method for detecting the malicious nodes in WSN	Bayesian Classification algorithm	High accuracy rate and low FPR
[49]	Suggested a fuzzy detector-based intrusion detection system to detect unknown attacks	Negative selection algorithmand Genetic algorithm	Reduces search space and yielded high accuracy in detecting intrusions
[50]	Proposed an improved intrusion detection system by using ML algorithms and signatures	Randomforest classifier	Reduced computational complexity and detect unknown attacks quite effectively
[60]	Proposed a data mining-based IDS approach to improve the accuracy rate using NSL-KDD dataset	SVM	High accuracy with decreased FPR
[61]	Studied and analyzed various ML based intrusion detection approaches and their relationship with protocols and intrusions by utilizing NSL-KDD dataset	DifferentML algorithms	Information about the relationship among protocols and intrusions were revealed

(*continued*)

Table 5. (*continued*)

References	Work done	Feature extraction/Classifier	Outcome
[62]	proposed a unique hybrid strategy for detecting intrusions and categorizing attacks	Min-max for pre-processing, RF for Extracting features, SVM for identifying intrusion and ANFIS for classifying attacks	Accuracy of 99.3% achieved for binary class and MSE of 0.084964 for training data, 0.0855203 for testing data and 0.084964 for multiclass
[63]	Proposed a network based IDS that was based on ML and data quality and used NSL-KDD and CICID2017 datasets	Decision tree	99.42% and 98.80% accuracy for NSL-KDD and CICID2017 datasets respectively
[64]	Studied and analyzed different ML algorithms and proposed their own ML model in which pre-processing was done	Random forest for extracting features	Reduced FPR and high accuracy
[65]	proposed a method for identifying the unusual network activity and tried to solve the problems faced in SIDS by using the prediction model	MultipleML algorithms including NB, KNN and DT	The suggested method was able to distinguish between the regular and abnormal traffic effectively

(*continued*)

Table 5. (*continued*)

References	Work done	Feature extraction/Classifier	Outcome
[66]	Proposed an Anomaly based IDS model that combined HIDS and NIDS was designed for the Internet of mobile things	Six different ML algorithms were analyzed	Reducedcomputational complexity
[67]	studied and analyzed a number of ML algorithms that were used for detecting the attacks like SYN-DOS in IoT. They used Public dataset consisting of 2 million samples	Different machine learning algorithm from Apache Spark MLib library	Detection rate of over 99% achieved in all Spark methods. RF gets highest accuracy of 1
[68]	Worked on enhancing the recall value in enhanced fuzzy min0max neural network (EFMN) by using standard datasets	Fuzzy min-max NN	Improved performance than traditional SVM, rule based and neuro-fuzzy based IDS
[69]	Proposed a ML based IDS by using low power devices. NSL-KDD dataset for Weka was used	Multilayer perceptron (MLP)	Better results in terms of accuracy,precision and coverage
[70]	Proposed a linear SVM based IDS	Linear SVM	Performedbetterthan traditional SVM and other ML algorithms

(*continued*)

Table 5. *(continued)*

References	Work done	Feature extraction/Classifier	Outcome
[71]	Proposed a Sugeno based ANFIS model for identifying intrusion in the network	ANFIS	Better results in terms of MSE, confusion matrix and accuracy
[72]	proposed a new online anomaly learning and detection approach for IoT large-scale services	Reversiblejump MCMC learning	Detectedanomalies effectively in the network
[73]	examined the implementation and role of ML algorithms in Smart Homes for detecting the abnormal actions	HiddenMarkov model (HMM)	Accuracyof97%in detecting anomalies
[74]	proposed a hybrid intrusion detection system (HIDS) for identifying intrusions. Dataset used NSL-KDD and ADFA	C5 DT and One class SVM (OC-SVM)	High accuracy and low false positive rates

distributes IoT traffic processing and performs calculations in fog nodes. The necessary parameters required for training the network were shared by the master node that is present in the fog layer.In addition, the suggested anomaly detection method divides IoT data into two categories: normal and malicious, and afterwards sends it to the cloud for attack neutralization. Investigations were performed out on the Bot-IoT database from UNSW, and the findings demonstrated that the new distributed deep learning model can manage scaling data more effectively than traditional centralized algorithms. When contrasted to traditional AIDS, experimental results showed that the suggested technique is much better in terms of accuracy, precision, and recall. Similarly, the authors in [77], introduced a fresh new IDS framework for IoT devices, that was meant to recognize device-or host-oriented assaults in a lightweight fashion while taking into account the restricted CPU resources that were available on IoT devices. Keeping this in mind, they proposed a stacking model that combines Extreme Gradient Boosting (XGBoost) and the Long Short-Term Memory (LSTM) models for identifying the malicious behavior in IoT environment. They used the system call sequence as a signal of anomalous activity in particular. The renowned -gram model which is one of the common technique for detecting hostbased intrusions, was also utilized to analyze the gathered system call patterns first. Followed by tis, the stacking model was then used for identifying the aberrant behavior in the system call patterns. The performance of the suggested approach was tested by setting up the real world IP camera system towards which numerous IoT attacks were launched. Experimental investigation assessments revealed that the stacking model outperforms other existing anomaly detection techniques, with a 0.983 AUC score achieved for actual data. The XGBoost-LSTM stacking model also provided good performance, reliability, and generalization ability as per the numerical tests. Moreover, in [78], a long short-term memory (LSTM)-Gauss-NBayes method was proposed for detecting intrusion in the IIoT systems. The suggested LSTM-NN model entirely relies on the normal time series, where the Gaussian Naïve Bayes was used to anticipate errors. In the proposed work, the major benefits on LSTM and Gaussian Naïve Bayes were taken into consideration which aids in improving the classification accuracy via predictive error in Gaussian Naïve Bayes as well as high prediction ability of the LSTM for future data points. The current method was tested on three real-world datasets with both long-term and short-term time dependence. The results revealed that the suggested technique was able to detect anomalies in IoT based system more effectively and accurately. Similarly, the authors in [79], introduced an autoencoder based NN in wireless sensor networks for detecting anomalies in networks. The suggested approach basically worked in two stages, in which various intrusion or anomalies were detected in a distributed way and that too without nay human intervention. In the second stage, the cloud system was used for handling the computationally complex tasks more effectively with less frequency. Moreover, in majority of the COTS sensors, the computational load on sensors is much lower and affordable. The authors used the real WSN based testbed and sensor information that was gathered from last 4 months and illustrated the effectiveness of suggested scheme in terms of high accuracy rate and low FPR. In [80], proposed a hybrid approach, named as, SCDNN, in which spectral clustering (SC) and DNN (Deep Neural network) were integrated so that intrusion can be detected. In the beginning, they partitioned the given dataset into k number of groups on the basis of the similarity samples. On the basis of these similarity points, the total distance between each data point in training and testing dataset is analyzed which was later passed on to the DNN as input. A total of 6 KDD-Cup99 and NSL-KDD databases and a sensor dataset were used in the proposed framework to validate its efficacy. To evaluate the supremacy of the suggested model, its performance was analyzed and compared with the traditional BPNN, SVM, RF and Bayes Tree model in terms of accuracy and types of attack. It also serves as a useful tool for researching and analyzing intrusion detection in huge systems.

In addition to this, the authors in [81], used the CIDDS database for developing an improved IDS based on DL techniques, named as, LSTM. The results demonstrated that an accuracy of 0.85 was achieved by the suggested scheme. In [82], an enhanced approach was developed for identifying intrusions in the IoT systems. Initially, a Negative Selection Algorithm (NSA) was used to build a training dataset based exclusively on understanding of regular network behavior. They trained the basic Neural Network to accomplish the real categorization depending on this information. The multilayered method enabled the training intricacy to be separated from the computing and power limitations of IoT devices. Additionally, the introduction of a Negative Selection layer allowed them to train the NN solely on the network's normal behavior and eliminated the necessity for non-self-attack knowledge. They called the suggested scheme as Negative Selection Neural Network or NSNN, whose performance was analyzed by using the KDD-NSL database. The test findings showed that the suggested method can be used as a network intrusion detection classifier.

Furthermore, an Artificial Neural Network (ANN) based IDS was proposed by the authors in [83], in order to examine the security vulnerabilities in an IoT environment. In their work, the potential of a multi-level perceptron which is basically a sort of supervised ANN, to prevent Distributed Denial of Service (DDoS/DoS) threats was tested by employing internet packet logs. The current technique majorly focuses on the classification of normal and attack patterns. A simulated IoT system was used to test the proposed ANN method. The experimental findings showed that the system can recognize multiple DDoS/DoS threats with an accuracy of 99.4%. Similarly, the authors in [84], proposed a conditional variational autoencoder based IDS. The suggested model comprises of the particular framework that combines intrusion labels in the decoder phase. The main benefit of this technique was that it was not complicated when compared with traditional variational autoencoders and generated a high accuracy result. Furthermore, the suggested approach was capable of feature reconstruction, which means it can restore missing features from partial training data. The authors also showcased that even for categorical attributes with a wide range of different entries, the restoration accuracy was very good. In [85], the researchers investigated the use of DL along with the RNN for detecting intrusions in IoT. The suggested technique analyzed the operational codes or OpCodes of ARM IoT applications by using the RNN. Moreover, they employed an IoT application database in which 281 malware and 270 benign ware were present, which were used for training the system. The performance of the suggested model was then analyzed by passing the 100 new IoT malware samples along with the 3 different LSTM configurations. The topology in which 2 layered neurons were present showcased the highest accuracy of 98.18% for recognizing new intrusions, as per the results of 10-fold cross validation. Furthermore, a comparison of the LSTM technique with some other ML classifiers also showcased the efficiency and effectiveness of suggested model (Table 6).

1.4 IDS Validation Techniques

Validation of an IDS can be defined as, when a system is developed and implemented in such a way that it performs reasonably well in accordance with the objectives. However, a number of validation methodologies exist, that may all be divided into two categories based on the source of information, one is experts and other is data. In expert system, a highly subjective and qualitative system is provided by expert sources of information, whilst objectivity and quantitative validation is provided by data sources of information. Validation strategies can be, hypothetical, empirical, simulation or theoretical.

Table 6. Comparison OF various DL based anomaly IDS techniques

References	DL technique	Mechanism	Dataset utilized	Advantages	Drawbacks
[75]	Deep neural network	Classification	NSL-KDD dataset	Produce highly accurate results in real time scenarios by using fog computing	High resources and time demand
[76]	Vector Convolutional Neural network or vector CNN	VCN extracting for features and FCN for classification purpose	Bot-IoT dataset	Consumes less time and is scalable in nature	Performs better with specific features
[77]	Long short Term memory (LSTM)	Classification	Utilized ADFA-LD and their own dataset	Accurately detect intrusions in real time systems	Takes more time
[78]	LSTM	Classification	Created own dataset	High accuracy	Time consuming and high resource demand
[79]	Auto-encoder	Classification	Create own dataset	Low FPR	It is not effective for datasets
[80]	Hybrid-DNN	Classification and prediction	NSL-KDD and KDD99 databases	Better performance than conventional ML algorithms and compatible with heterogenous system	Less training samples were included with average accuracy
[81]	LSTM	Classification	CIDDS	Performs better than conventional ML algorithms	It cannot detect zero day attacks
[82]	Basic network neural	Classification	NSL-KDD dataset	High accuracy, less processing time	Cannotdetect zero-day attacks

(continued)

Table 6. (*continued*)

References	DL technique	Mechanism	Dataset utilized	Advantages	Drawbacks
[83]	Basic NN	Classification	Created their own dataset	Detection accuracy is good	Less efficient
[84]	ID-CVAE	Classification	NSL-KDD dataset	Less complicated and latency	Huge FPR
[85]	Bi-RNN	Malware analysis and feature extraction	Created own dataset and Internet based applications	High detection rate	Computational complexityis high

i. **Hypothetical:**As it is understood by the name itself, that the observations and results attained are not real and related to the model.

ii. **Empirical:** In this type ofvalidation technique, the information is collected from multiple operational systems to perform systematic tests.

iii. **Simulation:**in this case, few of the IoT based IDS models are validated and simulated.

iv. **Theoretical**: these are the arguments which are extremely precise and have the ability to support the attained results. Major performance criteria in IDS are;

False positive rate (FPR) which can be mathematically calculated by using the formula given in equation 1,

$$FPR = \frac{FP}{FP + TN} \tag{1}$$

False Negative rate (FNR) can be calculated by using the formula given in 2,

$$FNR = \frac{FN}{FN + TN} \tag{2}$$

True positive rate (TPR), it can be defined as the ration of total accurately predicted attacks over total number of attacks. Mathematically,

$$TPR = \frac{TP}{TP + FN} \tag{3}$$

Accuracy: of a system defines how precise and accurately can a system detect intrusions and anomalies. Mathematically, it can be calculated as,

$$Accuracy = \frac{TP + TN}{TP + TN + FP + FN} \tag{4}$$

2 Key Outcomes

We compiled some significant points based on numerous aspects after evaluating the polls conducted by various academics over the previous few years. Table 7 lists the conclusions reached.

Table 7. Conclusions drawn from the literature review.

Factors	Key Findings/ outcomes
IDS system evolution	We discovered throughout our analysis of IDS that network-based and signature- based IDS were first used to identify intrusions in the IoT network. However, as a result of their exponential expansion, these systems were unable to deliver effective results, leading to the development of more sophisticated IDS, such as specification-based, profile-based, anomaly-based, statistical, knowledge-based, and machine learning (ML) based IDS
Attacks by Type	Another important observation was that, despite the fact that there are several assaults that can target IoT systems, the majority of researchers concentrated on preventing Denial of Service (DOS) attacks. At different stages of the IoT network, numerous more threats including node capturing, bogus data injection, and malicious virus/worm can also happen
utilised dataset	We also determined that there aren't many datasets available for model training and testing because of privacy and security concerns. There are a few datasets that are often used, though, including NSL-KDD, DARPA/KDD98/99, IoT botnet, CICID2017, CAIDA, CIDDS and ADFA However, we also found that NSL-KDD is mostly used by researchers because it produced effective results and aids in improving network performance as a whole
the IDS trends	Moving on to the following issue, we noticed that automation processes for detecting infiltration are being used as a result of the continual increase in data generated by the Internet of Things. For this, the majority of academics today use artificial intelligence-based (ML and DL-based)approaches since they produce useful results and manage big datasets relatively well
IDS with machine learning	Many researchers adopted the automatic method of discovering intrusions while working with ML algorithms One of the well-known and commonly employed classifiers is SVM The primary issue with ML approaches, however, was that they couldn't handle very big linear datasets and had significant false positive rates The most common problems with ML algorithms are overfitting and underfitting

(continued)

Table 7. (*continued*)

Factors	Key Findings/ outcomes
Using deep learning, IDS	IDS adopted DL approaches primarily because they are faster than traditional ML algorithms and can handle massive datasets
	The most popular DL methods for detecting intrusions are CNN and LSTM The problem with DL-based IDS was that it required a lot of resources yet didn't produce findings with a high degree of accuracy Their performance was also hampered by the architectural problems with DL approaches DL technique training is an expensive and challenging operation due to the complicated data model Small datasets don't work well for the DL-based IDS

3   Conclusion

Computer networks are becoming more susceptible to various cyberattacks as a result of the exponential growth in network data, and intrusion detection systems (IDS) were created to address these problems. To guarantee data security and integrity, these systems are in charge of identifying dangerous or threat content in communication networks. These systems must be continually updated because of their importance in computer networks and the daily emergence of new assaults, viruses, and intrusions. In order to protect against security threats, it is critical to create robust solutions as the IoT deals with user personal data and industry information. Due to the enormous amount of heterogeneous data that the Internet of Things (IoT) generates, this is conceivable using the ideas of machine learning and deep learning. In this review article, we examined several intrusion detection strategies put forth in recent years. After examining, we found that because traditional IDS are always evolving, they are unable to recognise incursions actively for new threats. Additionally, we evaluated the effectiveness of each of the main IDS types that were deployed in the past, as well as their advantages and disadvantages. Additionally, the use of AI algorithms, such as ML and DL techniques, in intrusion detection methods (IDS), is also examined. It was discovered that ML and DL techniques are utilised more specifically in modern IDS, increasing the general effectiveness of detection systems. However, there were some problems with those techniques as well because ML algorithms couldn't handle big, linear datasets and didn't produce a lot of positive results. However, given that the DL algorithms' detection accuracy is still poor, additional work needs to be done on them. DL-based approaches, however, are favoured to ML-based ones due to their capacity for handling big datasets and ability to lower the number of false positives in the detection system. The goal of this paper is to educate readers on various computer network attacks and to help them better understand the network infiltration issue, which is crucial for any internet user. Instead of being an evolution, the Internet of Things (IoT) is a revolution. The security challenges change as IoT changes. The Internet of Things is only beneficial to society when it is secure, which is made feasible by artificial intelligence.

References

1. Remesh, A., Muralidharan, D., Raj, N., Gopika, J., Binu, P.K.: Intrusion detection system for IoT devices. In: 2020 International Conference on Electronics and Sustainable Communication Systems (ICESC), pp. 826–830 (2020). https://doi.org/10.1109/ICESC48915.2020.9155999

2. Paul, S., Sarath, T.V.: End to End IoT based hazard monitoring system. In: 2018 International Conference on Inventive Research (ICIRCA), pp. 106–110 (2018)

3. Zarpelão, B.B., et al.: A survey of intrusion detection in Internet of Things. J. Netw. Comput. Appl. **84**, 25–37 (2017)

4. Cosson, A., et al.: Sentinel: a robust intrusion detection system for IoT Networks using kernel-level system information. In: Proceedings of the International Conference on Internet-of-Things Design and Implementation, pp. 53–66. Association for Computing Machinery, New York (2021)

5. Sikder, A.K., Acar, A., , H., , A.S., Akkaya, K., Conti, M.: IoT-enabled smart lighting systems for smart cities. In: 2018 IEEE 8th Annual Computing and Communication Workshop and Conference (CCWC), pp. 639–645. IEEE (2018)

6. Newaz, A.K.M., Sikder, A.K., Rahman, M.A., Uluagac, A.S.: A survey security and privacy issues in modern healthcare systems: attacks and defenses. arXiv preprint arXiv:2005.07359 (2020)

7. Newaz, A.K.M.I., Sikder, A.K., Babun, L., Uluagac, A.S.: Heka: a novel intrusion detection system for attacks to personal medical devices. In: IEEE Conference on Communications and Network Security (CNS). IEEE (2020)

8. Holst, A.: Number of Connected Devices Worldwide 2030. Statista, Hamburg ()2018

9. Pal, S., Hitchens, M., Rabehaja, T., Mukhopadhyay, S.: Security requirements for the internet of things: a systematic approach. Sensors **20**, 5897 (2020)

10. Lin, J., Yu, W., Zhang, N., Yang, X., Zhang, H., Zhao, W.: A survey on internet of things: architecture, enabling technologies, security and privacy, and applications. IEEE Internet Things J. **4**, 1125–1142 (2017)

11. Ferrag, M.A., Shu, L., Yang, X., Derhab, A., Maglaras, L.: Security and Privacy for green IoT-based agriculture: review, blockchain solutions, and challenges. IEEE Access **8**, 32031–32053 (2020)

12. Ghaleb, F.A., Maarof, M.A., Zainal, A., Rassam, M., Saeed, F., Alsaedi, M.: Context-aware data-centric misbehaviour detection scheme for vehicular ad hoc networks using sequential analysis of the temporal and spatial correlation of the consistency between the cooperative awareness messages. Veh. Commun. **20**, 100186 (2019)

13. Albulayhi, K., Smadi, A.A., Sheldon, F.T., Abercrombie, R.K.: IoT intrusion detection taxonomy, reference architecture, and analyses. Sensors (Basel). **21**(19), 6432 (2021). https://doi.org/10.3390/s21196432

14. Liao, H.-J., Lin, C.-H.R., Lin, Y.-C., Tung, K.-Y.: Intrusion detection system: a comprehensive review. J. Netw. Comput. Appl. **36**(1), 16–24 (2013)

15. Zhang, X., Li, C., Zheng, W.: Intrusion prevention system design. In: The Fourth International Conference on Computer and Information Technology, CIT'04, pp. 386–390. IEEE (2004)

16. Sridharan, R., Maiti, R.R., Tippenhauer, N.O.: Wadac: privacy-preserving anomaly detection and attack classification on wireless traffic. In: Proceedings of the 11th ACM Conference on Security & Privacy in Wireless and Mobile Networks, WiSec '18, pp. 51–62. Association for Computing Machinery, New York (2018)

17. Anthi, E., Williams, L., Sowinska, M., Theodorakopoulos, G., Burnap, P.: A supervised intrusion detection system for smart home iot devices. IEEE Internet Things J. **6**(5), 9042–9053 (2019)

18. Kasongo, S.M., Sun, Y.: A deep learning method with filter based feature engineering for wireless intrusion detection system. IEEE Access **7**, 38597–38607 (2019)
19. Khraisat, A., Gondal, I., Vamplew, P., Kamruzzaman, J.: Survey of intrusion detection systems: techniques, datasets and challenges. Cybersecurity **2**(1), 20 (2019)
20. Stallings, W., Brown, L., Bauer, M.D., Bhattacharjee, A.K.: Computer Security: Principles and Practice. Pearson Education, Upper Saddle River (2012)
21. Zarpelo, B.B., Miani, R.S., Kawakani, C.T., de Alvarenga, S.C.: A survey of intrusion detection in Internet of Things. J. Netw. Comput. Appl. **84**, 25–37 (2017)
22. Garfinkel, T., Rosenblum, M.: A virtual machine introspection based architecture for intrusion detection. NDSS **3**, 2003 (2003)
23. Park, D., Kim, S., Kwon, H., Shin, D., Shin, D.: Host-based intrusion detection model using siamese network. IEEE Access **9**, 76614–76623 (2021). https://doi.org/10.1109/ACCESS.2021.3082160
24. Khraisat, A., Gondal, I., Vamplew, P., Kamruzzaman, J., Alazab, A.: A Novel ensemble of hybrid intrusion detection system for detecting internet of things attacks. Electronics **8**, 1210 (2019)
25. Ioulianou, P., Vassilakis, V., Moscholios, I.: A signature-based intrusion detection system for the Internet of Things (2018)
26. Almutairi, A.H., Abdelmajeed, N.T.: Innovative signature based intrusion detection system: parallel processing and minimized database. In: 2017 International Conference on the Frontiers and Advances in Data Science (FADS), pp. 114–119 (2017). https://doi.org/10.1109/FADS.2017.8253208
27. Onyebuchi, O.B.: Signature based network intrusion detection system using feature selection on android. Signature **11**(6), 551–558 (2020)
28. Uddin, M., Rahman, A.A.: Dynamic multi layer signature based intrusion detection system using mobile agents. arXiv preprint arXiv:1010.5036 (2010)
29. Shah, S.N., Singh, M.: Signature-based network intrusion detection system using SNORT and WINPCAP. Int. J. Eng. Res. Technol. (IJERT) **1**(10), 1–7 (2012)
30. Kruegel, C., Toth, T.: Using decision trees to improve signature-based intrusion detection. In: Vigna, G., Kruegel, C., Jonsson, E. (eds.) Recent Advances in Intrusion Detection, pp. 173–191. Springer, Heidelberg (2003). https://doi.org/10.1007/978-3-540-45248-5_10
31. Patel, P.M., Rajput, P.H., Patel, P.H.: A parallelism technique to improve signature based intrusion detection system. Asian J. Converg. Technol. (AJCT) (2018). ISSN -2350–1146, 4(II). https://asianssr.org/index.php/ajct/article/view/607
32. Gaikwad, D., et al.: A proposal for implementation of signature based intrusion detection system using multithreading technique. Int. J. Comput. Eng. Res. **2**(7), 59 (2012)
33. Bronte, R., Shahriar, H., Haddad, H.M.: A signature-based intrusion detection system for web applications based on genetic algorithm. In: Proceedings of the 9th International Conference on Security of Information and Networks (SIN 2016), pp. 32–39. Association for Computing Machinery, New York (2016). https://doi.org/10.1145/2947626.2951964
34. Alazab, A., Hobbs, M., Abawajy, J., Khraisat, A., Alazab, M.: Using response action with intelligent intrusion detection and prevention system against web application malware. Inf. Manag. Comput. Secur. **22**, 431–449 (2014)
35. Butun, I., Morgera, S.D., Sankar, R.: A survey of intrusion detection systems in wireless sensor networks. IEEE Commun Survey Tutorial **16**(1), 266–282 (2014)
36. Qayyum, A., Islam, M.H., Jamil, M.: Taxonomy of statistical based anomaly detection techniques for intrusion detection. In: Proceedings of the IEEE Symposium on Emerging Technologies, pp. 270–276 (2005). https://doi.org/10.1109/ICET.2005.1558893
37. Agrawal, S., Agrawal, J.: Survey on anomaly detection using data mining techniques. Procedia Comput. Sci. **60**, 708–713 (2015)

38. Ding, Y., Zhou, X.-W., Cheng, Z.-M., Lin, F.-H.: A security differential game model for sensor networks in context of the Internet of Things. Wirel. Pers. Commun. **72**, 375–388 (2013). https://doi.org/10.1007/s11277-013-1018-y
39. Om, H., Hazra, T.: Statistical techniques in anomaly intrusion detection system (2012)
40. Herve, D., Marc, D., Andreas, W.: Towards a taxonomy of intrusion detection systems. Comput. Netw. **31**, 805–822 (1999)
41. LeventKoc, T.A., Mazzuchi, S.: A network intrusion detection system based on a Hidden Naïve Bayes multiclass classifier. Expert Syst. Appl. **39**, 13492–13500 (2012)
42. Shanthamallu, U.S., Spanias, A., Tepedelenlioglu, C., Stanley, M.: a brief survey of machine learning methods and their sensor and IoT applications. In: Proceedings of the 2017 8th International Conference on Information, Intelligence, Systems & Applications (IISA), Larnaca, Cyprus, 27–30 August 2017, pp. 1–8 (2017)
43. Bezerra, C.G., Costa, B.S.J., Guedes, L.A., Angelov, P.P.: An evolving approach to data streams clustering based on typicality and eccentricity data analytics. Inf. Sci. **518**, 13–28 (2020)
44. Jyothsna, V., Ramaprasad, V.V., Prasad, K.M.: A review of anomaly based intrusion. Int. J. Comput. Appl. **28**(7), 26–35 (2011)
45. da Costa, K.A., Papa, J.P., Lisboa, C.O., Munoz, R., de Albuquerque, V.H.C.: Internet of Things: a survey on machine learning-based intrusion detection approaches. Comput. Netw. **151**, 147–157 (2019)
46. Chalapathy, R., Chawla, S.: Deep learning for anomaly detection: a survey. arXiv (2019). arXiv:1901.03407
47. Tavallaee, M., Bagheri, E., Lu, W., Ghorbani, A.A.: A detailed analysis of the KDD CUP 99 data set. In: 2009 IEEE Symposium on Computational Intelligence for Security and Defense Applications, pp. 1–6 ((2009))
48. Xiao, Z., Liu, C., Chen, C.: An anomaly detection scheme based on machine learning for WSN. In: IEEE International Conference on Information Science and Engineering (2009)
49. Xu, J., You, J., Liu, F.: fuzzy rules based approach for performance anomaly detection. IEEE (2005)
50. Zhang, J., Zulkernine, M.: based network intrusion detection with unsupervised outlier detection. In: IEEE International Conference on Communications (2006)
51. Kasinathan, P., Pastrone, C., Spirito, M.A., Vinkovits, M.: Denial-of-service detection in 6LoWPAN based internet of things. In: 2013 IEEE 9th International Conference on Wireless and Mobile Computing, Networking and Communications (WiMob), pp 600–607. IEEE, Lyon (2013)
52. Kasinathan, P., Costamagna, G., Khaleel, H., Pastrone, C., Spirito, M.A.: DEMO: an IDS framework for internet of things empowered by 6LoWPAN. In: Proceedings of the 2013 ACM SIGSAC Conference on Computer; Communications Security, CCS 2013, Berlin, pp. 1337–1340 (2013)
53. Jun, C., Chi, C.: Design of complex event-processing IDS in internet of things. In: 2014 Sixth International Conference on Measuring Technology and Mechatronics Automation, pp 226–229. IEEE, Zhangjiajie (2014)
54. Zhou, C., et al.: Design and analysis of multimodel-based anomaly intrusion detection systems in industrial process automation. IEEE Trans. Syst. Man Cybern. Syst. **45**(10), 1345–1360 (2015)
55. Pan, S., Morris, T., Adhikari, U.: Developing a hybrid intrusion detection system using data mining for power systems. IEEE Trans. Smart Grid **6**(6), 3104–3113 (2015)
56. Surendar, M., Umamakeswari, A.: InDReS: an intrusion detection and response system for internet of things with 6LoWPAN. In: 2016 International Conference on Wireless Communications, Signal Processing and Networking (WiSPNET), Chennai, pp 1903–1908 (2016)

57. Ambusaidi, M.A., He, X., Nanda, P., Tan, Z.: Building an intrusion detection system using a filter- based feature selection algorithm. IEEE Trans. Comput. **65**(10), 2986–2998 (2016)
58. Bostani, H., Sheikhan, M.: Hybrid of anomaly-based and specification-based IDS for internet of things using unsupervised OPF based on MapReduce approach. Comput. Commun. (Supplement C), 52–71 (2017)
59. Bhavsar, Y.B., Waghmare, K.C.: Intrusion detection system using data mining technique: support vector machine. Int. J. Emerg. Technol. Adv. Eng. **3**, 581–586 (2013)
60. Dhanabal, L., Shantharajah, S.P.: A study on NSL-KDD dataset for intrusion detection system based on classification algorithms. Int. J. Adv. Res. Comput. Commun. Eng. **4**, 446–452 (2015)
61. Mehmood, M., et al.: A hybrid approach for network intrusion detection. CMC-Comput. Mater. Contin **70**, 91–107 (2022)
62. Guezzaz, A., Benkirane, S., Azrour, M., Khurram, S.: A reliable network intrusion detection approach using decision tree with enhanced data quality. Secur. Commun. Netw. **2021**(1), 1230593 (2021). https://doi.org/10.1155/2021/1230593
63. Wu, F., Li, T., Wu, Z., et al.: Research on network intrusion detection technology based on machine learning. Int. J. Wireless Inf. Networks **28**, 262–275 (2021). https://doi.org/10.1007/s10776-021-00520-z
64. Bhatta, J., Gajurel, K., Nepal, S., Pandey, S., Koirala, S.: Anomaly based intrusion detection system (2019). https://doi.org/10.13140/RG.2.2.25949.84960
65. Zachos, G., Essop, I., Mantas, G., Porfyrakis, K., Ribeiro, J.C., Rodriguez, J.: An anomaly-based intrusion detection system for internet of medical things networks. Electronics **10**, 2562 (2021). https://doi.org/10.3390/electronics10212562
66. Morfino, V., Rampone, S.: Towards near-real-time intrusion detection for IoT devices using supervised learning and apache spark. Electronics **9**, 444 (2020)
67. Upasani, N., Om, H.: A modified neuro-fuzzy classifier and its parallel implementation on modern GPUs for real time intrusion detection. Appl. Soft Comput. **82**, 105595 (2019). https://doi.org/10.1016/j.asoc.2019.105595
68. de Almeida Florencio, F., Moreno, E.D., Teixeira Macedo, H., de BrittoSalgueiro, R.J.P., Barreto do Nascimento, F., Oliveira Santos, F.A.: Intrusion detection via MLP neural network using an arduino embedded system. In: 2018 VIII Brazilian Symposium on Computing Systems Engineering (SBESC), Salvador, Brazil, pp. 190–195 (2018)
69. Ham, H.S., Kim, H.H., Kim, M.S., Choi, M.J.: Linear SVM-based android malware detection for reliable IoT services. J. Appl. Math. **2014**(1), 594501 (2014)
70. Rahman, S., Mamun, S.A., Ahmed, M.U., Kaiser, M.S.: PHY/MAC layer attack detection system using neuro-fuzzy algorithm for IoT network. In: 2016 International Conference on Electrical, Electronics, and Optimization Techniques (ICEEOT), pp. 2531–2536 (2016). https://doi.org/10.1109/ICEEOT.2016.7755150
71. Wang, J., Kuang, Q., Duan, S.: A new online anomaly learning and detection for large-scale service of Internet of Thing. Pers. Ubiquit. Comput. **19**, 1021–1031 (2015). https://doi.org/10.1007/s00779-015-0874-8
72. Kalnoor, G., Gowrishankar, S.: A model for intrusion detection system using hidden Markov and variational Bayesian model for IoT based wireless sensor network. Int. J. Inf. Tecnol. (2021). https://doi.org/10.1007/s41870-021-00748-1
73. Khraisat, A., Gondal, I., Vamplew, P., Kamruzzaman, J., Alazab, A.: Hybrid intrusion detection system based on the stacking ensemble of C5 decision tree classifier and one class support vector machine. Electronics **9**, 173 (2020). https://doi.org/10.3390/electronics9010173
74. Diro, A.A., Chilamkurti, N.: Distributed attack detection scheme using deep learning approach for Internet of Things. Future Gener. Comput. Syst. **82**, 761–768 (2018). https://doi.org/10.1016/j.future.2017.08.043

75. Ng, B.A., Selvakumar, S.: Anomaly detection framework for Internet of things traffic using vector convolutional deep learning approach in fog environment. Future Gener. Comput. Syst. **113**, 255–265 (2020)
76. Wang, X., Lu, X.: A host-based anomaly detection framework using XGBoost and LSTM for IoT devices. Wirel. Commun. Mob. Comput. **2020**, 8838571 (2020). https://doi.org/10.1155/2020/8838571
77. Wu, D., Jiang, Z., Xie, X., Wei, X., Yu, W., Li, R.: LSTM learning with bayesian and gaussian processing for anomaly detection in industrial IoT. IEEE Trans. Ind. Inf. **16**, 5244–5253 (2020). https://doi.org/10.1109/TII.2019.2952917
78. Luo, T., Nagarajan, S.G.: Distributed anomaly detection using autoencoder neural networks in WSN for IoT. In: Proceedings of the 2018 IEEE International Conference on Communications (ICC); Kansas City, KS, USA, 20–24 May 2018 (2018)
79. Ma, T., Wang, F., Cheng, J., Yu, Y., Chen, X.: A hybrid spectral clustering and deep neural network ensemble algorithm for intrusion detection in sensor networks. Sensors **16**, 1701 (2016). https://doi.org/10.3390/s16101701
80. Althubiti, S.A., Jones, E.M., Roy, K.: LSTM for anomaly-based network intrusion detection. In: Proceedings of the 2018 28th International Telecommunication Networks and Applications Conference (ITNAC), Sydney, NSW, Australia. 21–23 November 2018. IEEE, Manhattan (2018)
81. Pamukov, M.E., Poulkov, V.K., Shterev, V.A.: Negative selection and neural network based algorithm for intrusion detection in IoT. In: Proceedings of the 2018 41st International Conference on Telecommunications and Signal Processing (TSP); Athens, Greece. 4–6 July 2018. IEEE, Manhattan (2018)
82. Hodo E., et al.: Threat analysis of IoT networks using artificial neural network intrusion detection system. In: Proceedings of the 2016 International Symposium on Networks, Computers and Communications (ISNCC); Hammamet, Tunisia, 11–13 May 2016 (2016)
83. Lopez-Martin, M., Carro, B., Sanchez-Esguevillas, A., Lloret, J.: Conditional variational autoencoder for prediction and feature recovery applied to intrusion detection in IoT. Sensors **17**, 1967 (2017). https://doi.org/10.3390/s17091967
84. HaddadPajouh, H., Dehghantanha, A., Khayami, R., Choo, K.-K.R.: A deep recurrent neural network based approach for internet of things malware threat hunting. Future Generat. Comput. Syst. **85**, 88–96 (2018). https://doi.org/10.1016/j.future.2018.03.007

An Effective Framework for Gastrointestinal Disease Detection Using Hybrid Features

J. Sharmila Joseph[1]([✉]) [iD], Abhay Vidyarthi[1] [iD], and Vibhav Prakash Singh[2] [iD]

[1] VIT Bhopal University Bhopal-Indore Highway, Sehore, Madhya Pradesh 466114, India
{sharmila.joseph,abhay.vidyarthi}@vitbhopal.ac.in
[2] Motilal Nehru National Institute of Technology, Prayagraj, Allahabad 211004, India
vibhav@mnnit.ac.in

Abstract. Machine Learning (ML) techniques are widely used for the detection and classification of diseases in medical fields. These algorithms make decisions based on extracted features from images. Therefore, selecting appropriate feature extraction techniques together with appropriate Machine Learning (ML) algorithms is crucial for achieving high accuracy in classification. Through the literature review, we observed a very few information about ML techniques for the classification of Gastrointestinal (GI) diseases. To solve this issue, we collected five classes of GI images from a publicly available Kvasir dataset and extracted textural features like Local Binary Pattern (LBP) and Gray Level Cooccurrence Matrix (GLCM), color features from RGB (Red Green Blue) and HSV (Hue, Saturation, and Value) color spaces and shape information from HOG. These features are combined and dimension was reduced by Principal Component Analysis (PCA). Finally, the selected features are classified with well-known ML classifier One Against All-Support Vector Machine (OAA-SVM) and obtained classification Accuracy-96.8%, Precision-96.74%, Recall-96.75% and F1-Score-96.73%.

Keywords: Gastrointestinal (GI) diseases · Machine Learning (ML) · Feature extraction · One Against All-Support Vector Machine

1 Introduction

Machine Learning are applied in many fields, including face identification, finger vein recognition, tumor detection, segmentation and classification [1–5]. Recently, it is showing its remarkable performances in medical field. Automatic identification of disease and anatomical features from medical images is such a difficult task that could be highly assist in medical diagnostics and significantly speed up while also lower the expense of investigative procedures.

The extraction of relevant features is very essential part of classification task. We observed that, based on the particular categorization tasks, researchers have employed various types of features in prior studies. In paper [6], ultrasound images are classified as benign and malignant using uniform LBP features. [7] utilized GLCM features for the detection of lung cancer, finally SVM classifier was trained with those features and

© The Author(s), under exclusive license to Springer Nature Switzerland AG 2025
M. Gupta et al. (Eds.): MISS 2023, CCIS 1952, pp. 56–68, 2025.
https://doi.org/10.1007/978-3-031-69115-7_3

classified with accuracy 96.7% with less computational speed. Classification of liver CT images based on HOG-SVM algorithm was developed in [8] and provided the categorization accuracy of 94%. Then coming to Gastro intestinal diseases classification, [9] utilized geometric and deep CNN for extracting features and utilized KNN for classifying Healthy, Bleeding and Ulcer from Wireless Capsule Endoscopy (WCE) images and paper [10]. Paper [11] used a combination of GLCM, LBP and color features with KNN and SVM classifiers for the detection GI abnormalities. A new method was introduced [12] which classified the abnormalities from WCE images using Histogram Oriented Gradient (HOG), LBP and color features and Extreme Learning Machine. In paper [22] a CNN based CAD system was developed to identify Small-Bowel Diseases and Normal Variants from WCE images and obtained best results. In paper [23] an ensemble of Mobilenetv2 and customized CNN for the detection of bleedy images from normal WCE images. The methos used Mobile net for feature extraction and flattened the features and given as the input to the customized CNN for training and classification. A unified CAD model was presented in [24] for the classification of three classes from the WCE images. This method used pixel labeled images and trained using Linear discriminant analysis for ROI and feature extraction and finally classified with hierarchical classifier for classification and in paper [25] a model for detection of ulcerous from WCE images was developed. It segmented the abnormal region using saliency-based color and texture. The color LBP and Pyramid HOG features are extracted and converted to codebook. Eventually classified with hidden Markov model classifier. A model presented in [26] which utilized fine-tuned AlexNet and GoogLeNet architecture for the detection of abnormalities from WCE images and showed that AlexNet outperformed GoogLeNet in the detection of ulcer.

From the survey, it observed that only few handcrafted features-based Computer Aided Diagnostic (CAD) models were developed for GI disease classification. So, in our paper A new GI abnormality detection method using Handcrafted features was presented. Inspired by the previous studies fusion of GLCM, LBP, HOG and color features was done in our paper. The concatenation of these handcrafted features created most discriminative feature set. From the combined features, important features are selected and classified with OAA-SVM ML classifier.

The remaining sections are arranged as follows: The proposed methods for GI image classifications are described in Sect. 2. Experimental findings and a discussion are given in Sect. 3. The paper is concluded at Sect. 4.

2   Methodology

The ability to detect an individual's illness from medical images is a promising computation. The goal of the current work is to attain the highest accuracy in disease prediction. Here Kvasir dataset is utilized for image classification analysis. From these datasets five different GI diseases are classified by using proposed classification technique. The proposed technique consists of four stages.

Stage 1: Preprocessing: - Here image resizing and Contrast enhancement was done.
Stage 2: Feature extraction: - Here color, texture and shape features are extracted and fused together.

Stage 3: Feature selection: - Here irrelevant features are weed out by reducing dimension.
Stage 4: Classification: - The important features are classified as five different classes
of GI diseases. The framework of proposed model is depicted in Fig. 1.

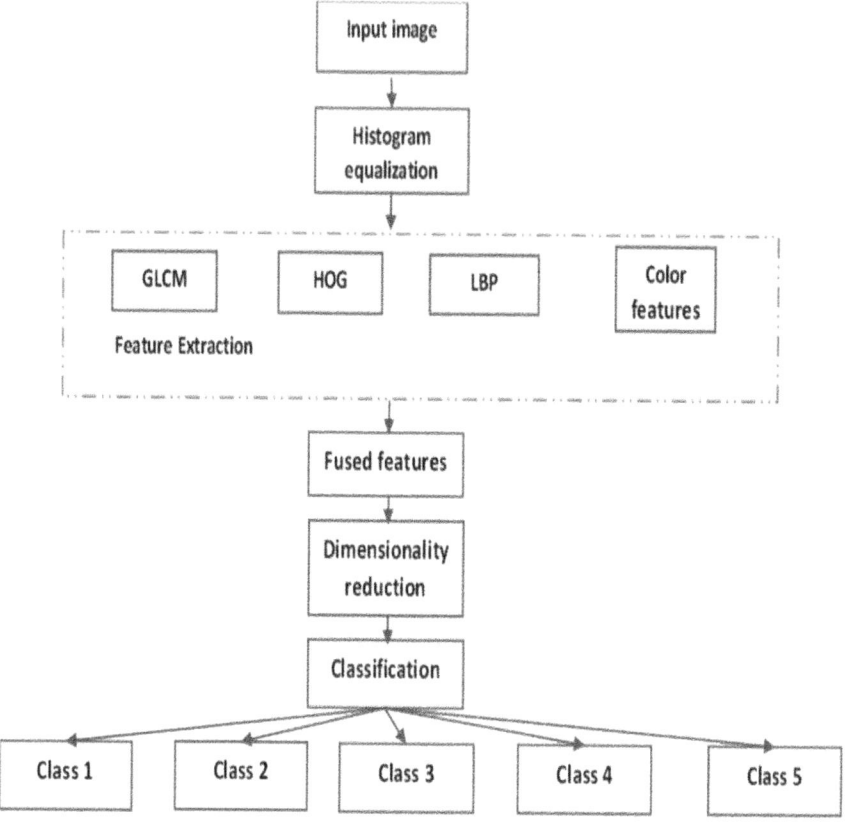

Fig. 1. Flow diagram of proposed work

2.1 Dataset

We used Kvasir dataset in our experiment. The Kvasir dataset comprises of expert analyzed images, comprising classes with images of gastrointestinal endoscopic operations
and anatomical markers. Image resolution ranging from 720 × 576 to 1920 × 1072
pixels and is in RGB color space. In our study, the dataset included 5,000 images from
five different classes are uniformly distributed: ulcerative colitis, normal cecum, normal
pylorus, polyps, and dyed-lifted polyps. The goal of this dataset collection is to help in
the early detection of lesions, which can stop the spread of cancer. In Fig. 2 some sample
images from the dataset are shown.

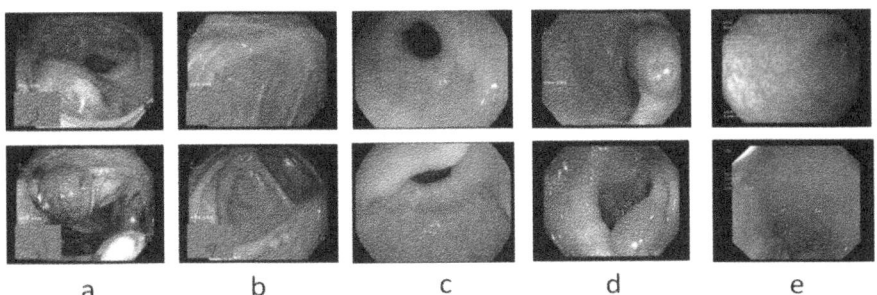

a b c d e

Fig. 2. a: Dyed-lifted polyps, b: Normal-cecum, c: Normal pylorus, d: Polyps, e: Ulcerative colitis.

2.2 Preprocessing

Image preprocessing is done to enhance the image quality. Here, raw images are first resized to 200 × 200 and improved using Histogram Equalization (HE) which adjust the image intensities to improve the contrast.

Histogram Equalization

The gray levels of images are expanded using this HE process [13]. The Probability Density Function (PDF), P(Y_l) of image I which has N no of pixels and G gray levels is given by,

$$P(Y_g) = \frac{m_g}{m} \tag{1}$$

For G = 0, 1...g-1, m_g indicates the total number of pixels at Y_g gray level and $Y_g \in \{Y_0, Y_1, ...Y_{G-1}\}$, Additionally, the Cumulative Density Function is defined as,

$$C(Y_g) = \sum_{s=0}^{g} P(Y_s) \tag{2}$$

where $(C(Y\_\{g-1\})) = 1$, The transform function f is derived using the CDF as

$$f(Y_g) = Y_0 + Y_{G-1} - Y_0)C(Y_g) \tag{3}$$

After histogram equalization, the improved image might be represented as

$$E = f(Y_g) \tag{4}$$

There would be 256 Gy levels for an 8-bit grayscale picture, or G = 256 and g = 0, 1..., 255.

As a result, the pixels of high intensities are stretched more than the pixels of low intensities using HE process.

2.3 Feature Extraction

Feature extraction is the crucial step of classification task. The extracted features are more associated to the classification accuracy. So, in order to identify discriminant features, texture descriptors like GLCM and LBP, color features and HOG features are considered. The feature descriptors used in our work are described below.

Color Feature Color features are the most important factors in disease diagnosis and they have a big impact on machine learning-based categorization methods [4]. In this paper, two statistical color features like Mean and Standard deviation of each channel of RGB and HSV color spaces are extracted. Totally 12 color features are obtained and list of features are given in Table1.

Table 1. Description of extracted Color Features

Feature Description	Color Features	Description
F1	Mean_H	mean value of Hue. Hue: The color is represented by the hue. The color value is between 0 and 360
F2	Mean _S	mean value of saturation channel. Saturation: The saturation value indicates how much of the appropriate color should be added. A 100% saturation indicates the addition of just pure color, whereas a 0% saturation indicates the absence of color resulting in grayscale
F3	Mean _Value	mean of Value channel. Value: The value depicts the brightness in relation to the color's saturation. Total blackness is represented by the value 0, whereas full brightness, which depends on the saturation, is represented by the value 100
F4	Std _H	standard deviation of Hue channel
F5	Std_S	standard deviation of Saturation channel
F6	Std_V	standard deviation of Value channel
F7	Mean_R	mean of Red channel
F8	Mean_G	mean of Green channel
F9	Mean_B	mean of Blue channel
F10	Std_R	standard deviation of Hue channel
F11	Std_G	standard deviation of Saturation channel
F12	Std_B	standard deviation of Value channel

Texture Features

Local Binary Pattern

LBP was first established by Ojala [5]. LBP dis the most frequently used feature descriptor for texture analysis. An enhanced local binary model called rotation invariance is capable of effectively describing the shallow texture patterns found in medical imaging. Traditional LBP identifies important points in image and creates a histogram which displays the distribution of those points. It uses sliding window to scan the pixels and creates a binary code that based on the variations between the center pixel and its evenly spaced circular neighbors. P represents the number of neighbors, whereas the radius parameter R denotes the distance [6]. The LBP operator is given by{t}_{c})){2}^{j}

$$LBP_{P,R} = \sum_{i=0}^{P-1} x(I(s_j, t_j) - I(s_c, t_c))2^j \tag{5}$$

$$x(s) == \begin{cases} 0, if s < 0 \\ 1, otherwise \end{cases} \tag{6}$$

where, s_j and t_j are the coordinates of neighbor j^{th} of central pixel (s_c, t_c).

Three phases are involved in the calculation procedure for each sliding window position: Firstly, the difference between the center and surrounding pixels must be calculated. Second, x(s) is set to 1 if it is larger than or equal to center pixel, otherwise to 0. Third, the central pixel is given as the summing of all values in a clockwise manner, and this binary code is then used as a distinctive local texture.

Gray-Level Co-occurrence Matrix

GLCM is a statistical technique for texture analysis that consider the spatial relationship between pixels. GLCM matrix is created by determining how frequently image contains pairings of pixels with certain values with given spatial relationship. It shows the probability that grey level t will be neighboring, either horizontally, vertically, or diagonally, at grey level s. The GLCM, indicated as G(s,t) here, is normalized by dividing each matrix element by the total of the pixel pairings. From the GLCM matrix following characteristics are extracted [14][17]. The equations of extracted GLCM features are given in Equations [6–11].

$$Contrast = \sum_s \sum_t (s - t)^2 G(s - t) \tag{7}$$

$$Energy = \sum_s \sum_t G(s, t)^2 \tag{8}$$

$$Entropy = - \sum_s \sum_t G(s, t) log(G(s, t)) \tag{9}$$

$$Homogeneity = \sum_s \sum_t \frac{G(s, t)}{1 + (s - t)^2} \tag{10}$$

$$Disimilarity = \sum_{s,t} G_{s,t}|s - t| \tag{11}$$

$$Angular\ Second\ Moment = \sum_{s,t} G_{s,t}^2 \tag{12}$$

Shape Features

Histogram Oriented Gradient

Histogram Oriented Gradient (HOG) is mainly used in object detection in computer vision and image processing. HOG is a feature descriptor that focus on shape and appearance of the object. The distribution of intensity gradients over a space is represented by HOG [7]. The HOG descriptor counts the instances of gradient orientation in specific areas of an image, such as region of interest (ROI). The input images are partitioned into blocks, and then each block is even further subdivided into cells, from which a histogram of the edge orientations or gradient directions of the pixels within each cell is produced. The gradient values calculation along horizontal H_x and vertical directions V_y using the kernel functions are given below.

$$H_x = [-1\ 0\ 1], V_y = \begin{bmatrix} -1 \\ 0 \\ 1 \end{bmatrix} \tag{13}$$

$$I_x = I \times H_x I_y = I \times V_y \tag{14}$$

By convolving the kernel function specified in Eq. 13, the horizontal and vertical derivative of the GI image (I) is produced. The magnitude (M) and orientation (θ) of the gradient is calculated by,

$$|M| = \sqrt{I_x{}^2 + I_y{}^2}\ and\theta = \tan^{-1}\frac{I_y}{I_x} \tag{15}$$

For signed gradients, the range of orientation is either ($0\text{-}360^0$) or ($0\text{-}180^0$) for unsigned gradients. For each and every pixel in the cells, the gradients' magnitude and direction are determined. To represent a block, a normalized histogram is produced for each cell and then concatenated.

2.4 Feature Selection

The next step of feature extraction is feature selection. The goal of feature selection is to reduce the dimensionality of data. In our work Principal Component Analysis (PCA) is used to transform the dimensionality from high to low based on most essential features. PCA is a statistical procedure that transforms the observations of correlated character-istics into a group of linearly uncorrelated characteristics with the use of orthogonal transformation. These newly altered features are called Principal Components. These PCs are either the same number or less than the original characteristics that were included in the dataset. Some properties of PCA are:

- The linear combination of the original characteristics must be the principal compo-nent.
- As these components are orthogonal, there is no association between any two variables.
- Going from 1 to n, the importance of each component declines, making PC-1 the most important and PC-N the least important.

2.5 Classifiers

SVM is an optimization problem defined as a maximum margin classifier [8]. For the feature vector X_i, class label Y_i weight vector W,

$$Minimize: \frac{1}{2}||W|| + C\sum_i \xi_i \tag{16}$$

$$Subjectto : Y_i(W^T X_i + W_0) \geq 1 \qquad (17)$$

C is the cost parameter that controls the maximizing of margin and minimizing of classification error, where i is the number of samples, and, where ξ stands for the training errors. SVM is easily adaptable to multi-class issues. In practice, the One-Against-All (OAA) technique [21] and the One-Against-One (OAO) method are common methods for multiclass classification [9]. For the classification of N no of classes, the OVA method requires N no of binary classifiers, whereas the requirement of OAO method is $\frac{N(N-1)}{2}$. The proposed study employs a one-versus-all strategy and a 5-class classifier model using five SVM binary classifiers.

3   Results and Discussions

This section outlines the results of the proposed collection of handmade features. Following a discussion of the comparative performance of and various combinations of feature descriptors, then we examine the performance of our classification model before and after feature selection. An Intel Xeon CPU E3-1245-v6 @3.70 GHz computer system with 32 GB of RAM was used to carry out the experiment successfully. The tool used is MATLAB 2020a.

Our dataset consists of 5000 images in which 70% for training and 30% using cross validation. In the preprocessing stage the images are resized to 200×200 pixels before contrast enhancement. As GI images comes with complex back grounds the images are enhanced by using the Histogram equalization method. The sample image after Histogram Equalization is shown in Fig. 3.

Original Image (left) and Contrast Enhanced Image (right)

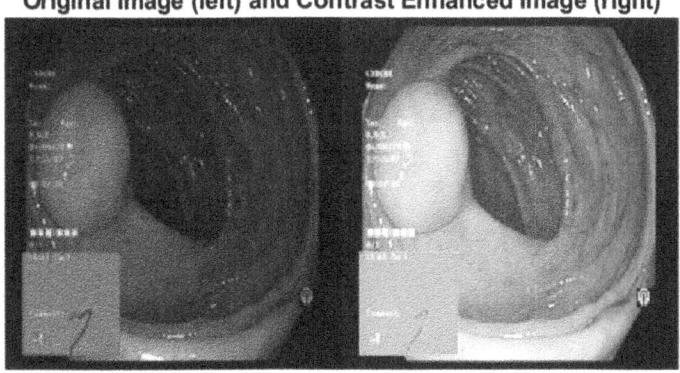

Fig. 3. A polyp sample image before and after Histogram Equalization

The enhanced images are fed to feature extraction step, Since the color information plays an important role in GI images, two important Statistical features like Standard deviation and Mean of RGB color space and HSV color space are calculated. But the color information is not sufficient to discriminate the five classes So, in addition Six GLCM features and LBP features for texture information and Hog descriptors for extracting the shape information are utilized. 4326 features are extracted using HOG. Then extracted hand crafted features are combined to form most powerful features with high dimension. The dimension was reduced to 1x1000 using PCA. Now the effective features are given to the popular OAA SVM classifier for classification.

The evaluation metrics like Precision, recall and F1-score are calculated for our proposed work are given in equations [17–20] below.

$$Accuracy = \frac{No\ of\ correct\ predictions}{Total\ no\ of\ predictions\ made} * 100 \qquad (18)$$

$$Recall/Sensitivity\frac{Number\ of\ truepositives}{Number\ of\ true\ positives + Number\ of\ false\ negatives} \qquad (19)$$

$$Precision = \frac{Number\ of\ true\ positives}{Number\ of\ true\ positives + Number\ of\ false\ positives} \qquad (20)$$

$$1 - score = 2 * \frac{(Precision * Recall)}{Precision + Recall} \qquad (21)$$

Individual feature descriptors are analyzed and the results are shown in figs [4–7]. The other performance metrics of handcrafted features are evaluated and illustrated in bar graph form. From the Fig. 4 we can recognize that the accuracy based non color features is very superior to other handcrafted features, next to color features HOG descriptor has moderate discrimination capability.

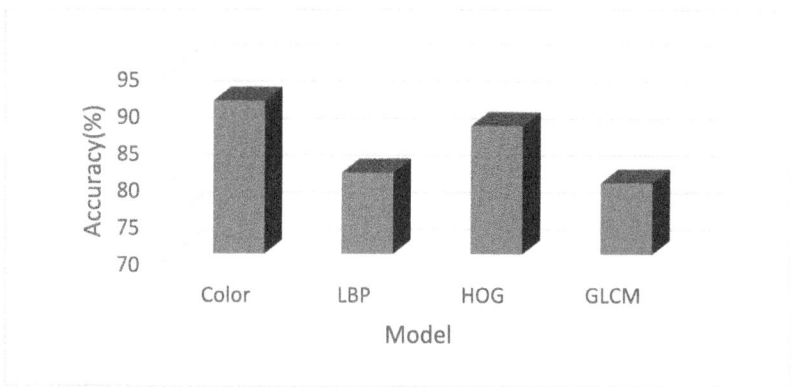

Fig. 4. Accuracy based on individual features

In addition to the accuracy metric, we also evaluated other performance metrics like precision, recall and F1-score. Figure 5 shows the precision is high when classified with the color features and incase of recall and f1-score also the performance improved with color features (Figs. 6 and 7).

Although the performance improved with color with color features, the accuracy rate is in moderate level. Inorder to enhance the prediction power of the handcrafted features

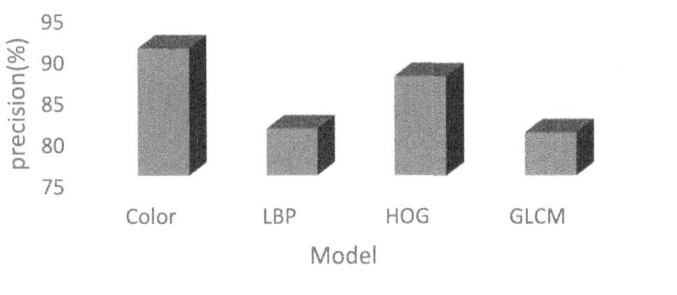

Fig. 5. Precision based on individual features

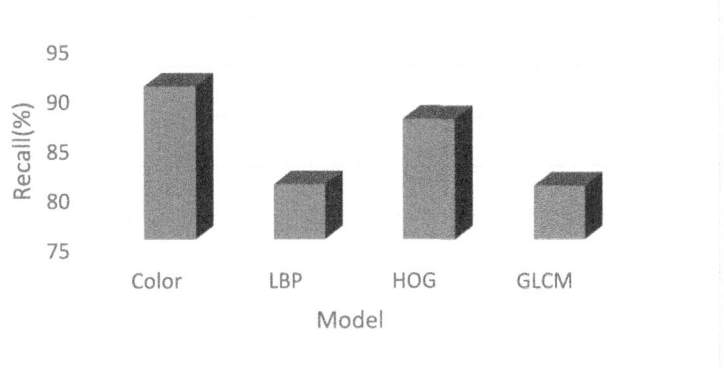

Fig. 6. Recall based on individual features

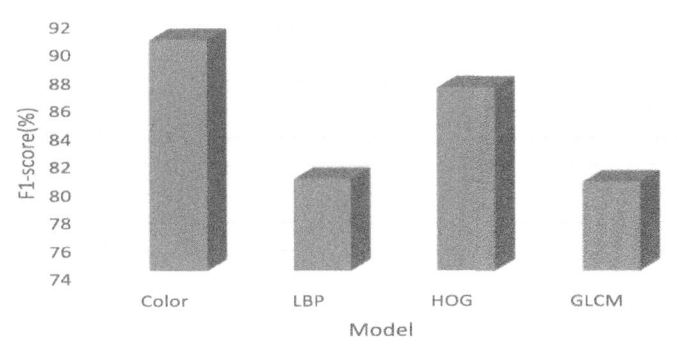

Fig. 7. F1-score based on individual features

we did some fusion tactics, that means we combined color, texture and shape features to form the effective features subset. The various combination of features used in this paper was shown in Table 2. Those features are finally classified with OAA-SVM classifier.

Table 2. Comparison of proposed model with various combinations of features before feature selection

Methods	Accuracy	Precision	Recall	F1-score
LBP + Color	92.13	91.86	91.9	91.88
HOG + Color	92.4	92.14	92.21	92.17
GLCM + Color	90.2	90.2	90.4	90.3
LBP + HOG	89.8	89.8	89.6	89.6
LBP + GLCM	84.4	84.2	84.2	83.1
HOG + GLCM	88	87.7	87.8	87.7
HOG + LBP + GLCM	89.73	89.81	89.9	89.84
HOG + LBP + GLCM + Color	94.4	94.14	94.21	94.17

From the Table 2 we identified that while fusing Color, HOG, LBP and GLCM the accuracy was raised to 94.4% in GI images classification. The minimum accuracy was obtained when classified with only the texture features that is 84.4%. In order to improve the accuracy further, PCA is adopted to select the important features. Because the features are extracted from the whole images, hence there might be some unwanted features that could slow decrease the classification accuracy.

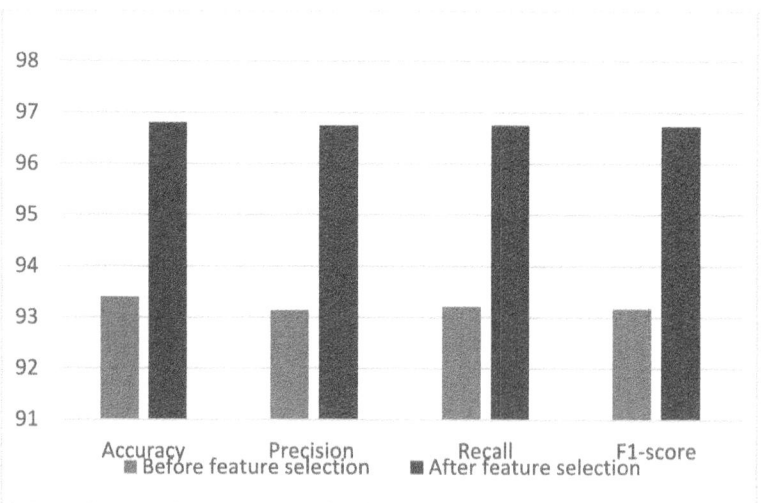

Fig. 8. Performance measures before and after feature selection.

After feature selection, the accuracy of our proposed model improved from 94.4% to 96.8%. The accuracy variations before and after feature selection is clearly illustrated in Fig. 8.

4   Conclusion

The main objective of this work is to create a reliable method for Gastro Intestinal diseases classification with less computational resources. Here Kvasir dataset was utilized for experimental analysis. In this paper handcrafted features like statistical color features, GLCM, LBP and HOG features and various combinations of these features are investigated. The fused high dimensional feature set is transformed to low dimensional optimal feature set using PCA. Finally, the OAA-SVM classifier was trained on selected features and achieved the classification accuracy of 96.8%. Two observations from this work are: Firstly, the prediction capability of ensemble handcrafted features is more effective than individual features. Secondly, the color information has a vital role in GI image classification.

References

1. Yaman, M.A., Subasi, A., Rattay, F.: Comparison of random subspace and voting ensemble machine learning methods for face recognition. Symmetry (Basel) **10**(11) (2018). https://doi.org/10.3390/sym10110651
2. Rosdi, B.A., Shing, C.W., Suandi, S.A.: Finger vein recognition using local line binary pattern. Sensors **11**(12), 11357–11371 (2011). https://doi.org/10.3390/s111211357
3. Chtihrakkannan, R., Dr, P., Kavitha, T., Mangayarkarasi, R.K.: Breast cancer detection using machine learning. Int. J. Innov. Technol. Explor. Eng. **8**(11), 3123–3126 (2019). https://doi.org/10.35940/ijitee.K2498.0981119
4. Azary, H., Abdoos, M.: A Semi-supervised method for tumor segmentation in mammogram images. J. Med. Signals Sens. **10**(1), 12–18 (2020). https://doi.org/10.4103/jmss.JMSS_62_18
5. Javaid, A., Sadiq, M., Akram, F.: Skin Cancer Classification Using Image Processing and Machine Learning (2021). https://doi.org/10.1109/IBCAST51254.2021.9393198
6. Zeebaree, D.Q., Abdulazeez, A.M., Zebari, D.A., Haron, H., Hamed, H.N.A.: Multi-level fusion in ultrasound for cancer detection based on uniform LBP features. Comput. Mater. Contin. **66**(3), 3363–3382 (2020). https://doi.org/10.32604/cmc.2021.013314
7. Ankita, R., Kumari, C.U., Mehdi, M.J., Tejashwini, N., Pavani, T.: Lung cancer image-feature extraction and classification using GLCM and SVM classifier. Int. J. Innov. Technol. Explor. Eng. **8**(11), 2211–2215 (2019). https://doi.org/10.35940/ijitee.K2044.0981119
8. Al Sadeque, Z., Khan, T.I., Hossain, Q.D., Turaba, M.Y.: Automated detection and classification of liver cancer from CT Images using HOG-SVM model (2019). https://doi.org/10.1109/ICAEE48663.2019.8975602
9. Sharif, M., Attique Khan, M., Rashid, M., Yasmin, M., Afza, F., Tanik, U.J.: Deep CNN and geometric features-based gastrointestinal tract diseases detection and classification from wireless capsule endoscopy images. J. Exp. Theor. Artif. Intell. **33**(4), 577–599 (2021). https://doi.org/10.1080/0952813X.2019.1572657
10. Amiri, Z., Hassanpour, H., Beghdadi, A.: A computer-aided method for digestive system abnormality detection in WCE images. J. Healthc. Eng. **2021**(1), 78631 (2021). https://doi.org/10.1155/2021/7863113

11. Khan, M.A., Rashid, M., Sharif, M., Javed, K., Akram, T.: Classification of gastrointestinal diseases of stomach from WCE using improved saliency-based method and discriminant features selection. Multimed. Tools Appl. **78**(19), 27743–27770 (2019). https://doi.org/10.1007/s11042-019-07875-9

12. Ellahyani, A., El Jaafari, I., Charfi, S., El Ansari, M.: Detection of abnormalities in wireless capsule endoscopy based on extreme learning machine. Signal Image Video Process. **15**, 877–884 (2020). https://doi.org/10.1007/s11760-020-01809-x

13. Saravanan, S., Karthigaivel, R.: A fuzzy and spline based dynamic histogram equalization for contrast enhancement of brain images. Int. J. Imaging Syst. Technol. **31**(2), 802–827 (2021). https://doi.org/10.1002/ima.22483

14. Hegde, P.R., Shenoy, M.M., Shekar, B.H.: Comparison of machine learning algorithms for skin disease classification using color and texture features (2018). https://doi.org/10.1109/ICACCI.2018.8554512

15. Ojala, T., Pietikäinen, M., Harwood, D.: A comparative study of texture measures with classification based on featured distributions. Pattern Recogn. **29**(1), 51–59 (1996). https://doi.org/10.1016/0031-3203(95)00067-4

16. Hazgui, M., Ghazouani, H., Barhoumi, W.: Genetic programming-based fusion of HOG and LBP features for fully automated texture classification. Vis. Comput. (2021). https://doi.org/10.1007/s00371-020-02028-8

17. Shrivastava, V.K., Londhe, N.D., Sonawane, R.S., Suri, J.S.: Reliable and accurate psoriasis disease classification in dermatology images using comprehensive feature space in machine learning paradigm. Expert Syst. Appl. **42**(15–16), 6184–6195 (2015). https://doi.org/10.1016/j.eswa.2015.03.014

18. Annalakshmi, M., Roomi, S.M.M., Naveedh, A.S.: A hybrid technique for gender classification with SLBP and HOG features. Cluster Comput. **22**, 11–20 (2019). https://doi.org/10.1007/s10586-017-1585-x

19. C. Cortes and V. Vapnik, "Support-Vector Networks," *Mach. Learn.*, vol. 20, no. 3, 1995, https://doi.org/10.1023/A:1022627411411

20. Rocha, A., Goldenstein, S.K.: Multiclass from binary: expanding one-versus-all, one-versus-one and ECOC-based approaches. IEEE Trans. Neural Networks Learn. Syst. **25**(2) (2014). https://doi.org/10.1109/TNNLS.2013.2274735

21. Sharmila Joseph, J., Vidyarthi, A., Singh, V.P.: Multiclass image classification using OAA-SVM. In: Agrawal, S., Gupta, K.K., Chan, J.H., Agrawal, J., Gupta, M. (eds.) Machine Intelligence and Smart Systems: Proceedings of MISS 2021, pp. 235–244. Springer Nature Singapore, Singapore (2022). https://doi.org/10.1007/978-981-16-9650-3_18

22. Saito, H., et al.: Automatic detection and classification of protruding lesions in wireless capsule endoscopy images based on a deep convolutional neural network. Gastrointest. Endosc. **92**(1), 144–151 (2020). https://doi.org/10.1016/j.gie.2020.01.054

23. Rustam, F., et al.: Wireless capsule endoscopy bleeding images classification using CNN based model. IEEE Access **9**, 33675–33688 (2021). https://doi.org/10.1109/ACCESS.2021.3061592

24. Kundu, A.K., Fattah, S.A., Wahid, K.A.: Multiple linear discriminant models for extracting salient characteristic patterns in capsule endoscopy images for multi-disease detection. IEEE J. Translat. Eng. Health Med. **8**, 1–11 (2020). https://doi.org/10.1109/JTEHM.2020.2964666

25. Charfi, S., El Ansari, M., Balasingham, I.: Computer-aided diagnosis system for ulcer detection in wireless capsule endoscopy images. IET Image Process. **13**(6), 1023–1030 (2019). https://doi.org/10.1049/iet-ipr.2018.6232

26. Alaskar, H., Hussain, A., Al-Aseem, N., Liatsis, P., Al-Jumeily, D.: Application of convolutional neural networks for automated ulcer detection in wireless capsule endoscopy images. Sensors **19**(6), 1265 (2019).

An Integrated Best-Worst-VIKOR Method for Evaluation and Selection of Luxury Hotels in India

Manini Chawla, Simarpreet Kaur, and Sugandha Aggarwal[✉]

Lal Bahadur Shastri Institue of Management, 11/07 Dwarka Sector 11, Near Metro Station, New Delhi 110075, India
{manini_chawla-r21,simarpreet_kaur-r21}@lbsim.ac.in, sugandha_or@yahoo.com

Abstract. Tourism is an important economic component in several countries. India, with its rich cultural heritage and natural beauty, has a promising reputation in the tourism business. With multiple options available, tourists often face problem in selecting the most suitable hotel for boarding, as they assess the hotels on their own different preferences. To assist in this decision, the primary objective of our research is to study the criteria based on which customers choose the luxury hotels in India. The second objective is to assess and select the luxury hotel based on the criteria identified in the first step. To attain these objectives, Multi Criteria Decision Making (MCDM) methods are utilized wherein Best Worst Method (BWM) is used to rank the criteria and VIKOR method is used to select the best luxury hotel. The results of this research can be used to drive future research on the performance assessment of hotels in general and can also serve as a guide for tourists looking to book the best hotel with their specific requirements.

Keywords: MCDM · Luxury Hotels · Selection · VIKOR · BWM

1 Introduction

Travel and tourism have been significant social activities for humans since the dawn of humanity. The aim is to discover new locations, whether inside one's own nation or elsewhere, and to seek a change of atmosphere.

Tourism is one of the most mobile industries in the twenty-first century. If we talk about India, the estimate is that India will contribute 250 billion USD GDP from Tourism, 137 million jobs will come from this sector and the foreign arrivals will be 25 million by 2030. By 2027, in India, the tourism market will reach $125 billion from $75 billion in 2020 [1].

It can be noted that the hotel sector, or more precisely, the "Hotel Section," plays a vital role in success and development of tourism since they contribute to the economic expansion of an area as a tourist destination. According to ceidcdata [2], the number of hotels reported in India in 2017 were 1784 among which 153 were above five-star deluxe hotels; 162 five-star hotels, 258 four-star, 488 three-star, and 43 two-star hotels apart

© The Author(s), under exclusive license to Springer Nature Switzerland AG 2025
M. Gupta et al. (Eds.): MISS 2023, CCIS 1952, pp. 69–82, 2025.
https://doi.org/10.1007/978-3-031-69115-7_4

from others. Thus, the hotel business in India has been growing at a reasonable pace in recent years and has the potential to develop much faster in the future. The nation's cultural and linguistic variety has attracted a great number of visitors from all over the world. India has also been regarded as a spiritual tourism destination for both foreign and local visitors.

In the hospitality market, the luxury hotel segment is a breed apart. These hotels provide top-notch service and lavish amenities, but they are also among the most difficult to manage and risky for developers and owners. Luxury hotels are among the most variable in the sector, doing well in good times and poorly in bad.

Since customers can choose from a wide variety of hotels available even in this segment [2], there is a strong competition in the market. Meeting the varying expectations of the tourists and maintaining a competitive edge over competitor hotels becomes quite challenging for the hotel management. They must deliver a high level of service and exceed their guests' expectations to sustain in the business.

Numerous MCDM methodologies have got developed, suggested, and effectively applied for selection purposes whether we talk about supplier selection, hotel selection, material selection, location selection, etc. MCDM is a method that integrates alternative performance across several contradicting qualitative and/or quantitative criteria, yielding a solution that requires consensus applied in a variety of application areas since the 1960s. The goal of MCDM is not to advise the optimal solution, but to assist decision makers in selecting option that meets their objectives and is consistent with their preferences stated early on. Analytical Hierarchy Process (AHP), Analytic Network Process (ANP), Technique for Order Preference by Similarity to Ideal Solution (TOPSIS), VIKOR, and Best Worst Method (BWM) are some of the MCDM methodologies popularly used as can be seen in Sect. 2 ahead.

This paper aims to identify the criteria that influence the selection of luxury hotel in India and choose the best hotel keeping the identified criteria in mind. In this regard, five of India's finest five-star hotels have been chosen to investigate the psychological and non-psychological factors that influence tourists' decisions. The performance of these hotels is assessed using VIKOR to determine which hotel is the best according to the data acquired. During the process of computation, BWM is used to determine the relative importance of each performance criterion by factoring in feedback from subject matter experts. And which explains why this is the best method. Thus, BWM computes weights for the selection criteria and VIKOR ranks the luxury hotels. This model can help the hotel authorities, visitors and the tourism industry to take more effective decisions that leads to satisfaction of the tourist.

In Sect. 2, a review of literature is conducted to discuss how different authors have applied MCDM techniques for selection purpose and identify the gap in the existing literature. Section 3 consists of the research methodology which includes the MCDM methods that have been used in the paper. In Sect. 4, the case description is discussed starting with the case introduction, the description of the five luxury hotels in India and the factor description. Section 5 includes the results and discussion wherein the results from applying BWM and VIKOR has been discussed, and Sect. 6 concludes the paper.

2   Literature Review

There has been a considerable work on hotel industry in different areas such as hotel selection, site selection, etc. Authors have used different techniques for selection of hotels like AHP, ANP, TOPSIS, OCRA, SWARAS, VIKOR etc. as discussed below.

Chou et al. (2008) [3] examined the variables that influence and help in hotel site selection in Taiwan using fuzzy MCDM technique. AHP with fuzzy sets theory and linguistic value concept is used that takes vagueness into account. The author explained how comparisons can be done when selecting the model, which provides direction to hotel authorities for making operating strategies.

Chen et al. (2012) [4] analyzed six important indices to assess the performance of four five-star hotels in Iran. They used ANP to determine the weights of the suggested indices and deduces the interdependence between the indices. Based on the results of ANP, rating of these hotels was done using TOPSIS approach.

Sohrabi et al. (2012) [5] identified the hotel selection aspects in Tehran hotels and used Exploratory factor analysis of the selected factors for effective decision making. The fuzzy sets were used for development purposes.

Li et al. (2013) [6] proposed the Choquet Integral to develop MCDM process designed for the tourists for Hong Kong hotels. The data was collected from TripAdvisor website. The ratings were taken on characteristics such as value for money, service, location, and overall. The research conducted by them can be used to find the preferences of travelers that impacts the decisions of hotel selections.

Zoraghi et al. (2013) [7] et al. proposed fuzzy MCDM technique to evaluate the service quality hotels. They integrated the subjective (based on judgment of decision makers) and objective (based on mathematical calculations) weights to rank the hotels. Five hotels were ranked using this approach and the model was validated through a real life case study.

Isik and Adali (2016) [8] developed a novel unified decision-making technique for handling a hotel selection issue that is built on stepwise weight assessment ratio analysis (SWARA) and operational competitiveness rating analysis (OCRA). The SWARA approach was used to establish the weights for the criterion, while the OCRA method was used to rank the alternatives and choose the best hotel. The decision-makers tasked with resolving the issue are chosen. Then, the problem's criteria and options are identified, and the appropriate data is acquired. They have detected an issue with hotel selection. To resolve the issue, five selection criteria and six hotel options were employed. They utilised a case study of the textile industry in Denizli to show the integrated method's applicability.

Aksoy and Ozbuk (2017) [9] employed the methods of MCDM and preference selection index to determine the elements that impact a tourist's hotel location choice. They have explored if there is a link between the MCDM findings and post-purchase customer evaluations, which suggests that visitors' hotel location selections are influenced by a high rationality feature. It was determined that travelers preferred a site that was within walking distance of their desired tourist attractions.

Kwok and Lau (2019) [10] employed methods to locate the finest hotel using a variety of factors. It implied that online travel firms should prefer the criteria used by clients to pick hotels. Additionally, it advises building a system for individualized hotel package recommendations. Vague Set TOPSIS is as decision support system that assists travellers in ranking hotel selections. A fundamental modification to standard TOPSIS models (including fuzzy TOPSIS, IF-TOPSIS, and VS-TOPSIS) since these approaches were unsuitable for hotel selection.

Sin and Sin (2019) [11] assessed the critical success factors (CSF) of quality management (QM) for hotels in Malaysia. The criteria for evaluating QM CSFs were defined using a literature search and MCDM approach aided by a DEMATEL-based analytical network process (ANP) (D-ANP). D-ANP is incorporated to calculate the significance level, weights for all dimensions and indicators, and significances. The results established a framework for evaluating and ranking hotels by examining the interactions between and among settings and characteristics.

Wang et al. (2020) [12] examined the variations in hotel choosing among various categories of travelers using internet hotel reviews. They intended to facilitate the customer's decision-making process. This research added to the pool of expertise on travellers' behavioural preferences and decision-making models. Multiple-criteria evaluations were converted to image fuzzy evaluation data. Finally, the fuzzy TODIM algorithm for images was developed. The text was pre-processed, which included collecting key factors from the TF-IDF and then using the Word2Vec method. The MCDM technique has been employed, i.e. the Fuzzy TODIM approach, which is then used to rate hotels.

Mahdi, A. and Esztergár-Kiss (2021) [13] used fuzzy analytical hierarchy process to analyse and prioritize hotel selection factors while booking hotels online. While there are several online booking websites and apps, booking.com is used to validate the model. By using the Fuzzy AHP approach, this research attempts to eliminate ambiguity and aid travellers in making the best hotel selection on the booking.com website. The weighting of the criteria is determined by comparing them to stakeholder preferences and using the FAHP method. The findings indicate that the most crucial factor in hotel selection is the price, while the free cancellation criterion has the least influence.

Tsai et al. (2021) [14] developed a model to aid hotel administrators in choosing bloggers using a unique mixed multiple-criteria decision-making approach that included IPA, AHP, and TOPSIS. This research interviewed hotel administrators, reviewed blogger selection literature, and used the stick concept to gather selection criteria. Then, using AHP, the IPA was used to screen each selection criteria for significance and performance (high importance and poor performance). Finally, to minimise the number of pairwise comparison matrices and improve decision-making efficiency, the AHP selection criterion weights were imported into TOPSIS.

Nuriyev and Baysal (2022) [15] developed a model for supplier selection for hotels in decision-making environments with high levels of uncertainty. ZVIKOR, MCDM technique based on Z numbers, was employed to pick alternatives (suppliers) for the hotels. Delphi analysis was used to define the supplier selection criterion. The study's findings demonstrate the approach's relevance for solving MCDM challenges in the tourist sector under situations of high uncertainty. The study demonstrated the significance of factors such as service quality, on-time delivery, and profile and rating in hotel suppliers a higher-level uncertainty formalism for decision-making in supply chain management and other fields under imperfect information selection. The results show that employing the Z-numbers paradigm as a higher-level uncertainty formalism for decision-making in supply chain management and other domains with imperfect information has a lot of potential.

Peng et al. (2022) [16] tackled the issue of hotel choices and designed a MULTI-MOORA approach and a pairwise evaluation-based MCDM method. This problem was solved by using two hotels. A multi-criteria decision-making strategy using Z-numbers and projection measures is developed. Certain Z-number Choquet integral projection operators are presented for fusing assessment data. In addition to the Multiplicative Form methodology, a new pairwise evaluation-based multi-criteria decision-making method is devised.

Pour and Hemati (2022) [17] prioritized the marketing efforts using FMCDM approaches and compared their effects for hotels in the world's second-largest religious city. As a result, two methodologies were used, combining the FAHP methodology with distance-based FMCDM methods. Distance-based approaches such as FTOPSIS and FVIKOR were employed. Based on the FTOPSIS, the most appropriate marketing approach was chosen, with an emphasis on targeted distinction. Three techniques were chosen as the best strategies using the FVIKOR approach. Two of these three tactics focus on targeted differentiation, while the third focuses on differentiation.

As can be seen from the above literature different authors have used several MCDM techniques individually or in integration but as per the authors knowledge none of the authors have used integrated VIKOR and BWM for the selection of the hotel. This gap has been filled in this paper.

3   Research Methodology

The research methodology is illustrated below in Fig. 1:

3.1   MCDM Techniques

MCDM techniques are used to find the best alternative by ranking the alternatives using a set of attributes (Gal et al. (2013) [18]). The solution so achieved is based on the consensus decision. Numerous MCDM techniques have been developed since 1960s, in different application areas. The aim is to help decision-makers in choosing the alternative that is in line with their preferences and fulfill their requirements. To apply MCDM methods efficiently proper knowledge of MCDM techniques is essential along with understanding the target group of decision makers who are the actual players in MCDM.

3.2   BWM

The best-worst method (BWM) (Hasan et al. (2022) [19]) is a type of MCDM technique that is used to solve different types of real life problems that needs decision making. The basic BWM arranges the criteria in order of importance. This is done by comparison of the best and worst criteria versus each other. This could lead to more reliable and consistent pairwise comparisons as we first identify the best and the worst criteria before doing the

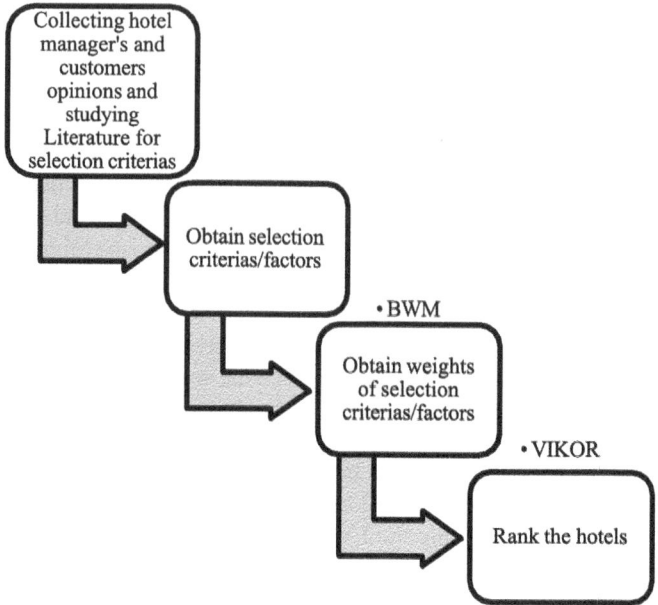

Fig. 1. Research methodology

pairwise comparisons among the criteria. This is due to a better understanding decision maker gets of the range of evaluation.

Suppose there are n criteria, and a pairwise comparison of these criteria using a 1/9 to 9 scale 1 has to be executed. The resulting matrix would be as in Fig. 2:

$$A = \begin{bmatrix} a_{11} & a_{12} & ... & a_{1n} \\ a_{21} & a_{22} & ... & a_{2n} \\ \vdots & \vdots & ... & \\ a_{m1} & a_{m2} & ... & a_{mn} \end{bmatrix}$$

Fig. 2. BWM Matrix

Where aij represents the relative preference of criteria i over criterion j, and $aij = 1$ indicate that i and j are equally important; $aij > 1$ indicates that i is more significant than j. $aij = 9$ indicate that i is extremely essential to j. aji demonstrates the significance of j to i. For matrix A to be reciprocal, aij must equal to $1/aji$ and aij must be equal to 1, for every i and j (Fig. 3).

3.3 VIKOR

VIKOR is a compromised ranking method and it starts by making a compromise ranking list, a compromise solution, and the weight stability intervals for the compromise solution (Mahmudova (2023) [20]). Then, it figures out the positive-ideal solution and the negative-ideal solution, which helps with ranking and choosing. In this approach we

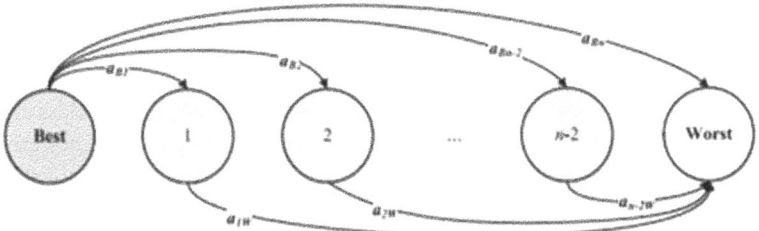

Fig. 3. Reference Comparison (Hasan et al. (2022) [19])

deal with the alternatives that have multiple conflicting or non-commensurable criteria. This method takes in to picture the subjectivity and data imprecision.

4   Case Description

4.1   Case Introduction

The Indian government has relaxed its hospitality-related investment policy in order to permit 100% foreign direct investment (FDI) in tourist building projects, hotel and resort developments, and the installation of recreational facilities. This has proven to be a significant catalyst for the growth of new hotel brands in the country.

People are also increasingly participating in staycations, extended weekend excursions, and social events such as weddings. This has led to the expansion of the hotel industry in tier 2 and 3 cities as well. In addition, improved road infrastructure throughout the nation, particularly in tourist locations such as hill regions, has offered a much needed boost to the growth.

To compete with domestic and international hotel chains, hotels in India are constantly innovating. Nowadays, five-star hotel chains serve not only the affluent and upper classes but also the upper middle and the middle market as well. They have started tying up with residencies and other chains to provide a more affordable option to people. Thus, hotels have become more competitive, inventive, and customer-centric, and have successfully established themselves in the hotel industry.

Some hotels are great in terms of food, some are more inclined towards architecture and interiors whereas some are location focused. As there is a variety of options available even in luxury segment, it becomes difficult for tourists to choose a hotel since they must take a variety of variables into account. To cater to this problem a model for selection of hotels is developed considering five hotels Hotel 1, Hotel 2, Hotel 3, Hotel 4, and Hotel 5. These are all luxury hotels in India, the names have not been disclosed due to confidentiality reasons. These hotels have been selected based on the number of properties they had and popularity in India. Hotel 5 has the largest number of properties around 100, followed by Hotel 1 around 80, Hotel 2 having around 20, Hotel 4 with around 15 and lastly Hotel 3 with around 10.

4.2 Hotels Description

The origins of Indian hospitality dates back to the ancient and mediaeval periods. Hotel 1 revives these traditions with a masterful touch of contemporary, world-class elegance, beginning with the Kings and Queens and their kingdoms and palaces. They are driven by feelings rather than logic, which means taking care of even the smallest of your daily needs and answering the simplest of questions in a way that has never been seen before. It is renowned for its extravagant Indian Interiors, Prime Location, and Indian-inspired luxury lifestyle. Hotel 2 family established the corporation that bears their name. Because of the warm and welcoming atmosphere, guests describe feeling extremely at home. Day in and day out, they prioritize their customers. It is recognized as one of the most reputable names in the hospitality industry due to the exceptional level of care that is embedded into every aspect of the company's operations.

In their daily transactions, Hotel 3 promotes fairness and confidence. Here, they are always trying to find ways to cut down on waste by getting rid of tasks that don't add value. A key component of its success is its dedication to staying current. Its nightclub is one example of how this has been accomplished. It stays true to its goal of being historically modern, delicately opulent, and unique. Modern Architecture, Strong Project Execution, and Quality Construction. Hotel 4 is devoted to adopting the best technology, equipment, and operational methods for the environment and ecology. They offer warm, genuine hospitality, opulent lodging, and original, individualized experiences on every trip. Hotel 5 is evolving and expanding, but the core values that guide its behavior and decisions remain the same: putting the needs of others before their own, striving for excellence, welcoming change, being honest and doing what is right while serving our planet.

4.3 Factor Description

After studying the literature and in discussion with the stakeholders from hotel industry, multiple factors like location, service quality, security, cost, sanitization and hygiene, food, interiors, hospitality, and room have been taken in to consideration for evaluation of the hotels as can be seen from Table 1.

Location includes distance from the airport, from the main city area, and from tourist hotspots and surroundings. Service quality includes the responsiveness of the hotel staff, individual attention given to the guest, benchmarking, reliability of the employees, attitude of the employees, and how they provide assurance to the guest. Security includes several sub-factors such as video surveillance, vehicle security, alert systems of the hotel, and what level of access control is given to visiting guests, and employees. Cost factor considers whether the hotel's prices are comparable to the competitors price and whether there are any hidden costs and taxes.

Sanitization and hygiene include sub-factors like the overall cleanliness of the hotel, the overall odor, washroom maintenance, and kitchen maintenance. In food, the factors considered are taste, the customization offered in food, variety, late-night availability, presentation of food, and timely delivery of food. In interiors, the factors considered are décor and artefacts, spacious layout and aesthetics. Hospitality includes spa and wellness, the events, entertainment and fun activities, cafeteria service, swimming pools, clubs and bars, play areas for children, and early check-in and late check-outs. Room includes room size, bathroom aesthetics, the variety of rooms, room interiors, view from the room, housekeeping, and electronic equipment that is available in the room.

Table 1. Criteria chosen for evaluation

Criteria No	Name of Criteria
Criteria 1 (C1)	Location
Criteria 2 (C2)	Service Quality
Criteria 3 (C3)	Security
Criteria 4 (C4)	Cost
Criteria 5 (C5)	Sanitization and Hygiene
Criteria 6 (C6)	Food
Criteria 7 (C7)	Interior
Criteria 8 (C8)	Rooms
Criteria 9 (C9)	Hospitality

5 Results and Discussion

5.1 BWM

The data on the factors stated in Table 1 was collected by visiting the five hotel chains. A random sample of 40–50 guests from each hotel were chosen and measure of central tendency, mode was taken to select the most frequently replied best and worst factors. BWM is used to determine the decision criteria weights wherein we decide the priority order through pairwise comparisons on the best & worst criteria.

Table 2. Best criteria for each hotel selected by respondents

Hotel No	Criteria No
H1	C1
H2	C4
H3	C8
H4	C4
H5	C1

Table 3. Worst criteria for each hotel selected by respondents

Hotel No	Criteria No
H1	C7
H2	C3
H3	C7
H4	C7
H5	C8

After finding the best and the worst criteria as shown in Tables 2 and 3 above, pairwise comparisons between the criteria was done. This gives more consistent pairwise comparisons, which have been shown in the original BWM study. The best and worst alternative along with the pairwise comparison is depicted in the below tables (Table 4).

Table 4. Ranking best criteria to other

Hotel No	Criteria No	C1	C2	C3	C4	C5	C6	C7	C8	C9
H1	C1	1	4	7	2	6	3	9	5	8
H2	C4	2	5	9	1	7	4	8	3	6
H3	C8	2	5	9	1	7	4	8	3	6
H4	C4	4	2	8	1	7	3	9	5	6
H5	C1	1	5	3	2	6	4	7	9	8

Table 5. Ranking worst criteria to others

Hotel No	H1	H2	H3	H4	H5
Criteria No	C7	C3	C7	C7	C8
C1	6	4	2	3	4
C2	3	3	4	5	5
C3	8	1	9	2	7
C4	7	8	3	7	3
C5	9	6	8	8	6
C6	2	9	5	4	2
C7	1	7	1	1	9
C8	4	5	6	9	1
C9	5	2	7	6	8

The solver on excel is used to find the weights based on the pairwise matrix. The weights of factors along with their average weight are depicted in the above Table 5.

Table 6. Weights and average weights

Criteria No. Hotel No	C1	C2	C3	C4	C5	C6	C7	C8	C9
H1	0.273	0.102	0.059	0.205	0.068	0.137	0.023	0.082	0.051
H2	0.200	0.080	0.023	0.291	0.057	0.100	0.050	0.133	0.067
H3	0.137	0.103	0.059	0.206	0.069	0.082	0.022	0.272	0.051
H4	0.101	0.202	0.051	0.282	0.058	0.135	0.023	0.081	0.067
H5	0.258	0.084	0.139	0.209	0.070	0.104	0.060	0.024	0.052
Average	0.194	0.114	0.066	0.238	0.064	0.112	0.035	0.118	0.058

From the above Table 6, it can be concluded that the Cost, Location, Rooms and service are the most substantial criteria when selecting a hotel by the potential customers with an overall weight of 23.8, 19.4, 11.8 and 11.4%, respectively.

5.2 VIKOR

The weights calculated for criteria from the best-worst method are further used in VIKOR. Decision matrix is formed by accumulating and modifying the individual linguistic preferences for necessities and alternatives.

First, we identify and select the criteria, options and a group of decision-makers. Then, for weights and ratings, the linguistic time duration sets are identified. After this, we will accumulate and modify the linguistic preferences individually for necessities and alternatives into weights and decision matrices. And lastly we will defuzzify the weights and decision matrix into crisp (Tables 7 and 8).

Table 7. Preferences for criteria's

Criteria No	C1	C2	C3	C4	C5	C6	C7	C8	C9
Weights Hotel No	0.194	0.114	0.066	0.238	0.064	0.112	0.035	0.118	0.058
H1	8	7	5	6	5	6	8	6	4
H2	6	6	5	5	4	5	6	4	6
H3	7	8	8	6	6	7	6	5	4
H4	8	8	7	7	6	8	7	8	5
H5	7	4	4	7	6	8	4	8	7
f*	8	8	8	7	6	8	8	8	7
f-	6	4	4	5	4	5	4	4	4

Table 8. Crisp Weights and ratings

Criteria No. Hotel No	C1	C2	C3	C4	C5	C6	C7	C8	C9
H1	0.000	0.029	0.049	0.119	0.032	0.074	0.000	0.059	0.058
H2	0.194	0.057	0.049	0.238	0.064	0.112	0.018	0.118	0.019
H3	0.097	0.000	0.000	0.119	0.000	0.037	0.018	0.089	0.058
H4	0.000	0.000	0.016	0.000	0.000	0.000	0.009	0.000	0.039
H5	0.097	0.114	0.066	0.000	0.000	0.000	0.035	0.000	0.000

To compute S_i, R_i, and Q_i index values we calculate the best and worst values and then rank the alternatives. For this, a compromised solution is recommended (Table 9).

Table 9. S and R values

Hotel No	S-Value	R-Value
H1	0.421	0.119
H2	0.870	0.238
H3	0.418	0.119
H4	0.064	0.039
H5	0.313	0.114
	$S^* = 0.064$	$R^* = 0.039$
	$S = 0.870$	$R = 0.238$

Table 10. Q values and Ranks

Hotel Number	Q-Value	Rank
H1	0.423	4
H2	1.000	5
H3	0.421	3
H4	0.000	1
H5	0.344	2

Using the VIKOR approach to rank the hotels, it can noted that Hotel 4 is the most desirable for consumers.

Two more conditions can be examined to see if the answer that is obtained is optimal or not. These conditions are: C1 = Acceptable advantages; C2 = Acceptable stability in decision making.

For C1: Acceptable advantage, $Q(A2)-Q(A1) >= DQ$, Where $DQ = 1/(j-1)$ ($j =$ number of alternatives). In this case, there are 5 hotels so $DQ = 0.25$. As seen from the

Table 10, the difference in Q value of rank 1 (0.000) and 2 (0.344) is more than DQ. So, condition 1 is satisfied.

For C2: Acceptable stability in decision-making, Alternatives must also be the best ranked by either R value or S value.

It is seen that for hotel 4, S and R values are ranked best which is minimum. Therefore, it can be said that Hotel 4 has added advantage over others. In other words, Hotel 4 is proposed as a compromise solution because both two conditions (C1 and C2) are satisfied.

The research findings from the integrated MCDM methods applied in this paper can aid policymakers in formulating rules mandating a quality assessment of hotels and quality management procedures in the Indian hospitality industry can be improved. Using the model will provide experts with information and data on best procedures, ensuring the accomplishment of hotel businesses. Using the BWM-VIKOR approach, an efficient model for the hotel selection has been developed. Wherein the overall results indicated that Cost, Location, Rooms and service are the most crucial factors when selecting a hotel respectively. Thus, the hotels have been ranked as follows: Hotel 4, Hotel 5, Hotel 3, Hotel 1, and Hotel 2. Hotel 4 has been ranked the best.

6 Conclusion

Evaluation of hotels has emerged as one of the most significant challenges that Indian hotels must surmount to develop new strategies and plans that have the potential to improve overall efficiency. There has been a lot of research in this area as stated in the Sect. 2 that concentrate on the assessment and choice of hotels. However, the integrated model developed in this paper is unique using two MCDM techniques BWM and VIKOR.

The findings of this paper present a true contextual contribution that may aid decision-makers in the hotel industries in comprehending and learning MCDM model integration for selection of hotels. The results can be used by customers to select the hotel that provides the greatest solution for meeting their requirements, while hotel management can utilize it to discover new clients and attract them. The method presented may lead to significant increase in the efficiency of the decision-making process regarding the hotel choosing. Researchers who are interested in a more practical understanding of the hotel selection problem, hotel stakeholders and tourists can benefit from this work.

References

1. Investindia. https://www.investindia.gov.in/sector/tourism-hospitality. Accessed 18 Jan 2023
2. Ceicdata. https://tels-and-hotel-rooms/number-of-hotels. Accessed 10 Oct 2022
3. Chou, T.Y., Hsu, C.L., Chen, M.C.: A fuzzy multi-criteria decision model for international tourist hotels location selection. Int. J. Hosp. Manag. 27(2), 293301 (2008)
4. Chen, I.S., Hashemkhani Zolfani, S., Rezaeiniya, N., Hasan Aghdaie, M.: The evaluation of five-star hotel: a case in Iran. Int. J. Bus. Soc. 13(3) (2012)
5. Sohrabi, B., Vanani, I.R., Tahmasebipur, K., Fazli, S.: An exploratory analysis of hotel selection factors: a comprehensive survey of Tehran hotels. Int. J. Hosp. Manag. 31(1), 96–106 (2012)

6. Li, G., Law, R., Vu, H.Q., Rong, J.: Discovering the hotel selection preferences of Hong Kong inbound travelers using the Choquet Integral. Tour. Manag. **36**, 321–330 (2013)
7. Zoraghi, N., Amiri, M., Talebi, G., Zowghi, M.: A fuzzy MCDM model with objective and subjective weights for evaluating service quality in hotel industries. J. Ind. Eng. Int. **9**(1), 1–13 (2013)
8. Işık, A.T., Adalı, E.A.: A new integrated decision-making approach based on SWARA and OCRA methods for the hotel selection problem. Int. J. Adv. Oper. Manag. **8**(2), 140–151 (2016)
9. Aksoy, S., Ozbuk, M.Y.: Multiple criteria decision making in hotel location: does it relate to postpurchase consumer evaluations? Tour. Manag. Perspect. **22**, 73–81 (2017)
10. Kwok, P.K., Lau, H.Y.: Hotel selection using a modified TOPSIS-based decision support algorithm. Decis. Support. Syst. **120**, 95–105 (2019)
11. Sin, K.Y., Sin, M.C.: Applying multi-criteria decision making technique in hospitality quality management system. Int. J. Acad. Res. Bus. Social Sci. **9**(11), 1–9 (2019)
12. Wang, L., Wang, X.K., Peng, J.J., Wang, J.Q.: The differences in hotel selection among various types of travellers: a comparative analysis with a useful bounded rationality behavioural decision support model. Tour. Manag. **76**, 103961 (2020)
13. Mahdi, A., Esztergár-Kiss, D.: Analysis of the effective factors for hotel selection by using the fuzzy AHP method. Industry 4.0 **6**(2), 79–82 (2021)
14. Tsai, J.F., Wang, C.P., Chang, K.L., Hu, Y.C.: Selecting bloggers for hotels via an innovative mixed MCDM model. Mathematics **9**(13), 1555 (2021)
15. Nuriyev, A., Baysal, A.B.: Selection of the hotel suppliers under high-level uncertainty. In: International European Conference on Interdisciplinary Scientific Research (2022)
16. Peng, H.G., Wang, X.K., Wang, J.Q.: New MULTIMOORA and pairwise evaluationbased MCDM methods for hotel selection based on the projection measure of Z-numbers. Int. J. Fuzzy Syst. **24**(1), 371–390 (2022)
17. Pour, M.K.S., Hemati, M.: Comparison of distance-based fuzzy MCDM techniques to evaluate marketing strategies for tourism-pilgrimage hotels during a pandemic. Fuzzy Optim. Model. J. **3**(3), 33–48 (2022)
18. Gal, T., Stewart, T., Hanne, T. (eds.): Multicriteria Decision Making: Advances in MCDM Models, Algorithms, Theory, and Applications, vol. 21. Springer, Heidelberg (2013)
19. Hasan, M.G., Ashraf, Z., Khan, M.F.: Multi-choice best-worst multi-criteria decision-making method and its applications. Int. J. Intell. Syst. **37**(2), 1129–1156 (2022)
20. Mahmudova, S.: Development of an algorithm using the VIKOR method to increase software reliability. In Predictive Analytics in System Reliability, pp. 229–246. Springer, Cham (2023)

IPCCH: Intrusion Prevention in Cloud Computing Using Honeypot

Lataben Gadhavi[1]([✉]), Vivek Prasad[2], and Manashri Patel[2]

[1] Information Technology Department, Government Polytechnic Gandhinagar, Gandhinagar, India
`latagpg@gmail.com`
[2] Computer Science and Engineering, Institute of Technology Nirma University, Ahmedabad, India

Abstract. In this modern era, there has been rapid growth of technology. Information and Data have also been growing exponentially with time. With advancement comes responsibilities; the management, storage, security, and maintenance of these data and information has also increased. Cloud Computing offers the solution for the above problems by providing the users unlimited storage, the flexibility of resources, and maintenance of these resources. The security aspects in cloud computing include Intrusion Prevention (preventive measures against the attacker) and Intrusion Detection (detects the attacker and provides proper response), which are yet to entirely mature. Honeypots are used by some IPS and IDS to collect information about the attacker's attack methods, as well as to trap the attacker. Here in this paper, we will mainly get to know about the usage of honeypot systems and how they provide security to Cloud Computing. Also, we will learn about a novel concept and model, together with its working, advantages, and disadvantages.

Keywords: Cloud Computing · Intrusion Prevention · Honeypot · Information security

1 Introduction

Security and storage of information have been a vital aspect in recent years due to the increase of information and data production exponentially with time. Storage of information and data, in a cloud- based environment, has solved the problem of increasing difficulty with data storage, its management, and maintenance. Security of information has been provided by many tools and software that has specifically developed to solve a particular security-related issue. Cloud-based also provides with its security, but it has not been up to the mark. Cloud computing usage has been increasing year after year, and due to this progressive increase, many additional services are required on the cloud platform. There are various aspects of data or resource which must be checked and analyzed by the cloud security, for example:

a) Risk analysis has to be performed before the data or resources are allocated any space in the cloud.

© The Author(s), under exclusive license to Springer Nature Switzerland AG 2025
M. Gupta et al. (Eds.): MISS 2023, CCIS 1952, pp. 83–96, 2025.
https://doi.org/10.1007/978-3-031-69115-7_5

b) The cloud service models (IaaS, PaaS, SaaS) must be checked at different levels, and security must be provided and checked at each service level.
c) There should be enough security provided for various cloud types (public, private, and hybrid) along with its risk analysis before transferring or allocating resources from one cloud type to another.

While we can increase the internal security of cloud computing, external security aspects can also help in improving the cloud security. Many times, we can observe the usage of a firewall along with IDS (Intrusion Detection System) and IPS (Intrusion Prevention System) for maintaining the security between the client and cloud connection. The firewalls have to be used to filter the network traffic according to the mentioned rules [1]. IDS are used to detect intrusion and provide necessary steps to fight against intrusion [2]. IPS is generally placed between the client and firewall, providing an extra line of defense and screening to the incoming packets on the network. Sometimes the attacker may bypass this amount of security and proceed further into the network and attack the servers and other services of the cloud. For providing add-on security in Cloud computing, there have been instances when Honeypot had been incorporated in the securities defense line. Honeypots are generally placed between the firewall and the client endpoint. This provides additional filtering of packets during their transmission over the network [3].

A honeypot is a decoy system or a decoy server used to attract the attackers, to let them attack. It behaves like a genuine system, that is part of the network, but in reality is isolated from the network and is used to trap and monitor attacker's attack patterns and methods. These attack patterns and methods are then used by Honeypot to learn and improve the security of other systems in the actual network. Some information can also be used by the Honeypot to retaliate against the attacker. Honeypots are found on the basis of their usage, and the types of honeypot that come in handy in different case scenes are listed here:

1. Research Honeypots – As the name suggests, it is deployed to gather attacker's attack patterns and learn to form them.
2. Production Honeypots – As the name suggests, it is deployed to detect and prevent the risks that an organization might face when under cyber- attack.

1.1 Motivation

We have already discussed and found that cloud computing itself has a lot of demand in the market due to the services it provides to its customers. Cloud computing is a package that offers the following facilities: a). External servers for the storage, maintenance, and retrieval of data, b). Service of the virtual desktop, that is used for accessing different OS by the customer, c). a platform for developing various software, d). Deployment of new applications and using the inbuilt applications for several purposes. Despite cloud computing's wide usage, we discovered some of the important drawbacks, which are already discussed in the literature survey section. Even though Cloud-Computing is known to be safe through different means, the attacker can always exploit the slightest vulnerabilities found. Here we have discussed one such possibility where the attacker who has avoided all the checks and firewalls arrives at the internal network due to lack of security methods can effortlessly divulge in the network. Due to which the attacker can gain access to the secrets that have been stored away by the enterprise and can use it for malicious reasons, hence by harming the third party. These downsides were the reason that motivated me to do some research and increase the security of the network.

1.2 Contribution

Here in the paper, the main goal of the proposed technique is to prevent the internal network from being attacked by the intruder, which has been regarded as one of the main risks to date. Here we have elaborated on a technique that offers a secure connection while working with the cloud environment. Whenever an attacker disguised as a legit user tries to break through the many defenses of the network and attacks the modules of the network, making it vulnerable; moreover, the requests from a legit user are not dealt with properly due to a shortage of resources. Thus, we propose an administrator who will oversee all the switches and firewalls and direct all the suspicious activities to the honeypot servers. This makes the network free from attackers, and allows the honeypot servers to collect the attack patterns of the attackers and enable them to counter attack the attacker. Also, with the increase of free resources, the network modules will efficiently provide the required resources to legitimate users. Here we provide a model that will increase the security over the network and prevent the attackers from attacking the significant modules of the network. So, we will have efficient use of a network that provides better confidentiality for the user's data.

1.3 Organization

The paper here-by follows the following schema. Section 1 gives introduction part of the paper inclusing motivation and contribution. Section 2 depicts the literature review done for the proposed topic. Section 3 portrays the issues and challenges that exist in the present mechanism for Intrusion Prevention System. In Sect. 4, we have talked about why it is advantageous to use the proposed solution that has been modeled. There is an in-depth working of the architectural model IPCCH, which is presented in this paper. Various algorithms have been discussed in Sect. 4 to get a better understanding of the architecture. Finally, we have described the analysis delivered by the IPCCH framework.

2   Literature Survey

Muhammet Baykara and Resul Das in 2018 [4] had proposed a novelty of honeypot for the security purpose of cloud computing. They had given a unique approach for intrusion detection and intrusion prevention systems, which is combing both the architects and avoid network bottleneck. This provided many benefits that had not been observed before, which was a reduction of configuration cost along with detection of a zero-day attack. It also maintained and managed the network traffic more effectively. The only drawback was it could not prevent the zero-day attack from happening, and could only give a warning if it was going to happen. Dina Moloja and Noluntu Mpekoa in 2017 [5] had proposed an intrusion detection and prevention system for the electronic voting that took place in South Africa. The model they proposed was a mixture of IDPs (Intrusion Detection and Prevention System) and HIDS (Host-based Intrusion Detection System). They had conducted rigorous research by collecting data and analyzing the alert pattern. They had developed a flexible and convenient Voting system with the help of cloud services, and had also incorporated enough security. The drawback being the Tipping point when too many people were gathered on the network, the load could not be handled well. Wided Ben Daoud, Amel Meddeb-Makhlouf, and Faouzi Zarai in 2018 [6] had created a model of intrusion prevention system based on risks involved in the cloud computing environment. They had introduced the model, which allowed the cloud services to discard non-authorized users, which greatly reduced the risks being involved with cloud security. This aided in to gaining a new foundation of trust between the user and service providers. The model was a mixture of Role-based access control (RBAC) and the risk analysis done by various machine learning techniques. There were a few suggestions to reduce the overheads and delays, that was the FOG computing model, which was a talk for the future. Muhammet Baykar and Resul Das in 2019 [7] proposed a centralized honeypot-based security system, which used a software-defined switch for securely managing the VLAN networks. They proposed a system which was operated by GNS3 simulation software, which helped in improving the security of VLAN by reducing false negatives by a certain level. Even though the performance level was increased, they failed to detect new attack patterns, thus failing to prevent the zero-day attack, which mostly occurs when signature-based detection methods are used. They were successful in reducing high false-negative alarms, which are often observed as a disadvantage in anomaly detection methods. Guangfeng Guo, Junxing Zhang, and Zhanfei Ma in 2020 [8] had proposed an intrusion prevention system, that trace-backs attacks and can control the campus networks with software specifically designed for this purpose. They made use of an SDN controller to control the traffic flow and manage the faulty nodes. They were able to reduce the overheads even when the environment was real-time. There can be more work done on improving the algorithm to support more stateful forwarding devices, even when the foremost network topology is not known.

Intrusion Prevention approaches in Cloud Computing defined by various researchers are presented in Table 1. Approaches are discussed with number of parameters. Based on the findings from these research papers, various issues and challenges are identified.

3   Identification of Issues and Challenges

As we know, because of the increase in information and data generation, the cloud environment has become the most trending platform. Cloud platform services extend their benefits not only to giant corporations and companies, but also to communal people. Thus, we can say that many people hand over their private data to the cloud, making it responsible for the security and management of those data and information. So providing a better amount of protection to these data and making the people feel secure is the responsibility of the cloud platform [21–23].

As we discovered earlier in our literature survey, we need to build a highly safe-guarded network for easy uploading and retrieving data by the user, without worrying about the attackers. But first, let's see what issues exist due to the insufficiency of security on the channel through which data will pass over.

Table 1: Intrusion Prevention Approach in Cloud Computing

Authors of Paper	Approach	Parameters	Issues
JunHo Jo et al. [12]	Microcontroller, Laser distance, sensor, AWS, Image recognition	Effective, prevention and inspection, damages, scheduling overhead	Accuracy of data accumulation
Ihsan H Abdulqadd et al. [1]	ML-IDPS proposed in SDN/NFV for 5G network	Mitigating IP spoofing, overflow, DDoS, host location hijacking	-
Farouq Aliyua et al. [3]	Light weight encryption	Low usage of processor, and memory	More latency and energy consumption if the traditional (fog device) is used
Muhamme t Baykara et al. [4]	Signature based; hardware constraints used to avoid the network bottleneck	Reduce cost of configuration n, maintenance e and management, can detect 0- day attack, network traffic can be monitored	0-day attack
Ahmed Patel et al. [13]	Network- based sys, host-based sys, CSIDPS (collaborative smart IDPS), cooperative smart soft computing, automatic computing components	Reliable, scalable and flexible design	Overcome vulnerabilities, giving better detection rate with low FP alarms
Yaping Chi et al. [10]	Virtual resource management technology, SDN series connection of IPS	Enhancing the efficiency and flexibility	Snort (detected) Bro and Tipping-point
Wided Ben Daoud et al. [6]	RBAC, risk, trust	Allows the discard of non- authorized users based in a computed and updated risk metrics	Fog model to reduce the overheads in term of delay
Nayyar Ahmed Khan et al. [14]	AWS cloud trail	User friendly, fast with high productivity	Avoid breach of data, exit strategy for DDoS attack

1. Can breach into the internal network and cause havoc
2. Can breach into the service network and cause the legit user to delay or time out the response.
3. Data can be lost, stolen, or altered.

Many other attacks are possible by the attacker, who might be an insider or an outsider [9, 10]. The cloud platform is explored more as time passes by, during which number of users expanded exponentially with time. More and more data flows in and out of the cloud, boosting the attacks happening. These attacks are growing much more advanced, to which the security is not able to follow up. The security fails to recognize the intentions of the users, which only leaves us with the option of increasing the general security. There must be a detection mechanism that detects the malicious attack, along with a prevention mechanism that prevents or diverts the attacks or any malicious intended user to a sacrificial server [11, 15]. These sacrificial servers play a vital role in counter attacking the malicious user.

4   Proposed Solution and Methodology Used

In this paper, we have discussed an improvement in the Honeypot deployment architectural model used traditionally. This novel deployment will track the malicious attackers that have made their way to the internal network, which will then be re-directed to the Honeypot in the network, reducing the risks brought by the attacker. Despite there being many ways in a cloud to aggravate the security parameters, here we have used the method of deploying Honeypot. Honeypots are sacrificial machines, having many fake data and techniques implanted in them to trap the attacker [16–20]. Later on, these data patterns are used to trace and counterattack the attacker [24]. The data gathered in these honeypots are updated in the IDS and IPS by instilling the newly found attack methods and attack patterns.

There are three types of honeypots based on their design factors, they are:

a. Pure honeypots: attacker's activities will be monitored with the help of a bug tap that has been pre-installed on the honeypots network linkage. The most important benefit of such a honeypot is when we surreptitiously want to gain remote control over the defense mechanisms.
b. High-Interaction honeypots: attackers are provided with plenty of fake services to play. With the help of a virtual machine, we can deploy multiple honeypots on a single device, which creates a backdoor to recover the device even when compromised. The prime example of this type is Honeynets.
c. Low-interaction honeypots: the attacker has limited services to play around with. The consumption of resources is relatively low, thus providing a chance to host multiple virtual machines on a system. The security here is relaxed, but it has a short response time. The prime example of this type is Honeyd.

Here in the proposed model, we have deployed three honeypots, which increases the efficiency of finding the faulty requests produced by either an inside attacker or an outside attacker. Here, the efficiency of an intrusion detected and prevented is written as per the Eq. (1):

$$Efficiency = \frac{Total\ number\ of\ True\ Requests * 100}{Total\ number\ of\ Requests} \tag{1}$$

The number of false requests (request made by attackers) that have incurred out of the total user requests is given in the Eq. (2):

$$Total\ number\ of\ False\ Requests = TNR - TNTR \tag{2}$$

TNR = *Total number of Requests*, TNTR = *Total number of True Requests*

Thus, with this, we can find the efficiency at which the requests will be handled at any given point.

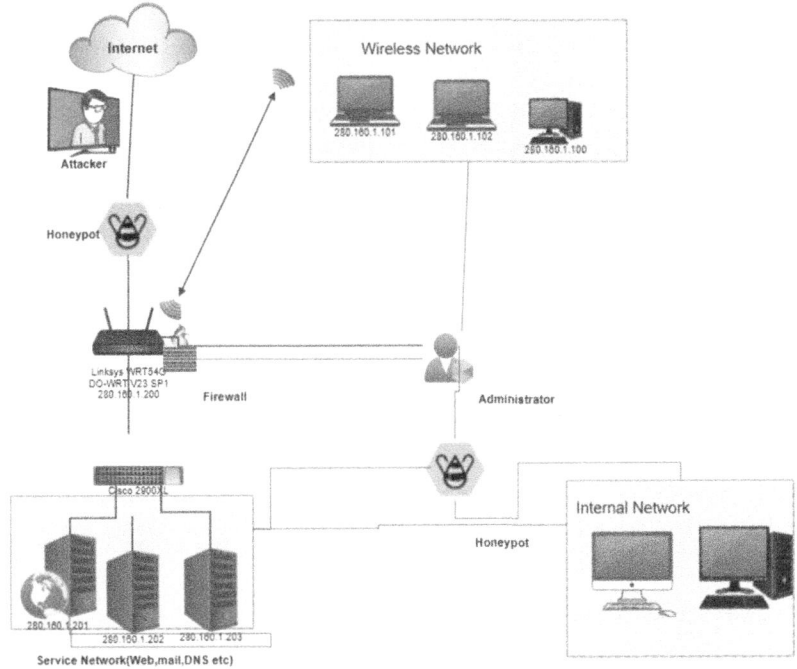

Fig. 1. Deployed honeypot architecture

As discussed in the earlier modules, the architecture as shown in the Fig. 1, proposed here consists of 3 honeypots for the Intrusion Prevention mechanism. Two honeypots can be high-level honeypots, honeypot3 and honeypot2, while honeypot1 can be a pure honeypot that enables the server to check the logs. The other vital component in this architecture is the administrator, whose duty is to respond and control the routers and external firewall if any problem arises. The administrator keeps a check and observes if any suspicious entity has intruded and affected the security system. The user bypasses all the components before it gets the required services. These back-end tasks are processed speedily, providing the user with the required output after just a few clicks. Here we will be explaining the back-end process that goes on. From the internet, a user requests services or internal network access from the cloud platform. It is not known to us if the user is requesting the services with all intent or not. It passes via Honeypot1 that checks the IP, port_no, whether there is any previous registration of the user, etc. It examines the log activities that are registered beforehand and monitors whether it is a new user or not. If the user is unknown, then all the information is logged. If we found out that the user is malicious, the request will not be processed further and will redirect to the last accessed page, which will block the user's access to the network services. If the user is not malicious, then the request is forwarded to the second component, which happens to be the firewall. The firewall gets administered by the administrator control. The details of the packets, port, IP, and other factors, will be examined by the firewall with the previously established security policy. If any problem is detected, it will be sent to Honeypot1 and will not have access to the services provided by the network. The user continues further and goes to the respective network, where it passes via routers and switches. These routers will route the user to its desired network, while the switches allocate a server, PC, or machine to access the required services. The routers have a security check placed in them, which cross-checks the user IP, port, and request destination with the administrator information, and allow users to pass through if they are legit. When a malicious user enters the LAN and begins to act maliciously, the malicious user will be redirected to the honeypots in their respective LAN, preventing the intrusion attack. Here, the user gets examined via the security aspects provided in the switch, and if the intentions are malicious, they get re-routed to the honeypot server, which is not on the original LAN network. When the ill-intended user gets caught in the honeypot, the honeypot can easily find the attack patterns and update the other honeypot data. Also, honeypot2 and honeypot3 being a high- interactive honeypot, the honeypots can create honeypots in themselves, which ultimately leads to a massive honeynet. By successfully getting the required information, these honeypots can also trace back to the attacker and counter attack as shown in the Fig. 2.

Here, various algorithms are designed based on the flowchart of proposed model. Algorithm 1, explains for the prevention of malicious user from accessing sensitive files available on the cloud-environment.

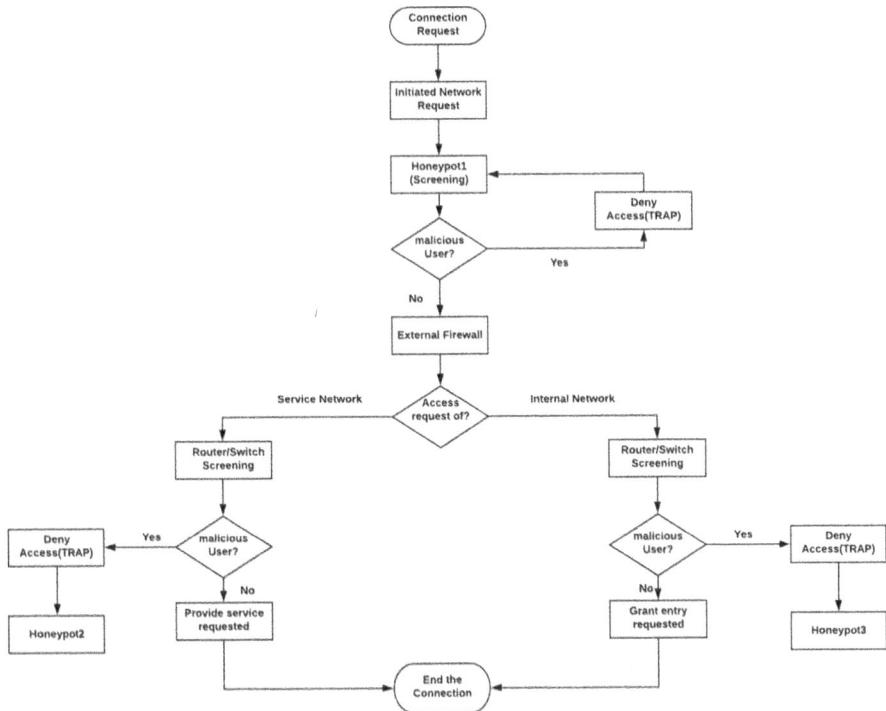

Fig. 2. Flowchart of proposed model

Algorithm 1 For the prevention of malicious user from accessing sensitive files available on the cloud-environment

inputs: User, Network access;
outputs: Alert and Report is generated if there is an attacker;

while *TRUE* **do**
    user goes to Honeypot1 for screening;
    Honeypot1 log checks;
    if *user == malicious user* **then**
        Generate alert;
        Block attack;

    else
        user can proceed to next check in network i.e External Firewall;

    end
    goto External firewall;
    External Firewall checks;
    ——This will be further explained in algorithm 2
end

Algorithm 2 gives details to provide the selection option before going to LAN after bypassing the firewall.

Algorithm 2 To provide the selection option before going to LAN after bypassing the firewall

---

inputs: User, Network access, firewall;
outputs: Alert and Report is generated if there is an attacker;

when the user choose the request type to be service request;
if *request-type* == *service-network-request* **then**
  | Goto SN router;
  | Goto SN switch;
  | —-The further explanation is given in algorithm-3

when the user choose the request type to be internal request;
else if *request-type* == *internal-network-request* **then**
  | Goto SN router;
  | Goto SN switch;
  | —-The further explanation is given in algorithm-4

else
  | Error request type not found, please goback to previous step;
end

Algorithm 3 and 4 explains to prevent the malicious user from entering the LAN of service's provided by cloud environment with the help of honeypot2 and to prevent the malicious user from entering the LAN of internal network provided by cloud environment with the help of honeypot3 respectively.

Algorithm 3 To prevent the malicious user from entering the LAN of service's provided by cloud environment with the help of honeypot2

---

inputs: user, LAN network
outputs: alert generated if there is an attacker

From algorithm2 it is redirected here to check the service-network-request;
Administrator control checks;
if *user* == *malicious user* **then**
  | Generate alert;
  | Block attack;
  | Redirect to Honeypot2;

else
  | User can proceed the request of legit user;

end

Algorithm 4 To prevent the malicious user from entering the LAN of internal network provided by cloud environment with the help of honeypot3

---

inputs: user, LAN network
outputs: alert generated if there is an attacker

From algorithm2 it is redirected here to check the internal-network-request;
Administrator control checks;
if *user == malicious user* then
 | Generate alert;
 | Block attack;
 | Redirect to Honeypot3;

else
 | User can proceed the request of legit user;

end

Performance Evaluations

For assessing the proposed model, a test was conducted by deploying a honeypot in a Linux system and observing the results obtained. We took a use case where we tested the system with and without the honeypot [25, 26]. To determine the consistency, we calculated the efficiency that is measured, with the increase in number of users. This calculation is done for the following situation: When there was no honeypot in the internal network, the intruder that attacked the network could easily gain access to sensitive files and information of the user, which could then be used for malicious intention. Initially, as the numbers of users were zero, the efficiency was maximum; but as the users increased, the efficiency kept on dropping until it stagnated at a point as shown in the Fig. 3.

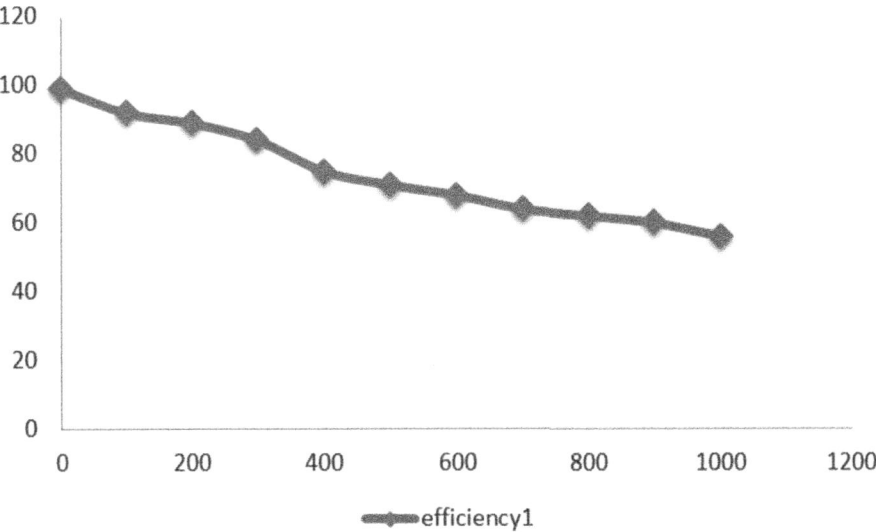

Fig. 3. Internal network deployment without honeypot

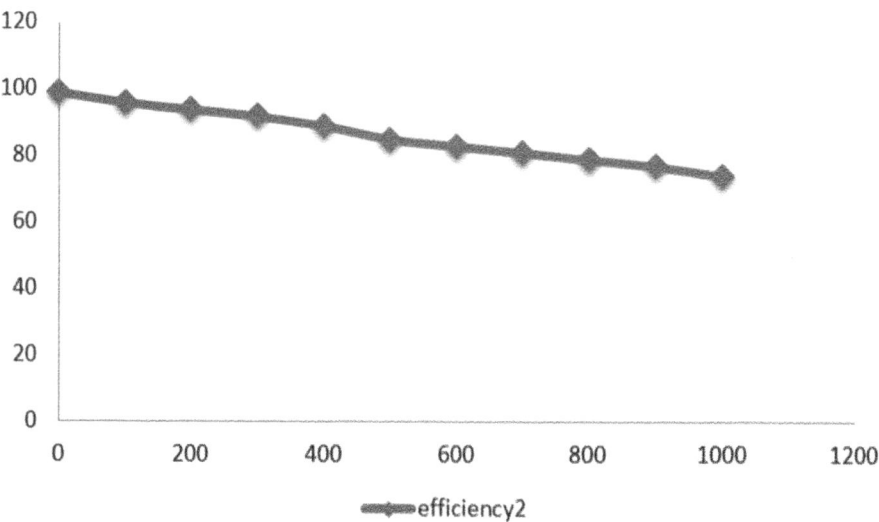

Fig. 4. Internal network deployment with honeypot

Whereas in the case a honeypot was added into the internal network, the intruder's attack would be diverted to the honeypot server, which would mitigate the threat of having the sensitive data being leaked. Initially, as the number of users were zero, the efficiency was maximum; but as the users increased, the efficiency kept on dropping until it stagnated at a point as shown in the Fig. 4. Still, when compared to the case where no honeypot in the Internal Network, we can observe an increase in efficiency of 18% which is shown on the Fig. 5.

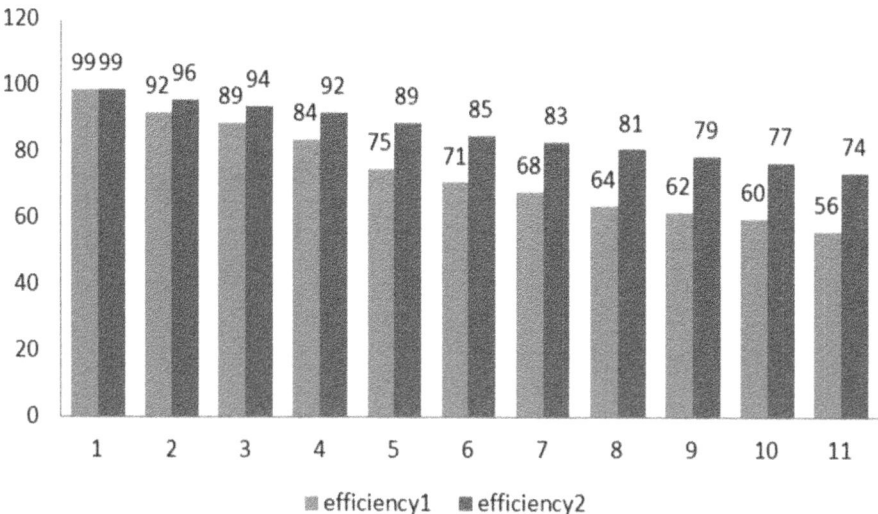

Fig. 5. Comparison of efficiency

The final outcomes that we found leaned towards having a honeypot, rather than not having a honeypot in the internal network area. Further work can be done to increase the efficiency of the honeypot in the internal network, by making it more secure and sturdy.

5 Conclusions

The security aspect of cloud computing was one of the most important features in protecting the data and information that was assembled on the cloud platform by the number of users around the world. To protect the users-data, and provide them with the proper and timely services, cloud security has adopted many distinctive features and models. They have also integrated many external modules to guarantee that the user's data stays safe. Here, we have used one such external module, which is the Intrusion Prevention System. We have integrated it with Honeypot deployment to prevent the attackers from gaining access to user's information and making all use of this information. We have successfully re-routed the malicious user to the honeypot user on a different network, which also helps in reducing the network traffic. When the attacker tries to intrude on any component of the network, alarm is generated, which notifies the other network components and takes preventive measures to stop the intrusion by directly redirecting the attacker to the honeypot system. To determine the consistency, we calculated the efficiency that is measured, with the increase in number of users. This calculation is done for the following situation: When there was no honeypot in the internal network, the intruder that attacked the network could easily gain access to sensitive files and information of the user, which could then be used for malicious intention. Thus, we can say that the results showcased here signify the importance of the proposed architecture.

References

1. Abdulqadder, I.H., Zhou, S., Zou, D., Aziz, I.T., Akber, S.M.A.: Multi-layered intrusion detection and prevention in the sdn/nfv enabled cloud of 5G networks using ai-based defense mechanisms. Comput. Netw. **179**, 107364 (2020)
2. Achbarou, O., El Kiram, M.A., Bourkoukou, O., Elbouanani, S.: A new distributed intrusion detection system based on multi-agent system for cloud environment. Int. J. Commun. Netw. Inf. Secur. **10**(3), 526 (2018)
3. Aliyu, F., Sheltami, T., Shakshuki, E.: A detection and prevention technique for man in the middle attack in fog computing. Procedia Comput. Sci. **141**, 24–31 (2018)
4. Baykara, M., Das, R.: A novel honeypot based security approach for real-time intrusion detection and prevention systems. J. Inf. Secur. Appl. **41**, 103–116 (2018)
5. Moloja, D., Mpekoa, N.: Towards a cloud intrusion detection and prevention system for m-voting in South Africa, PP. 34–39(2017)
6. Ben Daoud, W., Meddeb-Makhlouf, A., Zarai, F.: A model of role-risk based intrusion prevention for cloud environment. In: 14th International Wireless Communications Mobile Computing Conference (IWCMC), pp. 530–535 (2018)
7. Baykara, M., Das, R.: Softswitch: A centralized honeypot based security approach using software-defined switching for secure management of vlan networks. Turk. J. Electr. Eng. Comput. Sci. **27**, 3309–3325 (2019)
8. Guo, G., Zhang, J., Ma, Z.: Intrusion prevention with attack traceback and software-defined control plane for campus networks. Comput. Sci. Inf. Syst. **18**(3), 867–891 (2021)

9. Chauhan, K., Prasad, V.: Distributed denial of service (ddos) attack techniques and prevention on cloud environment. Int. J. Innov. Adv. Comput. Sci. **4**, 210–215 (2015)
10. Chi, Y., Jiang, T., Li, X., Gao, C.: Design and implementation of cloud platform intrusion prevention system based on sdn, pp. 847–852 (2017
11. Fan, W., Du, Z., Fernandez, D., Hui, X.: Dynamic hybrid honeypot system based transparent trafic redirection mechanism (2015)
12. Jo, J., Jo, B., Khan, A., Kim, J.: A cloud computing based damage prevention system for marine structures during berthing. Ocean Eng. **180**, 23–28 (2019)
13. Patel, A., Alhussian, H., Pedersen, J., Bounabat, B., Junior, J., Katsikas, S.: Nifty collaborative intrusion detection and prevention architecture for smart grid ecosystems. Comput. Secur. **63** (2016)
14. Khan, N., et al.: Intrusion management to avoid web-form spamming in cloud based architectures, pp. 437–442 (2019)
15. Ibrahim, N.M., Zainal, A.: A model for adaptive and distributed intrusion detection for cloud computing. In: 7th ICT International Student Project Conference (ICT-ISPC), pp.1–6. IEEE (2018)
16. Idhammad, M., Afdel, K., Belouch, M.: Distributed intrusion detection system for cloud environments based on data mining techniques. Procedia Comput. Sci. **127**, 35–41 (2018)
17. Jelidi, M., Ghourabi, A., Gasmi, K.: A hybrid intrusion detection system for cloud computing environments. In: International Conference on Computer and Information Sciences (ICCIS), pp. 1–6. IEEE (2019)
18. Gadhavi, L.J., Bhavsar, M.D.: Efficient resource provisioning through workload prediction in the cloud system. In: Zhang, Y.-D., Mandal, J.K., So-In, C., Thakur, N.V. (eds.) Smart Trends in Computing and Communications: Proceedings of SmartCom 2019, pp. 317–325. Springer, Singapore (2020). https://doi.org/10.1007/978-981-15-0077-0_33
19. Kumar, V., Sangwan, O.P.: Signature based intrusion detection system using snort. Int. J. Comput. Appl. Inf. Technol. **1**(3), 35–41 (2012)
20. Zhang, L.: Honeypot-based defense system research and design. In: 2nd IEEE International Conference on Computer Science and Information Technology, pp. 466–470 (2009)
21. Gadhavi, L.J., Bhavsar, M.D.: Adaptive cloud resource management through workload prediction. Energy Syst. **4**, 1–25 (2019)
22. Gadhavi, L.J., Bhavsar, M.D.: Prediction based efficient resource provisioning and its impact on QoS parameters in the cloud environment. Int. J. Electr. Comput. Eng. **8**(6), 5359–5370 (2018)
23. Gadhavi, L.J., Bhavsar, M.D. Efficient and dynamic resource provisioning strategy for data processing using cloud computing. Int. Rev. Comput. Softw. (I.RE.CO.S.) **11**, 1828–6003 (2016)
24. Mishra, A.K., Paliwal, S. Mitigating cyber threats through integration of feature selection and stacking ensemble learning: the LGBM and random forest intrusion detection perspective. Cluster Comput. (2022)
25. Farooqui, N., Banduni, A., Mehra, R.: Data mining and fusion techniques for wireless intelligent sensor networks (2020)
26. Singh, P., Bhargava, B., Paprzycki, M., Kaushal, N.,& Hong, W.: Handbook of Wireless Sensor Networks: Issues and Challenges in Current Scenario's (2020)

Entropy Based Transparent and Secure Watermarking Approach Using Arnold Transform

Sanjay Patsariya[1]([✉]) and Manish Dixit[2]

[1] Department of Computer Science and Engineering, RGPV, Bhopal, MP, India
sanjaypatsariya@gmail.com
[2] Department of Computer Science and Engineering, MITS, Gwalior, MP, India

Abstract. Due to speedy development in internet technology, peoples have usually dependent upon internet enabled digital equipment's in a wide range of application. As a result, different challenges relating to information security and authentication are brought up with multimedia objects in both the civil and defense sectors. The use of watermarking technology eliminates the need for an overheard as in traditional encryption by integrating it directly into various frames. As a result, it is effective when employed for data encryption that require restricted resources, as those engaged in the Internet of Things (IoT). The projected study illustrates a spatial domain approach using the concept of entropy for digital watermarking grounded on LSB substitution and Arnold transform. Original Image chunk keeping highest entropy value is employed for watermark implantation. In addition, the perceptual excellence, robustness and security of the watermarked images produced by the proposed framework have been compared to certain cutting-edge watermarking techniques using image quality parameters namely MSE, PSNR, SSIM NPCR, UACI and NCC.

Keywords: Entropy · MSE · PSNR · NPCR · UACI · NCC · Arnold Transform · Hill Cipher

1 Introduction

Internet usage as a medium for digital multimedia distribution is on the rise in modern life. Due to that, a strong reliable and secure copyright protection technique is required for authentication of digital multimedia objects [1]. Digital watermarks came in to picture to avoid this type of issues. The word "digital watermarking" refers to an unseen alteration of data, and it has grown in importance in the area of multimedia signal processing. It can also be viewed as an improved technique for preventing digital piracy and protecting the ownership of digital media. The process of digital watermarking involves embedding a watermark image within a host image in such a way so that small deviations in data values cannot be detected by human visual system [2–4]. Recently, because to its availability for numerous lightweight solutions for a variety of approaches and applications, such as cloud computing, electronic health, and IoT, digital watermarking technology has

© The Author(s), under exclusive license to Springer Nature Switzerland AG 2025
M. Gupta et al. (Eds.): MISS 2023, CCIS 1952, pp. 97–112, 2025.
https://doi.org/10.1007/978-3-031-69115-7_6

become indispensable for diverse industries. The spatial and transform domain are used to implement watermarking approach. The main issue with digital picture watermarking solutions is watermark security. The main characteristics of watermarking are imperceptibility, robustness, security, payload, and computing cost. Cryptographic approaches have been considered as a demanding aspect for security objectives, particularly for images in telecommunication, defense, and medical applications. This work presents a safe spatial domain watermarking based on entropy. Additionally, the performance of the suggested scheme is examined and contrasted with other state-of-the-art methods.

The block diagram of watermarking method is depicted in Fig. 1.

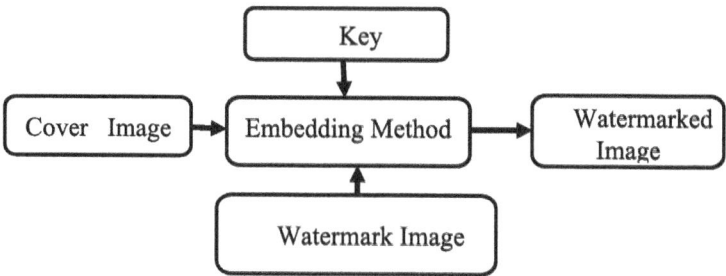

Fig. 1. Block diagram of normal watermarking Process

2   Literature Survey

Sanjay et al. [5] suggested watermarking method grounded on spatial domain to improve transparency. They used cover Image chunk with the maximum entropy level to implant confidential data. To strengthen the watermark's security, hill cipher encryption is employed. Evaluation of the imperceptibility by the presented method was tested by means of subjective and objective picture quality assessment techniques. Hill Cipher encryption is used for watermark security.

Parah et al. [6] Presented a least significant bit based blind image watermarking approach established on spatial domain. Pseudo-random keys were used for security of watermark. Pseudorandom address vector was used for security purpose. They had compared the suggested watermarking method perceptual quality with some cutting-edge watermarking methods. The robustness of the provided method is evaluated using a variety of assaults.

Bal et al. [7] suggested a watermarking approach established on bit pair similarity. The Symmetric key cryptography was employed for the security of watermark image.

Bouslimi et al. [8] projected message accessible in both domain namely encrypted and spatial. Here, instigator has proposed a new data concealing method to encrypted image which allow approaching message in both domains.

Soltani et al. [9] described semi-blind watermarking methods based on the ordered Hadamard Transform (HT). This method has a number of benefits over previous transform domain and hybrid domain approaches, including fast processing times and simpler equipment setup.

Patvardhan et al. [10] have suggested a reliable watermarking approach. Here, YCbCr color space watermarking is carried out by adopting SVD and DWT. A QR pattern was employed as confidential data in this instance. According to their experimental findings, a QR code's data can still be recovered even after being extensively altered by an image processing attack.

Singh et al. [11] has presented a watermarking methods grounded on DWT and SVD. Dual dissimilar watermarks were employed; these are logo image and the QR pattern.

Pandey et al. [12] projected a watermarking method built on the SWT-SVD. The Y plane of YCbCr color model was chosen to implant the confidential data. To strengthen the picture quality characteristics, hybrid technique was applied.

Pandey et al. [13] presented a GWO-based Non-blind LWT-SVD method. To obtain better results, the Y plane of YCbCr color model was altered to implant confidential digital materials.

Kumari et al. [14] presented a novel watermarking approach grounded on LSB substitution mechanics and also used matrix multiplication for data security and authentication. In their proposed work, the last three bits of each RGB component of the cover picture or watermark image are replaced using this technique by choosing a non-consecutive matrix of pixels. Each matrix block is subjected to the filtering process in order to choose the best block for the embedding procedure.

Faheem et al. [15] suggested watermarking method grounded on chaotic map and image gradient based on the least significant bit. The gradient of each non-correlated block in the image is computed after segmenting the image into blocks. The gradient of the picture expresses how quickly an image can change. The image gradient is a method for determining the ideal location to incorporate a watermark while preventing image degradation. The watermark signal is implanted in line with the gradient of the original image by altering its least significant bits.

Mashad et al. [16] suggested a hybrid digital watermarking and image processing strategy. A blob detection algorithm is used with variations of the popular Least-Significant Bit (LSB) watermarking approach in order to embed data into the boundary pixels of the largest blob in a digital image. The findings demonstrate that the suggested methods beat the conventional LSB algorithm in terms of time, assessment criteria, and the proportion of modified pixels.

3   Proposed Methodology

Proposed methodology consists of three parts i.e. (1) Watermark Insertion (2) Watermark Extraction and (3) Arnold Transform for security.

3.1   Watermark Insertion

Entropy aims to measure the level of randomness in an image. It represents how much ambiguity or unpredictability available in the image. The Entropy can be described as:

$$E(I_m) = -\sum P(I_m) \log P(I_m) \tag{1}$$

Figure 2 depicts the watermark embedding procedure and consists of following steps

1. Divide Cover (Host) image into equal size segments.
2. Calculate the entropy of each segment.
3. Watermark is encrypted using Arnold Transform prior to insertion
4. Select the image segments of cover image having highest entropy for watermark insertion.
5. Apply LSB method to obtain watermarked Image in which LSB bits are set to 0 in each cover image block and then apply right shift over encrypted watermark by 7 bits.

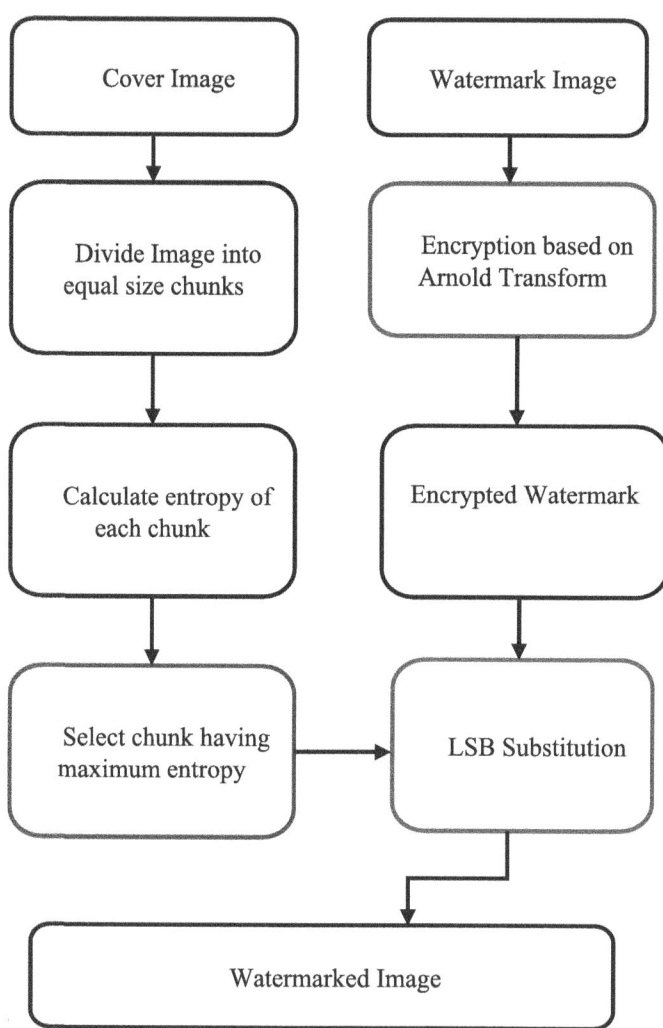

Fig. 2. Watermark Embedding Procedure

3.2 Watermark Extraction

The watermark pulling out process is shown in Fig. 3 that consists of following five steps.

1. Divide watermarked image into equal size segments.
2. Calculate the entropy of each segment.
3. Select block having highest entropy
4. Apply LSB extraction method in which reverse process of LSB substitution is applied.
5. Apply anti-Arnold transform to obtain watermark.

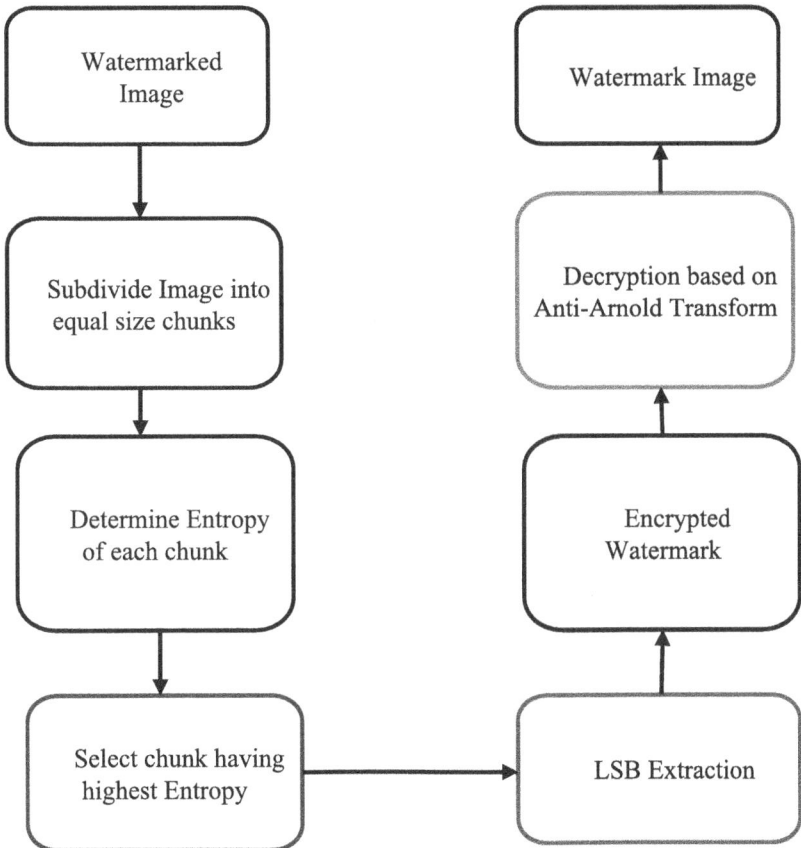

Fig. 3. Watermark extraction procedure

3.3 Watermark Security

Watermark security is a big concern in watermarking technique. To make the watermark secure, Arnold transform is used here. Arnold transform confuse the image by distortion the correlation among the pixels in the images [17]. For an image of size N *N, Arnold transforms and Anti –Arnold on the position Cr_x, Cr_y pixel is given by Eq. 2 and Eq. 3.

$$\begin{bmatrix} Cr_x' \\ Cr_y' \end{bmatrix} = \begin{bmatrix} 1 & 1 \\ 1 & 2 \end{bmatrix} \begin{bmatrix} Cr_x \\ Cr_y \end{bmatrix} \bmod N \tag{2}$$

$$\begin{bmatrix} Cr_x' \\ Cr_y' \end{bmatrix} = \begin{bmatrix} 2 & -1 \\ -1 & 1 \end{bmatrix} \begin{bmatrix} Cr_x \\ Cr_y \end{bmatrix} \bmod N \tag{3}$$

4   Performance Analysis

The effectiveness of proposed study has been investigated and same compared with other methods using various gray scale images. The various Cover and watermark image used for performance analysis are shown in Fig. 4 and Fig. 5, respectively.

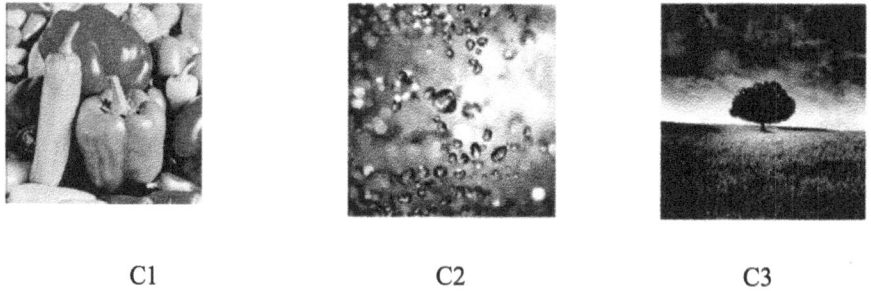

C1   C2   C3

Fig. 4. Considered Cover (Host) Images

W1   W2

Fig. 5. Watermark Images

4.1   Transparency Analysis

The transparency analysis measures the effectiveness of the invisible watermarking methods. The PSNR, MSE and SSIM are adopted to determine the transparency of proposed

Table 1. Comparison of PSNR with the considered methods

S. No	Author	PSNR
1	**Proposed**	**75.18**
2	Sanjay et al. [5]	69.22
3	Parah et al. [6]	68.72
4	Bal et al. [7]	52.00
5	Boushmi et al. [8]	62.00

Table 2. Estimation on MSE, SSIM and NCC(without attack)

Cover Image	Watermark	MSE	SSIM	NCC
C1	W1	0.0010	1.0000	1.0000
C1	W2	0.00067	1.0000	1.0000
C2	W1	0.00099	1.0000	1.0000
C2	W2	0.00067	1.0000	1.0000
C3	W1	0.0011	1.0000	1.0000
C3	W2	0.00079	1.0000	1.0000

work. SSIM is employed to decide the likeness between host and watermarked picture. Table 1 and Table 2 shows the transparency assessment parameters values obtained by proposed method.

From Table 1 and Table 2, it can be seen that the suggested picture watermarking method provides strong imperceptibility in terms of MSE, PSNR and SSIM as compare to the considered methods. The proposed method produces improved outcomes by maintaining high PSNR and SSIM values between the watermarked image and the cover image. The strategy is effective to hide the information at more appropriate location.

4.2 Security Analysis

Insecure watermarking cannot be used for authentication and copyright protection. The adjacent pixel values are correlated in the image and hacker can try to establish correlation among pixel values to find watermark image. The security algorithm destroys the relationship among the nearby pixel values. The Security of the encryption techniques has been evaluated using PSNR, Number of Pixel Changing Rate (NPCR) and Unified Average Changing Intensity (UACI) [18]. Investigation has been imposed over two watermark image namely W1 and W2 image. Experimental outcomes of the encryption techniques are shown in Table 3. The effectiveness of the encryption technique (Arnold Transform) is shown in Fig. 6.

To check the level of security, the performance analysis test of the scrambling/encryption algorithm is done through the parameters namely PSNR, NPCR and

Table 3. Estimation of Security assessment parameters

Watermark	Arnold Transform			Hill Cipher Encryption [5]		
	PSNR	NPCR	UACI	PSNR	NPCR	UACI
W1	4.57	86.16	22.28	13.89	15.57	16.49
W2	7.63	88.41	15.95	10.68	40.13	15.30

UACI on both gray and color images. High NPCR and UACI are commonly understood to indicate high resistance to differential attacks. The smaller value of the PSNR shows the more dissimilarity between two images. From Table 3, the PSNR, NPCR and UACI values obtained using Arnold transform is better than the values obtained using Hill cipher transform.

It has been found that in a plain image, there are highly strong correlations between adjacent pixels. For the optimal scrambling, there must be no association between the pixels. To examine the correlation, the correlation coefficient is computed for both original watermark and scrambled watermark images. Comparison of the horizontal, vertical, and diagonal correlation coefficients between original and scrambled images using Arnold transform and Hill cipher encryption schemes is illustrated in Table 4. Zero correlation coefficient indicates that no linear relationship exists between two images. Correlation plot is dispersed as illustrated in Fig. 7, means there is no association between the adjacent pixels in the scrambled image. The correlation between adjacent pixels should be near to 0 for a truly random image. The outcome shows that Arnold transforms establish better de-correlation among pixels of the images.

4.3 Robustness Analysis

The robust refer to the capability to recognize watermark after various image processing attacks [19]. The watermark will be intact after various attacks and hence used for authentication, copyright protection, broadcast monitoring, copy control and finger printing. The Normalized cross correlation (NCC) is used for the assessment of robustness [20–22]. The NCC values after various attacks are shown in Table 5.

From Table 5, it is observed the suggested technique preserves the watermark image's perceptual quality. The proposed study provides enough robustness against various images processing operation.

No. of Iterations	Arnold Transform	
	W1	W2
1		
2		
3		
4		
5		

Fig. 6. Visualization of encrypted watermark after number of iterations as a key

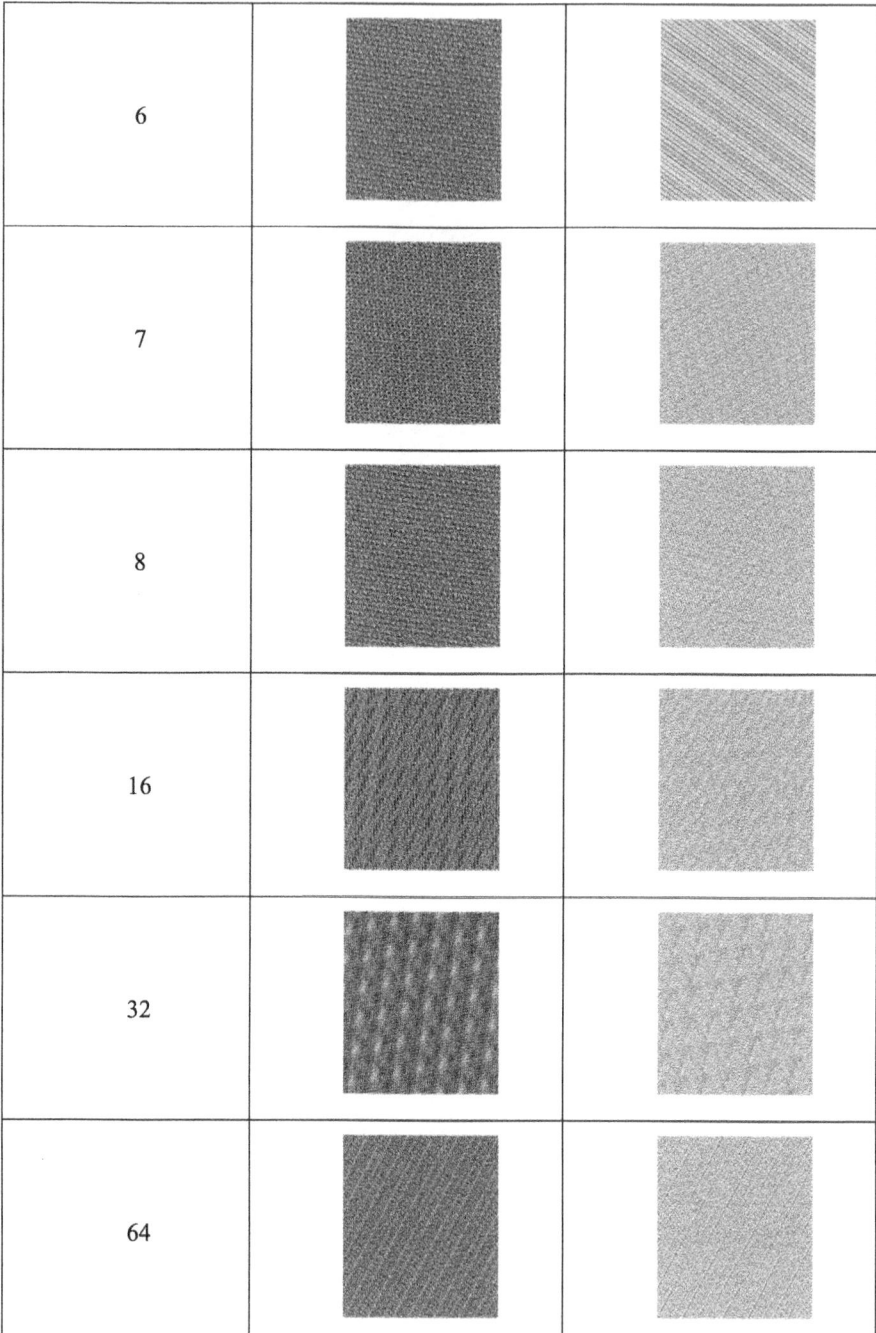

Fig. 6. (*continued*)

Table 4. Estimation of Horizontal, vertical and diagonal correlation coefficient using Arnold and Hill Cipher methods.

Image	Encryption /Scrambling Method	Horizontal Correlation	Vertical Correlation	Diagonal Correlation
	None	0.9905	0.9923	0.9871
	Hill Cipher	0.5542	0.6994	0.4795
	Arnold(k=8)	0.0034	-0.0536	0.0192
	None	0.9845	0.9806	0.9630
	Hill Cipher	0.1190	0.1040	0.0763
	Arnold(k=8)	-0.0514	0.0435	-0.0207

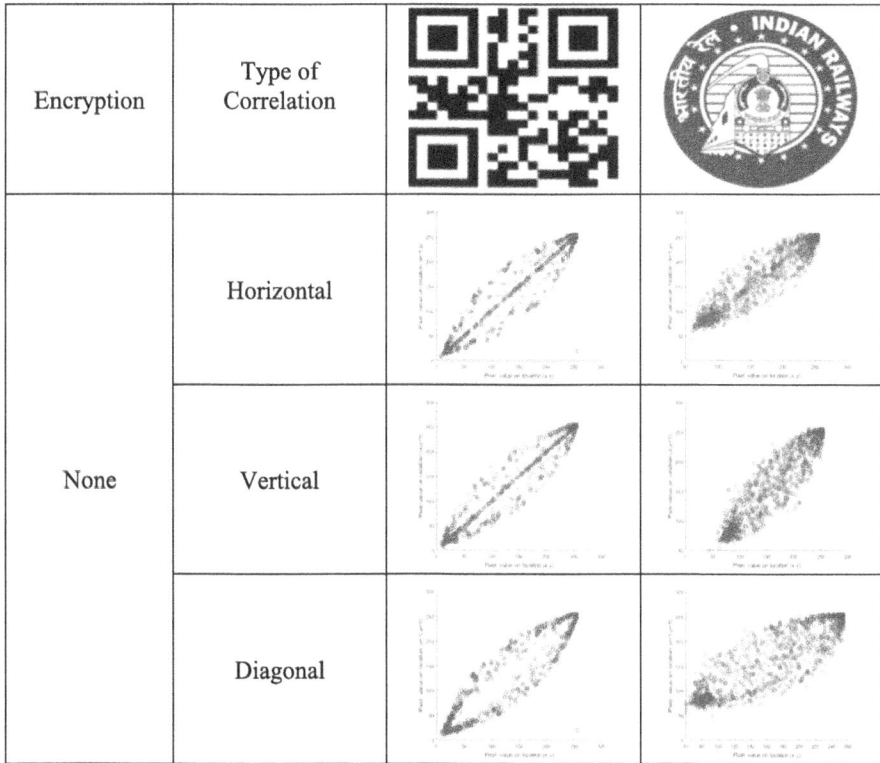

Fig. 7. Visualization of various correlations

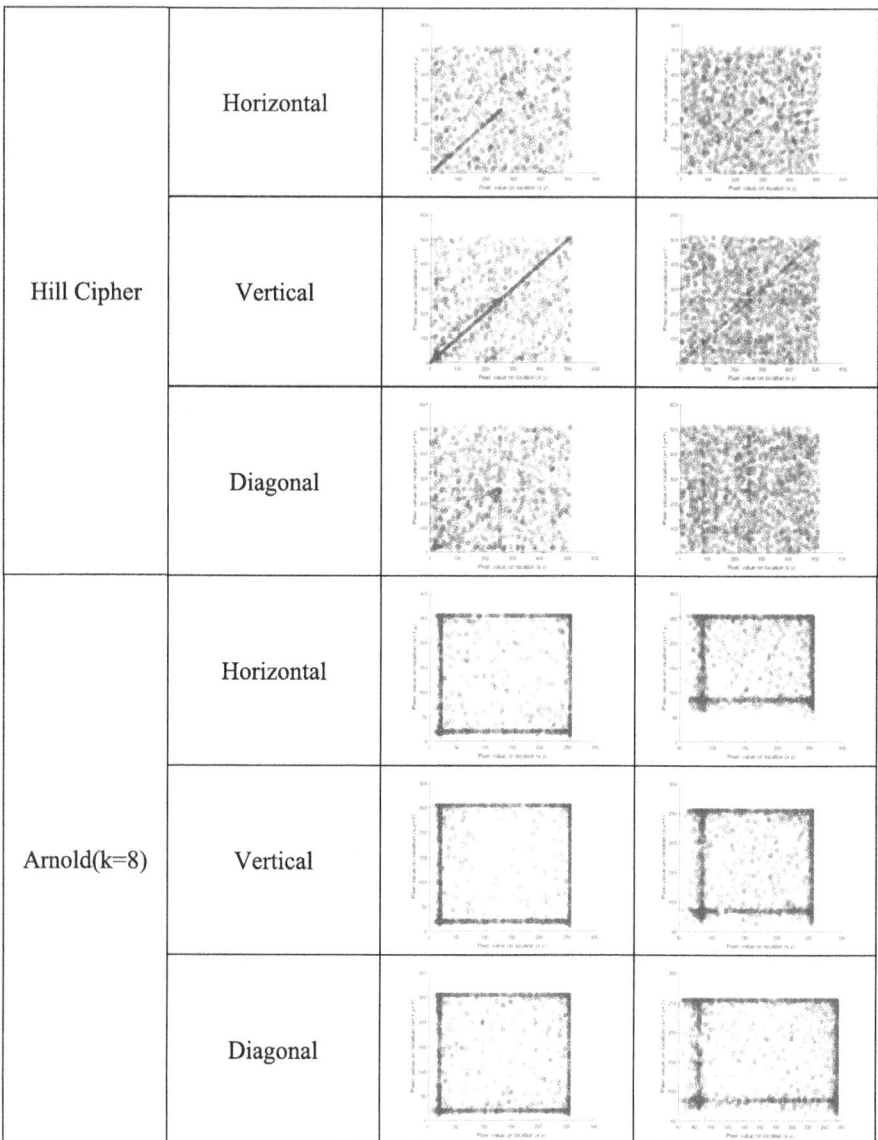

Fig. 7. (*continued*)

Table 5. Estimated NCC value and extracted watermark image (EWI) after various attacks

Name of attack	Salt and Pepper (d1=0.001)	Salt and Pepper (d1=0.002)	Salt and Pepper (d1=0.006)
EWI			
NCC	**0.9998**	**0.9994**	**0.9987**
Name of attack	Gaussian Noise (v1=0.001)	Gaussian Noise (v1=0.002)	Gaussian Noise (v1=0.006)
EWI			
NCC	**0.7772**	**0.7762**	**0.7760**
Name of attack	Speckle Noise (v1=0.001)	Speckle Noise (v1=0.002)	Speckle Noise (v1=0.006)
EWI			
NCC	**0.7787**	**0.7779**	**0.7814**
Name of attack	Poisson Noise	Resize attack (50%)	Median filter[3*3]
EWI			
NCC	**0.7822**	**0.7826**	**0.8779**

(*continued*)

Table 5. (*continued*)

Name of attack	Median filter [5*5]	Gaussian low pass filter[5*5]	Wiener filter-ing[3*3]with noise =.001
EWI			
NCC	0.8144	0.7935	0.7857
Name of attack	Hybrid attack(salt & Pepper followed by speckle)	Hybrid attack(Salt and pepper followed by scaling)	Hybrid at-tack(Gaussian followed by scaling)
EWI			
NCC	0.7872	0.7812	0.7885

5   Conclusion

The Spatial domain approach using the concept of entropy is proposed to achieve better transparency over existing methods. The image quality parameters namely MSE, PSNR and SSIM are used to evaluate the transparency. To secure watermark from unauthorized access, Arnold transform is employed due to simplicity and more effectiveness. The image is confused by changing the pixels position. The number of repetition is used as key and higher the number of iterations produces more distortions in watermark image. PSNR, NPCR and UACI are used to evaluate the security effectiveness of encryption method. The robustness of the proposed method is evaluated between the cover and extracted watermark (after attack) using NCC. It is clearly evident the presented study provides better transparency, security and better robustness in some cases. The proposed method can be employed in defense and civil areas.

References

1. Patsariya, S., Dixit, M.: A survey on watermarking and its techniques. In: Agrawal, S., Gupta, K.K., Chan, J.H., Agrawal, J., Gupta, M. (eds.) Machine Intelligence and Smart Systems. AIS, pp. 71–78. Springer, Singapore (2021). https://doi.org/10.1007/978-981-33-4893-6_7

2. Anand, A., Singh, A.K.: Watermarking techniques for medical data authentication: a survey. Multimedia Tools Appl. **80**, 30165–30197 (2020). https://doi.org/10.1007/s11042-020-088 01-0
3. Pandey, M.K., Parmar, G., Patsariya, S.: An effective way to hide the secret audio file using high frequency manipulation. In: Communications in Computer and Information Science, pp. 125–130 (2011). https://doi.org/10.1007/978-3-642-18440-6_15
4. Begum, M., Uddin, M.S.: Digital image watermarking techniques: a review. Information **11**, 110 (2020). https://doi.org/10.3390/info11020110
5. Kumar, S., Singh, B.K.: Entropy based spatial domain image watermarking and its performance analysis. Multimedia Tools Appl. **80**, 9315–9331 (2020). https://doi.org/10.1007/s11 042-020-09943-x
6. Parah, S.A., Sheikh, J.A., Assad, U.I., Bhat, G.M.: Realisation and robustness evaluation of a blind spatial domain watermarking technique. Int. J. Electron. **104**, 659–672 (2016). https:// doi.org/10.1080/00207217.2016.1242162
7. Bal, S.N., Nayak, M.R., Sarkar, S.K.: On the implementation of a secured watermarking mechanism based on cryptography and bit pairs matching. J. King Saud Univ. Comput. Inf. Sci. **33**, 552–561 (2021). https://doi.org/10.1016/j.jksuci.2018.04.006
8. Bouslimi, D., Coatrieux, G., Cozic, M., Roux, C.: Data hiding in encrypted images based on predefined watermark embedding before encryption process. Signal Process. Image Commun. **47**, 263–270 (2016). https://doi.org/10.1016/j.image.2016.06.012
9. Mansoori, E.G., Soltani, S.S.: A new semi-blind watermarking algorithm using ordered Hadamard transform. Imaging Sci. J. **64**, 204–214 (2016). https://doi.org/10.1080/13682199. 2016.1159816
10. Patvardhan, C., Kumar, P., Vasantha Lakshmi, C.: Effective Color image watermarking scheme using YCbCr color space and QR code. Multimedia Tools Appl. **77**, 12655–12677 (2017). https://doi.org/10.1007/s11042-017-4909-1
11. Singh, R.K., Shaw, D.K., Jha, S.K., Kumar, M.: A DWT-SVD based multiple watermarking scheme for image based data security. J. Inf. Optim. Sci. **39**, 67–81 (2017). https://doi.org/ 10.1080/02522667.2017.1372153
12. Pandey, M.K., Parmar, G., Gupta, R., Sikander, A.: Non-blind Arnold scrambled hybrid image watermarking in YCbCr color space. Microsyst. Technol. **25**, 3071–3081 (2018). https://doi. org/10.1007/s00542-018-4162-1
13. Pandey, M.K., Parmar, G., Gupta, R., Sikander, A.: Lossless robust color image watermarking using lifting scheme and GWO. Int. J. Syst. Assur. Eng. Manag. **11**, 320–331 (2019). https:// doi.org/10.1007/s13198-019-00859-w
14. Rinki, K., Verma, P., Singh, R.K.: A novel matrix multiplication based LSB substitution mechanism for data security and authentication. J. King Saud Univ. Comput. Inf. Sci. **34**, 5510–5524 (2021). https://doi.org/10.1016/j.jksuci.2021.01.013
15. Faheem, Z.B., et al.: Image watermarking scheme using LSB and image gradient. Appl. Sci. **12**, 4202 (2022). https://doi.org/10.3390/app12094202
16. El-Mashad, S.Y., Yassen, A.M., Alsammak, A.K., Elhalawany, B.M.: Local Features-Based Watermarking for Image Security in Social Media. Comput. Mater. Continua **69**, 3857–3870 (2021). https://doi.org/10.32604/cmc.2021.018660
17. Wu, L., Zhang, J., Deng, W., He, D.: Arnold transformation algorithm and anti-arnold transformation algorithm. In: 2009 First International Conference on Information Science and Engineering (2009). https://doi.org/10.1109/icise.2009.347
18. Dawahdeh, Z.E., Yaakob, S.N., Bin Othman, R.R.: A new image encryption technique combining elliptic curve cryptosystem with hill cipher. J. King Saud Univ. Comput. Inf. Sci. **30**, 349–355 (2018). https://doi.org/10.1016/j.jksuci.2017.06.004

19. Tao, H., Chongmin, L., Mohamad Zain, J., Abdalla, A.N.: Robust image watermarking theories and techniques: a review. J. Appl. Res. Technol. **12**, 122–138 (2014). https://doi.org/10.1016/S1665-6423(14)71612-8
20. Al-Najjar, Y., Soong, D.C.: Comparison of image quality assessment: PSNR, HVS, SSIM, UIQI. Int. J. Sci. Eng. Res. **3**(8), 1–15 (2012)
21. Patsariya, S., Dixit, M.: A new block based non-blind hybrid color image watermarking approach using lifting scheme and chaotic encryption based on arnold cat map. Traitement du Signal. **39**, 1159–1168 (2022). https://doi.org/10.18280/ts.390408
22. Patsariya, S., Dixit, M.: Entropy based secure and robust image watermarking using lifting wavelet transform and multi-level-multiple image scrambling technique. Traitement du Signal. **39**, 1751–1759 (2022). https://doi.org/10.18280/ts.390533

Dual Scrambling Based Non Blind Robust and Secure Color Watermarking Technique

Sanjay Patsariya[1(✉)] and Manish Dixit[2]

[1] Department of Computer Science and Engineering, RGPV, Bhopal, MP, India
sanjaypatsariya@gmail.com
[2] Department of Computer Science and Engineering, MITS, Gwalior, MP, India

Abstract. Due to advancement in network technology, digital data is very popular medium for communication but eased to duplication and manipulation. As a result, diverse challenges related to information security and authentications are brought up with multimedia objects in both the civil and defense sectors. Digital watermarking is the process of embedding digital evidence (also called watermark) into another digital object (also called cover image) to proof the content authenticity and copy right protection. The performance of watermarking technique can be assessed using various parameters like MSE, PSNR, SSIM and NCC. In this paper, Non-blind watermarking technique is proposed in frequency domain using dual scrambling techniques to make this method robust and secure.

Keywords: watermarking · frequency domain · scrambling · correlation · MSE · PSNR · SSIM · NCC

1 Introduction

Due to expeditious development in communication technology, Internet is most preferable medium for communication. The secret information is implanted into a digital material during watermarking process. In other words, the process of invisible digital watermarking involves embedding a watermark image within a host image in such a way so that small deviations in data values cannot be detected by human visual system. Digital watermarking are used to verify the authenticity, integrity of the digital data or to show the identity of its ownership [1, 2]. Basically, three techniques—steganography, cryptography, and watermarking can be used to conceal information. Digital materials like text, image, audio or video can be introduced as digital proof. According to the type of documents, the watermark can be differentiating into text, image and audio watermark [3]. As per human perception, the watermark can be visible, semi-visible and invisible. The watermark embedding process can be taken place in two domain namely (1) spatial (SD) and (2) transform domain (TS). SD refers to the image plane itself and directly manipulates the value of pixels [4–6]. Imperceptibility, capacity, robustness and security are the fundamental properties of digital watermarking algorithm. Strength of SD is easiness but robustness is the main concern, means that information concealing in this domain is more open to threats [7]. Instead of adjusting the pixel values, frequency

© The Author(s), under exclusive license to Springer Nature Switzerland AG 2025
M. Gupta et al. (Eds.): MISS 2023, CCIS 1952, pp. 113–127, 2025.
https://doi.org/10.1007/978-3-031-69115-7_7

domain inserts messages by altering the transform coefficients of the cover message. The use of watermarking in various fields is illustrated in Fig. 1.

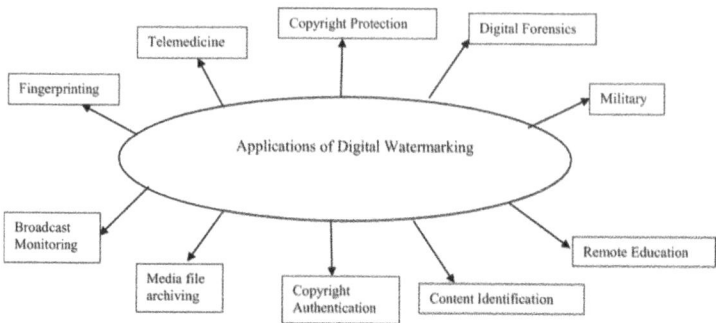

Fig. 1. Illustrate the fields of Watermarking

2 Literature Survey

Razak et al. [8] proposed a DWT- SVD grounded watermarking methods intended for RGB images. In their projected study, DWT is utilized on images i.e. host and watermark picture and select LL sub-band of both picture for implementation of SVD. PSNR and MSE measures are utilized to estimate the efficiency of proposed method.

Singh et al. [9] presented a watermarking method based on spatial domain to insert watermark into least significant bit. Due to implementation in spatial domain, the method has less complexity but more vulnerability against various types of attacks.

Mishra et al. [10] proposed method by means of bit plane slicing scheme by applies DWT (Haar wavelet transform). Firstly, they divided gray level cover (host) image into binary plane including 8 bits. DWT is applied to split plane 1 into four sub-bands. SVD is applied to the HH sub-band to conceal the evidence.

$$S1 = P1.Q1.R1^T \tag{1}$$

where, Q1 represent the diagonal matrix, S1 indicate the HH sub-band of image. The matrix Q1 is employed to implant secret digital materials. The proposed method's outcome gratifies the reversibility characteristics, but does not work with colorful images.

Saikrishna et al. [11] projected a novel invisible and safe watermarking method grounded on texturization. In their proposed work, they firstly locate out the place of embedding with some criteria. Black and white texture region of host image are separated. DWT transformation is employed to conceal digital materials and Arnold transform are used for security purpose. The proposed method applicable for gray scale images only.

Su et al. [12] provided a blind method based on QR decomposition for color pictures with non-overlapping chunks that demonstrated resilience against different attacks. The Arnold transform was used to encode watermarks.

Pandey et al.[13] projected non-blind watermarking method grounded on lifting wavelet transform and for security point of view, Arnold transform is used. Single level Arnold transform was adopted to destruct the correlation among the pixels of the image. They used the number of iterations as key.

Pandey et al. [14] introduced Non-blind, IWT-SVD grounded method using lifting scheme GWO. One level scrambling approach was used by means of Arnold transform to make watermark secure. They worked on color image and YCbCr color model was adopted to implant watermark.

3   Proposed Methodology

In proposed method, dual scrambling approach is adopted prior to implant watermark into the cover image. There are two parts i.e. embedding and extraction of watermark into/from color cover/watermarked image. To make watermarking robust against various attacks SVD and DWT transform is used.

3.1   Embedding Procedure

The following steps are adopted to insert watermark into cover image as illustrated in Fig. 2

Step 1: Select Cover image and watermark image.

Step 2: Select watermark image and employ block based scrambling technique via secret key (K1) followed by Arnold transform with number of iterations as a key (K2).

Step 3: The scrambled watermark captured in step 2 is used for embedding purpose.

Step 4: Apply DWT and select LL sub-band of both cover and watermark images. Apply SVD on LL sub-band of watermark and cover image and calculate singular matrix for watermarked image.

Step 5: Determine new singular matrix by grouping of singular matrix with scaling factor (α)

$$S1_{wm} = S1_h + \alpha \times S1_w \tag{2}$$

where, $S1_{wm}$, $S1_h$, $S1_W$ point out singular value of watermarked, Cover and Watermark images respectively.

Step 6. Use calculated singular value and apply inverse SVD and inverse DWT to merge new LL sub-band to the unchanged sub-band (HL, LH, HH) to obtained watermarked image.

3.2   Extraction Procedure

The watermark extraction is the inverse practices of watermark embedding process. Figure 3 shows the extraction procedure that consists of following steps.

Step 1: Select watermarked image and apply DWT transform and select LL sub-band to apply SVD.

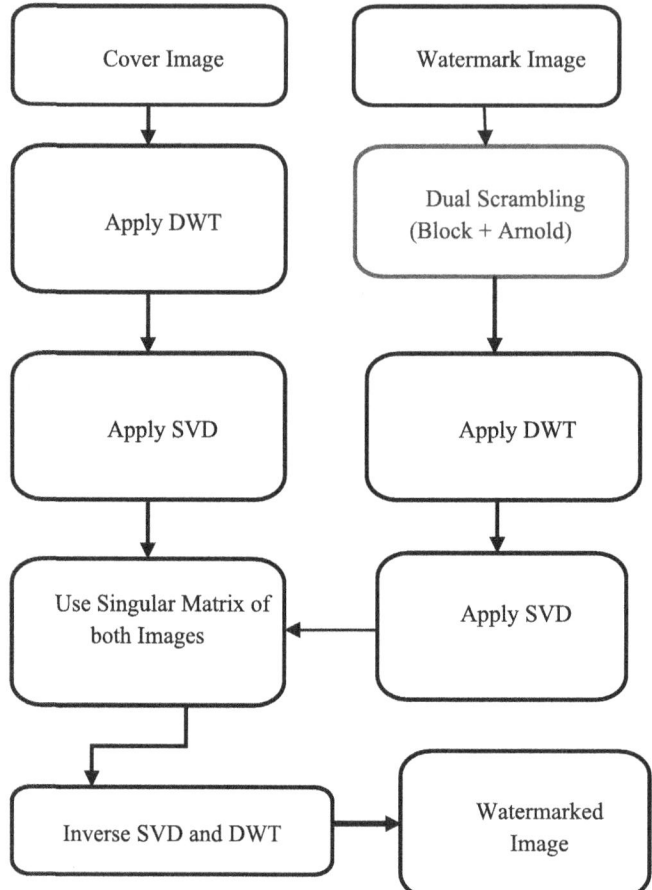

Fig. 2. Procedure of Watermark embedding

Step 2: Evaluate singular matrix via grouping of both singular matrix with scaling factor (α)

$$S1_{w\_n} = (S1_{wm} - S1_h)/\alpha \tag{3}$$

where, $S1_{w\_n}, S1_{wm}, S_w\ S1_h$ represent the watermark, watermarked and original singular value respectively.

Step3: Using evaluated singular value along with orthogonal matrix, the watermark can be obtained by

$$W = U1_w \times S1_{w\_n} \times V1'_m \tag{4}$$

Step 4: Apply dual descrambling method using the same secret keys (K2 &K1) to obtain the original watermark.

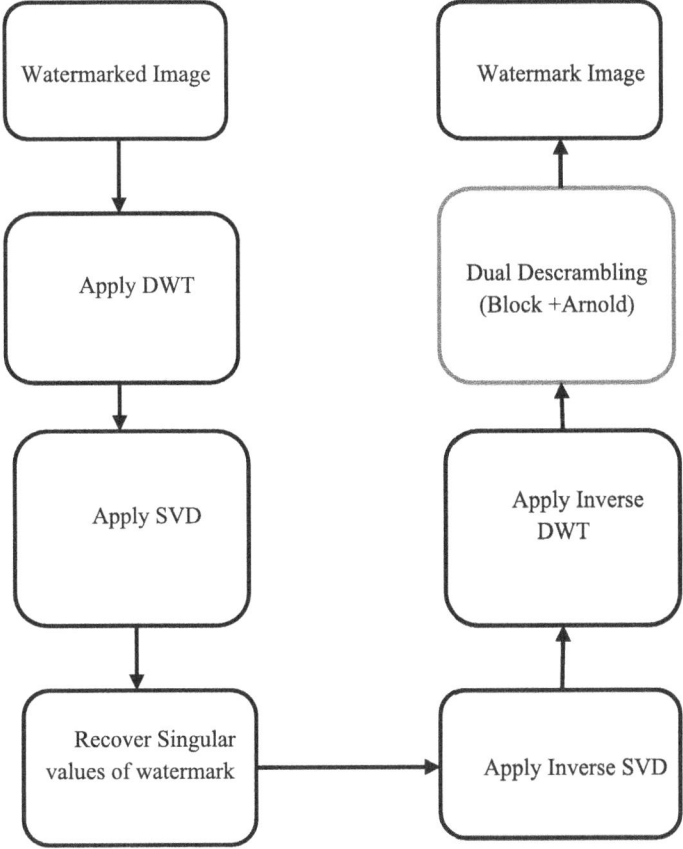

Fig. 3. Watermark extraction procedure

4 Experimental Results

The usefulness of watermarking methods can be assessed by the MSE, PSNR and NCC. All result obtained using MATLAB software by using color image of size 512×512.

4.1 Transparency Analysis

MSE, PSNR and SSIM image quality parameters are used to measure the imperceptibility level. High values of PSNR and SSIM indicate the high transparency level. The Table 1 shows the fidelity parameters using proposed algorithm.

4.2 Robustness Analysis

Robust refer to the ability of persistence after various image processing attacks [15, 16]. Normalized_Cross_Correlation (NCC) is used to assess the robustness of given

Table 1. Estimated image quality parameters for different values of strength factors (α).

Strength Factor(α)	MSE	PSNR	SSIM	NCC(With out Attack)
0.01	0.000069	41.5528	0.9998	1.0000
0.02	0.000279	35.5322	0.9993	1.0000
0.05	0.0017	27.5734	0.9958	1.0000
0.1	0.0070	21.5528	0.9840	1.0000
0.5	0.1748	7.5734	0.7433	1.0000
1.0	0.6994	1.5528	0.4595	1.0000

watermarking approach. Table 2 reflects the robustness by estimating NCC against various images processing operation and visualization of watermarked image and extracted watermark.

5 Security Analysis

In order to add confusion, scrambling is frequently employed in encryption methods. The some of the methods that are used for image scrambling are as under:

(A) Based on row and column swapping, secure image scrambling is performed via key-based method [17, 18]. The rows or column of the image are flipped around using certain random sequences.

(B) One of the most well-known methods of image scrambling is bit-plane based scrambling [19, 20]. It functions as one of the fundamental phases in numerous encryption algorithms. In this case, the rows are permuted using random sequences and the image is expanded to a bit plane binary image. Furthermore, the same random sequences are used to move the columns.

(C) A secret message can be encrypted using the Hill cipher encryption method [21]. It is a matrix based traditional block cipher that uses matrix multiplication. Both the sender and the receiver use the same key patterns for ciphering and decoding. Below is a description of the hill cipher encryption method.

$$K1 = \begin{bmatrix} 36 & 169 & 221 & 87 \\ 152 & 61 & 104 & 196 \\ 37 & 169 & 220 & 87 \\ 152 & 62 & 104 & 195 \end{bmatrix} \tag{5}$$

$$P1 = \begin{bmatrix} 84 \\ 96 \\ 54 \\ 111 \end{bmatrix} \tag{6}$$

where, K1 and P1 is the self- invertible matrix of size 4×4 and Plain text (Watermark), respectively.

Table 2. Estimated NCC value between watermark and watermark extracted after various attacks and visualization of watermarked Image after attacks and extracted watermark

Name of Attack	NCC	Watermarked Image after attack	Extracted Watermark
Gaussian attack with variance 0.001	0.8112		
Gaussian attack with variance 0.002	0.7802		
Salt and Pepper with noise density 0.001	0.9807		
Salt and Pepper with noise density 0.002	0.9559		
Poisson attack	0.9992		
Rotation 1^0	0.8208		
Rotation 5^0	0.6736		

(*continued*)

Table 2. (*continued*)

Histogram equalization attack	0.8196		
Contrast Adjustment	0.9959		
Image Intensity attack	0.8499		
Resize attack 50%	0.8710		
Resize attack 200%	0.9788		
Median Filtering[3×3]	0.8526		
Median Filtering[5×5]	0.7223		
Blur attack	0.9988		

The Plain text P1 (Watermark) is ciphered and deciphered using the Eq. (7) and Eq. (8), respectively.

$$D1 = (K1.P1) \bmod F \tag{7}$$

$$P1 = (K1.D1) \bmod F \tag{8}$$

The value of F dependent upon the type of Image.

(D) The Fibonacci transform can also be employed to perform the matrix-based scrambling [22]. The uniformity of the Fibonacci transform is a special quality. The pixels that are evenly spaced apart in the original image are also equally spaced apart in the scrambled image. The Fibonacci and Anti- Fibonacci transforms are illustrated by Eq. (9) and Eq. (10), respectively.

$$\begin{bmatrix} C1'_x \\ C1'_y \end{bmatrix} = \begin{bmatrix} 1 & 1 \\ 1 & 0 \end{bmatrix} \begin{bmatrix} C1_x \\ C1_y \end{bmatrix} \bmod N \tag{9}$$

$$\begin{bmatrix} C1'_x \\ C1'_y \end{bmatrix} = \begin{bmatrix} 0 & 1 \\ 1 & -1 \end{bmatrix} \begin{bmatrix} C1_x \\ C1_y \end{bmatrix} \bmod N \tag{10}$$

where, $C1'_x$, $C1'_y$ and $C1_x$, $C1_y$ are the new coordinate and original coordinates of the pixels, respectively.

(E) V.I. Arnold proposed the cat map, commonly known as the Arnold transform [14, 23]. The number of iterations used as key in Arnold based scrambling and visualization of distorted image using block and Arnold based scrambling are depicted in Fig. 4 and Fig. 5, respectively.

Scrambling (Arnold) and descrambling transform are illustrated by Eq. (11) and Eq. (12), respectively.

$$\begin{bmatrix} C1'_x \\ C1'_y \end{bmatrix} = \begin{bmatrix} 1 & 1 \\ 1 & 2 \end{bmatrix} \begin{bmatrix} C1_x \\ C1_y \end{bmatrix} \bmod N \tag{11}$$

$$\begin{bmatrix} C1'_x \\ C1'_y \end{bmatrix} = \begin{bmatrix} 2 & -1 \\ -1 & 1 \end{bmatrix} \begin{bmatrix} C1_x \\ C1_y \end{bmatrix} \bmod N \tag{12}$$

where, $C1'_x$, $C1'_y$ and $C1_x$, $C1_y$ are the new coordinate and original coordinates of the pixels, respectively.

The adjacent pixels are correlated among the image. Hacker can apply the concept of correlation to extract the embedded watermark [2, 23]. Horizontal (HC), Diagonal (DC) and Vertical (VC) Correlation coefficient are employed to determine the usefulness of the proposed scrambling techniques.

Block Size	First Level Image Scrambling
64×64	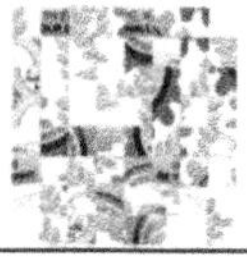
32×32	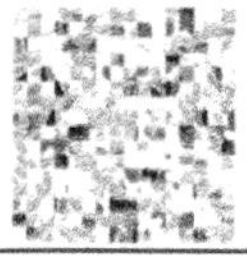
16×16	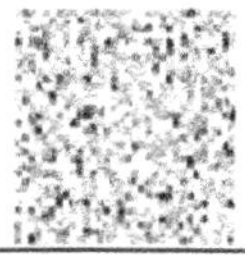
8×8	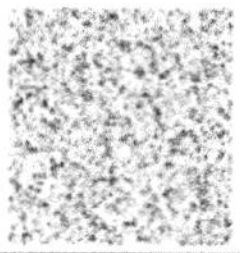

Fig. 4. Visualization of watermark after block based scrambling

The dual based scrambling is presented here. At first level, Block based scrambling is used while Arnold based scrambling employed at second level. The performance of proposed and considered scrambling methods is mentioned in Table 3.

The effect of block based scrambling using different block sizes are depicted in Fig. 4.

Figure 6 and Table 3 depict the effectiveness of scrambling approach using dual (Block + Arnold) scrambling over Fibonacci and Arnold transform in terms of horizontal, diagonal and vertical correlation. It is clearly apparent that projected dual scrambling style is more effective to demolish the correlation among pixels and provide more security.

No. of Iterations	Second Level Image Scrambling(Arnold Transform)
1	
2	
4	
8	
16	
32	
64	
128	

Fig. 5. Visualization of watermark after Arnold based scrambling

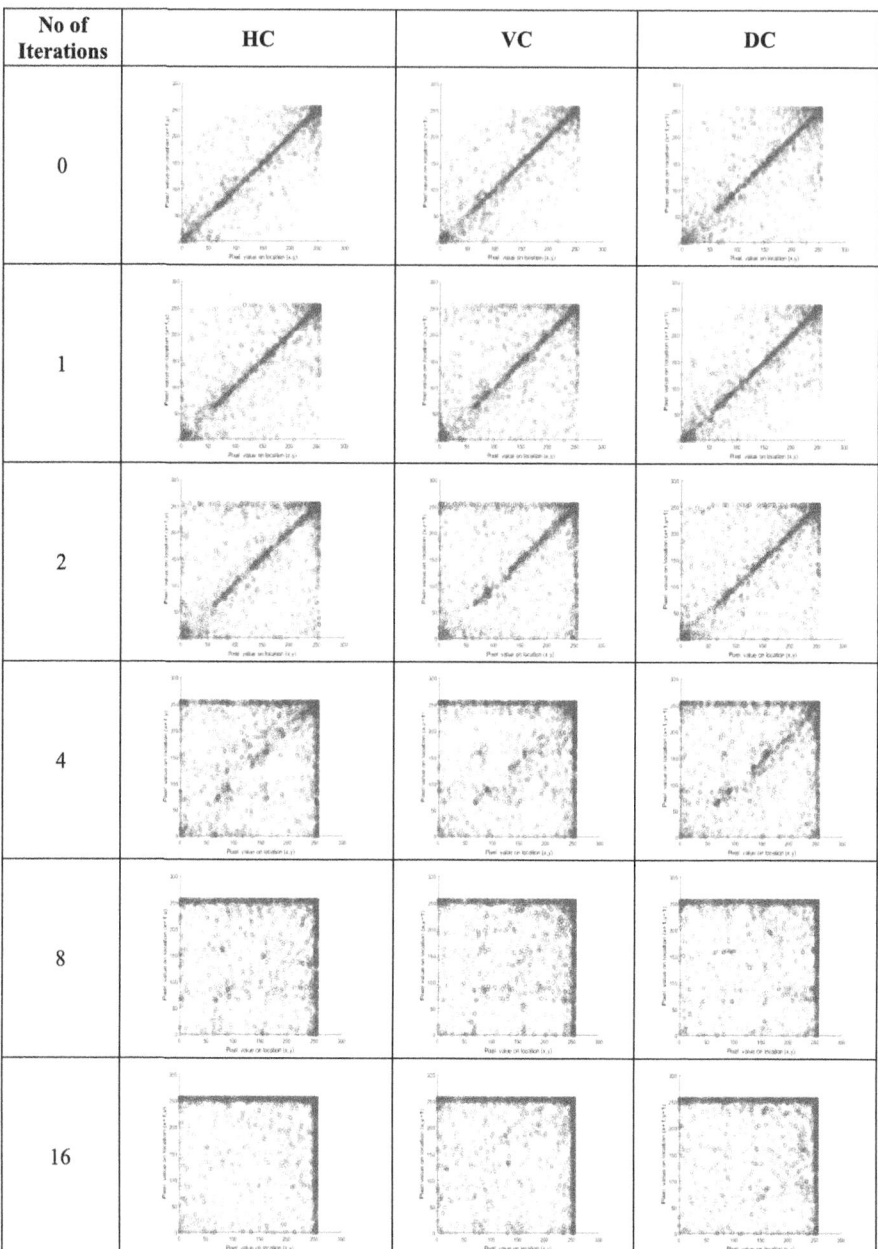

Fig. 6. Outcome of Horizontal, Vertical and Diagonal correlation of RJIT logo after the scrambling using Arnold transform.

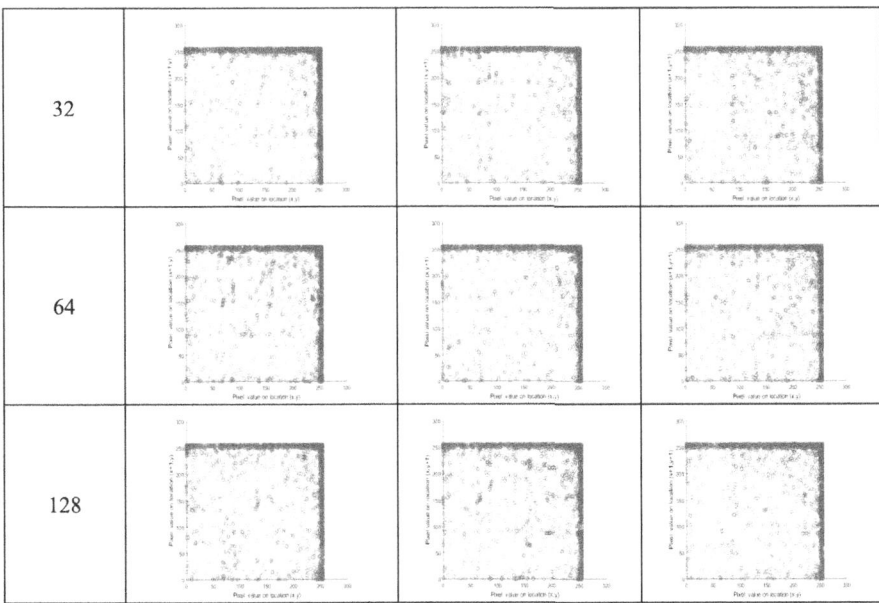

Fig. 6. (*continued*)

Table 3. Comparative Analysis of Correlation coefficient after applying Fibonacci, Arnold (Single) and hybrid (Block + Arnold based) Transform

No. of Iterations(N)	Fibonacci Based Scrambling [22]			Arnold Based Scrambling [14, 23]			Proposed (Block + Arnold Based) Scrambling		
	HC	VC	DC	HC	VC	DC	HC	VC	DC
None	0.94	0.94	0.90	0.94	0.94	0.90	0.94	0.94	0.90
1	0.92	0.96	0.95	0.90	0.82	0.94	0.68	0.52	0.81
2	0.84	0.89	0.95	0.77	0.64	0.82	0.39	0.12	0.56
4	0.65	0.77	0.85	0.35	0.14	0.35	−0.01	0.06	0.01
8	0.20	0.35	0.42	0.03	0.08	0.01	−0.05	0.08	−0.01
16	0.09	0.07	0.03	−0.11	−0.01	−0.01	0.05	−0.03	−0.01
32	0.01	−0.05	0.02	-0.06	−0.04	−0.03	0.02	0.04	0.01
64	−0.02	−0.07	0.01	0.04	−0.03	0.06	0.04	0.04	0.05
128	−0.06	0.01	−0.02	−0.02	0.03	−0.11	−0.05	0.03	0.01

6 Conclusion

Watermarking has turned up as an influential tool for copyright protection and to proof authenticity of digital materials. In presented method, dual scrambling approach is utilized to make watermark more secure. Block based and Arnold transform are used together to breach the correlation among pixels. Watermark is inserted using Transform domain (DWT) and SVD to make watermarking technique more robust against various attacks. The projected approach can be deployed in hypersensitive application like defense and civil areas.

References

1. Patsariya, S., Dixit, M.: A survey on watermarking and its techniques. In: Agrawal, S., Kumar Gupta, K.H., Chan, J., Agrawal, J., Gupta, M. (eds.) Machine Intelligence and Smart Systems. Algorithms for Intelligent Systems. Springer, Singapore, pp.71–78 (2021)
2. Bouslimi, D., Coatrieux, G., Cozic, M., Roux, C.: Data hiding in encrypted images based on predefined watermark embedding before encryption process. Sig. Process. Image Commun. **47**, 263–270 (2016). https://doi.org/10.1016/j.image.2016.06.012
3. Pandey, M.K., Parmar, G., Patsariya, S.: An effective way to hide the secret audio file using high frequency manipulation. Commun. Comput. Inf. Sci. 125–130 (2011). https://doi.org/10.1007/978-3-642-18440-6_15
4. Malik, S., Reddlapalli, R.K.: Histogram and entropy based digital image watermarking scheme. Int. J. Inf. Technol. **11**, 373–379 (2018). https://doi.org/10.1007/s41870-018-0259-0
5. Parah, S.A., Sheikh, J.A., Assad, U.I., Bhat, G.M.: Realisation and robustness evaluation of a blind spatial domain watermarking technique. Int. J. Electron. **104**, 659–672 (2016). https://doi.org/10.1080/00207217.2016.1242162
6. Begum, M., Uddin, M.S.: Digital image watermarking techniques: a review. Information **11**, 110 (2020). https://doi.org/10.3390/info11020110
7. Anand, A., Singh, A.K.: Watermarking techniques for medical data authentication: a survey. Multimedia Tools Appl. (2020). https://doi.org/10.1007/s11042-020-08801-0
8. Razak, N.A.: Digital Image watermarking base on DWT and SVD techniques. J. Netw. Commun. Emerg. Technol. (JNCET) **8**(2) (2018)
9. Singh, A.K., Sharma, N., Dave, M., Mohan, A.: A novel technique for digital image watermarking in spatial domain. In: 2nd IEEE International Conference on Parallel, Distributed and Grid Computing, pp. 497–501. Solan, India (2012)
10. Mishra, M., Rout, N.K., Budipi, N.R.: Bit plane slicing based digital watermarking technique in dwt domain. Int. J. Eng. Adv. Technol. **8**, 525–529 (2019). https://doi.org/10.35940/ijeat.e7301.088619
11. Saikrishna, N., Resmipriya, M.G.: An invisible logo watermarking using arnold transform. Procedia Comput. Sci. **93**, 808–815 (2016). https://doi.org/10.1016/j.procs.2016.07.299
12. Su, Q., Niu, Y., Wang, G., Jia, S., Yue, J.: Color image blind watermarking scheme based on QR decomposition. Signal Process. **94**, 219–235 (2014). https://doi.org/10.1016/j.sigpro.2013.06.025
13. Pandey, M.K., Parmar, G., Gupta, R., Sikander, A.: Non-blind Arnold scrambled hybrid image watermarking in YCbCr color space. Microsyst. Technol. **25**, 3071–3081 (2018). https://doi.org/10.1007/s00542-018-4162-1
14. Pandey, M.K., Parmar, G., Gupta, R., Sikander, A.: Lossless robust color image watermarking using lifting scheme and GWO. Int. J. Syst. Assurance Eng. Manage. **11**, 320–331 (2019). https://doi.org/10.1007/s13198-019-00859-w

15. Sunesh and Kumar, H.: Watermarks Attacks And Application in Watermarking. In: National Workshop-Cum Conference on Recent Trends in Mathematics and Computing (RTMC) (2011)
16. Tao, H., Chongmin, L., Mohamad Zain, J., Abdalla, A.N.: Robust image watermarking theories and techniques: a review. J. Appl. Res. Technol. **12**, 122–138 (2014). https://doi.org/10.1016/S1665-6423(14)71612-8
17. Liu, W., Sun, K., Zhu, C.: A fast image encryption algorithm based on chaotic map. Opt. Lasers Eng. **84**, 26–36 (2016). https://doi.org/10.1016/j.optlaseng.2016.03.019
18. Mondal, B., Mandal, T.: A light weight secure image encryption scheme based on chaos & DNA computing. J. King Saud Univ. – Comput. Inf. Sci. **29**, 499–504 (2017). https://doi.org/10.1016/j.jksuci.2016.02.003
19. Mondal, B., Mandal, T., Kumar, P., Biswas, N.: A secure partial encryption scheme based on bit plane manipulation. In: 2017 7th International Symposium on Embedded Computing and System Design (ISED) (2017). https://doi.org/10.1109/ised.2017.8303925
20. Chen, J., Zhu, Z., Fu, C., Yu, H., Zhang, L.: An efficient image encryption scheme using gray code based permutation approach. Opt. Lasers Eng. **67**, 191–204 (2015). https://doi.org/10.1016/j.optlaseng.2014.11.017
21. Kumar, S., Singh, B.K.: Entropy based spatial domain image watermarking and its performance analysis. Multimedia Tools Appl. (2020). https://doi.org/10.1007/s11042-020-09943-x
22. Zhou, Y., Panetta, K., Agaian, S., Chen, C.L.P.: Image encryption using P-Fibonacci transform and decomposition. Optics Commun. **285**, 594–608 (2012). https://doi.org/10.1016/j.optcom.2011.11.044
23. Wu, L., Zhang, J., Deng, W., He, D.: Arnold transformation algorithm and anti-arnold transformation algorithm. In: First International Conference on Information Science and Engineering, pp. 1164–1167. Nanjing, China (2009)

Efficient Technique for Image Enhancement Using Generative Adversarial Network

Anand Jawdekar[1][(✉)] and Manish Dixit[2]

[1] RGPV, Bhopal, India
anand.cs2007@gmail.com
[2] MITS, Gwalior, India

Abstract. Image processing is the recent trend in the computing, images play very crucial role in the field of engineering and technology. Many real-world applications used images as computation purpose. To find the high-resolution images still very challenging, due the capacity of the various acquisition devices, it is difficult to obtain the high-resolution images, still noise also present in the images. Various techniques for the image enhancement proposes by the various researchers to remove the noise present in the image. In this work generative adversarial network-based model is use here to improve image resolution. Adversarial network is the deep learning-based model which comprises various convolution layers, ReLU functions and normalization layers. GAN consists of discriminator and generator functions which is dedicated to perform different tasks. Single image super resolution-based GAN pretrained model used here to upscaling the images. This work upscale the different medical images with good PSNR values.

All simulation performed in python environment with tensor flow and keras. The model training performs of colab with virtual GPU support.

Keywords: Image enhancement · Medical Imaging · SRGAN · High Resolution image · Low resolution image · PSNR

1 Introduction

In recent years, digital imaging has grown rapidly and is becoming more popular as medical imaging technology advances. The most significant issue in the field of image processing is the preprocessing step, image enhancement. The basic purpose of image enhancement is to increase the clearer picture of any object with its surroundings. Image processing is used in all major fields such as medicine, satellite imagery, digital photography and underwater imaging [1].

Medical Imaging is the one of the promising areas for the interest of researchers. Medical images are very crucial and helpful for the disease detection and asymptotic treatment of disease. Due to the advancement in the technology healthcare system transformed to the smart healthcare.

Image enhancement is the first step in medical imaging. Medical Image Enhancement improves the image acquisition process or images acquired from digital cameras and devices used in primary healthcare such as MRI, PET SCAN.

© The Author(s), under exclusive license to Springer Nature Switzerland AG 2025
M. Gupta et al. (Eds.): MISS 2023, CCIS 1952, pp. 128–138, 2025.
https://doi.org/10.1007/978-3-031-69115-7_8

In medicine, radiation is of great importance in diagnosing diseases and their symptoms. Although many major technological advances have already been made in this area to achieve better results, uncertainties and deficiencies still exist. Radiologist scan the object and generate the scan image as per the requirement for the specific disease, so the images must be very clear to identify situations and predict outcomes. Even the slightest error distorts the image and the prediction is incorrect [2].

In addition, the amount, quality and clarity of image information is important. To meet these criteria, researchers have provided various image enhancement approaches to improve the quality of image perception [3].

In the Feature extraction process some specific feature extracted from image, which is further used for diagnostic and research purposes. Edge detection and image enhancement are key elements in this process. The three main aspects of image feature extraction methods are pixelated, local and global, and visual features of the image are mainly based on pixel values. Edge detection, on the other hand, reduces the amount of data and filters out meaningless data in the image [4].

Image enhancement also concerns for the increase the resolution of the image, primarily images taken from the various sources are always low-resolution image and for particular medical imaging each part of the body or organ should be clearly visible for the treatment and diagnosis purpose.

This paper is organized into Sect. 2, Related Work where previous related work is discussed in Image Enhancement, Sect. 3, Proposed Methods where key algorithms and recommendation architectures already discussed, Sect. 4 illustrates the experimental results and discussion, Sect. 5 finally concludes the conclusions and future work.

2   Related Work

2.1   Point Based Enhancement

A digital image comprises the crucial element known as pixels. You can improve pixel values by performing some basic operations on your digital images. This approach is generally categorized as the point-based technique. Major approaches listed here as:

2.1.1   Brightness Modification

In that approach fixed values could be added to the existing image. The constant values should be changed according to the diagram. In the Eq. (1) constant value of V added to the existing image Y (i, j) and the resulting image X (i, j) is the brightened image.

$$X(i, j) = Y(i, j) + V \tag{1}$$

In Eq. (2) constant value V could be subtracted from the existing image and the resulting image X(i, j) indicated less brightened image.

$$X(i, j) = Y(i, j) - V \tag{2}$$

2.1.2 Contrast Adjustment

It works only for contrast, to multiply some factor to the existing image.

$$X(I, j) = Y(i, j) * V \qquad (3)$$

Equation (3) shows the contrast modified image X(i, j) which is obtained with multiplied constant value of V to the existing image Y(i, j).

2.1.3 Image Negative or Inverse Transform Image

Negative or inverse transform is very useful in the medical image processing. Various types of negative images used in the various types of medical treatment. In some cases, image inversion is very important for disease identification. This is a very ancient procedure for enhancing digital images. Here dark and light shades are exchanged with each other. Inverted image of 255 minus the original image.

$$X(i, j) = 255 - Y(i, j) \qquad (4)$$

Equation (4) describes the inverse transformation process where X(i, j) obtained using the subtracting original image with maximum pixel value ie 255.

2.2 Histogram Based Techniques

Histogram equalisation is the most highly regarded and often used method of contrast enhancement because to its accuracy and simplicity. It is accomplished by employing its cumulative density function to normalize the intensity distributions, which increases the contrast of an input image and produces a consequent image with a uniform intensity distribution [3].

The foundation of many spatial domain processing methods is the image histogram. It can be used to provide related image statistics, segment images, compress images, and enhance images. It is easy to implement in hardware as well as calculation in software. Histogram equalization (HE) creates an image with equal probabilities of intensity levels across the image [5].

There are certain lighting effects with the histogram-based method. Several histogram-based methods have already been presented by different academics. In recent years, numerous researchers have developed various global histogram equalisation algorithms to address the issue of mean shift in output images. Another new approach, the HE brightness-preserving approach, was discussed to control the brightness present in the image and enhance the contrast [6].

Weng et al. proposed a similar technique called Dualistic Sub image Histogram Equalization (DSIHE). This approach is based on segmented medians [7, 8]. Experimental results show that DSIHE outperforms BBHE in terms of luminance preservation and entropy. Recursive Mean Discrete Histogram Equalization (RMSHE) and Recursive Sub image Histogram Equalization (RSIHE) are proposed to solve the problem of unwanted side effects [9, 10].

Both plans rely on recursive algorithms that use BBHE and DSIHE methods. Next, researchers examine a novel technique called recursive separated weighted histogram equalisation to enhance image contrast (RSWHE). This method is the same as RMSHE and RSIHE, except that RSWHE uses a normalized power law function [11].

These three of histogram high-enhancement (HE) techniques (RSIHE, RMSHE, and RSWHE) offer excellent contrast enhancement and reasonable brightness retention, but the resultant image has issues with over-enhancing. To address these shortcomings, the authors' group proposed a new automatic conversion technique, Weighted Distribution Adaptive Gamma Correction (AGCWD). Using a gamma correction approach that modifies the value of the pixel-wise probability distribution of the image, this technique boosts the brightness level of low-contrast images [12].

Another study combined adaptive gamma correction with two-level weighted histogram equalisation to successfully preserve brightness while enhancing contrast, but this approach is hindered by uneven lighting. Happens [13].

As a result, these new findings explore another effective His AGC-based method that trades off good enhancement and low computational cost. This method combines range-limited two-histogram equalization (RLBHE) with his AGC technique [14]. According to experimental data (AGCWD), compared with RLBHE and weighted distribution adaptive gamma correction, this proposed method is more suitable for enhancing low-contrast images.

As discussed by the various researcher's good quality images or high-resolution images have a good understanding and give more meaningful results. Small size images were very less for pixel data point of view, so these types of the images not suitable for the real time applications. While every time it is not possible to use the high-definition cameras to capture the image like ocean engineering, underwater imaging and also medical imaging, the image captured by the devices have low pixel and small in size. Various researchers already proposed various image enhancement techniques which discussed in the above literature. Also, some intelligent deep learning or neural network-based tech niques also suggested by the author to work for the image enhancement as well as noise reduction. Still some kind of intelligent techniques still require for the upscaling the images which is very useful in the various real time image processing applications (Table 1).

3   Proposed Work

As discussed in the literature review various researchers proposed different image enhancement methods for different types of the images. The strategy for improving images in this paper is based on a GAN (Generative Adversarial Network). This model improvises the image quality and also increase the image resolution.

Deep learning provides the better solution for the image enhancement and object recognition. It includes neural network and its different variations for the processing of the images. Generative adversarial network is one the deep learning-based solutions for the image enhancement. Good fellow et al. introduces Adversarial Nets Framework, generative models play against opponents. If the sample originates from the model distribution or the data distribution. A generative model can be thought of as analogous to a team of counterfeiters Create a fake currency and try to use it unnoticed while the discriminatory model is this It's like the police trying to spot counterfeit bills. This game

Table 1. Literature Summary

Author name and year	Approach	Remark
Kim & Chung (2008) [11]	The method is comparable to earlier ones, but it also incorporates a weighting process	PSNR and entropy level
S. Haung, F. Cheng, Y. Chiu (2013) [12]	Adaptive Gama Correction method	Useful for colour images
Wang, W et al. (2019) [14]	Low illumination method	Low performing over video images
Xiaodong Kuang et al. (2018) [15]	Deep Convolution Method	Work for specific images
Li-Wen Wang et al. (2020) [16]	Deep lightning network	More visibility of the low light images

race drives Both teams need to improve their methods until the fake is indistinguishable from the real thing paper.

GANs are a subset of AI algorithms used in unsupervised machine learning. A GAN is a deep neural network architecture consisting of two networks (a generator and a discriminator) that compete with each other (hence the "opponents"). A GAN is made up of a generator and a discriminator. Consider it as a game in which the Discriminator serves as the judge and the Generator attempts to produce some data from a probability distribution. The discriminator establishes whether the input is drawn from the genuine training dataset or synthetic data that was generated. Generator tries to adjust the data so that it more closely resembles the training data. Alternately, we may claim that the discriminator is instructing the generator to generate accurate data. Like encoders and decoders, they operate (Fig. 1).

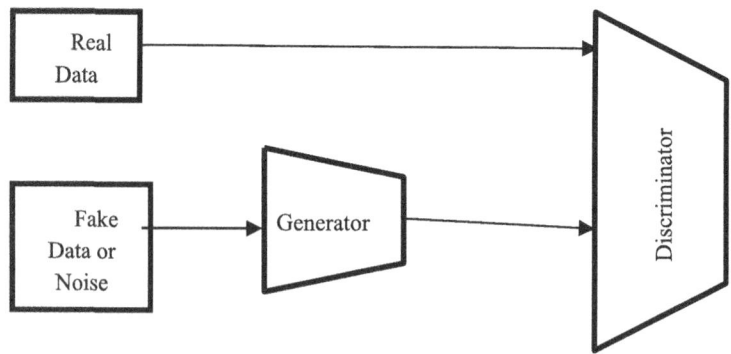

Fig. 1. GANs Architecture

Generator and discriminator both learning at the same time; however, generator slowly learns according to the input. Finally, when generator completely learns the process then generate the samples from the input which was very similar to the original data or real data.

3.1 SRGAN

SRGAN is the pretrained model of Generative Adversarial Network. As already various image enhancement techniques focus to the image enhancement also improves the parameters of image like PSNR, MSE and many more, but still some images basically have a very low-resolution images have a very difficult to understand and give the proper results, because good quality or high-resolution images continuously used in many fields of engineering and technology. SRGAN (Super Resolution GAN) is one of the promising pretrained model works perfectly over the low-resolution images.

Training Process

1. Separate the data for training and testing by separating the images into low reso lution (LR) and high resolution (HR).
2. Send the low-resolution images to the generator, which will produce the high-resolution image.
3. The discriminator trains the generator and discriminator using the high-resolu tion image and back-propagate to the GAN loss.

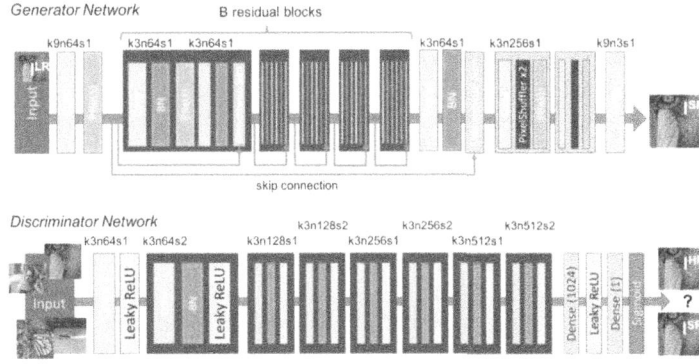

Fig. 2. Architecture of Generator and Discriminator Network with corresponding kernel size (k), number of feature maps (n) and stride (s) indicated for each convolutional layer [17]

Figure 2 shows the network architecture of SRGAN which includes generator and discriminator. This comprises majorly convolution layer, ReLU and batch normalization.

3.2 Mathematic and Terminologies

Performance of the SRGAN evaluated with the different parameters which gives the various mathematical equations and terminologies, we understand the mechanism of the model with the help of these parameters [17].

3.2.1 Perceptual Loss Function

This is a vital part of GAN's generator component. The weighted sum of the content loss and adversarial loss functions yields the percep tual loss function.

$$l^{sr} = l^{sr}_x + 10^{-3} l_{Gen}sr \qquad (5)$$

3.2.2 Content Loss Function

This is the Mean square error of the pixel

$$l^{SR}_{MSE} = \frac{1}{r^2 WH} \sum_{x=1}^{rW} \sum_{y=1}^{rH} \left(I^{HR}_{x,y} - G_{\theta G}\left(I^{LR}\right)x, y \right)^2 \qquad (6)$$

3.2.3 Adversarial Loss Function

This is the another very important characteristic, it defines the generative loss l^{SR}_{GEN} is defined based on the probabilities of the discriminator $D\theta D$ ($G\theta G$ (I^{LR})) of all samples

$$L^{SR}_{MSE} = \sum_{n=1}^{N} -log D_{\theta D}\left(G_{\theta D}\left(I^{LR}\right)\right) \qquad (7)$$

3.2.4 Mean Square Error

This is the error rate which should be calculated with respect to the image compres sion. This is the cumulative squared error with compressed and the original image.

$$MSE = \frac{\sum M, NI(m, n) - J(m, n)^2}{M \times N} \qquad (8)$$

3.2.5 Peak Signal to Noise Ratio

This is the very crucial parameter in image processing, it is the ratio of peak signal to noise power. Standard equation for PSNR as follow:

$$PSNR = 10 \log_{10} \frac{R^2}{MSE} \qquad (9)$$

3.2.6 Entropy

It is the statistical measure of the input image which represents randomness. It is useful to characterize the image.

$$sum(p. * log2(p)) \qquad (10)$$

All the above stated formulas and functions taken from MATLAB 2020 [18].

4 Experimental Result and Discussion

All simulation and experiments performed in python environment with tensor flow and Keras. SRGAN model was trained over 3,50000 over samples through NVIDIA Tesla M40 GPU. In this study we setup model over colab environment and train the model over colab with having virtual GPU.

4.1 Simulation Results of Image Enhancement Using SRGAN

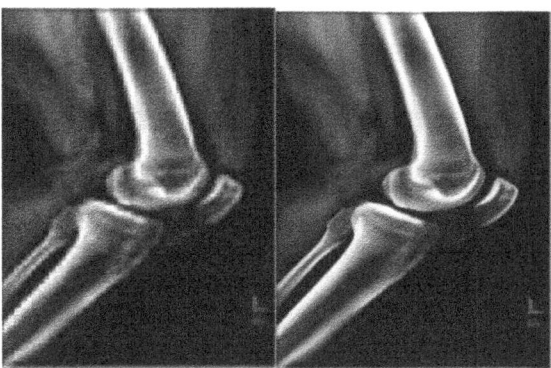

Fig. 3. Bone image low resolution image (Left) and High-Resolution image (Right)

In the Fig. 3 two images shown here left image is low resolution image and corresponding right image is high resolution bone image.

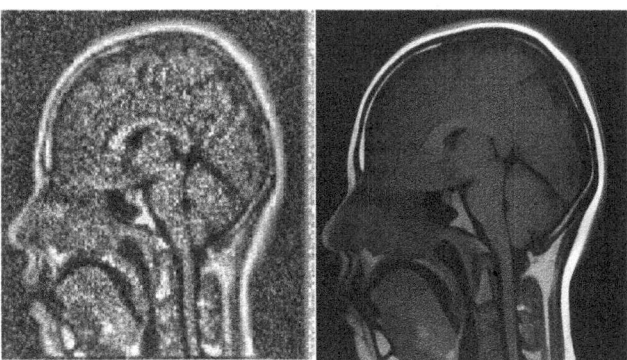

Fig. 4. Brain image low resolution image (Left) and High-Resolution image (Right)

In the Fig. 4 two images shown here left image is low resolution image and corresponding right image is high resolution brain image.

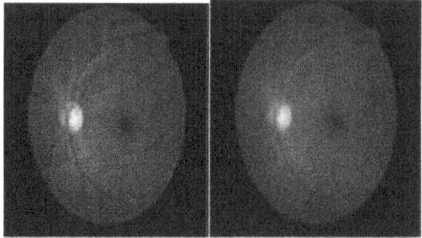

Fig. 5. Eye image low resolution image (Left) and High-Resolution image (Right)

In the Fig. 5 two images shown here left image is low resolution image and corresponding right image is high resolution eye image.

4.2 Performance Metric

Simulation performed over the medical images brain, eye and bone, these are the low resolution images captured through various medical devices. High resolution image generated through SRGAN, which was very high-resolution images.

Quantitative values of the images calculated for the performance evaluation of the model. Peak Signal to Noise Ratio is the perfectly describe the quantitative values of the image.

Table 2. PSNR Comparison

Image Type	PSNR SRResNet	PSNR SRGAN
Bone	32.03	35.95
Brain	31.32	34.42
Eye	29.23	32.39
Average	30.86	34.25

As shown in Table 2 the promising values of PSNR of the different medical images have calculated (Fig. 6).

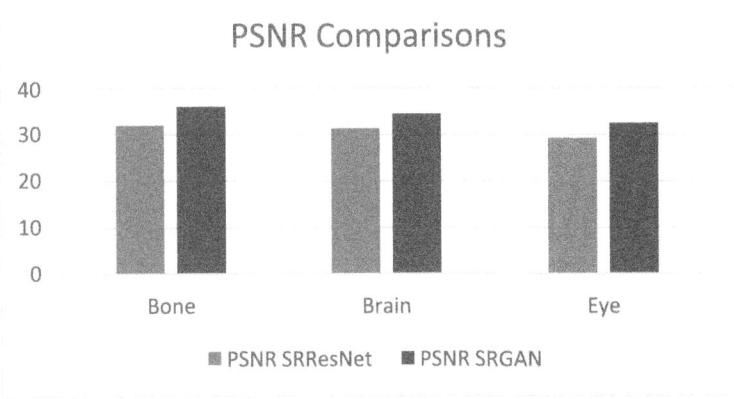

Fig. 6. Shows various values of the different images with PSNR values

5 Conclusion and Future Scope

This work focuses to image enhancement to increase the resolution of the correspond
ing images. This model works perfectly over various low-resolution images, in this work
some medical images were used for increase the image resolution. Here SRGAN (Single
Image super resolution GAN) used here. SRGAN perform well, there is a more efficient
model may use in future to increase the clearer quality of the image, also still more
improvement require in the PSNR values of the image.

References

1. Gonzalez, R.C., Woods, R.E.: Digital Image Processing, Third Edition (2008)
2. Salem, N., Malik, H., Shams, A.: Medical image enhancement based on histogramalgorithms.
   In: 16th International Learning & Technology Conference 2019, Published by Elsevier B.V.
   Procedia Computer Science, vol. 163, pp. 300–311 (2019)
3. Agarwal, M., Mahajan, R.: Medical image contrast enhancement using range limited
   weighted histogram equalization. In: 6th International Conference on Smart Computing and
   Communications, ICSCC 2017, 7–8, Published by Elsevier B.V, pp 149–156 (2018)
4. Wu, M., Sun, Q., Wang, J.: Medical image retrieval based on combination of visual semantic
   and local features. Int. J. Sig. Process. Image Process. Pattern Recogn. **5**(4), 43–55 (2012)
5. Krutsch, R., Tenorio, D.:Histogram equalization. Freescale Semiconductor, Document
   Number AN4318, Application Note (2011)
6. Kim, Y.T.L Contrast enhancement using brightness preserving bi-histogram equalization.
   IEEE Trans. Consumer Electron. 43(1), 1–8 (1997)
7. Wang, Y., Chen, Q., Zhang, B.M.: Image enhancement based on equal area dualistic sub-image
   histogram equalization method. IEEE Trans. Consumer Electron **45**(1), 68–75 (1999)
8. Chen, S.D., Ramli, A.R.: Minimum mean brightness error bi-histogram equalization in
   contrast enhancement. IEEE Trans. Consumer Electron. **49**(4), 1310–1319 (2003)
9. Chen, S.D., Ramli, A.R.: Contrast enhancement using recursive mean-separate histogram
   equalization for scalable brightness preservation. IEEE Trans. Consumer Electron **49**(4),
   1310–1309 (2003)

10. Sim, K.S., Tso, C.P., Tan, Y.Y.: Recursive sub-image histogram equalization applied to gray scale images. Pattern Recogn. Lett. **28**(10), 1209–1221 (2007)

11. Kim, M., Chung, G.C.: Recursively separated and weighted histogram equalization for brightness preservation and contrast enhancement. IEEE Trans. Consumer Electron. **54**(3), 1389–1397 (2008)

12. Haung, S., Cheng, F., Chiu, Y.: Efficient contrast enhancement Using Adaptive Gama Correction with Weighting Distribution. IEEE Trans. Image Process. **22**(3), 1032–1041 (2013)

13. Baby, J., Karunakaran, V.: Bi level weighted histogram equalization with adaptive gamma correction. Int. J. Comput. Eng. Res. (IJCER) **4**(3), 25–30 (2014)

14. Wanga, W., Chenb, Z., Yuanc, X., Wua, X.: Adaptive image enhancement method for correcting low-illumination images. Inf. Sci. 0020–0255 (2019)

15. Kuang, X., Sui, X., Liu, Y., Chen, Q., Gu, G.: Single infrared image enhancement using a deep convolutional neural network. Neurocomput. 0925–2312 (2018)

16. Wang, L.W., Liu, Z.S., Siu, W.C., Lun, D.P.: Lightening network for low-light image enhancement. IEEE Trans. Image Process. 1057–7149 (2020)

17. Ledig, C., et al.: Photo- realistic single image super-resolution using a generative adversarial network. arXiv:1609.04802v5 [cs.CV] 25 May 2017

18. Image processing toolbox, MATLAB R 2020, Mathworks. https://in.mathworks.com/help/

An Overview of Security Intelligence in IoT Applications with Learning Approaches

Tuhin Shukla$^{(\boxtimes)}$ (iD) and Nishchol Mishra (iD)

School of Information Technology, Rajiv Gandhi Proudyogiki Vishwavidalya, Bhopal,
MP 462033, India
shukla.tuhin87@gmail.com

Abstract. Traditional inanimate objects are given new life and interactivity thanks
to the Internet of Things. Sensors that can detect and identify their surroundings
are becoming more widespread in everyday electronics as we go further into the
Internet of Things (IoT) domain.They are able to process data, be connected to
a network, and interact with other electronic devices through the internet. The
Internet of Things aims to achieve its goal of connecting almost everything to
a centralized network. It gives us the ability to monitor and take control of our
various electronic devices. The framework of Iot systems consists of several layers,
sensors, controllers, protocols, as well as web services. The Internet of Things
design is crucial to the deployment of services required. Previously there are tens
of distinct Internet of Things designs available from the research community. The
research community has brought to light a number of difficulties, some of which
are interoperability, privacy and security, reliability, energy limit, scalability, and
a lack of agreed standards. However, in order to provide an acceptable design
for IoT, consideration should be given to requirement analysis since the degree
of need and the order of importance of demands might vary depending on the
context. The purpose of this paper research is to provide a methodical mapping that
focuses on security and application survey to Offer a well-organized summary of
current research tendencies, such as the design of the Internet of Things (IoT), the
difficulties associated with information security, functioning items that make use
of the technology, simulation tools, and so on. In addition, a technical taxonomy
for these challenges is provided, which is derived from the publications that were
researched.

Keywords: IoT Applications · Machine Learning Approaches · IoT · Internet
IoT Security · RFID · TCP/IP

1 Introduction

The Internet of Things (IoT) is a network of embedded devices that can be identified in
only one way and contain the embedded software that is necessary for them to interact
with one another. The Internet of Things has had a significant impact on the day-to-day
activities of end customers, significantly altering their lives. Individuals living spaces,
educational institutions, and places of employment are all linked to the Internet of Things,

© The Author(s), under exclusive license to Springer Nature Switzerland AG 2025
M. Gupta et al. (Eds.): MISS 2023, CCIS 1952, pp. 139–155, 2025.
https://doi.org/10.1007/978-3-031-69115-7_9

which enables them to make use of intelligent environments (both at home and in the city), e-Health services, and transportation networks [12, 39]. Enterprises and organizations are showing a growing interest in highly developed forms of automation and industrial production, information sharing, and data management, as well as intelligent and self-modifying mechanisms and systems [40]. They are creating a new digital world that is more connected and intelligent than the one we already live in, in which everything is linked to everything else, sharing information and collaborating with others. This vision is made possible by the Internet of Things, which is poised for extremely rapid expansion and offers enormous potential for all stakeholders, including corporations, IT professionals, application developers, academics, and students [41]. This vision is both driven and facilitated by the Internet of Things. It is rapidly gaining momentum as a game-changer that will have a huge effect on a number of sectors in India and abroad. Businesses of all stripes would feel the effects of this consequence to varying degrees. India is searching for designers, professionals, and executives that are well-versed in all things IoT [36]. (IoT),who comprehend its effect and potential, and who discover specialized possibilities in order to realize the enormous opportunities that the IoT provides and to benefit from those opportunities. They should also be able to take advantage of these opportunities by creating and releasing novel IoT apps or launching brand new companies that put a focus on IoT [37]. This means thatthe authors of this book are inspired to conduct an in-depth research of the principles of the Internet of Things, as well as the significance of this concept and the ways in which it may be used in our everyday lives. Additionally, it raises worries about the potential security problems that are associated with the IoT. Therefore, as the need of the hour, it is essential to do a full and systematised evaluation of current Internet of Things technology, applications, and security challenges, along with their limits, in order to encourage researchers to work in this sector [35].

In this part of the article, we will talk about the fundamental aspects of Internet of Things (IoT) applications, Featuring the newest innovations in healthcare, business, industry, the environment, and smart city applications: In the next part, "Sect. 2," we will explore the protocol and security of IoT applications, as well as their security goals, threats, approaches, and design issues. The explanation of machine learning may be found in Sect. 3 of the study. The difficulties encountered throughout the study are discussed in Sect. 4, followed by a review of the relevant literature in Sect. 5, and finally, a conclusion is presented and discussed in Sect. 6.

2   Protocols and Security

2.1   IoT Security Goal Available

Table 1. IoT Security Goal Available

S. No	Features	Protocols & Parameters	Remarks about Attacks
1	Confidentiality [19, 20]	Wireless-HART	Spoofing, Eavesdropping, Cloning
		6LoWPAN	–
		IEEE 802.15.4	Tempering, Replay attack, Eavesdropping
		IPSec	Eavesdropping, Spoofing
		Embedded Security	Temper-proofing
2	Integrity [21, 22]	Wireless-HART	Sybil Attack, DoS, Jamming, Tampering, Selective Forwarding De-Synchronization, Wormhole, Collision, Exhaustion
		6LoWPAN	Numerous schemes might categorize the assaults: outsider–insider skeptic source, passive-active, compromising, host-based approaches, or network-based
		IEEE 802.15.4	–
		IPSec	D-DoS attack
		Embedded Security	Software attacks, Cryptanalysis attacks
3	Availability [23, 24]	Wireless-HART	Failed access attempts, Denial of Service, Message failures, Authentication Failures, integrity check
		6LoWPAN	Sinkhole attack, routing attack, Flooding, Selective forwarding, Resource depleting attack
		IEEE 802.15.4	Wireless poses a security risk, sensor side of the network Security, and dangers from the online world
		IPSec	Botnets
		Embedded Security	Malware, Side Channel Attacks

2.2   IoT Security Threats

Table 2. IoT Security Threats

S. No	Features	Protocols & Parameters	Remarks about Threats
1	Confidentiality	Wireless-HART	A moderately bidirectional network gadget with great power, a focal point network administrator, and a controller Version 7 is now available. In addition, HART integrates an IEEE 802.15.4-compliant wireless-based mesh network as a possibility for the physical layer
		6LoWPAN	Per-hop protection, at the very least, integrity protection should be employed 6LoWPAN to networks prevent unauthorized access via listening to the radio medium, as well as protect from effortless attacks to be launched waste constrained resources
		IEEE 802.15.4	For message encryption and authentication on a per-hop basis, a foundation 6LoWPAN Networks
		IPSec	The most well-known algorithms include: MD5 and SHA are two cryptographic algorithms. Furthermore, Non-repudiation, accessibility, and authenticity are the assurance of communication procedures such as IPSec stands for Internet Protocol Security
		Embedded Security	The hash of information is formerly verifying the authenticity of the data—a person's integrity message from supplying a signature that's unique to each individual message. The most widely used algorithms, MD5 and SHA, are two types of hash algorithms
2	Integrity	Wireless-HART	Wireless-HART Security Manager
		6LoWPAN	Cryptography techniques Intrusion detection system techniques
		IPSec	
		Embedded Security	A number of processing cores that are independent of one another, a second bus masters such as DMA engines, as well as a significant number of both memory and peripheral bus saves
3	Availability	Wireless-HART	–
		6LoWPAN	IDS protection layers
		IEEE 802.15.4	For example, the author recommended that important management mechanisms for network sensors in the IoT the situation is appropriate layer-to-link security

(*continued*)

Table 2. (*continued*)

S. No	Features	Protocols & Parameters	Remarks about Threats
		IPSec	–
		Embedded Security	In order to provide physical or cyber security, it is at the execution level. A method based on execution that is a safe environment(SEE) must be constructed. A SEE is a term that refers to a group of people who come together unit of processing; it can securely execute various uses, implying the originating attacks from the outside SEE are not tamper-proof. With data and code about the SEE. The first is a fundamental component; a SEE is required. A protected processor– either a devoted or a semi-dedicated or a single processor- can promote a safe environment, which is hardware compartmentalized from the untrustworthy mode. Using a dedicated processor has the benefit of being easy. Separation, in addition to unloading the central processor, results from the handling duties relating to security—the drawback of a devoted. The processor is a device with the expansion of a silicon footprint

3   Machine Learning in Iot Applications

AI is a renowned procedure, the key element of it's machine learning [38] as well as deep learning. In most cases, the models for learning are a set of regulations, procedures, or intricate "transfer functions". These are the kinds of things that could be used to identify and predict behaviour in addition to detecting important security issue trends in IoT data. In this way, machine learning [39] and deep learning [40] may operate in dynamic IoT networks without the intervention of people or users. Figure 1 illustrates additional proof that demonstrates it may be advantageous to construct a data-driven model for the Internet of Things security intelligence utilizing deep and machine learning approaches (Tables 1 and 2).

Many different machine-learning methodologies can be used to train from IoT security data, including classification and regression analysis [41], clustering [42], rule-based methods [43], feature optimization approaches, and deep learning methods based on artificial neural networks like the multilayer perceptron system, the convolutional network, the recurrent network, etc. Therefore, many machine learning and deep learning approaches could be employed to solve the issue of securing the Internet of Things. Combinations of the following machine learning and deep learning methods may be used to safeguard IoT apps.

1. Regression Techniques,
2. Clustering Techniques,
3. 3.Rule Based Techniques,
4. Security Feature Optimization & PCA,
5. Deep Neural Network Learning-Based Approach
   a. Convolutional Neural Network
   b. Recurrent Neural Network

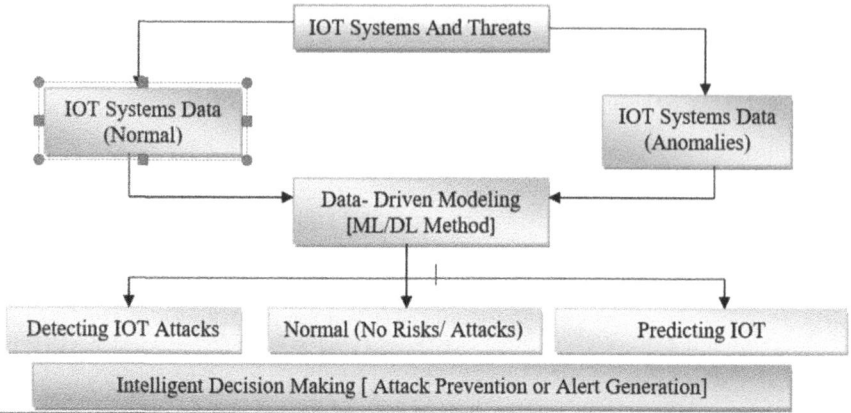

Fig. 1. Illustration of the Artificial intelligence and IoT security

4 Security Challenges in IoT Applications

4.1 IoT Security Challenges

Table 3. Security Issues and Challenges in the Internet of things

Services	Compressed IP-Sec		802.15.4 link-layer security	
	Mode	Overhead	Mode	Overhead
Integrity	HMac-SHAI-96	16-B	AES - CBCMac-96	12 B-X n frags
Confidentiality	AES-CBC	12-B	AES - CTR	5 B-X n frags
Integrity and Confidentiality	AES-CBC and HMAC-SHAI-96	26-B	AES – CCM - 128	21 - X n frags

4.2 Layer-Wise IoT Protocols and Security Protocols with Security Issues

Table 4. Layer-wise IoT Protocols and Security Protocols with Security Issues

IoT Layers	IoT Protocols	Security Protocols	Security Issues
Application Layer	MQTT, CoAP, HTTP	Specified by the user	Viruses, Trojans, malicious-code injection, and social engineering base attacks
Transport Layer	TCP, UDP	DTLS, TLS	Eavesdropping, man-in-the-middle(MITM)
Network Layer	IPv4/IPv6, RPL,6LoPAN	IP sec, RPL security	DoS Attack, exhaustion, collision, and spoofing
Data-link	IEEE 802.15.4 MAC	802.15.4 security	malware, collision, and unfairness attacks
Physical	802.15.4 PHY, Wi-Fi, LTE	–	Jamming, Radio interference, Tampering

5 Review of Literature

Abdullahi, M. et al. (2022), The fourth industrial revolution (Industry 4.0), made possible by technological breakthroughs, has seen the fast and widespread expansion of the IoT, fog computing, computer security, and cyberattacks. Massive amounts of data are being produced by the rapidly growing number of IoT devices and networks, which need solid means of identification and protection. Many experts agree that artificial intelligence (AI) is one of the best ways to lessen the severity of cyberattacks and improve security overall. Here, we suggest an SLR to organise, map, and assess the existing literature on artificial intelligence algorithms for detecting cybersecurity attacks against IoT devices. This SLR looks at solutions that are up-to-date as well as the most common artificial intelligence (AI) technologies used for cyber security right now [1].

Inayat, U. et al. (2022), An emerging technology is the Internet of Things, or IoT for short. Utilizing the cloud or wireless networks, it is simple to exchange information with other linked devices. It's possible, but not certain, that as the Internet of Things (IoT) ecosystem matures and grows, IoT systems may become more vulnerable to cyberattacks. This SLR looks at solutions that are up-to-date as well as the most common artificial intelligence (AI) technologies used for cyber security right now. There is a risk that material and monetary resources may be lost as a consequence of these incursions. In the wake of an attack, There are a number of challenges that IoT devices and systems must overcome, and this article focuses on those challenges as well as the system or framework that supports them, as well as learning-based approaches to those challenges [02].

Sarker, Iqbal H. et al. (2022), The authors of this article present a thorough analysis of previous works devoted to the issue of Information security intelligence. The Internet of Things, sensing devices built on the IoT, security issues, and solutions based on machine learning are all talked about.reseachersalso looked at recent studies on IoT security to get a feel for the topic and write this article from a well-rounded vantage point. In order to locate and secure IoT gadgets and infrastructure, extensive study of IoT architectures, particularly their layered cyber attacks, is necessary. Because of this, they have started to look into how machine learning and deep learning could be used to provide security solutions within the IoT framework [3].

Nikolskaia, K. et al. (2021), People's lives are becoming more permeable to various forms of artificial intelligence in recent years. There are instances when people don't even consider the possibility that some technologies that make our lives simpler and more pleasant are working in tandem with approaches to artificial intelligence. However, the fast rise in the penetration of such technology brings with it a variety of concerns that might potentially be disastrous. These dangers are connected to the safety of a person's private information. In this study, we investigate recent developments in people's usage of artificial intelligence in their day-to-day lives. The writers made an effort to draw a parallel between computer network security and artificial intelligence in their writing. Also taken into consideration are the potential dangers posed by cyberattacks that make use of artificial intelligence as well as assaults against AI itself [4].

Jahromi, A. N. et al. (2021), In this study, we used a two-stage ensemble deep learning model to come up with a new way to find attacks and figure out who was behind them. The attack detection step applies a DT to identify attack samples after using deep representation learning to map the data to a new higher dimensional space and then mapping the samples back to their original space. This level can find attacks that have never been seen before and can handle data sets that aren't perfect. The attack attribution step is comprised of a grouping of numerous one-versus-all classifiers, each of which has been trained on a particular attack attribute. As can be seen in the demonstration, the whole thing comes together to produce a complicated DNN that has both partly linked and completely connected parts that can appropriately identify cyberattacks [5].

Elsisi, M. et al. (2021), The cyber-physical system idea found in Industry 4.0 served as the inspiration for the architecture that was developed for the Internet of Things. Real-time monitoring and tracking of the GIS state through an Internet of Things topology offers great promise but faces significant challenges, the most significant of which are cyberattacks and the categorization of GIS insulation faults. In order to fulfil this objective, sophisticated methods of machine learning are applied to identify cyberattacks in order to carry out the paradigm and verification. To show that the proposed architecture for the IoT will work, multiple test cases that cover a wide range of GIS issues are run. The inputs for the Internet of Things architecture are represented by the characteristics that are retrieved from Each fault's unique partial discharge pulse sequence. According to the findings, the suggested IoT architecture that is built on the machine learning app-roach known as extreme gradient boosting (XG-Boost) is able to visualise all flaws in the GIS using a variety of warnings, in addition to efficiently displaying cyberattacks on the networks. In addition, the flaws in the GIS as well as the fake data that was caused by the cyberattacks are identified and shown with more precision on the dashboard of the proposed IoT platform, clearvisualisation in order to improve the decision-making process regarding the current state of the GIS [6].

Elbouchikhi, E. et al. (2021), Here, we'll examine many of the most common techniques for keeping tabs on the quality of the electricity being distributed via smart grids. Classification and decision-making techniques, as well as feature extraction algorithms, have been considered. Current breakthroughs in PQ analysis using advanced signal processing and machine learning have been brought to light. Attempts like this have failed so far. Determining the appropriate signal processing methods for use in defect identification and characterization may be challenging. Indeed, a brief data collection time is required when dealing with transitory events, off-nominal settings, and nonstationary operating conditions. This impacts the precision with which we can estimate frequencies and the precision with which we can predict their amplitudes [7].

Shah, Y. et al. (2020), Issues and concerns pertaining to the security of IoT and IIOT devices have been described in this article in the hopes that it may act as a guide in the process of building a secure network for such devices. The purpose of this research is to identify the many cyberattacks that are associated with Internet of Things and Industrial Internet of Things devices. The process of fully safeguarding IoT and IIOT devices is likely to be a lengthy one; but, in order to design new security mechanisms, we need to understand the many forms of cyber-attacks that an IoT/IIOT network is susceptible [8].

Kseniia, N. et al. (2020), When you look at the laws that govern the country, you'll notice that there is a very specific set of behaviours that are subject to legal scrutiny. Transportation, education, medicine, public safety, industry, and government administration are all examples of sectors that come under this umbrella. However, there is a clear inclination toward expanding the laws that regulate the use of AI into new fields. The quest for a solution to the difficult issue of streamlining interactions in the usage of cyber-physical systems and artificial intelligence is shown by the active use of organisational norms, ethical standards, self-regulation and co-regulation, and the establishment of transnational norms [9].

Abu Al-Haija, Q. et al. (2020), High-performance computing is used in the proposed IoT-IDCS-CNN, which makes use of fast Intel CPUs with I9 cores and potent Compute Unified Device Architectures (CUDA)-based Nvidia GPUs (Graphical Processing Units). The designed architecture is divided into three sections: a features extraction module, a feature learning module, and a traffic classification module. As a means of gauging the efficacy of the proposed method, we used the NSL-KDD dataset, which contains examples of all the most serious threats to IoT systems. According to the simulation results, the accuracy of the binary-class classifier (normal vs. anomalous) was 99.3%, while the accuracy of the multiclass classifier (five categories) was 98.2% [10].

Soe, Y. N. et al. (2020), They started out by highlighting a problem with the way Internet of Things detection is done today. Next, we used our new feature selection approach and implemented it on the Raspberry Pi to create a lightweight detection system for the IoT setting (named CST-GR). By running the CST-GR algorithm for each unique attack, the detection system only needs a handful of features to correctly identify each kind of attack, making it not only very lightweight but also incredibly fast. Plus, the lower weight of this detection system has almost no effect on its performance. In addition, many prominent machine learning techniques, such as J48, Random Forest, Logistic Model Trees (LMT), and Vega-Filter Decision Trees (VFDT) (RF), were tested to see whether they could be integrated into our system. The publicly available Bot-IoT dataset, which was created in a simulated version of an Internet of Things context, was used to measure the performance of our detection algorithm. Based on what we've learned from our studies, the CST-GR feature selection and the J48 classifier could help make a light detection system [11].

Muheidat, F. et al. (2020), IoT systems are employed in a wide variety of different applications, many of which have made our lives easier and more convenient. As a direct consequence of COVID-19, there has been an increase in the number of cyberattacks targeting Cloud-Enabled IoT health system services. New layered models for the Internet of Things have been suggested; these models have been broadened and extended to include privacy and security components as well as layer recognition. Each of the three layers— the edge, the cloud, and the Internet of Things—can have extra safety measures put in place in addition to the suggested approach [13].

Zia1, M. F. et al. (2019), A growing number of remote and rural areas are turning to microgrids as a viable option for addressing their electrical needs. Reliable renewable energy sources, energy storage, and microgrids powered by conventional generators can meet the energy demands of islands. To make the most optimal use of all of these different energy sources, a microgrid energy management system is necessary. There is a great deal of unrealized potential for marine renewable energy sources in the Bretagne region of France. Islands in this region may meet their energy demands using a combination of tidal turbines and other power generation methods. Specifically, this paper proposes a case study of an independent maritime microgrid system to be carried out on Ouessant Island. A solar array, a tidal turbine, a diesel generator, and a lithium-ion battery make up the microgrid under investigation. The architecture of the newly constructed microgrid system, as well as the optimal timeline for it, is discussed in detail with the goal of reducing operational and maintenance costs. In the not-too-distant future, the developed energy management architecture may aid in the creation of microgrid systems for islands [14].

Alsamiri, J. et al. (2019),The purpose of this research is to use machine learning techniques to detect assaults on IoT networks. The dynamic (continuous updates), expansive (vast attack variety), and deep characteristics of the bot IoT made it an ideal dataset for our investigation (various network protocols). With CIC Flow Meter's support, we were able to parse the raw traffic traces for flow-based features. CICFlowMeter produced all 84 of the network traffic components used to characterise the network flow in the dataset. The Random Forest Regressor technique was used to determine feature weights in order to determine which features should be included in the machine learning approaches. These numbers were arrived at in two different ways. In the original method, we accounted for the variety of attacks by computing individual significance weights. This second strategy involves, we grouped all the assaults together and calculated the importance weights for that set, i.e., we identified the shared characteristics that were crucial to each kind of attack. The original approach included individually calculating significance weights for each potential threat. In the end, seven different machine learning methods were utilised on the data, each with its own set of advantages. Here is a rundown of these algorithms, along with their corresponding F-measure performance ratios: Naive Bayes was at 0.77, QDA at 0.86, Random Forest at 0.97, ID3 at 0.97, AdaBoost at 0.97, MLP at 0.83, and K-Nearest Neighbors at 0.99. The F-measure was between 0 and 1[15].

Roopak, M. et al. (2019), In this study, deep learning models for the cyber security of Internet of Things (IoT) networks are suggested. The Internet of Items network is a potentially useful technology that links living and nonliving things located all over the planet. It is crucial for the success of any network that the network be entirely safe; otherwise, people may be hesitant to utilise this technology. IoT implementation is growing rapidly, but it is still deficient in cyber security, leaving it open to various threats. DDoS attacks (distributed denial of service) protocol have recently been launched against a number of Internet of Things networks, which has led to significant financial losses. For DDoS detection, researchers have proposed deep learning models as well as evaluated those models using the most current CICIDS2017 databases, which have provided the highest accuracy as 97.2%. The presented models are also contrasted using machine learning techniques. Furthermore, this research solves some of the most pressing open questions about the application of deep learning algorithms to the challenge of cyber security for the Internet of Things [16].

Khalid, H. M. et al. (2019), In this article, a new technique is described that may significantly enhance the precision with which a combined cycle power plant's water-cooling system infection is estimated. This is achieved by using an IMM-based fusion approach to take into account both the current and time-varying nonlinear dynamics. A sub-observer-based estimate is a local extension of the IMM structure that further computes the time-delay and cross-covariance between measurements in order to boost the system's observability. This means the proposed method can protect the system from the introduced attacks. One of the problems with the proposed method is the complexity of the calculations needed, which is exacerbated by the need to handle data in both local and central locations. But you can get around this problem by doing both computations at the same time on a high-performance machine [17].

Djenna, A. et al. (2018), Recent years have seen an uptick in interest in the Internet of Things (IoT) from a variety of sectors, including business and academia, due to its potential as a game-changing innovation. This technology provides a significant increase in value to many different industries, including home automation, transportation, energy, and most importantly, health care. On the other hand, Internet of Things (IoT) based health care suffers from a number of security issues that differ from those in other domains in terms of methodologies, motivation, and consequences. This is because of the complexity of the environment and the nature of the devices that are being deployed. The purpose of this article is to present an outline of the most current security concern affecting health care that is based on IoT [18] (Tables 5 and 6).

Table 5. A review of supervised machine learning techniques for identifying Internet of Things cyber attacks

Ref. No	Detection Method	Types of Attacks	Feature Selection/Classification Method	Datasets
[25]	Identifying Unauthorized Access	Using a combination of selective forward and blackhole attacks on the network layer	SVM	Information from an Internet of Things testbed
[26]	Identifying and analyzing irregularities	Black holes are classified into three types: black, selective, and sinkhole	c-SVM	KDD-99
[27]	Detection in all cases	R2L, R2L, U2R, and the Probe	Improbable Woods	Together, NSL-KDD and KDDCUP99
[28]	A Role for the Internet of Things in the Argo System	threat posed by something nobody knows about yet	To name a few: NN, LR, RF, NB, and KNN	–
[29]	Signalling of an intrusion	DDoS and spoofing	radio frequency, signal-to-noise ratio, and XG-Boost	IoTID20
[30]	Spark's ML library in Apache	Distributed denial of service using synchronous denial-of-service attacks	RF, DT, LR, SVM, and GBT to	SYNDOS2M

Table 6. Deep learning methods for cyber-attack detection in IoT.

Ref. No	Detection Method	Types of Attacks	Feature Selection/Classification Method	Data set
[31]	Synergy between a genetic algorithm and a deep belief network (GA)	Disk Operating System, Remote-to-Local, Probe, and User to Root	For example, Tax Deduction or Collection Account Number, Fuzzy Clustering and Artificial Neural Network, SA-DT-Support Vector Machines, and Back-Propagation Neural Network	KDDCUP
[32]	To model neural networks using the LM-BP framework	Disk Operating System, Remote-to-Local, Plankton Research On The Bay Environment, and User to Root	Both conventional and PSO-BP	KDDCUP99
[33]	CSSAE	Differential Operational Strategy, Reverse Operational Strategy, Plankton Research On The Bay Environment, and User to Root	Simultaneous Authentication Of Equals And Non-Symmetric Deep Auto-Encoder	A Comparison of KDDCUP99 with NSL-KDD
[34]	In a nutshell: an LSTM module ensemble	cyber attack	A Comparison of RF, KNN, MLP, and SVM	Activity on a Modular Bus Network
[44]	Protection against incursion using artificial intelligence	Distributed denial of service, sinkhole, blackhole, wormhole, and opportunistic assault	IWC	–

6   Conclusions and Future Work

The purpose of this study is to identify and characterize the new terminology, innovations, and problems in IoT application and Security concerns. The study covers a variety of Internet of Things applications, including significant domains such as health care, commercial, industrial, environmental, and smart city initiatives, among others. In this study, a variety of issues, such as data security and scalability, are examined in relation to the many different kinds of applications for the internet of things. It has been determined that the current issues are mostly centered on mobility and the structure of the system. As a consequence of this, the principal challenges associated with the Internet of Things (IoT) systems consist of the development of an ideal algorithm, the management of security issues, and the upkeep of an energy-efficient and cost-effective network in order to foster the expansion of the IoT system. In addition to that, we had a discussion on a few essential enabling technologies, as well as concerns about security and obstacles. We looked at how apps built for the Internet of Things may help enhance and simplify our lives on a day-to-day basis. Research on the Internet of Things is being conducted at a significant scale in this particular domain. It is necessary to find solutions to the problems and challenges that have surfaced as a direct consequence of adoption clearance. Examining and making it accessible to the user are both requirements for information security, privacy, and confidentiality. It is essential for the adoption of IoT technology to address concerns about the threats to confidentiality, security, and privacy. The problems and concerns that have been brought up in this article will serve as the basis for the scope of future research, with the goal being to find solutions to as many of these problems as possible across a variety of fields.

References

1. Abdullahi, M., et al.:Detecting cybersecurity attacks in internet of things using artificial intelligence methods: a systematic literature review. Electronics **11**(2), 198 (2022)
2. Inayat, U., Zia, M.F., Mahmood, S., Khalid, H.M., Benbouzid, M.:. Learning-based methods for cyber attacks detection in iot systems: a survey on methods, analysis, and future prospects. Electronics **11**(9), 1502 (2022)
3. Sarker, I.H., Khan, A.I., Abushark, Y.B., Alsolami, F.:Internet of things (iot) security intelligence: a comprehensive overview, machine learning solutions and research directions. Mobile Netw. Appl. 1–17 (2022)
4. Nikolskaia, K.Y., Naumov, V.B.:The Relationship between Cybersecurity and Artificial Intelligence. In: 2021 International Conference on Quality Management, Transport and Information Security, Information Technologies (IT&QM&IS), pp. 94–97. IEEE (2021)
5. Jahromi, A.N., Karimipour, H., Dehghantanha, A., Choo, K.K.R.: Toward detection and attribution of cyber-attacks in iot-enabled cyber–physical systems. IEEE Internet Things J. **8**(17), 13712–13722 (2021)
6. Elsisi, M., Tran, M.Q., Mahmoud, K., Mansour, D.E.A., Lehtonen, M., Darwish, M.M.: Towards secured online monitoring for digitalized GIS against cyber-attacks based on IoT and machine learning. IEEE Access **9**, 78415–78427 (2021)
7. Elbouchikhi, E., Zia, M.F., Benbouzid, M., El Hani, S.: Overview of signal processing and machine learning for smart grid condition monitoring. Electronics **10**(21), 2725 (2021)

8. Shah, Y., Sengupta, S.:A survey on classification of cyber-attacks on IoT and IIoT devices. In: 2020 11th IEEE Annual Ubiquitous Computing, Electronics & Mobile Communication Conference (UEMCON), pp. 0406–0413. IEEE (2020)

9. Kseniia, N., Minbaleev, A.:Legal support of cybersecurity in the field of application of artificial intelligence technology. In: 2020 International Conference Quality Management, Transport and Information Security, Information Technologies (IT&QM&IS), pp. 59–62. IEEE (2020)

10. Abu Al-Haija, Q. and Zein-Sabatto, S.: An efficient deep-learning-based detection and classification system for cyber-attacks in IoT communication networks. Electronics 9(12), 2152 (2020)

11. Soe, Y.N., Feng, Y., Santosa, P.I., Hartanto, R., Sakurai, K.:Towards a lightweight detection system for cyber-attacks in the IoT environment using corresponding features. Electronics 9(1), 144 (2020)

12. Souza, L.F.D.F., et al.:Internet of medical things: an effective and fully automatic IoT approach using deep learning and fine-tuning to lung CT segmentation. Sensors 20(23), 6711 (2020)

13. Muheidat, F., Tawalbeh, M., Quwaider, M., Saldamli, G.: Predicting and preventing cyber-attacks during covid-19 time using data analysis and proposed secure iot layered model. In: 2020 Fourth International Conference on Multimedia Computing, Networking and Applications (MCNA), pp. 113–118. IEEE (2020)

14. Zia, M.F., Elbouchikhi, E., Benbouzid, M.E.H.: An energy management system for hybrid energy sources- based stand-alone marine microgrid. IOP Conf. Ser. Earth Environ. Sci. 322, 012001 (2019)

15. Alsamiri, J., Alsubhi, K.: Internet of things cyber-attacks detection using machine learning. Int. J. Adv. Comput. Sci. Appl. 10(12) (2019)

16. Roopak, M., Tian, G.Y., Chambers, J.: Deep learning models for cyber security in IoT networks. In: 2019 IEEE 9th annual computing and communication workshop and conference (CCWC), pp. 0452–0457. IEEE (2019)

17. Khalid, H.M., Muyeen, S.M., Peng, J.C.H.: Cyber-attacks in a looped energy-water nexus: an inoculated sub-observer-based approach. IEEE Syst. J. 14(2), 2054–2065 (2019)

18. Djenna, A., Saïdouni, D.E.: Cyber-attacks classification in IoT-based-healthcare infrastructure. In: 2018 2nd Cyber Security in Networking Conference (CSNet), pp. 1–4. IEEE (2018)

19. Stellios, I., Kotzanikolaou, P., Psarakis, M., Alcaraz, C., Lopez, J.: A survey of iot-enabled cyberattacks: assessing attack paths to critical infrastructures and services. IEEE Commun. Surv. Tutorials 20(4), 3453–3495 (2018)

20. Berman, D.S., Buczak, A.L., Chavis, J.S., Corbett, C.L.: A survey of deep learning methods for cyber security. Information 10, 122 (2019)

21. Al-Garadi, M.A., Mohamed, A., Al-Ali, A.K., Du, X., Ali, I., Guizani, M.: A survey of machine and deep learning methods for internet of things (IoT) security. IEEE Commun. Surv. Tutor. 22, 1646–1685 (2020)

22. Tahsien, S.M., Karimipour, H., Spachos, P.: Machine learning based solutions for security of internet of things (IoT): a survey. J. Netw. Comput. Appl. 161, 102630 (2020)

23. Mohanta, B.K., Jena, D., Satapathy, U., Patnaik, S.: Survey on IoT security: challenges and solution using machine learning, artificial intelligence and blockchain technology. Internet Things 11, 100227 (2020)

24. Amanullah, M.A., et al.: Deep learning and big data technologies for IoT security. Comput. Commun. 151, 495–517 (2020)

25. Ioannou, C., Vassiliou, V.: Experimentation with local intrusion detection in IoT networks using supervised learning. In: Proceedings of the 16th International Conference on Distributed Computing in Sensor Systems (DCOSS), Marina del Rey, CA, USA, pp. 423–428, 25–27 May 2020

26. Ioannou, C., Vassiliou, V.: Classifying security attacks in IoT networks using supervised learning. In: Proceedings of the 15th International Conference on Distributed Computing in Sensor Systems (DCOSS), Santorini, Greece, pp. 652–658, 29–31 May 2019

27. Rani, D.; Kaushal, N.C.: Supervised machine learning based network intrusion detection system for internet of things. In: Proceedings of the 11th International Conference on Computing, Communication and Networking Technologies (ICCCNT) Kharagpur, India, pp. 1–7, 1–3 July 2020

28. Wan, Y., Xu, K., Xue, G., Wang, F.: IoTArgos: a multi-layer security monitoring system for internet-of-things in smart homes. In: Proceedings of the IEEE INFOCOM 2020-IEEE Conference on Computer Communications, Toronto, ON, Canada, pp. 874–883, 6–9 July 2020

29. Krishnan, S., Neyaz, A., Liu, Q.: IoT network attack detection using supervised machine learning. Int. J. Artif. Intell. Expert Syst. **10**, 18–32 (2021)

30. Morfino, V., Rampone, S.: Towards near-real-time intrusion detection for IoT devices using supervised learning and APACHE Spark. Electronics **9**, 444 (2020)

31. Li, P., Zhang, Y.: A novel intrusion detection method for internet of things. In: Proceedings of the Chinese Control Additionally, Decision Conference (CCDC), Nanchang, China, pp. 4761–4765, 3–5 June 2019

32. Yang, A., Zhuansun, Y., Liu, C., Li, J., Zhang, C.: Design of intrusion detection system for internet of things based on improved BP neural network. IEEE Access **7**, 106043–106052 (2019)

33. Telikani, A., Gandomi, A.H.: Cost-sensitive stacked auto-encoders for intrusion detection in the internet of things. Internet Things **14**, 100122 (2019)

34. Saharkhizan, M., Azmoodeh, A., Dehghantanha, A., Choo, K.K.R., Parizi, R.M.: An ensemble of deep recurrent neural networks for detecting IoT cyber-attacks using network traffic. IEEE Internet Things J. **7**, 8852–8859 (2020)

35. Li, F., Shi, Y., Shinde, A., Ye, J., Song, W.: Enhanced cyber-physical security in internet of things through energy auditing. IEEE Internet Things J. **6**, 5224–5231 (2019)

36. Maurya, S., Sharma, S., Yadav, P.: Internet of things based air pollution penetrating system using GSM and GPRS. In: 2018 International Conference on Advanced Computation and Telecommunication (ICACAT), pp. 1–5. IEEE (2018)

37. Shrivastava, R., Tiwary, A., Yadav, P.: Challenges block chain technology using IOT for improving personal and physical safety-review. In: 2021 International Conference on Advances in Technology, Management & Education (ICATME), pp. 238–243. IEEE (2021)

38. Kumar, A., Shukla, S.K., Sharma, A., Yadav, P.: A robust approach for image super-resolution using modified very deep convolution networks. In: 2022 International Conference on Applied Artificial Intelligence and Computing (ICAAIC), pp. 259–265. IEEE (2022)

39. Mishra, A., Shrivastava, R., Yadav, P.: A modified cascaded feed froward neural network distributed denial of service attack detection using improved regression based machine leaning approach. In: 2022 6th International Conference on Trends in Electronics and Informatics (ICOEI), pp. 1292–1299. IEEE (2022)

40. Tiwari, S., Gupta, N., Yadav, P.: Diabetes type2 patient detection using LASSO based CFFNN machine learning approach. In: 2021 8th International Conference on Signal Processing and Integrated Networks (SPIN), pp. 602–608. IEEE ,(2021)

41. Sharma, B., Kumar, S., Tiwari, P., Yadav, P., Nezhurina, M.I.: ANN based short-term traffic flow forecasting in undivided two lane highway. J. Big Data **5**(1), 1–16 (2018)

42. Tiwari, P., et al.: Sentiment analysis for airlines services based on Twitter dataset. Soc. Netw. Anal. Comput. Res. Methods Tech. **149** (2018)

43. Singh, J., et al.: Fuzzy logic hybrid model with semantic filtering approach for pseudo relevance feedback-based query expansion. In: 2017 IEEE Symposium Series on Computational Intelligence (SSCI), pp. 1–7. IEEE (2017)
44. Thamilarasu, G., Chawla, S.: Towards deep-learning-driven intrusion detection for the internet of things. Sensors **2019**, 19 (1977)

Natural Language Processing: Innovations, Recent Trends and Challenges

Padma Prasada[1,2(✉)] ⓘ, M. V. Panduranga Rao[1] ⓘ,
and Ujwala Vishwanatharao Suryawanshi[3] ⓘ

[1] Faculty of Engineering and Technology, JAIN (Deemed to be University), Kanakapura Road,
Bengaluru, India
ppjain15@gmail.com
[2] Department of Artificial Intelligence & Data Science, Sri Dharmasthala Manjunatheshwara
Institute of Technology, Ujire, Karnataka, India
[3] Rajarshi Shahu Mahavidyalaya (Autonomous), Latur, Maharashtra, India

Abstract. Natural Language Processing (NLP) is a rising technique for developing various sorts of Artificial Intelligence (AI), and it will continue to be a major emphasis for more cognitive application. This article covered NLP's most beneficial uses. Our objective of this study is to provide a conceptual understanding of the many different areas where NLP has the potential to make a significant impact and change the entire landscape using its automating methods. It's a popular topic right now, and everybody wants to get in on the action by investing there. The completion of these implementations is made possible by detailed and in-depth research on NLP as well as the field it covers. The review begins with an explanation of the tendencies of NLP and the elements of NLP, and afterwards moves forward to the application of NLP as well as its development and issues. Text to speech generation, liver cancer diagnosis, sentence generator, graph-based technique, and postoperative complications prediction for patients with chronic conditions are some of the features that we have included in our discussion.

Keywords: Natural Language Processing · Artificial intelligence · machine translation · POS Tagging · Text Summarization

1 Introduction

The term "natural language processing" (NLP) refers to both a field of research and an area of application that investigates the ways in which machines could be used to read and modify natural language, either in the form of voice or written text, as required to bring out meaningful activities. A hypothetical framework of computer algorithms that categorise activities or applications using human-like terms is called "common language processing". The use of NLP is intended to accomplish certain goals. A number of the NLP's bulk tasks, including programme designs, oral tests, machine interpretations, and on and on, are completed using these activities.

© The Author(s), under exclusive license to Springer Nature Switzerland AG 2025
M. Gupta et al. (Eds.): MISS 2023, CCIS 1952, pp. 156–168, 2025.
https://doi.org/10.1007/978-3-031-69115-7_10

NLP lets computers understand native languages and evaluate language-related commercial ventures. For instance, Natural Language Processing enables personal computers to grasp literature, listen into dialogue, comprehend it, analyse it, and determine and comprehend what aspects were crucial. The present generation of personal computers can see more language-based knowledge than individuals, without any downsides and with steady as well as impartial thinking. With the daily influx of unstructured data ranging from medical records to internet platforms, a computerised system will be accepted textual and conversational material as part of the offering. Individuals have their own languages, which are distinct from the vocabulary of astonishment. Individuals communicate with one another through a wide range of formats, including verbally and in writing. Every language, even though there can be relatively fewer varieties of accents, contains its own pattern of punctuation as well as grammatical structuring rules, sentences, and idioms. Our speech is characterised by local dialects, stuttering, mumbling, and the acquisition of new vocabulary via the use of informal idioms. Learning that is both administered and unintended, and especially strenuous learning, is presently being extensively used to illustrate natural language. This is a necessity for semantic and syntactic comprehension, as well as spatial prowess, all of which are fundamentally important components of human language. NLP is significant since it all helps to attempt to correct misunderstandings in language, and for some, it keeps adding features that characterise the layout of information, such as application areas, discussion acknowledgments, or text inspections.

2 Application Areas of NLP

According to studies, computer scientists do most NLP work, although linguists, psychologists, philosophers, and others are also interested. The study of NLP is concerned with a variety of ideas and methods that attempt to solve the challenge of using natural language to communicate with computers.

2.1 Composing Sentences

When linguistic constraints and the findings of the models are combined, it is possible to produce sentences that have sense and are coherent. It has a wide range of applications and can be used for tasks such as conversational programs, translating tools, and photo annotation [1]. Whereas an auto-regressive approach uses beam searching to create phrases from left to right, which is challenging to effectively incorporate lexical limitations. Such difficult circumstances would be tackled using the newly developed computational paradigm BFGAN. In this system, cohesive statements are generated by applying lexical limitations that use a reverse producer, a forward producer, and a discriminator. A discriminator is responsible for combining the backward and forward phrases by employing signalling incentives. [2] "Lexically limited sentence generation" describes a situation in which a lexical restriction is a predetermined group of terms that are allowed to appear inside the final product. This is the most pressing problem that must be addressed right now in the field of NLP.

RNN had already notably emerged as a major force in the field of NLP, thanks to its notable performance in a variety of activities, including the formation of neural networks

[3]. Tech blogs [4] written text summaries [5] table-to-text formation [6], as well as efficient document creation [7]. Beam Search (BS) is carried out, and auto-regressive models are often used to create left-to-right phrases [8]. Generating phrases that adhere to certain lexical requirements is indeed a challenging procedure. Replacing it with a random word in the outputs would disrupt the paragraph's flow. If more information about the word is provided, it does not guarantee that the intended phrase will appear in the results [9, 10]. In order to ensure that the forward generator maintains its integrity, the dynamic attentiveness mechanism is implemented. A recurrence occurs from the beginning all the way through to the finish, and the attentiveness function's range expands as a result.

2.2 Assessment of the Likelihood of Postoperative Complications

Patients get readmitted to a hospital for a variety of reasons, including the failure of medical rules and directives connected to the discharge, the return of infections, and inadequate care services when they are discharged from the hospital. There seem to be numerous requirements that patients have when they are discharged from the hospital. Comprehensive case management should be present for patients with COPD, and post-discharge requirements must be met. However, readmissions beyond 30 days would've been associated with chronic pulmonary disorders. The hospitals in the United States that have high readmissions are subject to a fine from the federal government. The clinical documentation serves as the basis for making predictions. Records of the laboratory findings are being compiled and evaluated based on the patient's medical history, allergens, and demographic data. OpenNLP and a Java-based NLP package are being utilised in the construction of this prediction method.

The objective of "OpenNLP" is to provide a framework that can support a wide range of NLP applications, including cryptocurrency mining, risk mitigation, chunk tagging, and named entity identification [11]. The pipeline pattern [12] is a very important step in the process in the realm of NLP, which is why frameworks are beneficial for NLP processing. NLP activities are typically arranged from the lowest level to the highest, with every query possibly depending on the one that came before it. [14] We are using unstructured data from the electronic healthcare records to make predictions about patients with chronic conditions who may be readmitted during the next 30 days. In spite of the fact that investigations using NLP to forecast health outcomes have started to emerge in recent times, hardly any research has assessed feature extraction methodologies for patient outcomes, and only a small number of articles are now being produced in the area of forecasting outcomes for patients [13]. As the use of patient health records becomes more widespread, clinical documentation will have a greater degree of significance; hence, natural language processing techniques will have to be included in the process of constructing a predictive model.

2.3 Word Extraction

The field of NLP is broadening its scope as a result of its accelerated rate of growth. Because of NLP, word extraction seems to be a possibility as well. Specialists can produce related words based on the words provided and then compile those. This programme takes a completely hands-off approach to the extraction of linked terms. It is composed of two stages: the collection of connection words and the creation of computerised correlation networks. A trigger response is a signalling message that is sent in response to a certain action or behavior, and association refers to the state of being attached to a person or an object. Word association assignments done in this manner require a significant investment of both time and money. There is a degree of consistency in the pattern of human interaction. Therefore, researchers anticipate that the machine association networks will be correct to a certain degree even when the attentiveness object is held constant [15]. Because of this, we may draw the conclusion that this framework, which is built on neural nets, extracts related phrases from textual form. Reading and understanding algorithms as well as an attention mechanism are used to complete the task.

2.4 Liver Cancer Prediction

Even with the gains achieved in the domain of NLP, EMR processing operations have remained to be constrained and difficult owing to insufficient information, databases, and semantic and syntactic features, notably for radiology reports. The random forest model offers the greatest effectiveness and accuracy rate regarding liver cancer speculation. In addition to this, the NLP pipeline may be utilised for clinical testing as well as other types of prediction work. EMR's electronic medical patient records constitute invaluable resources. In the modern digital age, approaches based on machine learning serve an important part in the process of data processing. These approaches also have applications in domains such as clinical decision-making, the treatment of chronic diseases, and record-keeping [16, 17].

There are a lot of benefits that come with using NLP-based extraction of features. As a mostly direct consequence of this, natural language processing-based features were thus effectively used for diagnosis observation, cohort building, performance measurement, and professional aid in radiography [18–21]. Due to the restricted corpora and databases that are available for EMR computation of radiology reports, researchers are needed to generate the corpus for their research in order to proceed. The radiographic specialists contribute to the construction of a vocabulary that is then annotated; this is followed by the processing of the data depending on the linguistic limitations and clinical knowledge of the specialists. Although we were only able to utilise a very small fraction of all radiological data to compile the vocabulary, it is possible that the pipeline would be used as a benchmark in other computerised clinical applications [22]. An analysis is being conducted on how neural network-based schemes are assisting radiologists, particularly radiologists with minimal expertise, in making judgements regarding the diagnosis. The natural language processing method that was presented could be applied to the process of creating lexicons for new diseases and forms of medical journals [22].

2.5 Text-To-Speech Conversion

Language is among the most effective ways to communicate; nonetheless, it may give rise to confusion and uncertainty, as well as serve as a possible cause of misunderstandings. It's possible for an individual to make statements that aren't quite clear, leaving room for many interpretations [23]. Because it does not need language translation or machine vision, Interactive Text to Pick Up can complete the assignment notwithstanding the ambiguity of verbal orders as well as the constraints that are placed upon that. The constructed Text to Pickup system can handle the job dynamically, which is helpful in situations where an incoming command is complicated or numerous things are grouped in various ways [24]. A question generator and the text-to-pickup system are the main components that make up this solution. In addition to this, the Text to Pickup system generates a heatmap of location information. The interactivity made a significant contribution to the overall clarity of the instructions. Interactive and solitary text-to-pickup systems are compared (the version that does not include any interactions) [24].

2.6 Graph-Based Technique

This study examines graph-based NLP algorithms' efficiency, functionality, and comprehension. The competence that can be extracted from the methodologies covers elements like summarizing, textual inference, elimination of duplication, measurement of resemblance, and tagging. NLP and NLU research has been driven by data and the demand for fast access to accurate and complete information. This has led to the invention of features including event resolution (ER), grammar annotation (GrA), information mining (IM), labelling (Lab), novelty recognition (ND), question/answer (QA), semantic relatedness (SR), and similarity measure (SM). Concepts, connections, and qualities can be used to indicate how a text might be read at a semantic and abstract level. One goal that might be possible with a higher degree of (or semantic) representations [26] is natural language understanding, which is also known as NLU.

2.7 Part of Speech Tagging

Part-of-speech (POS) is a key NLP concept and commonly utilised application [27]. POS Tagging [28–30] is a big problem in NLP that has to do with giving eevery word in a text the right syntactic tag based on where it appears. Words such as verbs, adjectives, adverbs, nouns, and adverbs are all examples of parts of speech (POS). Machine translation, figuring out what a word means, parsing questions, and other NLP tasks depend heavily on properly tagged POS data.

Tagging words with their parts of speech manually is a time-consuming, costly, and error-prone procedure, which is why there is growing interest in automating the process [31]. According to Pisceldo et al. [28], the most pressing issue in POS tagging is ambiguity. Words across most languages respond differently depending on the context in which they are used, making it difficult to determine the proper tag for a given word. Automatic POS tagging has been implemented using a variety of methods, including those based on transformations, rules, and probabilities. Words are categorised as nouns or verbs depending on the linguistic rules programmed into the rule-based part of speech taggers [32]. Deep learning (DL) and machine learning (ML) have been tested as POS taggers. Both ML and DL, which fall under the general heading of "artificial intelligence," attempt to glean useful knowledge from large corpora of text [33].

Artificial neural networks (ANNs) are a kind of learning algorithm that takes its cues from biological neural networks; ANNs are often used to estimate functions whose

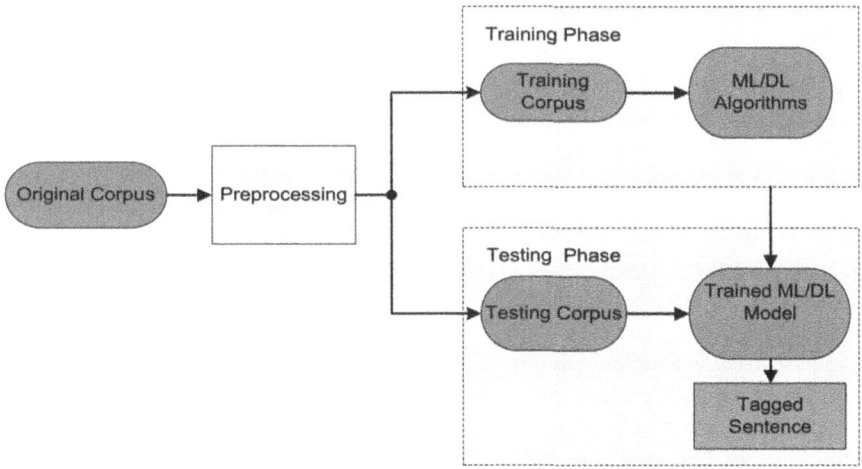

Fig. 1. Machine Learning/ Deep Learning Based POS Model [34].

exact inputs are unknown, but which may depend on many other variables [35]. These neural networks are adaptable to new information sources and ready to learn because the numerical weights on the connections between neurons may be altered based on experience. It's a collection of millions of interacting neurons for managing information and solving problems (Fig. 1).

2.8 Biomedical Named Entity Recognition

Named-entity recognition (NER) is a subtask of text analytics that finds and classifies nouns by name. Named Entity Recognition is often followed by Relation Detection (RD) [36] in the biomedical domain, which involves establishing connections between different biological entities in order to identify important relationships that may be investigated further. There is a combinatorial explosion among biological entities since there are so many distinct types of named entity classes. Therefore, it would be very expensive and time-consuming to use biological studies to identify which of these correlations is the most important. Nonetheless, by utilising computer methods to parse millions of biomedical research papers, millions of such relationships may be identified for use in developing networks. Specifically, protein-protein interaction networks may be built by tagging the connections between proteins.

Although "named entity recognition (NER)" and "relation detection (RD)" are necessary for several text mining-based machine learning applications, there is a dearth of survey publications that give in-depth explanations of these processes. Most reviews [37, 38] don't devote much attention to biomedical-specific methods for entity relation identification, or NER. A comprehensive system for documenting named entities must define NE types, provide correct class suggestions, overcome semantic challenges like metonymy and multi-class entities, and record NE restrictions [39]. Nevertheless, there are many more challenges to solve while creating a biomedical NER system than when working on generic NER [39]. Many of these problems go beyond feature extraction and system assessment since they are syntactic and semantic difficulties unique to certain domains.

2.9 Text Summarization

Abstractive summaries, as contrasted with extractive summaries, construct new sentences, or paraphrases, that use terms not found in the source text to summarise the material. Producing abstractive summaries involves considerable natural language processing, making them far more complicated and demanding than extractive summaries [40]. More recent studies on abstractive summarization have been influenced by the encoder-decoder paradigm, as shown by the work in [41, 42].

Because excellent summarization relies on careful text analysis, including semantic and lexical analysis, it presents a significant challenge to the discipline of NLP [43]. Non-redundancy, relevancy, completeness, consistency, and legibility are just a few of the other factors that need to be considered to create a well-written summary [44]. In 2018, there were 18 studies published on the topic of text summarization. In 2015, there was an enormous uptick in this study's research, with 15 papers (out of 85 total) making the cut. Extractive summaries remain a contentious topic in light of the review's conclusions that they are simpler than abstractive summaries (which are notoriously difficult). Figure 2 illustrates development roadmap of NLP form 2001 to present days.

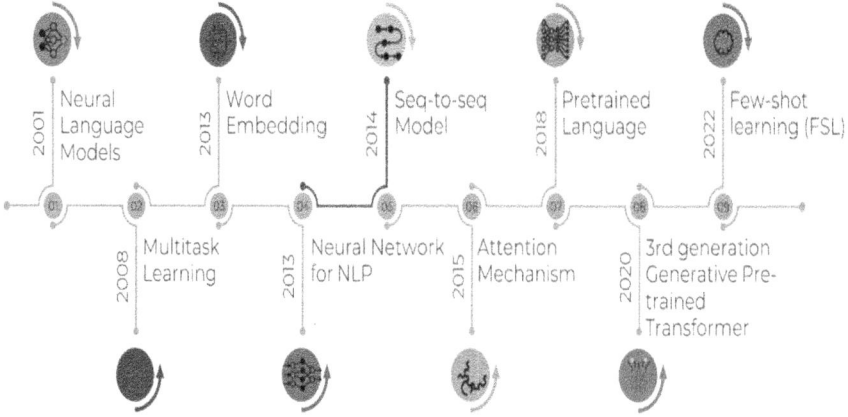

Fig. 2. NLP Development Roadmap

3 Challenges

Despite a great deal of recent progress, new difficulties keep cropping up as NLP finds more and more applications. Humans understand that words and phrases have different meanings based on context, but robots don't. Humans employ many words to indicate the same thing, making it hard to understand and design algorithms that can account for them all. "Large," "huge," and "giant" are used to express the same size, for instance. Question answering and speech-to-text applications also struggle with homophones, which sound same but signify different things.

Further, there is room for improvement in the degree to which phrases in the language are open to more than one interpretation [45]. Words have diverse meanings in different places, and languages with informal phrases, expressions, idioms, and culture-specific jargon make it hard to design models for general use. In addition to the challenges mentioned above, misspelt or misused words can cause problems. Despite continual improvements in auto-correct and grammar-correction systems, guessing the writer's purpose from a certain domain or geographic location by considering sarcasm, expressions, informal words, and so on is a significant difficulty. One may also consult [46] for a categorization of online harassment types and difficulties, [47] and [48] for public health issues and possibilities, [49] for a comprehensive literature survey and technological issues relevant to management research and NLP, and a recent review work by [50].

Models including visual commonsense reasoning [51] and NLP have recently attracted the attention of various academics, and this field of study is both promising and hard. These models use a visual reasoning paradigm to extrapolate information from images and videos beyond what is immediately apparent, such as the purposes of items, the motivations of individuals, and the emotional states of actors. Recently, [52] proposed a model to collect information from multiple viewpoints and anticipate common sense; the results of the tests done on the visual commonsense reasoning dataset VCR look highly satisfying and successful, suggesting that this route is fruitful.

4 Proposed Approach

Over the past few years, researchers in the field of machine translation (MT) have presented techniques to combat the issue of data sparsity and enhance the performance of neural machine translation (NMT) systems in low-resource circumstances. Context-aware machine translation techniques have not been thoroughly examined for low-resource language pairs.

4.1 Motivation

Machine translation has been studied extensively, however most of the Indian language research have focused on translating Hindi using traditional machine translation methods. In addition to this, it has been found that very little work is being done on Indian languages because of a lack of resources.

4.2 Kannada to English Machine Translation

To date, research on the use of machine translation for several Indian languages has been limited. Information and concepts provided in one language can only be understood through translation. Because of its syntactic and semantic diversity, Kannada, one of the Dravidian languages, has a rich historical literature but a lack of resources in computational linguistics. When compared to other Indian languages, Kannada has received less attention from MT researchers. Many studies have been conducted with the goal of improving machine translation from English to South Dravidian languages like Kannada and Malayalam.

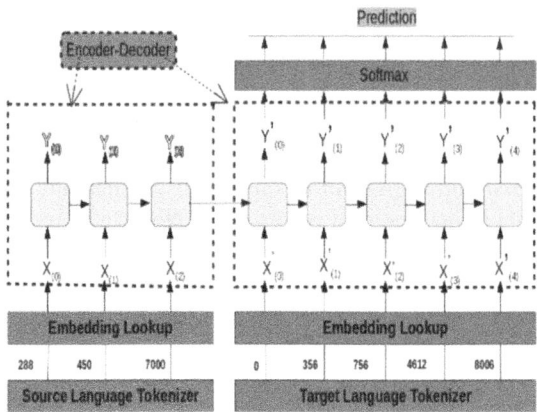

Fig. 3. Proposed Architecture for "Kannada – English" Machine Translation.

The suggested architecture for the MT procedure is shown as a block diagram in Fig. 3. Tokenizers break down sentences into individual words and assign a numeric value to each distinct word in the input or target language. For use as input by the LSTM cell, these integers are transformed into vectors. The encoder and decoder components of "Seq2Seq" mode are shown by the yellow blocks.

MT using a Seq2Seq model works well with an encoder-decoder architecture where both the encoder and the decoder are made of LSTM subunits. An encoder vector is first generated by connecting a sequence of LSTMs in the encoder unit. The encoder's final LSTM's internal state value is the encoder's vector, which contains data about the input components that came before it. The first LSTM decoder will use this vector as its starting state value, improving its prediction accuracy. Both the training and inference processes use encoders in the same way.

The platform on which the model must be trained will be determined by the size of the corpus. In machine learning, the model training may be done on several different platforms. Our plan is to conduct our training on Google Collab (an open-source platform) so that we may take advantage of its powerful graphics processing unit and tensor processing unit at no cost to our machine learning research.

A proposed metric for comparing anticipated sentences to their intended targets is the Bi-Lingual Evaluation Study (BLEU) score. If there is an exact match, the outcome will be 1, and otherwise it will be 0.

5   Conclusion

The research and use of training computers to comprehend and make sense of input data from human conversations is referred to as "natural language processing." Establishing channels of communication among people and computers may be accomplished via the use of NLP that is built on machine learning. Even though it is always undergoing development, NLP has already demonstrated its value in a variety of areas. NLP may be used in a variety of ways, each of which can spare organisations and people time and help them become more efficient while also raising customer satisfaction levels. The relative success of natural language processing, in particular in applied areas, may be attributed both to the ongoing development of technology and the present-day historical context.

Emotional intelligence and opinion mining are becoming indispensable tools for professionals in marketing strategy as well as communications managers for public figures in the arts and politics. Natural language processing exhibits interdisciplinarity. It draws from computer science, artificial intelligence, linguistics, cognitive psychology, and neuroscience. The linguistic models that are produced because of this will serve as a catalyst for the development of subsequent generations of natural language processing systems that will evolve to be much more modular and, therefore, less dependent on any specific language.

References

1. Liu, D., Fu, J., Qu, Q., Lv, J.: BFGAN: backward and forward generative adversarial networks for lexically constrained sentence generation (2019)
2. Hokamp, C., Liu, Q.: Lexically constrained decoding for sequence generation using grid beam search. In: Proceedings of 55th Annual Meeting of the Association for Computational Linguistics, pp. 1535–1546 (2017)
3. Wu, Y., et al.: Google's neural machine translation system: bridging the gap between human and machine translation. arXiv preprint arXiv:1609.08144 (2016)
4. Dong, L., Huang, S., Wei, F., Lapata, M., Zhou, M., Xu, K.: Learning to generate product reviews from attributes. In: Proceedings of Annual Meeting of the Association for Computational Linguistics, pp. 623–632 (2017)
5. See, A., Liu, P.J., Manning, C.D.: Get to the point: summarization with pointer-generator networks. In: Proceedings of Annual Meeting of the Association for Computational Linguistics, pp. 1073–1083 (2017)
6. Liu, T., Wang, K., Sha, L., Chang, B., Sui, Z.: Table-to-text generation by structure-aware seq2seq learning. In: Proceedings of AAAI Conference on Artificial Intelligence, pp. 4881–4888 (2018)
7. Ghosh, S., Chollet, M., Laksana, E., Morency, L.-P., Scherer, S.: Affect- LM: a neural language model for customizable affective text generation. In: Proceedings of Annual Meeting of the Association for Computational Linguistics, pp. 634–642 (2017)
8. Och, F.J., Ney, H.: The alignment template approach to statistical machine translation. Comput. Linguist. **30**, 417–449 (2004)
9. Wen, T.-H., et al.: Stochastic language generation in dialogue using recurrent neural networks with convolutional sentence reranking. In: Proceedings of Annual Meeting Special Interest Group Discourse Dialogue, pp. 275–284 (2015)

10. Yin, J., Jiang, X., Lu, Z., Shang, L., Li, H., Li, X.: Neural generative question answering. In: Proceedings of International Joint Conference on Artificial Intelligence, pp. 2972–2978 (2016)
11. Agarwal, A., Baechle, C., Behara, R., Zhu, X.: A natural language processing framework for assessing hospital readmissions for patients with COPD. IEEE J. Biomed Health Inform **22**, 588–596 (2018)
12. Ferrucci, D., Lally, A.: UIMA: an architectural approach to unstructured information processing in the corporate research environment. Nat. Lang. Eng. **10**(3/4), 327–348 (2004)
13. Agarwal, A., Behara, R.S., Mulpura, S., Tyagi, V.: Domain independent natural language processing – a case study for hospital readmission with COPD. In: Proceedings of the 2014 International Conference on IEEE Bioinformatics and Bioengineering, pp. 399–404 (2014)
14. Wasfy, J.H., et al.: Enhancing the prediction of 30-day readmission after percutaneous coronary intervention using data extracted by querying of the electronic health record. Circ. Cardiovas. Q. Outcomes **8**(50), 477–485 (2015)
15. Hu, Z., et al.: A natural language process-based framework for automatic association word extraction. IEEE Access (2020)
16. Jensen, P.B., Jensen, L.J., Brunak, S.: Mining electronic health records: towards better research applications and clinical care. Nat. Rev. Genet. **13**(6), 395–405 (2012)
17. Kourou, K., Exarchos, T.P., Exarchos, K.P., Karamouzis, M.V., Fotiadis, D.I.: Machine learning applications in cancer prognosis and prediction. Comput. Struct. Biotechnol. J. **13**, 8–17 (2015)
18. Goff, D.J., Loehfelm, T.W.: Automated radiology report summarization using an open-source natural language processing pipeline. J. Digit. Image. **31**(2), 185–192 (2018)
19. Huhdanpaa, H.T., et al.: Using natural language processing of free-text radiology reports to identify type 1 modic endplate changes. In: J. Digit. Image. **31**(1), 84–90 (2018)
20. Kang, S.K., et al.: Natural language processing for identification of incidental pulmonary nodules in radiology reports. J. Amer. College Radiol. **16**(11), 1587–1594 (2019)
21. Kehl, K.L., et al.: Assessment of deep natural language processing in ascertaining oncologic outcomes from radiology reports. JAMA Oncol. **5**(10), 1421 (2019)
22. Liu, H., et el.: A natural language processing pipeline of Chinese free-text radiology reports for liver cancer diagnosis. IEEE Access **8**, 159110–159119 (2020)
23. Loebner, S.: Understanding Semantics. 2nd edition. Routledge, UK (2013)
24. Mills, M.T., Bourbakis, N.G.: Graph-based methods for natural language processing and understanding—a survey and analysis. In: IEEE (2013)
25. Ambwani, G., Davis, A.R.: Contextually mediated semantic similarity graphs for topic segmentation. In: Proceedings of Workshop Graph based Methods Natural Language Process, pp. 60–68 (2010)
26. Duarte, R.M., Du Bois, A.R., Pilla, M.L, Cavalheiro, G.G.H., Reiser, R.H.S.: Comparing the performance of concurrent hash tables implemented in Haskell. Sci. Comput. Program. **173**, 56–70 (2019)
27. Demilie, W.B.: Analysis of implemented part of speech tagger approaches: the case of Ethiopian languages. Indian J. Sci. Technol. **13**, 4661–4671 (2020)
28. Singh, J., Joshi, N., Mathur, I.: Part of speech tagging of marathi text using trigram method. Int. J. Adv. Inf. Technol. **3**(2), 35–41 (2013)
29. Chungku, C., Rabgay, J., Faaß, G.: Building NLP resources for Dzongkha: a tagset and a tagged corpus. In: Proceedings of the 8th Workshop on Asian Language Resources, pp. 103–110 (2010)
30. Singh, J., Joshi, N., Mathur, I.: Development of Marathi part of speech tagger using statistical approach. In: Proceedings of 2013 International Conference on Advances in Computing, Communications and Informatics, ICACCI, pp. 1554–1559 (2013)

31. Pisceldo, F., Adriani, M.R., Manurung R.: Probabilistic part of speech tagging for Bahasa Indonesia. In: Proceedings of the 3rd International MALINDO Work. Coloca. Event ACL-IJCNLP (2009)
32. Kurniawan, K., Aji, A.F.: Toward a standardized and more accurate Indonesian part-of-speech tagging. In: Proceedings of International Conference on Asian Language Processing (IALP) (2018)
33. Alzubaidi, L., et al.: Review of deep learning: concepts, CNN architectures, challenges, applications. Future Direct **8**, 1–8 (2021)
34. Chiche, A., Yitagesu, B.: Part of speech tagging: a systematic review of deep learning and machine learning approaches. J. Big Data **9**, 1–25 (2022)
35. Chen, T.: An innovative fuzzy and artificial neural network approach for forecasting yield under an uncertain learning environment. J. Ambient Intell. Humaniz Comput. **9**(4), 1013–1025 (2018)
36. Bach, N., Badaskar, S.: a review of relation extractions (2007)
37. Goyal, A., Gupta, V., Kumar, M.: Recent named entity recognition and classification techniques: a systematic review. Comput. Sci. Rev. **29**, 21–43 (2018)
38. Song, Q.: An overview of reciprocal l 1-regularization for high dimensional regression data. Wiley Interdiscipl. Rev. Comput. Stat. **10**, e1416 (2018)
39. Marrero, M., Urbano, J., Sánchez-Cuadrado, S., Morato, J., Gómez-Berbís, J.M.: Named entity recognition: fallacies, challenges and opportunities. Comput. Stand. Interf. **35**, 482–489 (2013)
40. Gambhir, M., Gupta, V.: Recent automatic text summarization techniques: a survey. Artif. Intell. Rev. **47**, 1–66 (2017)
41. Xu, W., Li, C., Lee, M., Zhang, C.: Multi-task learning for abstractive text summarization with key information guide network. EURASIP J. Adv. Signal Process. (2020)
42. Lee, H., Choi, Y., Lee, J.H.: Attention history-based attention for abstractive text summarization. In: Proceedings of the ACM Symposium on Applied Computing, pp. 1075–1081 (2020)
43. Rane, N., Govilkar, S.: Recent trends in deep learning based abstractive text summarization. In: Int. J. Recent Technol. Eng. **8**, 3108–3115 (2019)
44. Verma, P., Om, H.: MCRMR: Maximum coverage and relevancy with minimal redundancy based multi-document summarization. Expert Syst. Appl. **120**, 43–56 (2019)
45. Khurana, D., et el.: Natural language processing: state of the art, current trends and challenges. Multimedia Tools Appl. 1–32 (2022)
46. Sharifirad, S., Matwin, S.: When a tweet is sexist. A more comprehensive classification of different online harassment categories and the challenges in NLP. arXiv preprint arXiv:1902.10584 (2019)
47. Oliver, B., et al.: Challenges and opportunities for public health made possible by advances in natural language processing. Can Commun. Dis. Rep. 161–168 (2020)
48. Wong, A., Plasek, J.M., Montecalvo, S.P, .Zhou L.: Natural language processing and its implications for the future of medication safety: a narrative review of recent advances and challenges. Pharmacotherapy. J. Hum. Pharmacol. Drug Ther. **38**, 822–841 (2018)
49. Kang, Y., Cai, Z., Tan, C.W., Huang, Q., Liu, H.: Natural language processing (NLP) in management research: a literature review. J. Manage. Anal. **7**, 139–172 (2020)
50. Alshemali, B., Kalita, J.: Improving the reliability of deep neural networks in NLP: a review. Knowl. Based Syst. 1–19 (2020)

51. Davis, E., Marcus, G.: Commonsense reasoning and commonsense knowledge in artificial intelligence. Commun ACM **58**, 92–103 (2015)
52. Wen, Z., Peng, Y.: Multi-level knowledge injecting for visual commonsense reasoning. IEEE Trans. Circ. Syst. Video Technol. **31**(3), 1042–1054 (2020)

Neural Machine Translation in Low-Resource Context: Survey

Padma Prasada[1,2]([⊠]) [iD], M. V. Panduranga Rao[1] [iD],
and Ujwala Vishwanatharao Suryawanshi[3] [iD]

[1] Faculty of Engineering and Technology, JAIN (Deemed to be University), Kanakapura Road,
Bengaluru, India
ppjain15@gmail.com
[2] Department of Artificial Intelligence & Data Science, Sri Dharmasthala Manjunatheshwara
Institute of Technology, Ujire, Karnataka, India
[3] Rajarshi Shahu Mahavidyalaya (Autonomous), Latur, Maharashtra, India

Abstract. Several approaches to solving challenges with limited resources in neural machine translation are explored in this survey. In particular, the case of English-Kannada NMT is used as an example. For NMT systems to make good translations, they need a lot of parallel corpora. The English-Hindi parallel corpus is used by the one-to-many (English to Hindi and Malayalam) multilingual model to enhance the quality of English-Kannada translations. For improving parallel data, techniques like phrase table injection, back-translation, and combined corpus methods are used. For improving transfer learning, techniques like pivoting and multilingual embeddings are used. When translating from English to Kannada, Hindi can be used to help with pivoting. In this article, a detailed case study is done with respect to machine translation between English and a Kannada language. The various Machine translation evaluation metrics are reviewed.

Keywords: Low resource language · Combined Corpus method · Neural machine translation · MT evaluation metrics · Transfer learning

1 Introduction

In addition to its linguistic diversity, India also has a wide variety of cultural traditions and landscapes. It is home to 780 different languages, 22 of which are officially recognised (as per the Constitution of India). Over a million people speak one of the 22 scheduled languages and 10 of the 99 non-scheduled ones. Most of these languages are classified as low-resource languages; however, this variety may be a result of, or an explanation for, the lack of a centralised language resource.

Five distinct language families can be identified here: the Indo-Aryan group (which includes languages like Hindi, Punjabi, Gujarati, Marathi, and Bengali) the Dravidian group (which includes languages like Tamil, Telugu, Malayalam, and Kannada), the Austro-Asian group (which includes languages like "Khasi" and "Munda"), and the Sino-Tibetan group (e.g., "Manipuri" and "Bodo") [18]. Such language variety necessitates

© The Author(s), under exclusive license to Springer Nature Switzerland AG 2025
M. Gupta et al. (Eds.): MISS 2023, CCIS 1952, pp. 169–180, 2025.
https://doi.org/10.1007/978-3-031-69115-7_11

the need for translation services to facilitate dialogue across India's many different states, which has never been a topic the administration has aspired to avoid. The advent of neural techniques in machine translation (MT) has led to substantial development for European languages (MT) [1, 2].

Researchers have shown that translation quality across Indic languages may be improved by using lexical and orthographic parallels despite the scarcity of available parallel corpora [3]. While Hindi and Marathi are connected, English and Kannada have no common heritage and are therefore unrelated. As a bonus, the English-Hindi corpus is far larger than the other English-to-Indic language corpora. In this work, we seek to leverage the English-Telugu language combination in several possibilities to improve English-Kannada translation as part of our multilingual experiments. These include transfer learning based on pivots, merged corpora, and combined resources.

1.1 Characteristics of Indian Languages

In 2011, 78.05% of Indians spoke an Indo-Aryan language, whereas 19.64% spoke a Dravidian language, according to the Census. The remaining 2.31% of the population speaks languages from families such as Austroasiatic, Sino-Tibetan, Tai-Kadai, and some small language isolates. Languages spoken in India (henceforth IL) have their own unique traits that impact their capacity to translate amongst one another and to other languages. Most Indian languages are SOV (subject-object-verb), but others, like English, are SVO (subject-verb-object). The VO-OV transformation must thus be applied by the translator, whether it be human or machine.

2 Related Work

When compared to earlier recurrent neural network (RNN)-based techniques, the Transformer model presented in 2017 significantly enhanced the translation quality [4]. Using self-awareness and doing away with recurrent layers allowed for more rapid model training and improved results. In a low-resource context, however, this did not enhance translation quality. Several solutions to the problem of limited NMT resources have been proposed over time. Techniques that work with monolingual data include utilising an autoencoding target [5], supplementing pseudo-parallel data with back-translation [6], or incorporating a separately learned language model into the decoder.

After performing research on a large number of language pairings in a constrained environment, Renduchintala et al. (2018) found that there is no classic ideal Byte-Pair Encoding (BPE) positioning for their long short-term memory networks architecture [7]. No one seems to agree on the optimal value for the merging operation count hyperparameter when it comes to BPE implementation. Researchers tested data from the LSTM and Transformer and shared their work at IWSLT 2016, which included translations from and into English from Arabic, Czech, French, and German. They found that there is no commonly accepted best BPE configuration for LSTMs but that for transformer designs, reducing the number of BPE merging operations is usually best [8]. The authors conclude that for Indian languages, BPE is preferable to the orthographic syllable as translation element for SMT across various language groups [9].

The author presented a full-stack learning technique that makes use of an external phrase memory to keep accurate translations between words. The data is pre-processed by using their phrase table, which is effectively a set of criteria for separating phrases from non-phrases in both the source and target sentences. A redesigned decoder that could switch between word mode and phrase mode was presented [10]. The research team proposed the phrase "table" as a recommendation memory for NMT systems.

To help NMT systems generate more accurate predictions, they introduce a new method for identifying phrase-level target words that merit recommendation and computing and using those recommendation scores. At each decoding stage, they start by creating a word suggestion set by matching a source sentence to a phrase translation table. Next, they include each suggested term into the NMT procedure and provide it a weighted bonus [11]. They have shown a considerable improvement in their abilities to translate between Chinese and English and between English and Japanese. The authors discuss their findings from experiments using morpheme-based and BPE-based classification, concluding that former yields better results for very dissimilar language pairings [12]. Researchers proposed a M-BPE, in which Morfessor is used for morpheme segmentation, and after that BPE is applied to the segmented morphemes as text elements. Three language pairs are used as test cases by the researchers.

In ILNMT, which comprises Bengali, Gujarati, Marathi, Malayalam, and Tamil, transfer learning is substantially impaired when there is no or little parallel corpus between the source and target languages [13]. To reconcile this discrepancy, the authors pre-order the phrase in the helping language so that its word order is identical to that of the source language before training the parent model. Despite of a neural net's shallowness, [14] perform exceptionally well on Urdu-Hindi, Punjabi-Hindi, and Gujarati-Hindi pairings. Similarly, [15] shows that using backtranslation with a filtered back-translation technique and then fine-tuning on to a small pair-wise language corpus may greatly boost translation model performance. Tamil, Malayalam, and Mizo are also being examined.

3 Approaches

Different approaches to the issue of low-resource language pairs are described in length in this section.

3.1 Phrase Table Injection (PTI)

Recently, neural machine translation (NMT) has demonstrated impressive performance on publicly accessible benchmark datasets, leading to its fast adoption in a wide range of industrial applications. Due to time, resources, and competence, a big, high-quality parallel corpus is not always attainable [16]. Sennrich et al. incorporated target-side monolingual data to test two ways for filling the source side. The first strategy replaced a fake source sentence for each target text sentence, whereas a second approach utilised back-translation to construct a synthetic source phrase [17]. Several approaches to incorporating vast amounts of source-side monolingual data into NMT were investigated. The first method takes its cue from [18], in which a baseline model is constructed before parallel synthetic data is created through the translation of monolingual dataset.

This supplementary data was utilised in conjunction with the primary data to train an attention-based GRU model [18]. The author suggests a natural language translation (NMT) paradigm that uses SMT phrase feedback to improve. Because the phrase table, created during SMT model training, is so important in SMT translation, It includes a probabilistic phrase-to-phrase mapping from the input language to the output language [19]. Combining phrase table phrases with parallel corpora increases NMT model training data. This aids to model's efforts in learning the right translation of both lengthy and short phrases.

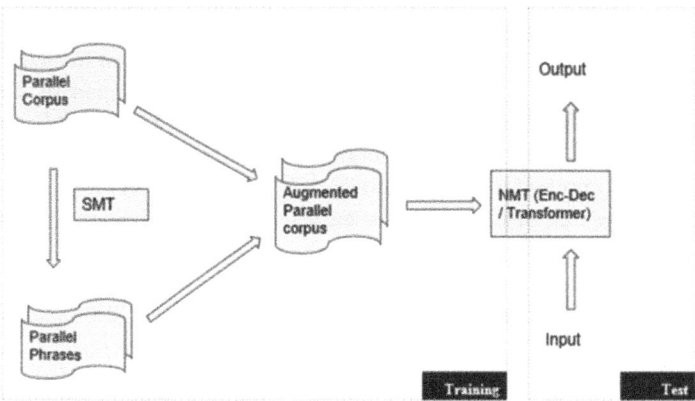

Fig. 1. Phrase Table Injection Approach [17].

Figure 1 depicts the overall architectural process flow. The primary concept is to supply more data on the degree to which the source and target terms agree with one another. Before passing on a sentence pair, encoder-decoders delete all metadata about its source and destination phrases. The approach naturally predicts and corrects translation mappings over a massive parallel corpus. Nonetheless, when the size of the corpus is limited, the model is unable to pick up on the connection between the words. To create the impression of a bigger corpus, feeding phrase pairs as the training examples has a significant effect.

SMT translation technique relies on the phrase table established during SMT model training. A probabilistic phrase-to-sentence mapping between the two languages is included. When the phrase table's phrases are added to those from parallel corpora, more data is made available for the NMT model to learn from. This aids the model in learning to translate both lengthy and short statements correctly.

3.2 Expanding Data Using Back-Translation

To boost the quality of translations for low-resource language pairings, back-translation makes use of monolingual data in the target language. Most target language monolingual data outnumber parallel data. This monolingual data can train a statistical machine translation (SMT) language model to create natural-sounding translations. Using back-translation, NMT may contain the capability of exploiting the monolingual data for training.

These monolingual data may train an SMT language model for natural-sounding translations. Back-translation is a procedure that may be used to include the use of monolingual data for training into NMT.

An encoder-decoder model embedded within a transformer design is the basis of back translation. The method of paraphrasing employs machine translation models (as seen in Fig. 2), which convert material into an intermediary language before returning it to its original context [20].

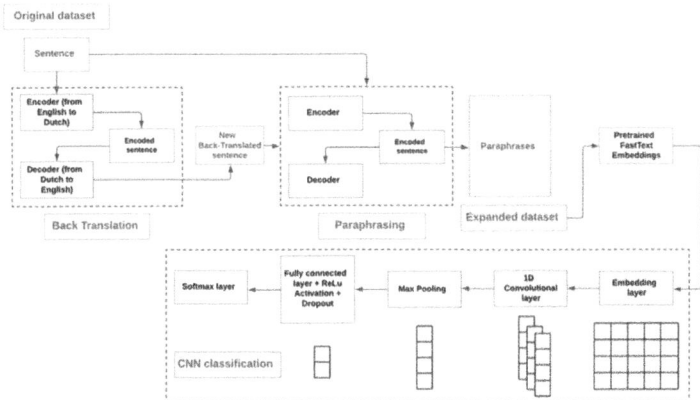

Fig. 2. Back translation and paraphrase are the general methods used to create an extended dataset from the original dataset [20].

The author used the pretrained English-Dutch translation model to perform back-translation-based data augmentation. New, enhanced data is created by translating data from assessed datasets into Dutch and back into English. Data augmentation was carried out with a paraphrase strategy based on a combination of experts and the transformer model. Using this method, it was able to produce several different yet semantically similar paraphrases of a given statement.

Data augmentation, based on visual or semantic invariance, is the easiest and most straightforward way to improve model performance. The model's generalisation ability and performance can improve while utilising fewer data sets, thanks to the data augmentation approach [21]. The author suggests using back-translation to enlarge the text data by mimicking the appropriate processing methods for picture enlargement [22].

A Kannada-English NMT model is first trained using the accessible parallel corpus. After this model is developed, it is utilised to transform the Kannada monolingual data into a pseudo-parallel corpus, which is then used in conjunction with the existing parallel corpora to train the NMT model.

3.3 Combined Corpus Method

The quantity of parallel corpora available for training deep neural network-based models classifies various Indian languages spoken on the subcontinent as low-resource languages. Unfortunately, there are no adequate automated translation systems for these languages due to a lack of resources. The acquisition of a parallel corpus allows for the creation of additional resources, such as cross-lingual word embeddings [27] and sentence embeddings [28].

In the combined corpus method, we draw upon information from related target languages. In Fig. 3, we see the NMT model being trained with a hybrid set of English-Marathi and English-Hindi corpora before being fine-tuned using solely the English-Marathi parallel corpora. The hypothesis is that a model that can translate between different languages right from the beginning would do better than one whose weights were randomly assigned to start training.

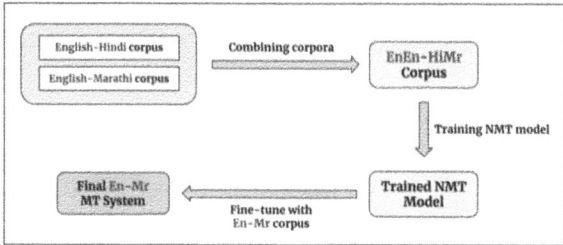

Fig. 3. Combined corpus model for English-Marathi Translation [29].

If the languages on the target side are comparable, there will be a greater chance of a vocabulary overlap, making this approach more useful. The two languages being spoken with here, Kannada and Telugu, are related because they have a common etymological root (Indo-Aryan) and use some of the same letters in their alphabets.

4 Machine Translation Approaches for Kannada

The many studies conducted to achieve machine translation of the Kannada language are covered here.

4.1 Dictionary Based Method

For languages like Kannada and Telugu, which have fewer linguistic resources, a dictionary-based approach is the ideal method for machine translation. This approach uses a dictionary to accomplish translations word by word, with little regard for the semantic connections between words. The morph analyzer, dictionary, transliteration, transfer grammar, and morph generator are all subprocesses that make up dictionary-based machine translation [23].

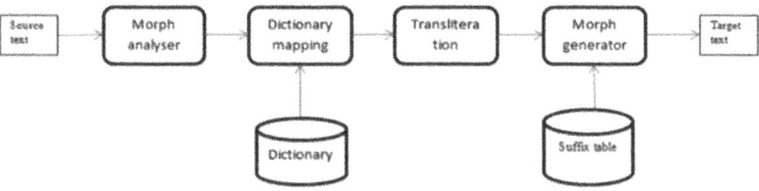

Fig. 4. Dictionary Based Machine Translation Model.

Due to limitations in dictionary coverage, dictionary mapping cannot provide an accurate translation in all circumstances. Using a combination of the dictionary as well as transfer grammar principles, machine translation can create more natural-sounding results. The target word is generated using a suffix mapping table and transfer grammar rules specified at the POS (parts of speech) tag level. Morph analyzer, bilingual dictionary, suffix mapping table, and morph generator are the components of the dictionary-based method, as shown in Fig. 4.

A word is fed into a morphological analyzer, and its analysis is what comes out the other end. Word morphology is captured in the output. The accuracy of a machine translation system improves with a larger number of entries in a bilingual lexicon, which is why dictionaries including both languages are essential.

4.2 Phrase Mapping Technique

The approach begins with a lexical analysis of the transliterated Kannada phrase as input, utilising the recently developed source language lexicon Kannada Word Net to identify and categorise tokens. The Kannada phrase is then checked against the syntactic constructions and grammatical rules using a domain-specific recursive descent parser. Additionally, the output generating techniques and phrase mapper are used to obtain the English sentence translation [24]. The suggested technique translates basic transliterated Kannada sentences into English using a WordNet lexicon, a recursive descent parser, a phrase mapper, and output-generating processes.

When given a transliterated Kannada sentence, the lexical analyzer breaks it down into its component lexical elements. Nouns, adjectives, conjunctions, and punctuation make up the lexical elements of Kannada sentences. The Kannada sentences that have been transliterated are checked using a recursive descent parser. For each syntactic construction in the language, the recursive descent parser employs a unique method to validate sentence syntax phrase by phrase. Parsers also specify the right sentence by rearranging words in the sentence to adhere to the grammatical rules supplied by the user. After the recursive descent parser parses the transliterated Kannada text, the next step is to construct an English sentence. The Phrase Mapper takes a parsed sentence and maps each phrase (adjective, noun, etc.) to a grammatically compatible phrase in the target English language.

4.3 Word Embedding

Word embeddings, in a brief, are a high-density representation of words as vectors in a 2- or 3-dimensional space. For each of these embeddings, there is a corresponding learnable vector, or parametrized function. During backpropagation, they use a loss function to be constantly updated as it searches for an appropriate pair of words to pair together while still maintaining their semantic and synaptic qualities.

As more and more languages become available online, cross-lingual information retrieval has become a hot issue in the field of natural language processing and data retrieval. One study [25] tried to convert inquiries in Kannada and English scripts into standard English queries.

5 Evaluation Metrics

In its most basic sense, "evaluation" refers to the process of determining the accuracy of a claim. Each new technological advancement requires some sort of testing or evaluation. The evaluation of an MT system is just as crucial as the MT itself since it provides answers to critical questions concerning the quality of the translation (its accuracy, fluency, and acceptance) and, in doing so, humanises the underlying MT algorithm. Testing and grading MT systems has been a challenging task for a long time. Because translation is more of an art than a science, there are often several viable options for rendering a given statement into another language.

When comparing machine translations to reference translations created by humans, a parameter called the **BLEU** score is used to determine overall quality. Accuracy in terms of n-grams relative to the reference translation is calculated, but grammatical accuracy is ignored. Those translations that are too brief incur a brevity penalty. Since it was developed to resemble human judgement on a corpus level, the BLEU score metric fares badly when applied to judging individual translated sentences. It disregards the semantics of language. When communicating with another human being, it is fine to use a different term with the same meaning. However, according to Bleu Score, that is the wrong term to use. The lines "The guard came late because of the rain" and "The rain arrived late because of the guard" would both receive the same (unigram) score, despite the obvious differences in meaning between the two. Even though the BLEU score is technically a number between 0 and 1, it is most commonly presented as a percentage score obtained by multiplying the value by 100. In terms of translation quality, a higher value is preferable. One of our assessment metrics is the BLEU score, which is majorly adopted by the MT society. For out-of-vocabulary (OOV) terms, the machine translation output frequently displays Unknown Word (UNK). It is often agreed that the quality of the translated sentences improves when fewer UNK terms are included in the final product.

BLEU uses a modified n-gram precision metric, which penalises translation candidates whose texts are longer than their sources. To allow for some flexibility at the sentence level, the brevity penalty is assessed throughout the whole corpus. Following is a [30] calculation of the brevity penalty (BP):

$$BP = \begin{cases} 1, & r < c \\ e^{(1-r/c)}, & \text{otherwise} \end{cases} \qquad (1)$$

Furthermore, the BLEU score for the whole test set is determined by applying following equation [30] to the BLEU score for each individual test set.

$$BLEU = BP \times \exp\left(\sum_{n=1}^{N} w_n \log P_n\right) \tag{2}$$

In this case, w_n's are taken to be the positive weights adding up to one, and P_n is defined as follows:

$$P_n = \frac{\sum_{C \in \text{Candidates}} \sum_{n-gram \in C} \text{Count}_{clip}(n\text{-gram})}{\sum_{C' \in \text{Candidates}} \sum_{n\text{-gram} \in C} \text{Count}(n - gram')} \tag{3}$$

The first step in determining the BLEU score is to determine the precision of the modified n-grams throughout the full test corpus. This may be done by solving Eq. 3 to determine the modified n-gram precision score, P_n.

Word Error Rate (WER) is a statistics-based evaluation metrics employed to evaluate the accuracy of the MT system. It compares the quality of machine translation and human translation. The lower the WER, the greater the translation. It does a word-by-word (or character-by-character) comparison of the anticipated output and the target transcript to count the number of discrepancies, illustrated in Fig. 5. Words that exist in the transcript but aren't in the prediction (counted as a deletion), words that don't exist in the transcript but are in the prediction (counted as an insertion), and words that are changed between the prediction and the transcript all count as differences (a substitution).

Fig. 5. Word Error Rate counting deletion, insertion, and substitution.

However, WER does not differentiate between words that are crucial to the content of the sentence and those that are not, making it difficult to use as a measure for speech recognition.

The F-score is the mathematical middle ground between accuracy and recall. Contrast this with recall, which evaluates how comprehensively remembered information is. A higher F-measure indicates a more accurate translation. If you're interested in computing the F-measure, the source code to do so is publicly available. This formula may be used to get an F-score, which is the harmonic mean of the system's precision and recall value; $2 \times [(\text{precision} \times \text{recall})/(\text{precision} + \text{recall})]$.

The use of F-scores to evaluate the efficacy of a predictive system has been criticised on the grounds that they may not provide a complete picture of the system's performance because even a relatively high F-score may come from an insufficient emphasis on either precision or recall. However, after quality has been established, most systems run into trouble when attempting to boost either the precision or the recall metric without sacrificing the other.

METEOR considers a harmonic mean of unigram recalls and accuracy. In contrast to BLEU, which aims for correlation between sentences, this one focuses on the whole text. Increasing METEOR height: the greater the value, the more accurate the translation. The machine translation literature was the primary user of the Meteor metric. In recent years, however, it has found widespread use in a variety of other text-related tasks, including summarization, captioning images, speech recognition, and so on. Also, METEOR may easily be modified to include other, superior matching methods.

METEOR determines a translation's quality by computing a score based on the number of precise unigram matches between the source and target texts. In order to compute METEOR, we first generate an alignment between the two phrases in a pair of translations. In other words, an alignment is a set of relationships between unigrams, where each unigram in one phrase corresponds to either one or no unigram in the other sentence.

Evaluations of MT on humans are comprehensive but prohibitively costly. Despite the fact that automatic systems are not intended to take the place of human translators and assessors, they are surprisingly effective at doing so. For a more thorough and accurate assessment, these metrics require careful attention and improvement.

6 Conclusion

In this work we discussed neural machine translation algorithms for a variety of Indo-Aryan language pairings using a variety of automated metrics. It was shown that the back-translation method benefits less successful MT models more than it does successful ones. Based on the survey, we understand that the SMT-trained phrases improve the NMT model's accuracy. As we found, there is a disconnect between BLEU and human judgement. While automatic systems are not intended to replace human translators or assessors, they are quite effective at saving time and resources. For a more thorough and accurate assessment, these metrics require close consideration and revision. Sentence structure and semantics are ignored by BLEU. As a result, it gives SMT systems a higher rating than RBMT systems. METEOR primarily depends on assessing accuracy and recall just considering unigrams. And whereas BLEU takes an n-gram method to surface forms, METEOR focuses on matching unigrams via stemming and synonyms. The most effective technique for improving machine translation quality is the use of rank and scale theory and the development of corpora. Additionally, the English-Telugu parallel corpus can be employed in the combined corpus experiment to enhance the quality of the English-Kannada translation.

References

1. Sennrich, R., Haddow, B.: Linguistic input features improve neural machine translation. In: Proceedings of the First Conference on Machine Translation, vol. 1, pp. 83–91 (2016)
2. Ding, S., Renduchintala, A., Duh, K.: A call for prudent choice of sub word merge operations in neural machine translation. arXiv preprint arXiv:1905.10453 (2019)
3. Kunchukuttan, A., Bhattacharyya, P.: Utilizing language relatedness to improve machine translation: a case study on languages of the Indian subcontinent. arXiv preprint arXiv:2003. 08925 (2020)

4. Vaswani, A., et al.: Attention is all you need. Adv. Neural. Inf. Process. Syst. **30**, 1–11 (2017)
5. Gülçehre, Ç., et al.: On using monolingual corpora in neural machine translation. Mach. Learn. (2015)
6. Luong, M.-T., Le, Q.V., Sutskever, I., Vinyals, O., Kaiser, L.: Multi-task sequence to sequence learning. In: International Conference on Learning Representations (2016)
7. Renduchintala, A., Shapiro, P., Duh, K., Koehn, P.: Character-aware decoder for neural machine translation. In: MT Summit 2019 (2019)
8. Ding, S., Renduchintala, A., Duh, K.: A call for prudent choice of sub word merge operations. In: Proceedings of Machine Translation Summit XVII, pp. 204–213 (2019)
9. Kunchukuttan, A., Bhattacharyya, P.: Learning variable length units for SMT between related languages via byte pair encoding. In: Proceedings of the First Workshop on Subword and Character Level Models in NLP, pp. 14–24 (2017)
10. Tang, Y., Meng, F., Lu, Z., Li, H., Yu, P.L.: Neural machine translation with external phrase memory. arXiv:1606.01792 [cs. CL] (2016)
11. Zhao, Y., Wang, Y., Zhang, J., Zong, C.: Phrase table as recommendation memory for neural machine translation. In: Proceedings of the Twenty-Seventh International Joint Conference on Artificial Intelligence, pp. 4609–4615 (2018)
12. Banerjee, T., Bhattacharyya, P.: Meaningless yet meaningful: morphology grounded subword-level NMT. In: Proceedings of the Second Workshop on Subword/Character Level Models, Association for Computational Linguistics, New Orleans, pp. 55–60 (2018)
13. Murthy, R., Kunchukuttan, A., Bhattacharyya, P.: Addressing word-order divergence in multilingual neural machine translation for extremely low resource languages. In: Proceedings of the Annual Conference of the North American Chapter of the Association for Computational Linguistics (2019)
14. Revanuru, K., Turlapaty, K., Rao, S.: Neural machine translation of Indian languages. In: Proceedings of the 10th annual ACM India Computer Conference, Bhopal (2017)
15. Akella, K., et al.: Exploring pairwise NMT for Indian languages. In: Proceedings of International Conference Natural Language Processing, Patna (2020)
16. Sen, S., et al.: Neural machine translation of low-resource languages using SMT phrase pair injection. Nat. Lang. Eng. 1–27 (2018)
17. Sennrich, R., Haddow, B., Birch, A.: Improving neural machine translation models with monolingual data. In: Proceedings of the 54th Annual Meeting of the Association for Computational Linguistics (2016)
18. Dewangan, S., Alva, S., Joshi, N., Bhattacharyya, P.: Experience of neural machine translation between Indian languages. Mach. Transl. 1–29 (2021)
19. Zhang, J., Zong, C.: Exploiting source-side monolingual data in neural machine translation. In: Proceedings of the 2016 Conference on Empirical Methods in Natural Language Processing, pp. 1535–1545 (2016)
20. Beddiar, D.R., et al.: Data expansion using back translation and paraphrasing for hate speech detection. Online Soc. Netw. Media **24**, 1–13 (2021)
21. Wong, S., Gatt, A., Stamatescu, V., Mcdonnell, M.D.: Understanding data augmentation for classification: when to warp. In: Digital Image Computing Techniques and Applications, pp. 1–6 (2016)
22. Yu, A.W., et al.: QANet: combining local convolution with global self-attention for reading comprehension. In: International Conference on Learning Representations, pp. 1–12 (2018)
23. Sindhu, D.V., Sagar, B.M.: Dictionary based machine translation from Kannada to Telugu. In: IOP Conference Series: Materials Science and Engineering, vol. 225, pp. 1–8 (2017)
24. Mallikarjun, M., Kodabagi, Angadi, S.A.: A methodology for machine translation of simple sentences from Kannada to English language. In: Proceedings of IEEE, pp. 237–241 (2016)
25. Lakshmi, S., Shambhavi: Learning to translate Kannada and English queries for mixed script information retrieval. Comput. Inform. **40**, 628–647 (2021)

26. Post, M., Callison-Burch, C., Osborne, M.: Constructing parallel corpora for six Indian languages via crowdsourcing. In: Proceedings of the Seventh Workshop on Statistical Machine Translation, pp. 401–409 (2012)
27. Ruder, S., Vulić, I., Søgaard, A.: A survey of cross-lingual word embedding models. J. Artif. Intell. Res. **65**, 569–631 (2017)
28. Conneau, A., et al.: XNLI: evaluating cross-lingual sentence representations. In: Proceedings of the 2018 Conference on Empirical Methods in Natural Language Processing, pp. 2475–2485 (2018)
29. Banerjee, A., Jain, A., Mhaskar, S., Deoghare, S., Sehgal, A., Bhattacharyya, P.: Neural machine translation in low-resource setting: a case study in English-Marathi pair. In: Proceedings of the 18th Biennial Machine Translation Summit, pp. 35–47 (2021)
30. Papineni, K., Roukos, S., Ward, T., Zh, W.-J.: Bleu: a method for automatic evaluation of machine translation. In: Proceedings of the 40th Annual Meeting of the Association for Computational Linguistics, pp. 311–318 (2002)

Recommendation System for Movies Using Improved Version of SOM with Hybrid Filtering Methods

Saurabh Sharma[✉] and Harish Shakya

Amity University, Gwalior, India
saurabhgyangit@gmail.com

Abstract. Recommendation systems (RS) are used by many businesses to identify product recommendations made by consumers that interact with e-commerce websites. Massive growth in both commodities and consumers has recently encountered significant difficulties. Many websites overwhelm the user with alternatives at once, which causes a lot of confusion. Additionally, a crucial component of RS is locating the appropriate product or active user. Based on consumer preferences and sociodemographic trends, products are already recommended. In order to enhance the user behaviour matrix, a hybrid-actionrelated recommendation based on K-Nearest Neighbor Similarity (HAR-KNN) combines the simplicity of hybrid filtering with the creation of feature vectors. It uses both quality and quantity classifiers to categorise properties. The suggested methodology also addresses weaknesses in past methods of feature analysis and user preference evaluation. The SOM AND KNN classification technique has been authorised for the purpose of locating data about user behaviour online and in real time for a particular user group that contains a huge amount of data in connection to the commonalities among many users and target users. Highly predictive metrics such as Precision (P), Recall (R), and F, as well as Mean Absolute Error (MAE), Mean Square Error (MSE), and Root Mean Square Error, are used to assess a test result (RMSE).

Keywords: Recommendation System (RS) · User behaviour data · Hybrid filtering · KNN · behavioural matrix

1 Introduction

In the past ten years, artificial intelligence technology has advanced quickly, making it increasingly vital to have systems that suggest tailored materials. Recommendation engines frequently examine user behaviour in order to deliver valuable information. But if some essential user data is not taken into account during the recommendation process, the suggested material might not be of high enough quality. In order to remedy this issue, additional data analysis is required, along with the addition of user log data to the suggested material upgrades.

The judicious use of data is a crucial issue in today's society because as networks are utilised more often, people's daily activities generate large volumes of data. The

© The Author(s), under exclusive license to Springer Nature Switzerland AG 2025
M. Gupta et al. (Eds.): MISS 2023, CCIS 1952, pp. 181–197, 2025.
https://doi.org/10.1007/978-3-031-69115-7_12

recommendation system (RS) is a crucial technique for impartially assessing data and automatically producing customer recommendations [1]. Almost all applications at current time use RS to provide more accurate recommendations for users in an effort to attract users to use their applications for a long time. These applications have an impact on almost every part of people's lives, including music [3], movies [4], books [5], the news [6], travel [7], and more. The role of the RS in people's online lives has progressively increased. Recommendation systems are frequently employed in business to help users discover things they haven't yet looked at and to promote sales to potential customers based on their past purchasing behaviour over similar data. If a recommendation system can present clients to the products that are most relevant to them, it is deemed effective. Because of this, recommendation systems are now the smart choice for advertising. The user's preferences, the item's features, the user's interactions with the item, and a few additional elements, such as temporal and location data, are typically used to make suggestions.

The main categories of RS approaches include Collaborative Filtering (CF), Content-Based Filtering (CBF), and hybrid recommender systems based on the input data [8]. The Collaborative Filtering method concentrates its suggestions to the active client on goods/products that other users with comparable preferences have previously enjoyed. To determine how similar users' likes are, their browsing histories or similar ratings (on a scale of 1 to 5 for movies) are used. The Content-Based model recommends products to the engaged user based on things he has previously appreciated. How similar the products are to one another is determined by the characteristics of the compared commodities, such as the setting, title, or even the product appearance. Hybrid approaches [9, 10] combine two or more current strategies to maximise their advantages while minimising their disadvantages. Collaborative Filtering in conjunction with other techniques is the most well-known and frequently used solution. Common categories include heuristic techniques, neighborhood-based collaborative filtering techniques, matrix factorization-based collaborative filtering techniques, and machine learning techniques. In order to solve the cold start problem, a variety of machine learning techniques, such as regularised regression and tree-based approaches, employ the CB features that are collected for people and items as inputs. Given the strategic importance of recommender systems in research and business, as well as algorithmic advancements, increasing prediction accuracy is a major problem. Using a hybrid technique that combines collaborative filtering and content-based analysis and is supervised by the ranking list generated by the well-known unsupervised learning artificial neural network method, we present a powerful recommendation system in this study.

Users can find the content they're seeking for with the aid of recommender systems, which employ both user information and data. Systems that make product recommendations consider consumer behaviour to identify the goods that are most likely to appeal to customers. Companies that sell electricity online, like Amazon, eBay, Netflix, and Youtube, have customised their websites and fundamentally altered how people shop. Collective filtering (CF) allows users to receive complicated suggestions while excluding specified users and item files. As a result, CF is evolving into a promising technological idea.

For instance, Netflix utilises CF to forecast movie viewership.

Two of the most popular collaborative filtering methods are the neighbourhood algorithm and the latent component model [1, 13]. It is simple to assess user or object similitude using neighbourhood algorithms. From raw customer rating data, late factor models use matrix factorization to identify the core commonalities between items and users. It is obvious that some characteristics connect to both users and products. Matrix factorization models have received a lot of attention recently because of their advantages, including maintaining accuracy in data scaling, having minimal computation costs, and minimising the issue brought on by high sparsity levels [6, 7]. Modern matrix factorization techniques frequently leverage database-stored data and operate in batch mode. Since it learns in a single session, this kind of batch mode matrix recommendation system does not require updating input data. The initial batch mode recommendation model will become out of date and provide subpar ideas as the recommendation system expands and changes since there will be a constant stream of new people and things.

The design of a recommendation system must take a variety of incoming data types into account, which is a crucial factor to take into account. The majority of matrix factorization methods that we employ today were created using transparent and explicit feedback data. However, since user interactions necessitate poor data collection, service providers must develop feedback websites. Implicit inputs, on the other hand, such past searches and browser histories, are simpler to gather. Implicit feedback recommendation systems need to overcome a number of obstacles in order to be successful [14]. Primarily, implicit feedback is a good thing. Due to its inherent noise, implicit feedback presents a second issue that makes it challenging to transform into meaningful preferences. A third issue is that it can be challenging to evaluate the implicit feedback recommender objectively. In order to incorporate implicit feedback information, the factorization of the matrix must be changed as a result. Matrix factorization techniques are studied from two perspectives as a result of the literature:

In order to address the issue of implicit feedback, [3] has developed a variety of kernel functions to enhance the suggestion's correctness in the analysis of implicit feedback. In [7], several kernel functions were put forth as a solution to the implicit feedback problem. [9] presented a local weighted matrix factorization for the top N suggestions using a local weighted matrix factorization with implicit feedback.

The weight function was applied to express user preferences, and the kernel function was employed to improve the local property of submatrices. Despite the fact that [8, 9] are generally accurate, the real-time update of the model has not yet been solved.

A stochastic gradient descent (SGD) technique for real-time updates was presented, but only for explicit feed-back suggestions, based on the real-time update issue aspect [10]. To solve a current issue, the authors in [2013] employed implicit feedback with binary values. However, they just translated implicit feedback into binary numbers rather of utilising explicit feedback with actual numerical values. As a result, [7]'s strategy isn't extremely exact. This technique effectively optimises an MF model using variable weighted missing data and elemental alternating low squares (eALS). [3] proposes utilising a single-sided, smaller number of squares to learn ALS integration more quickly (One-sided LS). When used in real-time applications, ALS is worse than SGD because it takes longer to update a single element [5, 17].

The advantages of ALS and SGD are combined in our hybrid real-time incremental stochastic gradient descent (RI-SGD) technique to update the MF recommendation systems with implicit feedback. We use a real-time, streaming data analysis platform to assess RI- SGD and other upgrading techniques. In comparison to all upgrade strategies, RI-SGD can surpass them by obtaining competitive accuracy with only 0.02 percent of the overall retraining time. In our prior work [11], we suggested a simple method for updating models with adaptive data. The architecture of RI-SGD recommendation systems, the time-dependent cost feature, and the progressively updating model based on earlier research are all presented in this study. Additionally disclosed are the number of elements and the attenuation parameter. Additionally, we show that RI-suggestion SGD is superior to other techniques using standardised decreased cumulative gains (NDCG).

2   Background and Related Work

A multimodal information source model (Multimodal IRIS) was established by Moon et al. [7] by incorporating information-sharing units amongst various modalities in its component module. Visual and textual representations can be blended to better understand how different modalities interact. The interaction between historical components and the goal objects may be represented by the multimodal aspect. The list of recommendations is also influenced by the expansion of multimodal data and the numerous fascinating significances of the recorded objects.

[14] In order to evaluate the platform for a variety of products, Huang et al. created a novel RS process configuration approach that makes use of the reusable qualities of product lines. The platform has a dimension-ality issue as a result of the huge feature models. This strategy is compatible with six proposing algorithms to configure the product line context. Three of the six presently in use recommendation systems were utilised by the creators.

Find the relevant characteristics. As a result, this approach can aid clients in understanding their preferences. An experimental study demonstrates that the suggested technique enhances two real-world data sets' quality and effectiveness and chooses effective traits to help decision-makers make well-informed choices.

Pilato et al. [3] developed the K-RecSys framework to enhance the framework for collaborative filtering proposals by defining space features. To gauge consumer preferences for the various components of the plan, this structure integrates information from online product clicks with information from abandoned product discussions. Less often than they do, clients would prefer to evaluate bids. Users want to buy things utilising item data as a result. The collaborative filtering method currently used to validate exhibitions is contrasted with K-RecSys. Tests are constrained and can only signal a success or failure if the results are being paid attention to.

Artificial neural networks can boost customer happiness, claim Duan et al. [5]. This strategy suggests similar items for other student groups while taking into account students' purchase patterns in order to satisfy each customer's wants. The prediction error is calculated using a NN model on relevant time series data. Data for prediction and forecasting could be produced from primary data. A variety of online tests could be utilised to collect information. The undesirable data could then be processed beforehand. Following that, the model technique was fed-forward trained. Finally, MSE-related training and testing can be used to assess the model. Thus, many datasets show that the expected result supports a specific purpose.

A theory-based suggestion on structural balance is a substructure-specific prescription for the ideal item for the intended user, according to Kyaw et al. [6] (SBT-Rec). Although the electronics sector has a limited number of rating information, comparable products could be overlooked on the internet. SBT-Rec combines based and item-driven collaborative filtering in the display % of its proposals, SBT-Recproposals.

A hybrid method to customer ratings, assessments, and demographic data was put out by Jiang et al. [12]. This model comprises the development of features, evaluation of the data, model training, community detection, blending of features, prediction, and evaluation. The experimental model separates and breaks audit material to discover social networks using convolutional suggestion models, such as network models. The ideas may be more accurate as a result of reading the text and social media posts.

Jayasudha et al. suggested a phased research strategy. In [4], the location of irregular assessments is found using the similarity between the product and the extracted features. To generate a rating matrix, a number of samples are taken from the user profile. You must thoroughly examine things' and users' fundamental and priceless mistakes in order to develop a grading matrix.

Determines the distribution by analysing a vast amount of data, rating intent, and time series analysis.

Bizimis and other [2] suggested a hybrid recommender system for the proposal process that is built on reading reviews and gathering user feedback. With the aid of complex algorithms, they produced suggestions for people who could provide supplementary information, such as review sentiment. Several research have found that ratings and user sentiment are only sporadically substantially related (for example, in the music domain).

A custom dataset that combines a well-known dataset for collaborative filtering with movie metadata was given by Wang et al. [9]. Then, for both regression and ranking scenarios, we provide models of the three architectures to be trained on it. Based on proper regression and rating criteria, we compare and assess them.

Many portions of the prior study in Custom RS, which included both structured and unstructured data, needed to be more insightful in order to produce improvement ideas. The architecture of the RS system is strained by the several attributes that control its input matrices. The service technique generates projected ratings using multi-attribute conduct and actual e-commerce data. Numerous changes have already been made to this procedure, and fixes have already been made for the problems with the recommendation service process.

3 Hybrid Action-Related System

Based on a variety of behavioural data, the suggested hybrid recommendation strategy. The entire data set is subjected to the hybrid approach in order to assess the information gathered as input and process explicit feedback on the sorted data frame. Based on user behaviour, data with user information qualities rank different products. The initial database has all the data required to process documents and vast volumes of relevant data beforehand. The hybrid method makes it simple to find users and others who are actively looking for them. Based on behaviour prediction and propensity, filtering hybrids are used to choose target products as recommendations. The present user is also given a feature description via a categorization approach. By choosing the KNN of the current user, this enhances the relationship between the user at hand and the selected product. Recommendations are provided using this method based on changing user behaviour. For all products connected to understanding behaviour based on data distribution, pattern information is offered. Choosing the nearest neighbours is the first step in the categorization procedure since it allows us to identify the vector features of the neighbour and the product. As a result, the degree to which the intended user and the number of users are similar is assessed. The outcomes can be paired with knowledge of user behaviour and prior purchases.

Recommendation Systems Relation
In this method, the browser may be suggested and defined by the behavioural scenario recommendation model. Based on these assumptions, we discuss how we can reach a relationship in the information system.

Take into account the product set P = p1, p2, ..., pi and U = u1, u2, ... To save space, the database saves user and product information. The equation R by U by P can express the relationship between U and P.

Users Recommend Compliance
The login user and RS work together to control the behavior of the recommender. The first user checks in to offer one or more particular items when they first sign up. The user selects a product or disconnects from the connection. User behavior indicates both greater and lower interest, of course. The behavior-based data is known as "three tuples", or X: U, L, and M >, if the user has only provided the clickstream or any other data for the product (1) Where L is a purchased product ID, M is a purchased product, and X is the sequence of time-ordered user activities.

4   Experiments and Discussions

With the use of Group Lens, we were able to obtain the dataset that we did. On grouplens.org, you may access the movielens/20m data sets. It demonstrates how the five-star rating and free-text labelling procedures on MovieLens operate (ML-20 m). 2000263 films have been rated as of today, and there are currently 465564 tags. Between January 9, 1995, and March 31, 2015, 138493 users contributed the data that makes up this report. This information gathering was started on October 17, 2016. A total of 188 participants were chosen at random from among those who expressed interest in taking part in the study. Each user had given at least 20 films a rating. The demographics are not broken down. Additional than the special ID that has been allocated to that person, no other information is provided to identify that person. Genome-scores.csv, Genome-tags.csv, Links.csv, Movies.csv, Ratings.csv, and Tags.csv are the CSV files that contain the information. Two files, "ratings.csv" and "movies.csv", will be needed to complete our goal in order to achieve our purpose.

Data Preparation and Processing To prepare the data for Machine Learning (ML) modelling, we'll start by importing the dataset and getting to know it.

5 Observations

– The data spans over 20 million records.
– UserId, movie, rating, date, title, and genres are among the six features.

Cleaning of Data
In order to handle missing values, outliers, and rare values and remove extraneous features that do not convey valuable information, we will start by cleaning the data.

Data columns (total 6 columns):
Column Dtype
— —— ——

0 userId int64
1 movieId int64
2 rating float64
3 date object
4 title object
5 genres object
memory usage: 915.5+ MB

Handling Missing Values. Identifying the features that have some missing values and imputing them.
  userId 0
  movieId 0
  rating 0
  date 0
  title 0
  genres 0 dtype: int64
  Observations:

– Since there are no missing numbers, the dataset appears well-maintained, which is good.

Analyzing Exploratory Data
Once the data has been cleaned, we can use EDA to find patterns and relation- ships that will improve our knowledge of the data.

Analysis by One Variable. Examining each feature separately to draw con clusions from the data and identify any outliers
  The count of unique userID in the dataset is: 138493
  The top 5 userID in the dataset are:
  118205 9254

8405 7515
82418 5646
121535 5520
125794 5491
Observations:

– "userId" refers to the Users whose ids have been made anonymous and who were chosen randomly for inclusion.
– The dataset has 138K+ unique users.
– The dataset for userId 118205 contains about 9K records.

The count of unique movieID in the dataset is: 26744
The top 5 movieID in the dataset are:
294 67310
353 66172
316 63366
588 63299
477 59715
Name: movieId, dtype: int64
Observations:

– The term "movieId" refers to movies in the dataset with at least one rating or tag.
– The collection contains around 26K + distinct movies.
– movieIds 294, 353, 316, and 588 are a few well-liked films that have received over 60K ratings.

Observations in Fig. 1

– The range of user ratings for movies is from 0.5 to 5.
– Most movies have received ratings of 3, 3.5, or 4 from the people.
– Given that most ratings fall between three and five, the ratings distribution appears to be slightly left-skewed.

The count of unique date in the dataset is: 6911
The first rating was given on: 1995-01-09
The latest rating was given on: 2015-03-31
The top 5 date in the dataset are:
2000-11-20 91753
2005-03-22 76568
1999-12-11 65077
2008-10-29 55163
2000-11-21 54131
Name: date, dtype: int64
Observations:

– There are around 7K distinct dates where users have rated movies.
– The initial rating was provided on January 9, 1995, and the most recent was given on March 31, 2015.

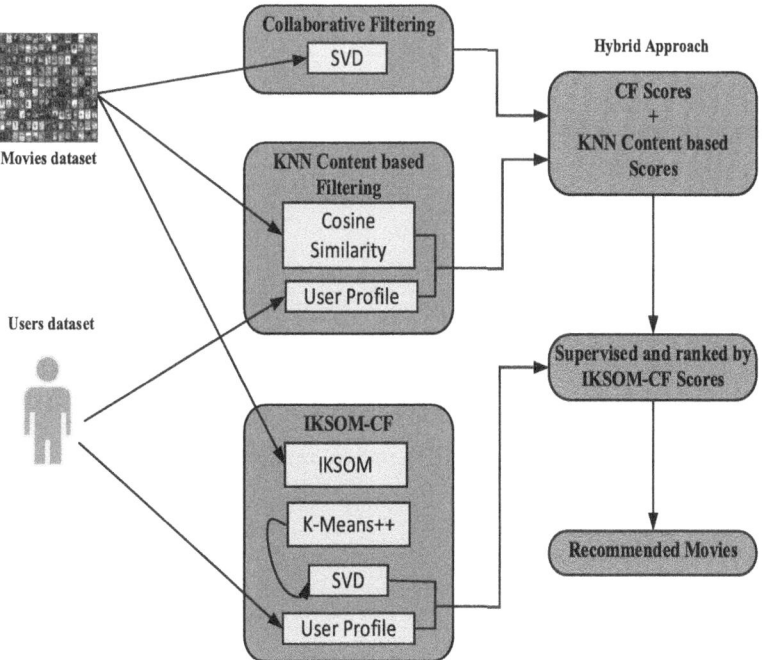

Fig. 1. Proposed Architecture

- On 2000–11-20, 91K + ratings were recorded.

  The count of unique title in the dataset is: 26729
  The top 5 title in the dataset are:
  Pulp Fiction (1994) 67310
  Forrest Gump (1994) 66172
  Shawshank Redemption, The (1994) 63366
  Silence of the Lambs, The (1991) 63299 Jurassic Park (1993) 59715
  Name: title, dtype: int64
  Observations:

- The collection contains about 26K distinct movie titles.
- The top four films in terms of the number of reviews that have over 60K + are Pulp
  Fiction, Forrest Gump, Shawshank Redemption, and Silence of the Lambs (Table 1).

  Split Training and Testing
  Before moving on to more EDA and feature engineering, split the data into train and
  test sets.

Table 1. Dataset Description

	userId	movieId	rating	date	title	genres
0	131160	1058	3.0	1995-01-09	Fish Called Wanda, A (1988)	Comedy|Crime
1	131160	47	5.0	1995-01-09	Seven (a.k.a. Se7en) (1995)	Mystery|Thriller
2	28507	1154	4.0	1995-01-09	Double Life of Veronique, The (Double Vie de V...	Drama|Fantasy|Romane e
3	131160	21	3.0	1995-01-09	Get Shorty (1995)	Comedy|Crime|Thriller
4	85252	7	5.0	1996-01-29	Sabrina (1995)	Comedy|Romance

> *# Checking the basic statistics for the training data*
> Total Train Data..
> Total number of movie ratings in train data : 16000210
> Number of unique users in train data : 112466
> Number of unique movies in train data : 12387

6   Bi-variate Analysis

Combining analysis of several features to find relationships, correlations, and patterns

Analyzing the Distribution of Ratings. The basic statistics for the feature is:
   count 1.600021e+07
   mean 3.512613e+00
   std 1.059931e+00
   min 5.000000e-01
   25% 3.000000e+00
   50% 3.500000e+00
   75% 4.000000e+00
   max 5.000000e+00

2. Analyzing the number of ratings with date.

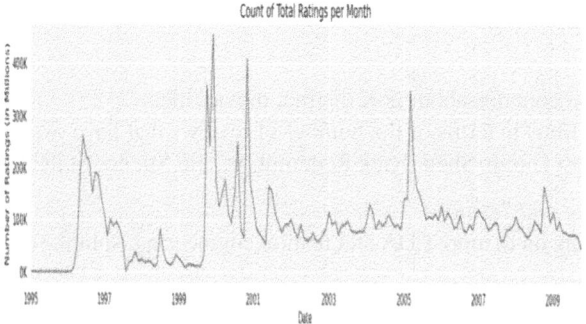

Fig. 2. Distribution of total rating in the dataset per month

Observations: Fig. 2

– In a handful of the months between 1996 and 1998, the number of ratings per month was extremely high.
– After the 2000s, there have also been a few months with a particularly high number of ratings.
– After 2001, the count remains constant until 2010, with a surge in a few months of 2006.

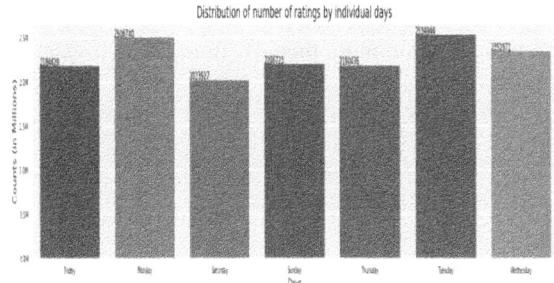

Fig. 3. Visualizing the count of ratings by individual days of the week

Observations: Fig. 3

– The number of ratings is not significantly affected by the day of the week.
– Undoubtedly, "Monday" and "Tuesday" receive higher ratings than any other day.
– The number of ratings over the weekend is much lower than during the workweek.
– Analyzing the average ratings by date.

7   Visualizing the Average Ratings by Weekday

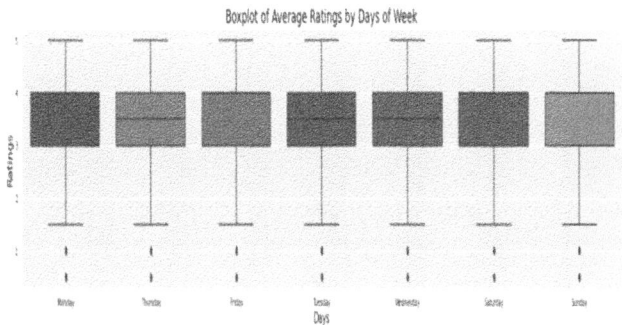

Fig. 4. Visualizing using box plot of average ratings by Days of week

Observations: Fig. 4

– The user's average ratings do not appear to vary between weekdays and weekends.

– The average ratings appear to remain consistent over all of the various days, even when we plot them by individual days.

  Analyzing the Ratings given by Top 15 Users.

8   Visualizing the Count of Ratings by Individual Users

Observations: Fig. 6

– The top 10 users typically rate more than 4K times, which seems like an excessive amount of usage (Fig. 5).

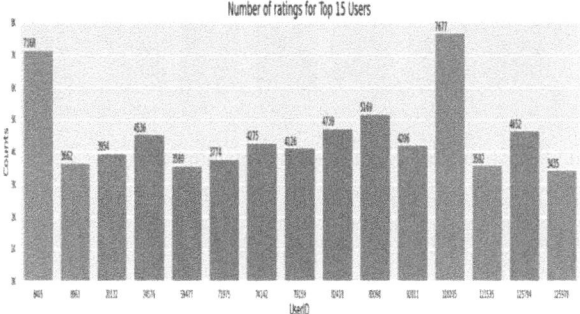

Fig. 5. Analyzing the Ratings given by Top 15 Users

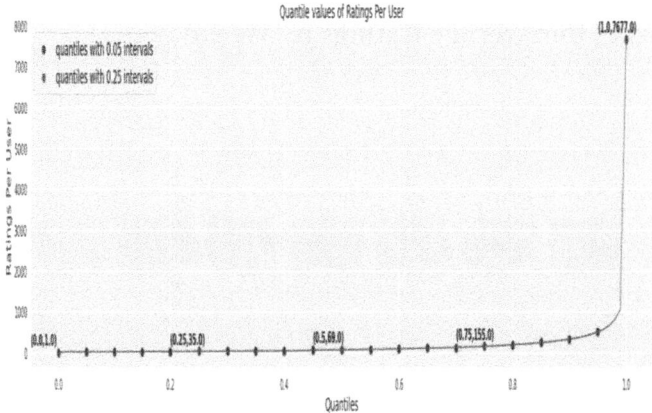

Fig. 6. Plotting the quantile values of ratings given by users

– It's astonishing that the userId 118205 has been rated over 7K + times.
– The KDE figure makes it abundantly clear that the number of ratings is significantly right-skewed and that the majority of user ratings fall between 0 and 1000.
– The CDF graph above also demonstrates that over 99% of users provide very few ratings.
– The median number of ratings given by users is 69, while the mean number is 142.
– From the 90th percentile, the number of movies starts to rise sharply.

9   Visualizing the Average Ratings by Individual Days of the Week

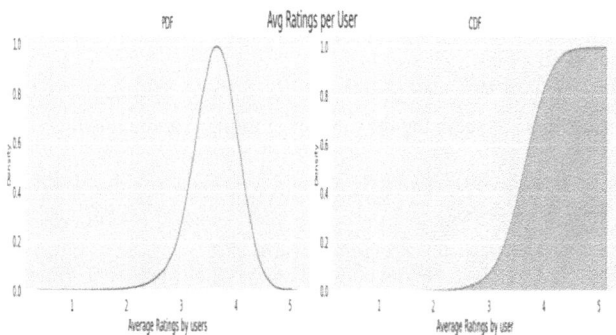

Fig. 7. Plotting the PDF and CDF for Avg. Rating by Users

Observations: Fig. 7

– The average rating for user ID 125794 is 3.81.
– The user's average rating PDF appears slightly left-skewed, with most num- bers falling between 3.5 and 4.
– The CDF also demonstrates that the most common average rating ranges between 3 and 5.

Analyzing the Ratings given to the Movies.
Observations: Fig. 8
Some films have skewed plots because they are particularly well-liked and have high user ratings compared to other films.

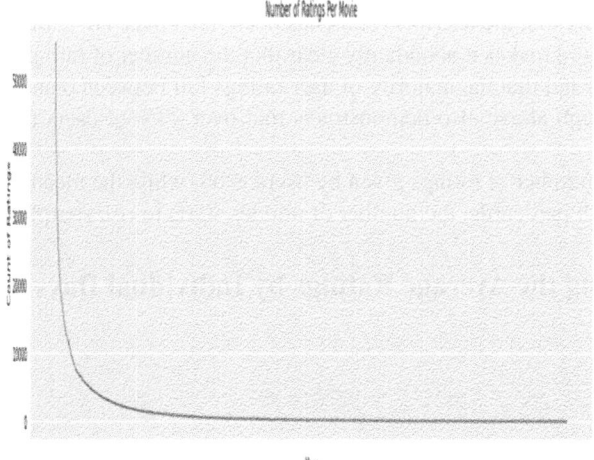

Fig. 8. Visualizing the number of ratings for the movies

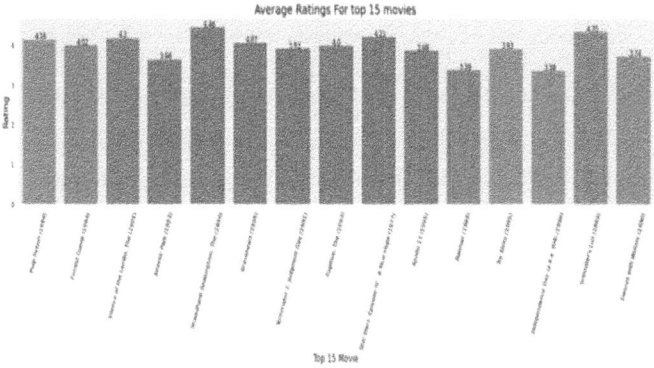

Fig. 9. Visualizing average ratings for top 15 movies

Observations: Figure 9

- The 1990s cult films have received the highest ratings.
- Over 50K people have rated Pulp Fiction, Forrest Gump, Shawshank Redemption, and Silence of the Lambs.
- Based on more than 50K ratings, Shawshank Redemption gets the highest average rating, 4.56.

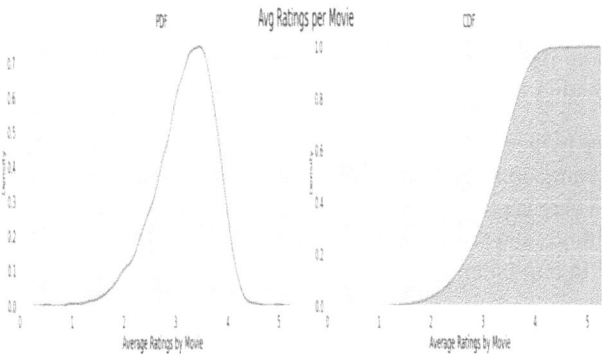

Fig. 10. Plotting the PDF and CDF for Avg. Rating by Movies

Observations: Fig. 10

- The average movie rating is distributed in a very typical way.
- The CDF demonstrates that the average rating increases after 3.

Table 2. Tabular Values of Errors

S.No	Model	Train RMSE	Test RMSE
0	XGBoost_13	0.807228	1.004849
1	BaselineOnly	0.827371	0.999581
2	XGB_BSL	0.807346	1.003603
3	KNNBaseline_User	0.337843	0.999480
4	KNNBaseline_Item	0.274160	0.999480
5	XGB_BSL_KNN	0.807366	1.004184
6	SOM	0.760572	0.999146
7	SVD	0.831105	0.999488
8	SVDpp	0.763126	0.999498
9	XGB_BSL_KNN_MF	0.807147	1.005231
10	XGB_KNN_MF_SO	1.066477	0.998114

10 Observations in Table 2

- The performance of all algorithms appears to be excellent, with the differ- ences staying quite near one another. As we can see, combining and stacking several rating prediction algorithms before utilising final methods appears to produce the lowest Testing RMSE. For instance, Baseline Only by Surprise used with KNN Baseline,

SVD, SVDpp, SOM, and Xgboost. The Testing RMSE of SlopeOne appears to be the lowest of all the algorithms. Except SlopeOne, the SVDpp and SVD algorithms have lower Testing RMSE than the other predictors.

Conclusion. Due to the fact that this is still an unresolved research problem, there is still a tonne of study being done on the topic of recommending cold start items for recommendation systems. There are two types of cold start items: a complete cold start item (CCS), which gets no ratings, and an incomplete cold start item (ICS), which mostly gets ratings that aren't zero. In particular, this work suggests two recommendation models that, by addressing the respective recommendation difficulties, can address the recommendation challenges related to CCS and ICS items. It solves the issue by combining a deep learning architecture called SDAE with a time-aware collaborative filtering (CF) model called time SVD++. The SDAE deep learning neural network is in charge of extracting the features of the item content. As a result of this research, a deep-learning neural network was utilised to predict unknown ratings using the temporal SVD++ model. The temporal dynamics of user preferences and item attributes were investigated in this study in relation to the prediction of unknown ratings. It has been shown that a number of tests were run to assess how well the suggested models performed when they were used to analyse Netflix datasets in terms of RMSE and recommendation prediction errors. Our experiments' findings were used to show that our cold start item recommendation models performed better than the baseline methods already in use. The results of our study and testing were significantly impacted by the inclusion of time and item content information, it turned out. The latent factor model of CF can be successfully used to our model to provide considerable performance improvements, especially when dealing with CCS issues. Our suggested models have also been contrasted with various levels of the sparsity of the rating matrix sparsity when it comes to ICS new item suggestions. According to some study, the ICS item-based approach does not generate useful recommendations for items that have earned a small number of ratings (i.e., three or fewer).

References

1. Nithya, B.N., Geetha, D.E., Kumar, M..: Optimization assisted personalized event recommendation for event-based social networks. Adv. Eng. Softw. **176**, 103368 (2023)
2. Gianluigi, F., Carla, G., Francesco, P.: An ensemble-based framework for user behaviour anomaly detection hybrid real-time implicit feedback based movie recommendation systems 15 and classification for cybersecurity. J. Supercomput. 1–24 (2023)
3. Giovanni, P., Filippo, V.: A survey on quantum computing for recommendation systems. Information **14**, 20 (2023)
4. Jayasudha, J., Thilagu, M.: A survey on sentimental analysis of student reviews using natural language processing (NLP) and text mining. In: Innovations in Intelligent Computing and Communication: First International Conference (2022)
5. Jin-Juan, D.: A modeling design method for complex products based on LSTM neural network and Kansei engineering. Appl. Sci. **13**, 710 (2023)
6. Khin, K.: Sandar: Business intelligent framework using sentiment analysis for smart digital marketing in the e-commerce era. Asia Soc. Issues **16**, e252965–e252965 (2023)

7. Moon, M.J., Lee, S.: The future of informational tools with big data informatics: opportunities and challenges for evidence-based policymaking. In: The Routledge Handbook of Policy Tools, pp. 559–569 (2023)
8. Mehrbakhsh, N.: The impact of multi-criteria ratings in social networking sites on the performance of online recommendation agents. Telemat. Inform. **76**, 101919 (2023)
9. Michael, B.: Hybrid recommendation systems using neural networks (2023)
10. Qianrun, Z.: Exploring the effects of overload and stress factors on wechat moments users' discontinuous usage intention: an SSO perspective. In: SHS Web of Conferences, vol. 155 (2023)
11. Sangwon, H.: Timed pattern-based analysis of collaboration failures in system-of-systems. J. Syst. Softw. 111613 (2023)
12. Tao, J.: Guest editorial: Special issue on integrated local energy systems. Int. J. Electr. Power Energy Syst. **148**, 108929 (2023)
13. Xiang, C.: Exploring science-technology linkages: A deep learning-empowered solution. Inf. Process. Manage. **60**, 103255 (2023)
14. Xiaojing, S.: Adaptive curriculum sequencing and education management system via group-theoretic particle swarm optimization. Systems **11**, 34 (2023)

Hierarchical Attention with Time Information Based Healthcare System for Drug Recommendation and ADR Detection

Swati Dongre$^{(\boxtimes)}$ and Jitendra Agrawal

Department of Computer Science, School of Information Technology,
RGPV, Bhopal, India
dswati31@gmail.com, jitendra@rgtu.net

Abstract. The data sparsity problem caused by information overload restricts the recommendation performance of the matrix factorization model based on clinical rating data. The recommendation model integrating the clinical suggestions can effectively alleviate the sparsity of the rating data. When current recommender systems use clinical data to model patients, diseases and ADR, they only use patient clinical queries/comments on diseases and ADR as a data source. In contrast, ignoring the impact of time information on patients, diseases and ADR attributes. Aiming at this problem, a recommendation method is proposed to integrate clinical comments-level attention and temporal information (RHATR), which can fully mine the latent semantic information of clinical queries/ comments and model the dynamic changes in patients' diseases and ADR features. By applying word-level attention to a single comment text, helpful mining information such as sentiment words and keywords in a single comment text, learning about patients' diseases and ADR representations and applying comment-level attention to patient comment sets and disease sets with time factors, respectively. Extract practical clinical suggestions and learn dynamic representations of patients' diseases and ADR features. The patients' illnesses and ADR representations were known from the clinical comments and ID-based clinical details. Patient embeddings are used as final features to capture the latent factors of each patient's diseases and ADR. Experimental results show that the proposed method achieves better results than the current baseline methods in root mean square error (RMSE) on Medhelp, Medline, Diego Laboratory and Daily Strength datasets.

Keywords: Adverse Drug Reaction · Collaborative filtering · Classification Technique · Drug recommendation · Drug Repositioning · Healthcare Recommendation System · Recommendation System · Social Media

1 Introduction

The rapid development of the Internet and information technology has led to a large-scale increase in the number and types of users and items. At the same

© The Author(s), under exclusive license to Springer Nature Switzerland AG 2025
M. Gupta et al. (Eds.): MISS 2023, CCIS 1952, pp. 198–216, 2025.
https://doi.org/10.1007/978-3-031-69115-7_13

time, the problem of information overload has become increasingly severe. The emergence of recommender systems can effectively alleviate the problem of information overload and is widely used in shopping, medical care and education.

In recommender systems, how accurately to learn user and item representations is very important for recommending items (Sulthana and Ramasamy, 2019). Classic collaborative filtering recommendation algorithms model users and items based on user history (explicit scores, click-through rates, etc.) (Jiang and Yang, 2017; Yang and Jiang, 2018; Zhou et al., 2018). However, the collaborative filtering algorithm is not interpretable in the recommendation process and has a cold start problem. Therefore, it isn't easy to accurately model users and items through information such as users' explicit ratings of items. The development of the Internet and information technology enables more and more data to be perceived and acquired. Multi-source heterogeneous data, including images, texts, and labels, contain rich user behaviour and personalised demand information. The hybrid recommendation method that integrates multi-source heterogeneous auxiliary information can alleviate traditional recommendation systems' data sparsity and complexity.

The cold start problem has been paid more and more attention (Miah et al., 2017). User comments on items contain rich information, which can reflect some features of items and express user preferences. The comments of different users under the same item can reflect the item's characteristics; the comment texts of the same user on different items potentially contain the user's preference and provide rich information for the user. Therefore, when the user's explicit rating data for the item is sparse, User comment texts on items can significantly enhance the representation learning of users and items, which is very helpful to improve the performance of recommender systems (Zhou et al., 2018).

The neural network can deeply understand the comment text and retain the contextual semantic information of the comment text. Therefore, deep learning techniques that integrate review texts for the user and item modelling have attracted more and more attention (Forestier et al., 2017; Miah et al., 2017; Yang and Yang, 2018).

DeepCoNN (deep cooperative neural networks) model (Ma et al., 2019), the NARRE (neural attentional regression model with review-level explanations) model (Zhang et al., 2021),, and the D-Attn model (Chen et al., 2019)(dual attention-based model) and WCN (Park et al., 2021) model (word-driven and context-aware networks) are some superior clinical recommendation system. But still have some shortcomings: (1) DeepCoNN (Ma et al., 2019), NARRE (Zhang et al., 2021), D-Attn (Chen et al., 2019), and WCN (Park et al., 2021) concatenate multiple comment texts into a lengthy document for encoding, and many long-distance features are lost. (2) The DeepCoNN (Ma et al., 2019), D-Attn (Chen et al., 2019), and WCN (Park et al.,2021) models regard different words and comments in the document as equally important; the NARRE (Zhang et al.,2021) model applies comment-level attention to distinguish the impact of different comments on users and comments. As a result, different words and comments have different contributions when modelling users and items. (3) Existing neural network recommendation models only consider the low-order interaction between user and item attributes through simple connection or element-wise product of user embedding and item embedding. (4) The above models all ignore the influence of time factors on user preference prediction and item feature extraction. Users' interests, preferences, and item characteristics will also change continuously with the change time.

Based on the above reasons, this paper proposes a recommendation model based on deep learning, which integrates clinical queries/comments hierarchical attention with time information (RHATR). The specific content of the process is as follows: (1) Make full use of the user's rating of the item and clinical comments information. The clinical comments information can show whether the user likes or dislikes a specific feature among all the project features. The rating information can reflect the user's preference for a particular item and its level of interest.

By applying word-level attention to a single comment text, mining useful information such as sentiment words and keywords in a single comment text, learning user and item representations; applying comment-level attention to user comment sets and item comment sets with time factors, respectively, Extract valid reviews and further learn user and item dynamic representations. (2) Model the interaction as the dot product of the user embedding and item embedding and replace it with the outer product. In the process of outer product operation, the inner product result and more features are included. It fully considers the high-order interaction between user attributes and item features, making the interaction more expressive and semantically reasonable.

2   Related Work

In recent years, there have been two methods for recommending users based on the user's rating information on items: matrix factorisation and deep learning.

Matrix decomposition is the mainstream collaborative filtering method (Chen et al., 2019; Liu and Aickelin, 2012a, 2012b; Park et al., 2021; Zhang et al., 2021). For example, Ma et al. (2019) proposed a matrix decomposition based on singular value decomposition (SVD) to learn from user and item rating matrices Potential representations of users and items. Yang and Yang (2013) proposed a probability-based matrix factorisation model (PMF), which decomposes the rating matrix into the product of two low-dimensional matrices, representing user attributes and item attributes, respectively. Since the relationship between users and items is determined by a simple linear combination of several factors, only shallow features of the model can be extracted. Rho et al. (2013) proposed a method to normalise the user's historical behaviour to the user's item rating, thereby alleviating the problem of explicit rating sparsity, using an improved Pearson correlation coefficient (PCC). Calculate the similarity between users, and use the incremental update algorithm to calculate the similarity between the current user and other users, not all users. This method works well on personalised websites. However, the user-based incremental collaborative filtering algorithm can only extract shallow features of user preferences.

With the development of deep learning, to mine the deep features of the model, recommendation methods based on deep learning are widely used in recommendation systems. Lin and Lo (2013) proposed a collaborative denoising auto-encoder (CDAE), which utilises two-layer neural networks of visible and hidden layers to learn distributed representations of users and items. Rho et al. (2013) proposed a marginalised denoising auto-encoder (MDA), which combines deep learning and matrix factorisation to extract deep features of users and items.

The above methods only use the user's rating data for items to learn the representation of users and items from the rating matrix, ignoring the review texts that contain rich semantic information. The development of information technology has caused the sparsity problem of scoring data, which restricts the recommendation performance of collaborative filtering models. The number of users and items on large-scale business websites is vast. The user's rating on items generally does not exceed 1% of the total number of items, and the items scored by two users are very few. The problem of data sparsity is the key to the performance of the recommendation system. To alleviate the problem of data sparsity, using the comment text to strengthen the user further and item representation can effectively improve the performance of the recommender system.

Modelling users and items by fusing review texts has become a research hotspot in recent years. As a result, learning user and item representations from review texts has attracted increasing attention (Lin and Lo, 2013; Pourebrahim and Keyvanpour, 2020; Sun et al., 2019; Zhang et al., 2020). Many existing methods extract topics from reviews to model users and items. For example, Yang and Yang (2016) proposed a latent factor-as-topic approach, using a topic model technique, the Dirichlet probability model, to discover latent aspects of users and items from reviews. In recent years, some deep learning-based methods learn user and item representations from review texts for recommendation (Pandit and Dubey, 2021, ?; Wen-ge and Jian-xiong, 2008).

The Deep Cooperative Neural Network (DeepCoNN) model (Ma et al., 2019) consists of two parallel convolutional neural networks (CNNs), which mine user preferences and item features from the user review set and an item review set respectively. Thereby, learning user and item representations dramatically improves the performance of recommender systems. The user (item) comment set of the DeepCoNN model is a lengthy document obtained by connecting the comments.

When the target user's rating on the target item is to be fitted during training, the target user's comment on the target item is included in the document. In practice, when predicting the target user's rating on the target item, it is generally not possible to obtain the target user's comment on the target item, which is unreasonable to a certain extent. Nan et al. (2015) proposed the TransNets method (transformational neural networks) based on the DeepCoNN model to learn user and item representations. When fitting the target user's rating on the target item, the TransNets method converts the target user's comments on the target item from Deleted from the document, using a convolutional neural network to obtain user feature vector and item feature vector representation. These models concatenate the comment texts of the same user/item into a document, treat other comments as equally important, and ignore the importance of different words and comments in the document for semantic representation. Different words and comments tend to contribute when modelling users and items.

The NARRE model (Zhang et al., 2021) uses the attention mechanism to mine effective comments for the comment set of the same user/same item. Still, when using a convolutional neural network (CNN) for text convolution processing, the maximum pooling operation loses some vital information in the comment text, so the semantic information in the comment text cannot be effectively extracted. In addition, Song et al. (2021) proposed a recommendation method based on the user's historical behaviour that integrates TextRank and Word2Vec and uses deep learning technology to mine the similar relationship between user behaviour sequences. However, when Word2Vec trains user behaviour, similar contexts. The words int have similar semantic information, and the training results are not affected by the context order of the words, regardless of syntactic and grammatical information. Meng et al. (2022) proposed the RHAOR model (recommendation method integrating review text hierarchical attention with the outer product) based on DeepCoNN and NARRE, which utilises two parallel neural networks to pair topic-level and comment-level attention networks with User and item comment sets are processed. The above methods ignore that user preferences and item characteristics can change over time.

User interests and item characteristics are not static. Schiza et al. (2019) pointed out that users' future interest preferences are mainly affected by their recent interests. Zaman and Li (2014) pointed out that some of the users' preferences are seasonal or cyclical under the influence of time. It is helpful to improve the performance of the recommender system by dynamically recommending users according to the time and historical behaviour of users. Therefore, considering the time information characteristics of users, it is necessary to make dynamic recommendations based on time and user status.

The continuous updating of information types makes user preferences change over time, including long-term and short-term preferences. Long-term preferences reflect the real interests of users (Wen et al., 2020), and short-term preferences are often associated with newly updated items. The time factor has a specific influence on user preferences and item characteristics.

To further strengthen the representation learning of users and items, capture the dynamic changes of user preferences and item features, and thus improve the performance of recommender systems, this paper proposes a recommendation method that integrates short-text-level attention and temporal information (RHATR), which applies words to a single review text. Level of attention to extracting effective information such as sentiment words and keywords in a single comment text. A review-level attention network is applied to the user review set with the time factor and the item review set with the time factor, respectively, to focus on recent valid review texts. The interaction is modelled as the dot product of user embedding, and the outer product replaces item embedding. The higher-order interaction relationship between user and item attributes is mined.

3   A Recommended Approach to Fusing Clinical Comments-Level Attention and Temporal Information (RHATR)

3.1   Hierarchical Attention for Comment Text

The hierarchical attention of comment text consists of two parts: the first part is to apply word-level attention to the words in a single comment text and finally get the encoding of the single comment text; the second part is to use the comment level to the user comment set and item comment set respectively.

Attention, to obtain the user's preference for the comment text and the correlation between the item and the comment text, and through the time information to mine the influence of the time factor on the user's preference and the item and feature, and finally obtain the user and item codes. The overall network architecture diagram is shown in Fig. 2.

3.2   Comment Text Encoding

The comment text encoding unit is used to learn comment text representations from words. As can be seen from Fig. 2, the comment text coding unit mainly has three layers. The first layer is word embedding, which converts words in a single comment text into sequence representations; the second layer is a word-level attention network, which uses a self-attention mechanism to perform attention on each word vector representation in a single comment text, and obtain a single The word-level attention vector representation of the comment text; the third layer is two unidirectional long short-term memory (LSTM) networks, to bring some between adjacent words in the word-level attention vector representation Dependencies, and finally, get the comment text representation vector.

The first layer is the word embedding layer, which converts word sequences into low-dimensional dense vectors containing the semantic information of these words. A single review text r consists of M words $(w_1, w_2, ..., w_M)$, where w_M represents the Mth word. By pre-training embeddings on the Wikipedia corpus using GloVe, a single comment text r consisting of M words is represented by word embeddings:

$$E = [e_1, e_2, ..., e_m] \tag{1}$$

Among them, eM is the D-dimensional embedding vector, and the word embedding matrix $E \in RM \times D$ is represented as a 2D matrix, where M represents the number of words in each comment text, and D represents the word embedding dimension.

The second layer is a word-level attention network. Different words contain different amounts of information and importance when modelling users and items in each comment text.

For example, in a comment text in the Amazon Instant Video dataset "Enjoyed some of the comedians, it was a joy to laugh after losing my father

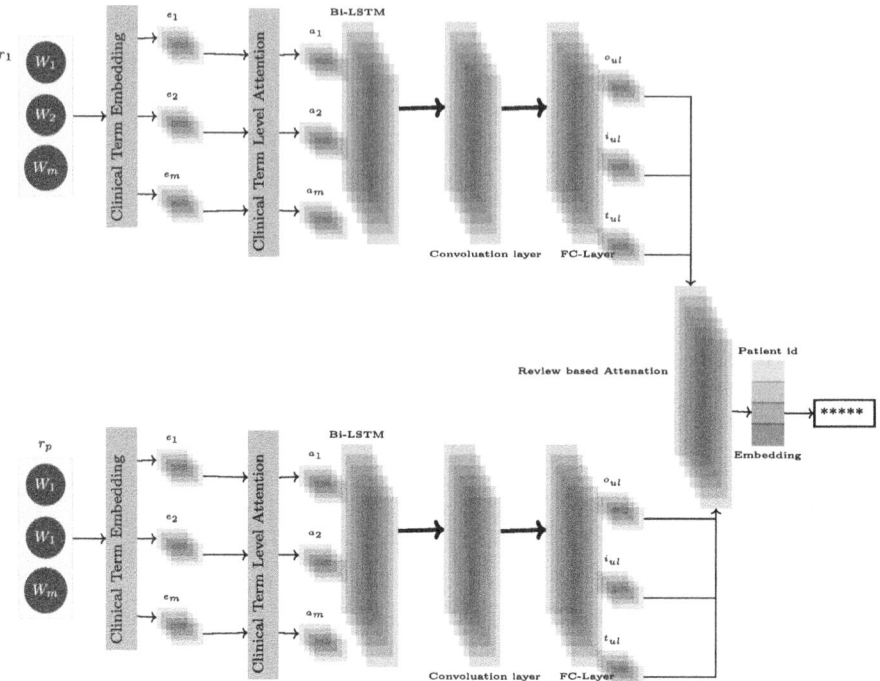

Fig. 1. Proposed Hybrid Model

whom I was a caregiver for", the words "Enjoyed" and "comedians" were signifi-cantly more Words such as "I" and "for" are more informative and better reflect the user's subjective opinion on the item and the characteristics of the item. Therefore, for the more effective user and item modelling, word-level attention mechanisms are utilised to select and focus on essential words (Fig. 1).

Each word has different importance for modelling users and items for a single review text consisting of M words. To extract the important information from the review text, the self-attention mechanism is used to perform R times on the phrase embedding representation vector E, and the word-level attention vector representation A of a single review text is obtained:

$$A = softmax(W_2 tanh(W_1 E^T)) \tag{2}$$

$$F = A \times E \tag{3}$$

Among them, A represents the importance of each word in each comment text obtained through the calculation of the attention network, and the size is $R \times M$. $W_1 \times RV \times D$, $W_2 \times RR \times V$ is the weight matrix, and V is a hyperparameter that can be set arbitrarily. Both tanh() and softmax() functions are nonlinear activation functions. F means to reduce the attention to irrelevant words while keeping the attention to each word unchanged, $F \in RR \times D$.

The third layer consists of two unidirectional long short-term memory (LSTM) networks. Aiming at the problems of gradient explosion and gradient disappearance in the long sequence training process of recurrent neural network (RNN), LSTM is suitable for modelling text data with time series and dependencies due to its design characteristics. Each word in the comment text does not exist in isolation but depends on the context information of the word. LSTM can effectively capture the context information of the entire sentence and strengthen the understanding of the context.

This paper uses LSTM to obtain a certain dependency between adjacent words in a single comment text. The matrix F is represented as $[f_1, f_2, ..., f_R]$, and two unidirectional long short-term memory networks are used to process the comment text:

$$\overrightarrow{h_t} = \overrightarrow{LSTM}(f_1, \overrightarrow{h_{t-1}}) \tag{4}$$

$$\overleftarrow{h_t} = \overleftarrow{LSTM}(f_1, \overleftarrow{h_{t-1}}) \tag{5}$$

Connect each group of $\overrightarrow{h_t}$ and $\overleftarrow{h_t}$ to obtain the hidden state h_t , and obtain Eq. 6. Each hidden unit of a unidirectional LSTM is marked as I, and each hidden unit of a bidirectional LSTM is 2I. Denote the R h_t as the vector H, as in Eq. 7:

$$h_t = (\overrightarrow{h_t}, \overleftarrow{h_t}) \tag{6}$$

$$H = (h_1, h_2, h_3, ..., h_R) \tag{7}$$

Among them, H is the comment text representation vector, $H \in RR \times 2I$.

BI-LSTM makes good use of the contextual information of the review text for processing. Still, the resulting feature matrix H has a large column dimension and contains too much information. A convolutional neural network (CNN) performs the dimensionality reduction process to find key features. Therefore, using CNN to process can maintain the context information of the embedded matrix and perform local convolution on the matrix H. Reduction of redundant dimensions impact of non-critical information on the model.

In recent years, much deep learning-based text processing methods have achieved better performance than traditional methods, such as TextCNN and TextRNN. This paper improves the DeepCoNN model. In DeepCoNN, CNN adopts the pooling layer. However, since the pooling layer only retains the maximum value of the local receptive field in the process of downsampling the embedding matrix, the key information of the feature matrix will be lost. Therefore, this paper uses CNN to further extract features from the comment text vector H, instead of using a pooling layer, but using a convolutional layer.

Suppose the convolutional layer consists of m neurons. Each neuron is associated with a convolution kernel $K \in R\tau \times 2I$, and performs a convolution operation on the embedding matrix H. The local feature z_j generated by the jth neuron is expressed as 8

$$z_j = relu(H_{R \times 2I} \times K_j + b_j) \tag{8}$$

where b_j is the bias term, $\times$ is the convolution operation, and relu is the nonlinear activation function.

Let $z_1, z_2, ..., z_j^{R-\tau+1}$ be the features produced by neuron j over the sliding window τ. Then, the final feature corresponding to this neuron is calculated using a fully connected layer, expressed as:

$$Z_r = [z_1, z_2, ..., z_j^{R-\tau+1}]^T \tag{9}$$

$$o_j = relu(WZ_r + b_j) \tag{10}$$

$W \in R_1 \times (R - \tau + 1)$ represents the weight matrix, and $b_j \in R_1$ represents the bias term. Concatenate the outputs of m neurons to obtain the feature vector $[o_1, o_2, ..., o_m]$ of a single comment text. In the same way, finally, get user comments The eigenvector α of the set:

$$\alpha = [o_{u1}, o_{u2}, ..., o_{up}] \tag{11}$$

Among them, o_{uP} represents the feature vector of the Pth comment of user u.

3.3   User/item Code

The user/item coding unit is based on the user comment/item comment representation and is used to enhance the user/item representation further. When modelling users and items, different comments from the same user/item can often reflect other user preferences and item characteristics. And user preferences and correlations between items and review texts change over time.

Therefore, this paper further strengthens the user and item representations by adopting a review-level attention network on the temporally factored set of user reviews and item reviews. The review-level attention method proposed in [8] is used to select valid review texts in the review set. Taking the user review set as an example, the goal of user/item representation is to select key review features in the reviews of user u and linearly combine these features to represent user u. The input to the attention network includes the feature vector of user u's P^{th} comment, the time of the Pth comment, and the item ID. In addition, add an item ID to mark items that the user has commented on.

To measure the influence of the time factor on user preference, the novelty of user comments is defined using the adaptive exponential forgetting function proposed in [28], as shown in Eq. 12

$$l_up = exp(-\alpha \times (u, r)) \tag{12}$$

where α is a hyperparameter that adjusts the rate of novelty decline, and times(u,r) returns a non-negative integer. For the user comment set considering time information, the attention network is expressed as Eq. 13, and the user's preference for each different comment text in the user comment set is obtained:

$$x_{up} = softmax(h^T relu(W_o o_{uP}^T + W_i i_{uP}^T + b_1) + b_2 + \beta l_{uP}) \tag{13}$$

Among them, $W_o \in R_l \times m$, $W_i \in R_l \times m$, $b_1 \in R_l \times 1$, $h \in R_l \times 1$ and $b_2 \in R_l \times 1$ belong to the model parameters, l is the size of the middle layer of the attention network, β is used for adjustment Hyperparameter for the influence of time factor on the relevance of user and comment text, X_{uP} represents the feature set contribution of the Pth comment to user u.

After obtaining the attention of each comment, the weighted sum of feature vectors of user u is expressed as Eq. 14

$$o_u = \sum_{p=1}^{p} X_{up} Oup \tag{14}$$

Among them, P is the number of comment texts in the user comment set. Finally, O_u is an output vector based on review-level attention by distinguishing the contribution of each review and representing the features of user u in the vector space.

Pass O_u to the fully connected layer to compute the k-dimensional feature vector representation X_u of the user comment set:

$$X_u = W_o O_u^T + b_o \tag{15}$$

Among them, the weight matrix $W_o \in R_k \times m$, $b_0 \times R_k$.

3.4   Rating Prediction

The outer product interaction hierarchy [29] is used to get high-order interaction information between different feature dimensions for the obtained user and item representations X_u and X_i. First, the exterior product feature interaction graph of the user feature vector and the item feature vector is obtained. Then, multi-convolutional hidden layers are used on the graph to extract feature representations at a higher level of abstraction. Finally, a latent factor model based on a matrix factorisation algorithm is used for scoring prediction.

Outer product interaction layer: The obtained user comment set feature vector X_u and item comment set feature vector X_i is known. In the same vector space, the user ID embedding and item ID embedding are combined with user comment set feature X_u and item comment set feature X_i. Combined to represent user feature embedding P_u and item feature embedding Q_i, such as Eqs. 16 and 17

$$P_u = p_u + X_u \tag{16}$$

$$Q_i = q_i + X_i \tag{17}$$

Among them, p_u and q_i represent the user ID embedding and item ID embedding, respectively, used to uniquely identify the target users and items participating in the scoring prediction. X_u and X_i denote user review set features and item review set features, respectively.

Perform outer product interaction on user feature embedding P_u and item feature embedding Q_i to obtain an external product interaction graph E, as shown in Eq. 18

$$E = P_u \bigotimes Q_i = P_u Q_i^T \tag{18}$$

where E is a $k \times k$ matrix.

Multi-convolutional Hidden Layers: This paper adopts multi-convolutional hidden layers to mine feature representations from the above matrices at a more abstract level. Convolution processing is performed on E, and the i-th layer is as follows:

$$E_1 = relu(E \times K'|b') \tag{19}$$

$$E_i = relu(E_{i-1} \times K'|b_i') \tag{20}$$

Among them, K' represents the convolution kernel, $\times$ represents the convolution operation, and bi' represents the bias term of the i-th layer. The final output tensor size is $1 \times 1 \times n$, and the vector V is obtained by adjusting the dimension. Finally, the output z of the multi-convolution hidden layer is calculated through Eq. (21).

$$z = W_e V + b_e \tag{21}$$

Among them, W_e represent the weight matrix with a size of $1 \times n$; b_e represent the bias term with a size of 1×1.

Rating Prediction Layer: In this paper, the latent factor model based on the matrix factorisation algorithm is used to predict user u's rating Ru, i for item i, as shown in formula (22):

$$R_{u,i} = z + b_u + b_i + \mu \tag{22}$$

Among them, z represents the output of the multi-convolution layer, reflecting the user's local preference for the item. bu and bi represent the user and item bias items, respectively, reflecting the influence of different users and items on the rating data. Finally, μ represents the global bias item, the average of all rating data, reflecting other datasets' differences in user ratings.

4   Network Model Training

The main goal of this paper is to perform score prediction, which is regarded as a regression task, and the mean square error (MSE) is used for network model training. The objective function is as Eq. 23.

$$L_r = \sum R^u, i \in D(R_u, i - R^u, i)^2 \tag{23}$$

where R^u, i is the given score on dataset D and $R_{u,i}$ is the predicted score. To alleviate the problem of overfitting, this paper introduces the L2 regular term into the objective function. That is, the L2 norm is added to the root mean square error of the loss function, as shown in Eq. 24.

$$L = L_r + \lambda \left\| \Delta \right\|^2 \tag{24}$$

When the model is under-fitting, it is only necessary to adjust the regularisation factor λ, reduce or even set it to zero; when over-fitting, increase it.

5   Experimental Setup and Result Analysis

5.1   Data Set

This paper used four publically available corpus, namely MedHelp (https://www.medhelp.org/ Accessed on Feb 12, 2022), Medline (https://pubmed.ncbi.nlm.nih.gov/ Accessed on Feb 2, 2022), Diego laboratory (http://diego.asu.edu/downloads/ Accessed on Feb 18, 2022.), Daily Strength (https://www.dailystrength.org/ Accessed on Feb 8, 2022) for crawled adverse drug reactions data set. Table 1 summarises the total number of patients, disease ADR, and total clinical query/advice for each dataset. The ratings of these datasets are in the interval [1, 5], and each clinical query has a corresponding patient comment time.

5.2   Evaluation Parameter

The root means square error (RMSE) was used as the evaluation index in the experiment. When the rating prediction result is a real value, the root mean square error is the square root of the ratio of the squared sum of the error between the rating prediction value $R_{u,i}$ and the true value $\hat{R}_{ui}$ and the ratio of the number N of all test instances, as shown in Eq. 25. The root mean square error is used to reflect the accuracy of the recommendation algorithm in rating prediction. The smaller the root mean square error, the better the performance of the model; on the contrary, the larger the root mean square error, the higher the dispersion of the error, and the worse the performance of the model, which reflects that in the scoring prediction, the predicted score deviates from the true value.

$$RMSE = \sqrt{\frac{1}{n} \sum_{u,i} (R_{ui} - \hat{R}_{ui})^2} \tag{25}$$

5.3   Performance Evaluation

The proposed model is compared with the following conventional models.

1. **Probabilistic Matrix Factorization Model (PMF) (Yang and Zhao, 2019):** It only uses the patient/expert rating information for the clinical query/comment to model the latent factors of patients and disease through matrix decomposition.

2. **Non-negative matrix factorization (NMF) (Meaney et al., 2022):** Only using the rating matrix information, considering that the patient/expert ratings for the clinical query/comment are all positive, the main feature is that the decomposed matrix elements are all positive, to learn patients, disease and ADR representations.

3. **HFT model (hidden factors as topics) (Yahyavi et al., 2021):** It uses review texts and ratings to jointly recommend and models patients, disease and ADR through rating matrix decomposition and LDA (latent Dirichlet allocation) without capturing the contextual semantics and word order of words.

4. Convolutional matrix factorization (ConvMF) (Chen et al., 2022): Effectively utilise the contextual information of clinical comments, fuse Convolutional Neural Network (CNN) into Probabilistic Matrix Factorization (PMF), use CNN to Extract contextual features from the clinical comment repository.

5. **Deep Collaborative Neural Network Model (DeepCoNN) (Yang et al., 2021):** Using all patient, disease and ADR clinical comment information, two parallel CNN networks are used to learn user preferences and item features from the patient query set, disease and ADR comment set, respectively.

6. Attn+CNN model (Usama et al., 2020): Attention-based CNN used patient query/comment text and rating information for the recommendation, uses CNN and applies attention to words in clinical comment documents to learn patient, disease and ADR representations.

7. **NARRE model (Su et al., 2022):** Use the patient ratings of disease and ADR information to make recommendations. Based on the DeepCONN model, this model considers the importance of different clinical comments for patient, disease and ADR modelling, introduces an attention mechanism into the model to explore the validity of clinical comment/ query, and selects valid ADR to provide model interpretability.

5.4   Experimental Setup and Parameter Selection

In the experiments, the clinical term embeddings are initialised by GloVe's pre-trained embeddings in the SIDER and Pharma -GKD corpus to initialise the clinical term embedding matrix. 80% of the patient, disease and ADR pairs are randomly selected for training, 10% for validation, and 10% for testing for each dataset. Each experiment was repeated three times independently, and the optimal value of RMSE was taken each time. The final result is the average of the optimal RMSE values of three experiments. The adjustment range of the parameters is shown in Table 2.

In the experiment, the grid search method is used to find the parameters that can make the model reach the optimum. The parameters are adjusted many times on the validation set, and the optimum parameters obtained are shown in Table 3.

Table 1. Data Set Description

Data Set	No of Patient	ADR	No of Clinical Query
Medhelp	5623	1658	30256
Medline	7568	2536	45236
Diego Laboratory	8562	2631	56234
Daily Strength	7526	2502	46235

6   Experimental Results and Analysis

1. **Experiment 1** Overall recommendation accuracy comparison Table 4 shows the RMSE values of the proposed model and existing models on four datasets. Through the analysis, the following conclusions can be drawn:

    (a) The traditional recommendation methods (such as PMF and NMF models) that only consider the patient rating information for disease/ADR and other methods that do not use the clinical data are recommended by HFT, DeepCoNN, Attn+CNN, and NARRE. However, it can be seen that commenting on clinical comments with clinical data can provide rich information on Drug -ADR preferences and disease attributes, which is very important for modelling patients, disease and ADR.

    (b) Among the models that consider both Drug rating and clinical comment information, DeepCoNN, Attn+CNN, and NARRE models based on deep learning technology are better than HFT models that cannot capture contextual semantics and word order.

    (c) Whether it is the Attn+CNN model that only applies the attention mechanism to the words in the clinical comment or the NARRE model that only uses the attention mechanism to the comment text in the review document, its recommendation performance is better than other methods without the attention mechanism.

2. **Experiment 2:** The effectiveness of different levels of attention This paper uses Experiment 2 to verify the effectiveness of different levels of attention, including word-level attention and comment-level attention. Compare the RMSE of the proposed model RHATR and its variants $RHATR_W$ (with only clinical word-level attention) and $RHATR_R$ (with only clinical comment-level attention) on four datasets as shown in Fig. 2.

    Figure 2 verifies that clinical word-level attention and clinical comment level attention effectively improve recommender system performance. Different clinical words contribute differently to learning patient, disease, and ADR representations in a single clinical comment. Applying clinical word-level attention to the clinical comment helps identify critical clinical terms and is effective for encoding clinical comment. The degree of preference for clinical comment differs, and the correlation between patient, disease and ADR is also distinct. Applying clinical comment set attention for patient query/ comment sets and ADR sets is helpful for the performance improvement of

Table 2. Parameter of Model

Model Parameter	Range
Word Embedding Dimension	100; 200; 300
Word-Level Attention Hyper-Parameter R	15; 20; 25;30
Word-Level Attention Hyper- Parameter V	20; 30; 40;50
Number of Neurons in the Hidden Layer of BI-LSTM	16; 32; 64; 128
L2 Regularization Parameter	0.001; 0.01; 0.1; 1

Table 3. Selected Parameter's

Parameter	Value
Word Embedding Dimension	200
Word-Level Attention Hyperparameters R and V	40; 50
Number of Neurons in the Hidden Layer of BI-LSTM	324
Kernel Size and Number of Kernels in Word-Level Attention CNN Networks	3,32; 16
Size and Number of Convolution Kernels in the Multi-Convolution Hidden Layer	2,2; 16
L2 Regularization Parameter λ	0; 1

recommender systems. Compared to $RHATR_W$, the importance of clinical comment is better than the importance of clinical term for the patient, disease and ADR modelling accuracy.

3. **Experiment 3:** To verify the validity of the time factor, Eq. 12 is improved, indicating that patient comments at different times are equally important for modelling clinical preferences and disease/ ADR features. Equation 13 is changed to $x_{up} = softmax(h^T relu(W_o o_{uP}^T + W_i i_{uP}^T + b_1) + b_2)$ for experiment represented by $RHATR_{NT}$. Table 5 shows the experimental results.

As shown in Table 5, the model that introduces time information has a smaller result on the root mean square error than the model that does not consider the time factor. This is because for patients, as patients grow older, their interests and preferences will continue to change. Equations 12 and 13 illustrate the importance of the influence of time information on patient preferences and disease/ADR characteristics. Taking the time information into account, making dynamic recommendations for patients is necessary. Therefore, it is adequate to model patients' preferences and disease/ADR features by combining patient comments and time factors to improve recommendation performance.

Table 4. Performance Evaluation on RMSE of Different Model

Data Set	Model							
	PMF	NMT	HFT	ConvMF	DeepCoNN	Attenation + CNN	NARRE	RHATR
Medhelp	0.9664	0.9085	0.9157	0.9385	0.9156	0.9256	0.8924	0.8534
Medline	0.7937	0.7426	0.7698	0.7938	0.7828	0.7634	0.7442	0.6245
Diego Laboratory	0.8658	0.8256	0.8345	0.8647	0.8568	0.8226	0.8138	0.7628
Daily Strength	0.9423	0.9289	0.9312	0.9456	0.9278	0.9372	0.9124	0.8924

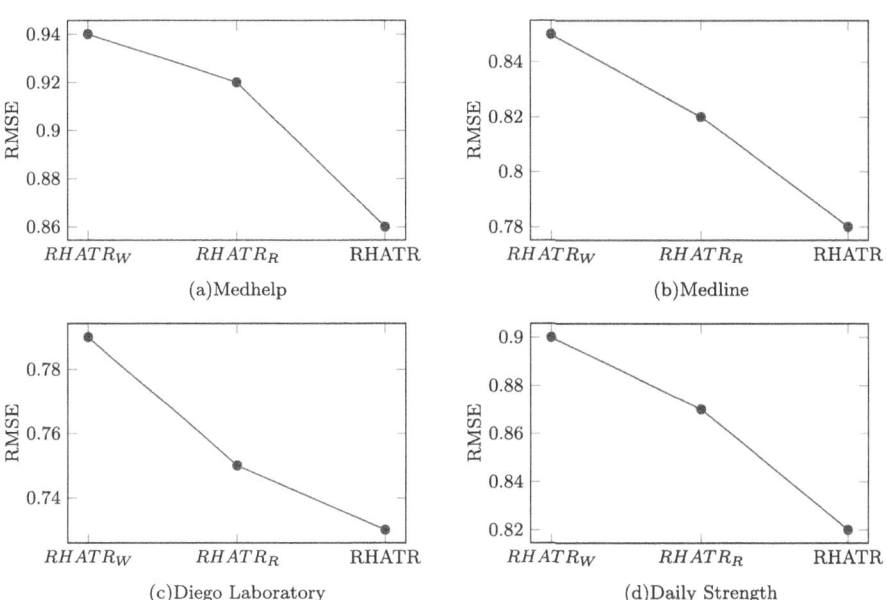

Fig. 2. Evaluation of effectiveness of Attention layer on Different Dataset

Table 5. Time Effectiveness on Different Dataset

Data Set	RHATR	RHATR_NT
Medhelp	0.9256	0.9452
Medline	0.8068	0.8526
Diego Laboratory	0.7264	0.7648
Daily Strength	0.9389	0.9562

7   Conclusion

This paper proposes a recommendation method that fuses clinical comments-level attention and time information by using patient ratings for ADR, clinical suggestions, and time factors. RHATR models patients, diseases and ADR by extracting clinical keywords, medically sentimental words and other information from a single clinical tweet; and applies review-level attention to patient, disease and ADR review sets that introduce temporal factors further to learn dynamic representations of patient, illness and ADR. This work Mining the dynamic attributes of patients, diseases and ADR to make accurate patient recommendations. Various comparative experiments show that the method in this paper is lower than other methods in the prediction score error. However, considering that the GloVe pre-trained word vectors used in this paper belong to static encoding, the same word has similar semantics in different clinical tweets, which may distort the meaning of the clinical word in the context. Future work will consider that the representation of each clinical word needs to change dynamically according to the different contexts of the disease to improve the recommendation performance.

References

Chen, Y., Wang, Y., Ding, Y., Xi, S., Wang, C.: RGCNCDA: relational graph convolutional network improves circRNA-disease association prediction by incorporating micrornas. Comput. Biol. Med. **143**, 105322 (2022)

Chen, Y., Lin, W., Wang, J.Z.: A dual-attention-based stock price trend prediction model with dual features. IEEE Access **7**, 148047–148058 (2019)

Forestier, G., Petitjean, F., Riffaud, L., Jannin, P.: Automatic matching of surgeries to predict surgeons' next actions. Artif. Intell. Med. **81**, 3–11 (2017)

Jiang, L., Yang, C.C.: User recommendation in healthcare social media by assessing user similarity in heterogeneous network. Artif. Intell. Med. **81**, 63–77 (2017)

Lin, W.-Y., Lo, C.-F.: Co-training and ensemble based duplicate detection in adverse drug event reporting systems. In: 2013 IEEE International Conference on Bioinformatics and Biomedicine, pp. 7–8 (2013)

Liu, Y., Aickelin, U.: Detect adverse drug reactions for drug alendronate. In: 2012 Second International Conference on Business Computing and Global Informatization, pp. 820–823 (2012a)

Liu, Y., Aickelin, U.: Detect adverse drug reactions for drug simvastatin. In: 2012 Fourth International Conference on Multimedia Information Networking and Security, pp. 246–249 (2012b)

Ma, L., Zhang, H., Li, T., Yuan, D.: Deep learning and social relationship based cooperative caching strategy for D2D communications. In: 2019 11th International Conference on Wireless Communications and Signal Processing (WCSP), pp. 1–6 (2019)

Meaney, C., et al.: Non-negative matrix factorization temporal topic models and clinical text data identify COVID-19 pandemic effects on primary healthcare and community health in toronto, canada. J. Biomed. Inform. **128**, 104034 (2022)

Meng, S., et al.: Privacy-aware factorization-based hybrid recommendation method for healthcare services. IEEE Trans. Industr. Inform. (2022)

Miah, S.J., Hasan, N., Hasan, R., Gammack, J.: Healthcare support for underserved communities using a mobile social media platform. Inf. Syst. **66**, 1–12 (2017)

Nan, S., Hou, J., Liu, P., Han, H., Liu, Z., Zhang, R.: An intelligent and personalized tobacco brand recommendation method. In: 2015 International Conference on Intelligent Transportation, Big Data and Smart City, pp. 98–101 (2015)

Pandit, A.A., Dubey, S.A.: A comprehensive review on adverse drug reactions (ADRS) detection and prediction models. In: 2021 13th International Conference on Computational Intelligence and Communication Networks (CICN), pp. 123–127 (2021)

Park, J., Lee, S., Kim, J.: CHNE: context-aware heterogeneous network embedding. In: 2021 IEEE International Conference on Big Data and Smart Computing (BigComp), pp. 342–345 (2021)

Pourebrahim, B., Keyvanpour, M.: Adverse drug reaction detection using data mining techniques: a review article. In: 2020 10th International Conference on Computer and Knowledge Engineering (ICCKE), pp. 118–123 (2020)

Rho, M.J., Kim, S.R., Park, S.H., Jang, K.S., Park, B.J., Choi, I.Y.: Development common data model for adverse drug signal detection based on multi-center EMR systems. In: 2013 International Conference on Information Science and Applications (ICISA), pp. 1–7 (2013)

Schiza, E.C., Kyprianou, T.C., Petkov, N., Schizas, C.N.: Proposal for an eHealth based ecosystem serving national healthcare. IEEE J. Biomed. Health Inform. **23**(3), 1346–1357 (2019)

Song, Y., et al.: Federated learning application on telecommunication-joint healthcare recommendation. In: 2021 IEEE 21st International Conference on Communication Technology (ICCT), pp. 1443–1448 (2021)

Su, Y., Yuliang Shi, W., Lee, L.C., Guo, H.: TahdNet: time-aware hierarchical dependency network for medication recommendation. J. Biomed. Inform. **129**, 104069 (2022)

Sulthana, A.R., Ramasamy, S.: Ontology and context based recommendation system using neuro-fuzzy classification. Comput. Electr. Eng. **74**, 498–510 (2019)

Sun, L., Sun, S., Wang, T., Li, J., Lin, J.: Parallel ADR detection based on spark and BCPNN. Tsinghua Sci. Technol. **24**(2), 195–206 (2019)

Usama, M., Ahmad, B., Wenjing Xiao, M., Hossain, S., Muhammad, G.: Self-attention based recurrent convolutional neural network for disease prediction using healthcare data. Comput. Methods Programs Biomed. **190**, 105191 (2020)

Wen, H., Song, J., Pan, X.: Physician recommendation on healthcare appointment platforms considering patient choice. IEEE Trans. Autom. Sci. Eng. **17**(2), 886–899 (2020)

Chen, W., Deng, J.: A study on signal detection and automatic warning algorithm for adverse drug reaction. In: 2008 International Conference on Computer Science and Software Engineering, vol. 4, pp. 202–205 (2008)

Yahyavi, S.T., Hoobehfekr, S., Tabatabaee, M.: Exploring the hidden curriculum of professionalism and medical ethics in a psychiatry emergency department. Asian J. Psychiatr. **66**, 102885 (2021)

Yang, C.C., Jiang, L.: Enriching user experience in online health communities through thread recommendations and heterogeneous information network mining. IEEE Trans. Comput. Soc. Syst. **5**(4), 1049–1060 (2018)

Yang, C.C., Yang, H.: Mining heterogeneous networks with topological features constructed from patient-contributed content for pharmacovigilance. Artif. Intell. Med. **90**, 42–52 (2018)

Yang, C.C., Zhao, M.: Mining heterogeneous network for drug repositioning using phenotypic information extracted from social media and pharmaceutical databases. Artif. Intell. Med. **96**, 80–92 (2019)

Yang, H., Yang, C.C.: Harnessing social media for drug-drug interactions detection. In: 2013 IEEE International Conference on Healthcare Informatics, pp. 22–29 (2013)

Yang, H., Yang, C.C.: Discovering drug-drug interactions and associated adverse drug reactions with triad prediction in heterogeneous healthcare networks. In: 2016 IEEE International Conference on Healthcare Informatics (ICHI), pp. 244–254 (2016)

Yang, Z., Wei, X., Chen, R.: A deep learning-based multi-turn conversation modeling for diagnostic Q&A document recommendation. Inf. Process. Manage. **58**(3), 102485 (2021)

Zaman, N., Li, J.: Semantics-enhanced recommendation system for social healthcare. In: 2014 IEEE 28th International Conference on Advanced Information Networking and Applications, pp. 765–770 (2014)

Zhang, T., et al.: Gated iterative capsule network for adverse drug reaction detection from social media. In: 2020 IEEE International Conference on Bioinformatics and Biomedicine (BIBM), pp. 387–390 (2020)

Zhang, Y., Ning, Y., Huan, Z.: An intelligent attentional-GRU-based model for dynamic blood glucose prediction. In: 2021 2nd International Conference on Artificial Intelligence and Education (ICAIE), pp. 10–14 (2021)

Zhou, L., Zhang, D., Yang, C.C., Wang, Y.: Harnessing social media for health information management. Electron. Commer. Res. Appl. **27**, 139–151 (2018)

Evaluation of Different Mapping Schemes for Detection of Tandem Repeats in DNA Sequences

Yashpal Yadav[1], Sanjeev Narayan Sharma[2], and Devendra Kumar Shakya[3(✉)]

[1] Department of Electronics and Instrumentation Engineering, Samrat Ashok Technological Institute, Vidisha, M.P., India

[2] Department of Electronics and Communication Engineering, PDPM, IIITDM Jabalpur, Jabalpur, M.P., India
snsharma@iiitdmj.ac.in

[3] Department of Electronics and Communication Engineering, Samrat Ashok Technological Institute, Vidisha, M.P., India
dkshakya.ec@satiengg.in

Abstract. Genomic sequences have a significant proportion of repetitive DNA nucleotides. The repetitive regions are functionally important and can be broadly classified as tandem repeats (TRs) and dispersed repeats. Computational methods have been used to identify the presence of these regions. Signal Processing (SP) based computational methods require the mapping of DNA sequences into numeric ones. Large numbers of mapping schemes have been reported for this purpose. The mapping strategy employed affects the sensitivity and specificity of the SP-based algorithms. In this work performance of Ramanujan Fourier transform based TR detection algorithm with different mapping schemes has been studied. Complex mapping scheme has been identified as the best mapping scheme for the TR detection.

Keywords: Ramanujan Fourier Transform · DNA sequences · Tandem Repeats

1 Introduction

Genomics is among the most incredible scientific innovation of our times. It contains deoxyribonucleic acid (DNA) which contains instructions for life. The DNA sequences consist of four nucleotides namely, Cytosine (C), Guanine (G), Adenine (A) and Thymine (T). Approximately 50% of the human genome is made up of repetitive sequences. Commonly occurring repeats in DNA are tandem repeats (TRs) and dispersed repeats. If in a DNA sequence two or more adjacent copies of repeating nucleotide pattern are present than it is considered as TRs. Further, TRs can be classified into approximate tandem repeats (ATRs) and exact tandem repeats (ETRs). Due to insertion, deletion and mutation in DNA sequences TRs are mostly ATRs. TRs in DNA are very important because their knowledge is essential for phylogenic studies, disease diagnosis, DNA mapping and fingerprinting [1]. Repeated DNA sequences are found in abundance in

© The Author(s), under exclusive license to Springer Nature Switzerland AG 2025
M. Gupta et al. (Eds.): MISS 2023, CCIS 1952, pp. 217–228, 2025.
https://doi.org/10.1007/978-3-031-69115-7_14

the eukaryotic genome. One of the characteristics of these repeated DNA sequences is the length of repeat pattern and number of copies of repeat pattern. Long repeat units referred to as satellite DNA have more than a hundred bases in the repeat pattern and also can go up to several thousand bases. For a medium-length repeat, the repeat unit is of the order of 9–80 bases and is referred to as minisatellites. DNA regions with repeat units of one to six base pairs are called microsatellites or short tandem repeats (STRs) [2]. Certain neurological, neuromuscular and heredity diseases such as Myotonic dystrophy, Fragile X-syndrome and Ataxia, [3] are caused due to the abnormal expansion of the STRs. Also, intron retention [4] takes place due to microsatellite expansion.

Since, TRs in DNA have important biological applications; many tools for detections of TRs along with the accurate prediction of structure, start position, end position, pattern size and number of copies have been developed. Broadly, TR detection tools are of two types; the first category of tools which are based on the experimental techniques use DNA amplification methodologies [5], and are expensive and time consuming. The second class of tools is based on computational methods and can reduce the experimental search. They either use string matching or signal processing algorithms. Benson developed tandem repeat finder (TRF) which is widely used string matching algorithm based on probability model [6]. Markov additive process [7] is another probability based method which finds ATRs as local sum. These two methods use string matching algorithms. The reported TR detection signal processing tools include discrete Fourier transform [8], short-time periodicity transform [9], exact periodicity subspace decomposition (EPSD) [10], quarternion periodicity transform [11], auto regression model (AR) [2] and wavelet based empirical model [12]. Adaptive S-transform has been applied in [13] to identify micro satellites in DNA, but large periods cannot be detected efficiently using this method. By replacing Gaussian window with Kaiser window larger periods can be accurately detected up to some extent [14]. Ramanujan filter bank [15] is a filter based technique that has also been used for TRs detections in DNA sequences. Recently reported short-time Ramanujan Fourier transform (ST-RFT) [16] can detect both smaller and larger periods more accurately and efficiently.

In Genomic Signal Processing, the first and foremost step is to convert DNA characters into corresponding numeric sequence and this conversion is commonly known as mapping. The effect of different mapping schemes for exonic detections in DNA [17] have been reported, but similar studies on TRs have not been conducted. This work primarily focuses on the study of different mapping schemes in TR detection, using recently reported ST-RFT, with an objective to identify best mapping scheme for the detection of TRs.

2   Mapping Schemes

In recent years, with the advancement in the area of genomics, a large number of mapping schemes have been studied and compared [17–21]. In [22], the mapping schemes which are based on DNA structural properties have been used. Table 1 summarizes different types of mapping schemes used in this work. The last four mapping schemes which are based on DNA structural properties are described in Table 2.

Table 1. Different Mapping Schemes

S.No	Mapping Schemes	DNA Representation	S(n) = [CGAT]
1	Complex	A = 1 + j, C = −1 + j, G = −1−j, T = 1−j	[−1 + j, −1−j, 1 + j, 1−j]
2	EIIP	A = 0.1260, C = 0.1340, G = 0.0806, T = 0.1335	[0.1340, 0.0806, 0.1260, 0.1335]
3	Paired Numeric	A or T = 1, C or G = -1	[−1, −1, 1, 1]
4	Integer	C = 1, G = 3, A = 2, T = 0	[1, 3, 2, 0]
5	Real	C = 0.5, G = −0.5, A = −1.5, T = 1.5	[0.5, −0.5, 1.5, −1.5]
6	Atomic Number	A = 70, C = 58, G = 78, T = 66	[58, 78, 70, 66]
7	Voss	$X_n = 1$ for S(n) = x $X_n = 0$ for S(n) ≠ x X_n applies to any C_n, G_n, A_n, and T_n	$C_n = [1, 0, 0, 0]$ $G_n = [0, 1, 0, 0]$ $A_n = [0, 0, 1, 0]$ $T_n = [0, 0, 0, 1]$
8	Z-curve	$$\begin{bmatrix} x_n \\ y_n \\ z_n \end{bmatrix} = 2 \times \begin{bmatrix} 1\ 0\ 1\ 0 \\ 1\ 1\ 0\ 0 \\ 1\ 0\ 0\ 1 \end{bmatrix} \times \begin{bmatrix} x_A[n] \\ x_C[n] \\ x_G[n] \\ x_T[n] \end{bmatrix} - \begin{bmatrix} 1 \\ 1 \\ 1 \end{bmatrix}$$	$$\begin{bmatrix} x_n \\ y_n \\ z_n \end{bmatrix} = \begin{bmatrix} -1\ \ 1\ \ 1\ -1 \\ 1\ -1\ 1\ -1 \\ -1\ -1\ 1\ \ 1 \end{bmatrix}$$
9	DNA bending stiffness	Using Table 2	[85, 60, 20]
10	Duplex free energy	Using Table 2	[−28, −15, −9]
11	Duplex disrupt energy	Using Table 2	[36, 16, 9]
12	Propeller twist mapping	Using Table 2	[−1003, −1348, −1501]

3 Short-Time Ramanujan Fourier Transform

The legendary Indian mathematician, Shrinivasa Ramanujan in 1918 introduced a summation called Ramanujan Sums (RS) [23]. The qth RS is for any positive integer q -

$$c_q(n) = \sum_{\substack{k=1 \\ (k,\, q)=1}}^{q} e^{j2\pi kn/q} \qquad (1)$$

Table 2. Numeric Conversion Using Structural Features

Dinucleotide	Propeller Twist	Duplex Disrupt Energy	Duplex Free Energy	DNA Bending Stiffness
CG	−1003	36	−28	85
GT	−1310	13	−25	60
GC	−1106	31	−23	85
CC	−810	31	−23	130
GG	−810	31	−23	130
CA	−945	19	−17	60
TG	−945	19	−17	60
AC	−1310	13	−15	60
AG	−1400	16	−15	60
CT	−1400	16	−15	60
GA	−1348	16	−15	60
TC	−1348	16	−15	60
AA	−1866	19	−12	35
TT	−1866	19	−12	35
AT	−1501	9	−9	20
TA	−1185	15	−9	20

where, notation $(k, q) = 1$ represents the greatest common divisor (gcd) for k and q.

Using linear combination of RS several standard arithmetic functions can be expressed by –

$$x(n) = \sum_{q=1}^{\infty} \alpha_q \cdot c_q(n) \tag{2}$$

where, the coefficients of the Ramanujan Fourier transform (RFT) are α_q.

The RFT coefficients α_q, can be calculated using Eq. (3) that has been used in many applications [24–27] –

$$\alpha_q = \frac{1}{\varphi(q)} \left(\underset{N \to \infty}{Lim} \frac{1}{N} \sum_{n=1}^{N} x(n) \cdot c_q(n) \right) \tag{3}$$

where, $\phi(q)$ denotes Euler's totient function which is equal to the number of integers k, with $1 \leq k \leq q$ for $(k, q) = 1$. Since most of the signal processing applications involve non-stationary signals, therefore ST-RFT computation [28] is more effective. Using the concept of windowing, ST-RFT can be obtained from RFT and is given by –

$$ST - RFT_s(p, q) = \frac{1}{\varphi(q)} \lim_{N \to \infty} \frac{1}{N} \sum_{n=1}^{N} x(n) \, w^*(n - p) \, c_q(n) \tag{4}$$

where, p & q are time and period domain indices respectively, and $w(n)$ represents window function. The modulus square of the ST-RFT, obtained using (4), provides the ST-RFT spectrogram. Using ST-RFT spectrogram time-period analysis of genomic sequences has been performed to identify regions where tandem repeats with different periodicities are present.

4   Methodology

The flow chart for TR detection using ST-RFT is shown in Fig. 1 [16].

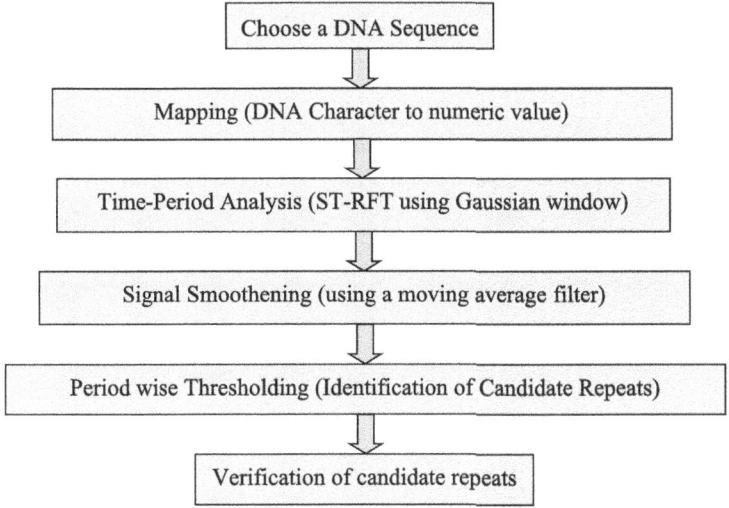

Fig.1. Step by step procedure used for TR detection

For demonstration purposes, a DNA sequence having accession number X64775 is used as an example sequence. The process of conversion of DNA character sequences into numeric form to use SP based tools is known as mapping. The results for various mapping schemes as discussed in Sect. 2 have been computed and the comparative study of different mapping schemes is discussed in the later part of this paper, but for demonstration purposes, electronic ion interaction potential (EIIP) mapping has been used here on the example sequence.

The ST-RFT spectrogram computation for sequence X64775 is done using Gaussian window of length 16. Figure 2 shows period-3 spectrum values for X64775.

Since the unprocessed or raw spectrum shown in Fig. 2 does not reveals accurate TR detections and may lead to inaccuracies, so signal smoothing has to be done for each period individually using Moving Average (MA) filter described by –

$$y(n) = \frac{1}{L} \sum_{m=0}^{L-1} x[n-m] \tag{5}$$

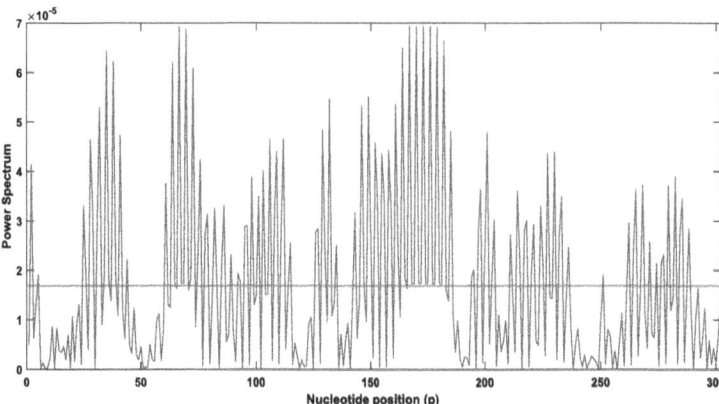

Fig. 2. Unprocessed ST-RFT spectrum for period-3

The value of L is so selected that it should be four times the period to be identified [16]. Figure 3 shows period-3 spectrum of X64775, filtered using an MA filter of length 12.

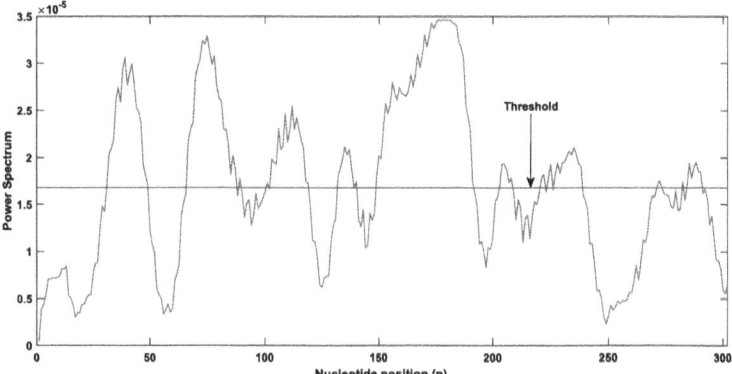

Fig. 3. Smoothened ST-RFT spectra for period-3

Similarly, raw spectrums for the other periods can also be filtered using MA filter. The time-period plot shown in Fig. 4 has been obtained using smoothened spectra for period-2 to period-8.

From the spectra obtained in Fig. 4 it is nearly impossible to identify TRs, so thresholding is required. The ST-RFT spectra are threshold individually period-wise with average or mean value used as the threshold limit for each period. For period-3 the threshold value is shown in Fig. 3 and for other periods it can be obtained in similar manner. Now, those values in spectrums which are above threshold limit are assigned a value of one, otherwise zero. After thresholding the binary time-period plot is shown by Fig. 5. The advantage of thresholding is that the possible candidate repeats get highlighted with unit value.

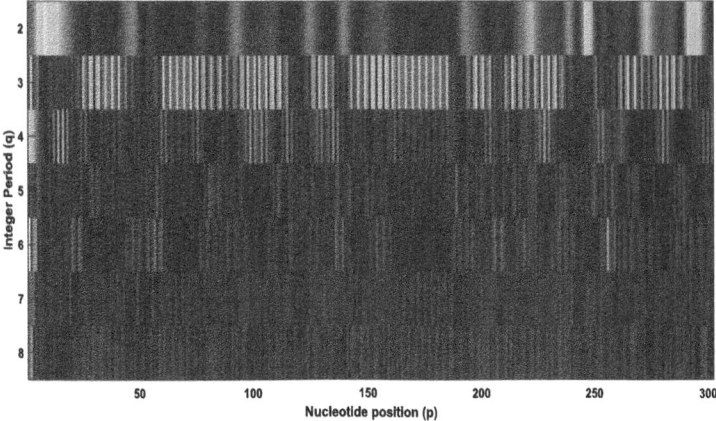

Fig. 4. Time-period plot of X64775 after signal smoothening

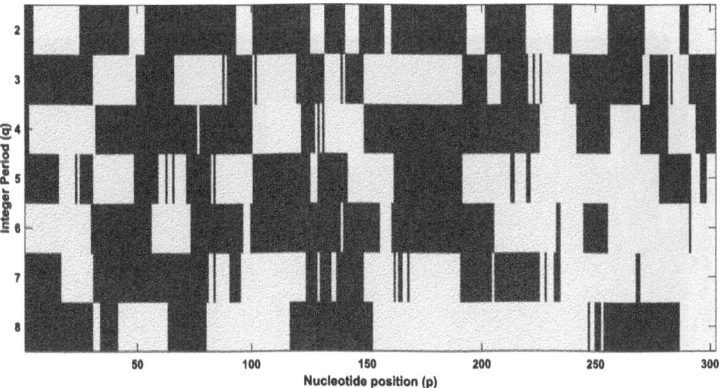

Fig. 5. Binary time-period plot obtained after thresholding for X64775

Once thresholding is done, the possible candidate repeats in Fig. 5 are subjected to verification phase. The post-processing algorithm proposed by Boeva *et.al.* [29] is used for verification of candidate repeats. The verification algorithm [2, 13, 16, 29] uses a mismatch ratio of $\delta = 0.6$ for deciding the number of possible mutations [2, 13, 29]. For reducing the computational load, only those regions of candidate repeats are considered whose length is twice the period to be computed. Using the above steps, candidate repeats of the sequence X64775 have been verified and the plot obtained is shown in Fig. 6. Table 3 shows candidate repeats (S'), verified repeats (Ŝ) and corresponding candidate repeats along with the consensus pattern, number of copies of the repeats and percentage match for sequence X64775 using EIIP Mapping.

The DNA sequences used in this work have been compiled in Table 4 along with their accession numbers and can be downloaded from NCBI database [30]. Selection of window length is an important parameter which significantly affects TRs detections. The criteria used for selection of suitable window length has been discussed in [16]. For

Table 3. TRs detection for X64775 by ST-RFT (EIIP Mapping)

Period	S′	Ŝ	Consensus Pattern	No. of Copies	Percentage Match
2	5–24	5–8	GA	02	100%
	240–255	246–249	CG	02	100%
	272–286	274–277	GA	02	100%
3	31–49	31–42	GCC/GCA	04	91.6%
	67–87	67–75	GGC/GGT	03	88.8%
		77–82	GAC/GAG	02	83.3%
	103–119	103–108	GGT/GGC	02	83.3%
	132–138	132–137	CCG/CTG	02	83.3%
	149–191	149–190	GGC/GAC	14	90.4%
4	3–31	5–12	GAGA/GCGA	02	87.5%

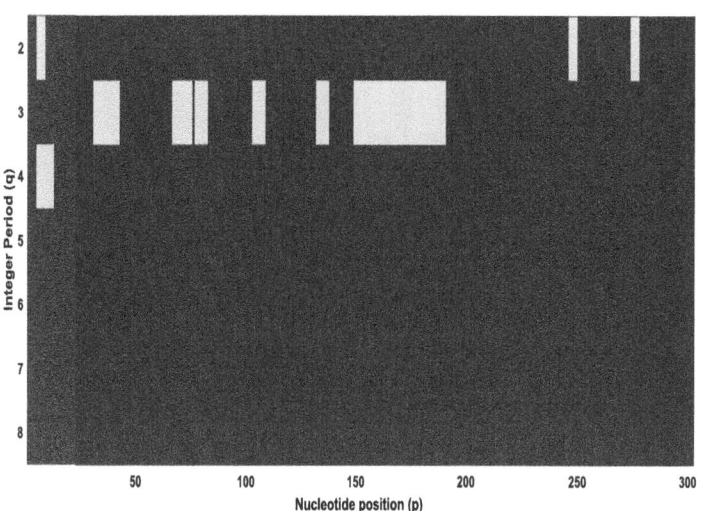

Fig. 6. TRs detected for X64775 after verification phase

all the DNA sequences Gaussian window has been used for the ST-RFT computation and the different window lengths used are given in Table 4.

Table 4. DNA sequences used for TRs detection

S. No	Accession Number	Sequence Length	Description	TR's Location	Window length used
(1)	X64775	303	O. sativa short highly repeated, interspersed DNA	1421186	16
(2)	M65145	1072	Human microsatellite repeat KLK1 AC DNA (Homosapiens)	8601900	32
(3)	AC004848	2000	Homo sapiens PAC clone RP4-649P17 from 7 (Homosapiens)	2611298, 1403–1433	32

5 Results

In this work, the effect of various mapping schemes on TRs detection using the ST-RFT have been studied for DNA sequences having accession numbers X64775, M65145 and AC004848. Table 5 shows total numbers of TR detections at each period for these sequences with 12 different mapping schemes described in this work. Also, in this study another parameter called match ratio (*MR*) [11] has been used –

$$MR = \frac{N_C}{L_R} \times 100 \tag{6}$$

Here, N_c denotes total number of matched characters between the fitted and the actual segment, and L_R stands for the repeat segment's overall length. In Table 5, all the TR detections which have *MR* of less than 80% have been discarded. Thus, for period-3 onwards only those repeats which have *MR* of more than 80% are being included in Table 5 (for period-2, the *MR* is taken as 100%).

In Table 5 total detections for periodicities 1 to 16 are given for 12 mapping schemes for the three DNA sequences. Complex mapping scheme identifies 703 periodic regions which is maximum, closely followed by Voss and Z-curve mapping. Complex mapping scheme is having an additional advantage of less computational complexity as compared to other two mappings. Among the four mapping schemes which are based on DNA structural properties described in Table 2, propeller twist mapping identifies the maximum periodic regions but its detection is less compared to complex mapping scheme. Duplex free energy identifies least number of repeat patterns among all 12 mapping schemes.

Table 5. Comparative Study for Different Mapping Schemes using ST-RFT

S. No	Mapping Scheme	Sequence	Detections at Period															Detections for each sequence	Total Detections
			2	3	4	5	6	7	8	9	10	11	12	13	14	15	16		
1	Complex Mapping	X64775	12	44	06	–	04	–	04	–	–	–	–	–	–	–	–	70	703
		M65145	66	56	25	22	30	06	10	06	08	12	04	08	02	02	02	259	
		AC004848	83	109	46	42	26	12	20	06	06	04	02	06	10	02	–	374	
2	EIIP	X64775	06	27	02	–	–	–	–	–	–	–	–	–	–	–	–	35	553
		M65145	53	40	24	22	26	11	10	12	06	12	06	06	04	–	02	234	
		AC004848	68	80	38	28	20	04	14	28	10	02	–	04	06	–	02	284	
3	Paired Numeric	X64775	16	21	06	–	03	–	04	–	–	–	–	–	–	–	–	50	620
		M65145	61	54	30	24	20	06	10	06	08	12	06	06	02	02	–	247	
		AC004848	81	90	48	28	22	06	16	08	10	04	02	02	04	–	02	323	
4	Integer Mapping	X64775	10	18	02	–	08	–	02	–	–	–	–	–	–	–	–	40	621
		M65145	55	58	20	24	14	12	06	12	06	14	06	06	04	–	04	241	
		AC004848	85	75	38	40	22	08	16	10	14	02	02	08	10	08	02	340	
5	Real Mapping	X64775	14	35	04	–	10	–	02	–	–	–	–	–	–	–	–	65	615
		M65145	41	59	20	20	15	08	14	06	10	06	04	04	04	-	02	213	
		AC004848	58	92	46	50	22	09	20	04	10	02	02	08	06	08	–	337	
6	Atomic Number	X64775	10	12	04	02	–	–	02	–	–	–	–	–	–	–	–	30	553
		M65145	51	46	28	18	22	07	12	06	10	10	04	06	04	–	02	226	
		AC004848	109	78	32	24	16	06	14	04	08	–	–	04	–	–	02	297	
7	Voss Mapping	X64775	16	21	02	–	08	–	06	–	–	–	–	–	–	–	–	53	692
		M65145	61	50	31	22	30	08	14	10	10	12	06	08	04	02	04	272	
		AC004848	87	110	38	38	14	12	18	12	14	04	02	06	08	02	02	367	
8	Z curve	X64775	16	21	02	–	10	–	04	–	–	–	–	–	–	–	–	53	692
		M65145	61	50	31	22	30	08	14	10	10	12	06	08	04	02	04	272	
		AC004848	87	110	38	38	14	12	18	12	14	04	02	06	08	02	02	367	
9	DNA Bending Stiffness	X64775	16	23	06	02	12	–	–	–	–	–	–	–	–	–	–	63	573
		M65145	30	52	29	24	18	06	12	06	08	12	04	04	06	–	02	213	
		AC004848	70	84	42	30	20	06	12	04	07	06	–	04	08	–	04	297	
10	Duplex Free Energy	X64775	10	17	08	02	08	–	–	–	–	–	–	–	–	–	–	45	505
		M65145	43	40	24	18	20	04	16	08	08	12	04	06	04	–	02	209	
		AC004848	57	47	44	30	18	13	12	08	08	02	–	04	02	02	04	251	
11	Duplex Disrupt	X64775	10	15	04	02	06	–	02	–	–	–	–	–	–	–	–	39	557
		M65145	50	42	24	14	24	08	10	06	10	04	04	06	04	–	–	206	
		AC004848	73	92	31	30	14	08	26	06	10	04	–	04	06	04	04	312	
12	Propeller Twist	X64775	06	25	08	04	12	–	02	–	–	–	–	–	–	–	–	57	651
		M65145	45	49	31	18	16	04	10	02	10	12	06	04	04	–	–	211	
		AC004848	73	129	53	40	26	12	20	06	06	06	–	02	06	04	–	383	

6 Conclusion

In this work effect of selection of mapping scheme on the number of TR detections has been studied. TR detection has been done using the ST-RFT. Twelve different mapping schemes have been included in the study. Simulation studies establish that complex mapping scheme identifies maximum number of TRs. Also, the complex mapping scheme is computationally efficient, as after mapping only single sequence is obtained which reduces the number of computations significantly.

Declaration of Originality. This work is original, and authors have no conflict of interest. All authors are equally responsible and contributed to the work.

References

1. Jeffreys, A.J., Wileon, V., Thein, S.L.: Individual-specific 'fingerprints' of human DNA. Nature **316**, 76–79 (1985)
2. Zhou, H., Du, L., Yan, H.: Detection of tandem repeats in DNA sequences based on parametric spectral estimation. IEEE Trans. Inf. Technol. Biomed. **13**(5), 747–755 (2009)
3. McMurray, C.T.: Mechanisms of trinucleotide repeat instability during human development. Nat. Rev. Genet. **11**, 786–799 (2010)
4. Sznajder, L.J., et al.: Intron retention induced by microsatellite expansions as a disease biomarker. Proc. Natl. Acad. Sci. U.S.A. **115**(16), 4234–4239 (2018)
5. Weber, J.L., Wong, C.: Mutation of human short tandem repeats. Hum. Mol. Genet. **2**(8), 1123–1128 (1993)
6. Benson, G.: Tandem repeats finder: a program to analyze DNA sequences. Nucleic Acids Res. **27**(2), 573–580 (1999)
7. Spouge, J.L.: Markov additive processes and repeats in sequences. J. Appl. Probab. **44**, 514–527 (2007)
8. Sharma, D., Issac, B., Raghava, G.P.S., Ramaswamy, R.: Spectral repeat finder (SRF): identification of repetitive sequences using Fourier transforms. Bioinformatics **20**(9), 1405–1412 (2004)
9. Buchner, M., Janjarasjitt, S.: Detection and visualization of tandem repeats in DNA sequences. IEEE Trans. Signal Process. **51**(9), 2280–2287 (2003)
10. Gupta, R., Sarthi, D., Mittal, A., Singh, K.: A novel signal processing measure to identify exact and inexact tandem repeat patterns in DNA sequences. EURASIP J. Bioinf. Syst. Biol. **2007**, 1–8 (2007)
11. Brodzik, A.K.: Quaternion periodicity transform: An algebraic solution to the tandem repeat detection problem. Bioinformatics **23**(6), 694–700 (2007)
12. Jiang, R., Yan, H.: Detection and 2–dimensional display of short tandem repeats based on signal decomposition. Int. J. Data Min. Bioinform. **5**(6), 661–690 (2011)
13. Sharma, S.D., Saxena, R., Sharma, S.N.: Identification of microsatellites in DNA using adaptive S-transform. IEEE J. Biomed. Health Inform. **19**(3), 1097–1105 (2015)
14. Sharma, S.D., Saxena, R., Sharma, S.N.: Tandem repeats detection in DNA sequences using Kaiser window based adaptive S-transform. Bio-Algorithms Med-Syst. **13**(3), 167–173 (2017)
15. Tenneti, S.V., Vaidyanathan, P.P.: Detecting tandem repeats in DNA using Ramanujan filter bank. In: Proceedings of IEEE International Symposium on Circuits and Systems (ISCAS), Montréal, QC, Canada, pp. 21–24 (2016)
16. Yadav, Y., Sharma, S.N., Shakya, D.K.: Detection of tandem repeats in DNA sequences using short-time Ramanujan Fourier transform. IEEE/ACM Trans. Comput. Biol. Bioinf. **19**(3), 1583–1591 (2022)
17. Sharma, S.D., Shakya, D.K., Sharma, S.N.: Evaluation of DNA mapping schemes for exon detection. In: Proceeding of IEEE International Conference on Computer, Communication and Electrical Technology, pp. 71–74. IEEE, Tirunelveli (2011)

18. Kwan, H.K., Arniker, S.B.: Numerical representation of DNA sequences. In: Proceeding of IEEE International conference of Electro/Information Technology, pp. 307–31. IEEE, Windsor (2009)

19. Akhtar, M., Epps, J., Ambikairajah, E.: On DNA numerical representations for period-3 based exon prediction. In: 5th IEEE International Workshop on Genomic Signal Processing and Statistics (GENSIPS 2007), pp.1–4. IEEE, Tuusula (2007)

20. Akhtar, M., Epps, J., Ambikairajah, E.: Signal processing in sequence analysis: advances in eukaryotic gene prediction. IEEE J. Sel. Top. Signal Process. **2**(3), 310–321 (2008)

21. Mendizabal-Ruiz, G., Román-Godínez, I., Torres-Ramos, S., Salido-Ruiz, R.A., Morales, J.A.: On DNA numerical representations for genomic similarity computation. PLoS One **12**(3) (2017)

22. Zhang, W.F., Yan, H.: Exon prediction using empirical mode decomposition and Fourier transform of structural profiles of DNA sequences. Pattern Recogn. **45**(3), 947–955 (2012)

23. Hardy, G.H., Atgar, P.V.S., Wilson, B.M.: Collected Papers of Srinivasa Ramanujan. Cambridge University Press (1927)

24. Planat, M., Rosu, H.C., Perrine, S.: Ramanujan sums for signal processing of low-frequency noise. Phys. Rev. E **66**(5), 1–7 (2002)

25. Carmichael, R. D.: Expansion of arithmetic functions in infinite series. In: Proceedings of London Mathematical Society, Landon, pp. 1–32 (1932)

26. Gadiyar, H.G., Padma, R.: Ramamanujan Fourier series, the Winer-Khintchine formula, and the distribution of prime numbers. Phys. A **269**, 503–510 (1999)

27. Vaidyanathan, P.P.: Ramanujan sums in the context of signal processing - part II: FIR representations & applications. IEEE Trans. Signal Process. **62**(16), 4158–4172 (2014)

28. Ma, X., Liu, D., Shan, Y.: Intra-pulse modulation recognition using short-time Ramanujan Fourier transform spectrogram. EURASIP J. Adv. Signal Process. **2017**(42) (2017)

29. Boeva, V., Regnier, M., Papatsenkoand, D., Makeev, V.: Short fuzzy tandem repeats in genomic sequences, identification, and possible role in regulation of gene expression. Bioinformatics **22**(6), 676–684 (2006)

30. National Centre for Biotechnology Information (NCBI) Homepage. http://www.ncbi.nlm.nih.gov. Accessed 12 Dec 2022

A Survey on Mammogram Datasets to Develop Breast CAD System

Shaila Chugh[1]([✉]), Sachin Goyal[2], Anjana Pandey[2], and Sunil Joshi[1]

[1] Computer Science and Engineering, Samrat Ashok Technological Institute,
Vidisha, India
lShailachugh.cse@satiengg.in

[2] Rajiv Gandhi Proudyogiki Viswavidyalaya, Bhopal 462033, MP, India

Abstract. The researchers are working on various types of abnormalities classification on mammogram images such as breast density, mass segmentation mass classification, and mico-calcification. The selection of a mammogram image dataset is an important step for developing advanced classification, segmentation, and lesion detection techniques using deep learning techniques. So in this manuscript, we summarise some of the public and private mammogram datasets such as MIAS Inbreast, KAU-BCMD, VinDr-Mammo, CBIS-DDSM, and many subsets of mass mammogram patch datasets are available in various repositories. The various properties of mammogram image datasets are discussed in detail of updations and compared them in the context of deep learning (DL) driven computer-aided diagnosis (CAD) for breast cancer.

Keywords: Mammogram image Dataset · Mass dataset · Breast Density · Deep learning

1 Introduction

For breast examination, various imaging techniques, such as magnetic resonance imaging, ultrasound imaging, and X-ray imaging mammogram. It has been found that breast cancer screening with mammogram imaging is the most effective tool for early diagnosis. Conventional or diagnostic mammography may help to identify eight types of abnormalities of the breast as Asymmetric Breast Tissue, Asymmetric Density, Architectural Distortion Mass, Micro-calcification, Interval Changes in comparison to previous films, Adenopathy, and Other miscellaneous Findings. Sometimes, there is a chance for the physicians to let pass the abnormality in a mammogram. All these abnormalities factors are affected the classification accuracy for correct diagnosis of cancerous or non-cancerous breast mammograms. A major concern with the manual detection of cancer is the chances of FP (false positive) and FN (false negative) results. Therefore, machine learning and a more legitimate algorithm base Computer-Aided Diagnosis system (CAD) help to increase the effectiveness and efficiency of screening procedures of mammogram that reduces the errors, FP, and FN. Before the

© The Author(s), under exclusive license to Springer Nature Switzerland AG 2025
M. Gupta et al. (Eds.): MISS 2023, CCIS 1952, pp. 229–241, 2025.
https://doi.org/10.1007/978-3-031-69115-7_15

implementation of any classification algorithm, it needs to find an appropriate dataset and problem for algorithm valuation.

A typical CAD system is designed with the following modules; Mammogram image capturing, Mammogram Pre-processing & Enhancement, Breast Region Segmentation, Mass Segmentation, Microcalcification detection, Feature Extraction, and mass classification in two levels (density, normal/abnormal, benign/malignant). Early, various segmentation techniques have been applied for mass segmentation but those results are more subjective and vary from mammogram to mammogram images [17]. The most of research articles on mammogram classification are focused on Automatic Mass Segmentation, Mass Classification (benign/malignant and cancerous and non-cancerous), detection microcalcification, Breast Density Classification, and Asymmetric distortion [7,15].

Recently, Deep learning (DL) is gaining popularity method for classifying medical images as it requires less pre-processing and segmentation and gives higher accuracy [6]. Implementing a Deep learning-based CAD system for mammograms requires a large collection of mammograms for various classes of abnormalities. Therefore, deep learning approaches such as CNN, U-Net, AUNet, UNet, Connected-UNets, SegNet, and Connected-SegNets are also adopted for mass segmentation and classification [2]. The majority of the mammogram databases have been published with an inadequate number of samples for deep learning, which is considered to be a substantial issue. Image augmentation techniques are adopted to increase the number of input images to solve the shortage. But it is not a good practice to solve real-life challenge problems [2]. The performance of CAD systems depends on the problem domain and the selection of the appropriate mammogram database.

The manuscript gives a brief review of mammogram datasets that are available for research purpose. Section 2 explains the objective for reviewing the various mammogram data sets. Section 3 briefs the visible properties of breast mammograms that are considered during the analysis the level of abnormalities. Section 4 reviews the available mammogram data sets with summary of mammogram images for each type annotation that are extracted from the .CSV or excel file which is given with each datasets. It is helpful for researchers to adopt appropriate data set as a benchmark for particular classification problem. Section Finally, Sect. 6 concludes the selection of datasets for various problem domains.

2   Objective

Implementing a Deep learning-driven CAD system for mammograms requires a large collection of various classes of mammograms for different abnormalities. Early no such public data set are available for research purpose but recently some of the preprocessed mammogram datasets are added in kaggle repository. Breast density and BIRADS classification are correlated because, in the higher dense mammogram, the diagnosis and classification of mass tissues is critical task for radiologists [12]. However, the mass tissue also has identified by high-intensity value and high dense mammogram has a bright region for the normal case. Therefore, the performance of CAD system is also affected by breast density. This issue can be handled CAD system developed for separate BIRADs classification techniques that will apply to individual Breast Density.

Till now most of the work is proposed on mass classification without considering breast density. The performance of CAD systems depends on the problem domain and the selection of the appropriate dataset. Therefore presented manuscript gives a brief review on mammogram Databases that are available for research purposes in terms of the number of distribution images for various classes.

Early, the most of articles are published on traditional feature extraction and classification using conventional machine learning approaches which do not require a large number of training samples. Early, the performance of mammogram classification algorithms are tested on MIAS and DDSM datasets which is published in 1996. Earlier, only the DDSM mammogram dataset was available as public data set for research purpose but in this data-set annotation of each mammogram are given in a separate folder and file which make it difficult to analyze. Therefore, some of the pre-processed mammogram datasets are added to kaggle repository to simplify the structure of the mammogram data-set. Therefore, recently some research groups have published mammogram data sets of a large number of mammogram images with an annotation that can be helpful for implementing deep learning-based fully automated mass segmentation and classification system.

3   Breast Mammogram Parameters

It is necessary to understand various breast mammogram measures parameters such as breast density and BIRADS classes before the Implementation of the mammogram-based CAD system.

A breast mammogram is examined several factors such as breast composition mostly- fatty, scattered-density, consistent-density, extremely dense, lesion or Mass (shape, margin, and density), calcification and architecture distortion. Breast density is a measure of the surrounding glandular or fibrous connective tissues. Mammogram composition density is treated as a decision risk factor for breast cancer. High-dense breast tissue might added the risk of breast cancer or malignancy. The risk of women with extreme breast density is high compared with women with scattered density [4, 13]. Masses and micro-calcification appear on mammograms in the early stage of the disease. Therefore, a well-trained diagnostic technique can help the patient to detect the disease early and can be curable.

The American College of Radiology created the risk assessment and quality control tool known as BI-RADS (Breast Imaging-Reporting and Data System), which delivers a broadly utilized lexicon and reporting scheme for breast imaging. BIRADS standard from 0 to 6 is the mass region category based on the cancer risk level. Table 1 Shows some of the mammogram samples for various type densities along with the annotated mass region.

4   Breast Mammogram Databases

For the optimization of deep learning-based approaches, the dimension of mammographic databases plays an important part in the research. A large number of images are required to train a deep learning-based architecture. From the above literature survey, we found that there is limited public data available for mammogram images in curated format.

4.1   MIAS

MIAS and mini-MIAS are also favorite datasets among researchers working in the breast cancer diagnosis area as it categorizes the images in a greater number of different labels. The Mammographic Image Analysis Society (MIAS) Digital Mammogram Database consists of 322 MLO digitized mammograms of size 1024×1024 corresponding to the left and right breasts of 161 women. The description of each image is mentioned such as breast density (fatty, glandular, or dense), abnormality types, the severity of abnormality (B Benign /M Malignant), and the coordinates of the mass center along with the approximate radius of the circle enclosing them. This mammogram gives images for seven different types of abnormalities (Calcification, Well-defined/circumscribed masses, Spiculated masses, Other, ill-defined masses, ARCH Architectural distortion, ASYM Asymmetry, NORM Normal). Table 2 summarizes information MIAS Mammogram dataset with a various numbers of images for breast density, mass type, and type of abnormalities.

4.2   Inbreast

The strengths of the actually presented INbreast database are focused on the fact that it was created with full-field digital mammograms (as compared to digitized mammograms), provides a broad range of instances, and is made accessible to the public along with correct annotations. We assume this database may be a guide for potential works based on breast cancer imaging or linked to it.

INbreast dataset containing 410 images for 115 cases of women. Each woman has 4 images for 90 cases and 2 images for 25 cases. Various cases of malignancies are masses, calcification, asymmetric and distortions where exact boundaries made by radiologists are also given in XML extension. Dataset provides rich presentation with full-field digital mammograms (as compared to digitized mammograms), provides a broad range of diversity, and is made accessible to the public along with correct annotations [16]. Each mammogram also gives breast density level for (1 to 5) along with mass information, micro-calcification (Table 3).

Table 1. Sample image of BIRADS

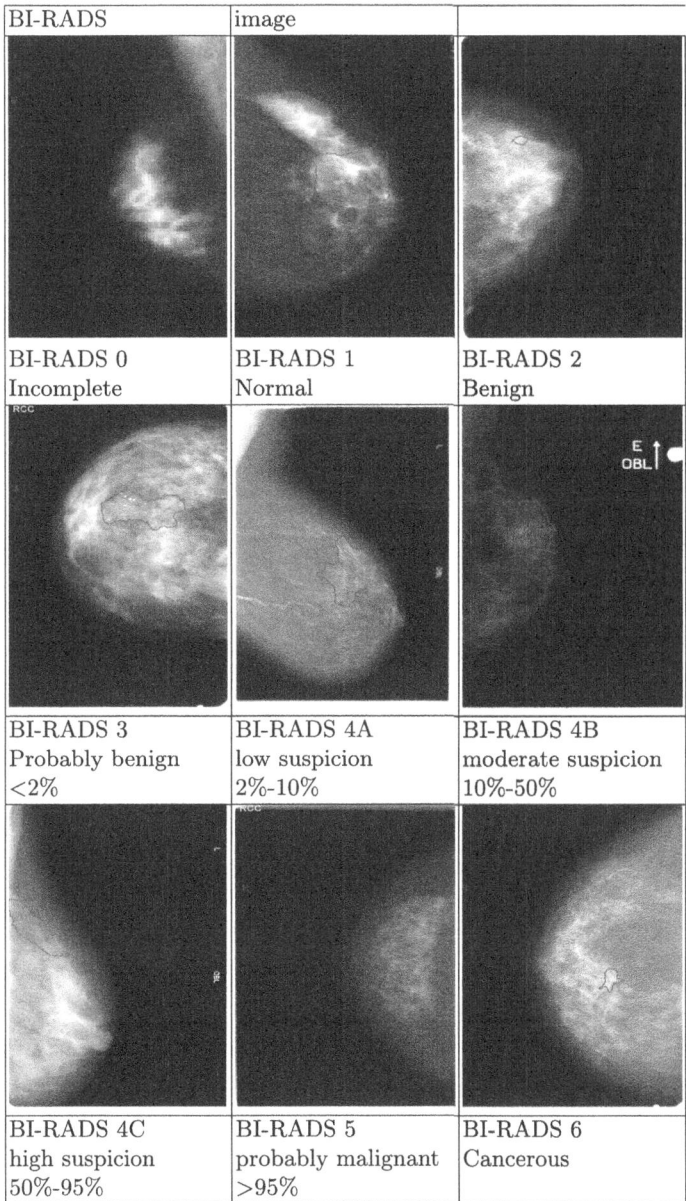

BI-RADS	image	
BI-RADS 0 Incomplete	BI-RADS 1 Normal	BI-RADS 2 Benign
BI-RADS 3 Probably benign <2%	BI-RADS 4A low suspicion 2%-10%	BI-RADS 4B moderate suspicion 10%-50%
BI-RADS 4C high suspicion 50%-95%	BI-RADS 5 probably malignant >95%	BI-RADS 6 Cancerous

Table 2. Summary of MIAS mammogram data set

Database Size	2.5 GB	# Benign Mass	66
# Cases	161	# Malignant Mass	56
# Mammogram	322	# Mass with CIRC	25
# Breast Density G	104	# Mass with MISC	15
# Breast Density D	117	# Mass with ASYM	15
# Breast Density F	109	# Mass with ARCH	19
# Breast with Mass	100	# Mass with SPIC	19
#Normal Case	209	#Microcalcification	28

Table 3. Summary of Inbreast mammogram data set

# Patient	115	# Mammograms	412
# BI-RADS- 1	67	#ACR 1	136
# Bi-Rads - 1	23	#ACR 2	146
# Bi-Rads - 2	220	#ACR 3	99
# Bi-Rads - 4	43	#ACR 4	28
# Bi-Rads - 5	49	#Micro-calcification	14
# Bi-Rads - 6	8	# Mass	108

4.3   BCDR

The objective of BREAST CANCER DIGITAL REPOSITORY (BCDR) is to develop a cutting-edge resource to investigate computer-based detection and diagnosis methods, and engage medical students, people diagnosed, and other professionals who work in the healthcare field. It comprises of two repositories: the Full-Field Digital Mammography-based Repository and the Film Mammography-based Repository (BCDR-FM) (BCDR-DM). Current BCDR is Composed by 1734 patients in 1010 cases of BCDR-FM and 734 cases of BCDR-DM. BCDR provides the mammogram dataset annotated with various parameters, Gender, Age, Breast Density, Breast Location, Biopsy results, BIRADS Classes for research purposes [14].

BCDR dataset contains annotated digitized film mammogram images and related data dictionaries. BCDR contain detected anomalies, density feature, BIRADS categories, and lesion segmentation outlines for both CC and MLO view. The beauty of this dataset is that mass regions are annotated with boundary pixels provided by the expert radiographer.

Table 4 summarizes the number of mammogram and mass images in both datasets for various classes.

4.4   CBIS-DDSM

The DDSM dataset contains the highest value in the case of a number of images but due to issues like image format, contention over some images made it less

Table 4. Summary of BCDR mammogram data set

Dataset	BCDR-DM	BCDR-FM
# Patient	1010	734
# Mammograms	3703	
# ROI	1493	
# Mass	1044	
#masses	639	
# micro-calcifications	341	
# calcifications	145	
# stromal-distortions	102	
#architectural-distortions	66	
#axillary-adenopathy	2	

suitable for getting accurate results. So, CBIS-DDSM was introduced as a refined form or filtered subset of the DDSM dataset.

Digital Database for Screening Mammography(DDSM) consists of a total contains 2620 cases in 43 volumes and 4024 mammograms divided into four types breast normal, cancer, begin and begin without callback. Annotations for the ROI of DDSM images are not accurately given the position of abnormalities for segmentation purposes for exact feature calculation. Further DDSM has a large scale of images as compared to any available public mammogram dataset still making it a strong dataset resource for researchers [10]. Some of the codes were available to convert LJPEG DDSM images into PNG, JPEG, TIF, GIF, and other formats [19].

While the DDSM dataset is much larger in size and in the improved version of the DDSM dataset i.e. CBIS-DDSM, data is selected and curated by a trained radiologist. In the CBIS-DDSM has the single comma-separated values (CSV) file*.CSV that are compiled from metadata of *.ics file for each mammogram which is given in the DDSM Dataset that helps researchers to separate mammograms for different classes. This dataset provides Mass-Training-Description.csv, Mass-Test-Description.csv, Calc-Training-Description.csv and Calc-Test-Description.csv files as the benchmark for CAD system evaluation [11]. This Dataset can be accessed from the cancer archive imaging website it is a collection of 6671 images that consists of full mammogram images and Mass ROIs. It has boundary information for each of the converted images and displays them. Three popular deep-CNN architectures (VGG16, ResNet50, InceptionV3 have been successfully trained by annotated CBIS-DDSM Mammogram dataset [1].

4.5 VinDr-Mammo

Recently, VinDr-Mammo is the largest mammogram dataset that are published by PHieu et al. [18] with The Institutional Review Board of Hanoi Medical University Hospital (HMUH) and Hospital 108 (H108).

It contains full-field digital mammograms with a sufficient dataset with 5000 cases having MLO and CC view for both left and right breast therefore it has 20,000 mammograms. These images are viewed twice by an experienced radiologist to assess the breast density, BIRADS classes and cancerous tissues. For the deep learning-based CAD, the dataset stratifies into standard 4000 training and 1000 test case. Table 5 summarizes the number distribution of mammograms for BIRADS and Breast Density combination. From this table, we found that breast density C and BIRADS 2 have a large number of mammograms.

Table 5. Summary of VinDr-Mammo mammogram data set

Density→ BIRADS ↓	A	B	C	D	Total
1	80	1326	10088	1912	13406
2	8	404	3624	640	4676
3	6	48	794	82	930
4	6	60	636	60	762
5	0	70	150	6	226
Total	100	1908	15292	2700	20,000

Other abnormal regions are also marked by rectangular bounding boxes by radiologists. Dataset can be used for 4 basic abnormalities - microcalcification, breast asymmetry, mass, and architectural distortion. Only BIRADS 3,BIRADS 4 and BIRADS 5 categories were marked by bounding boxes.

4.6   KAU-BCMD Data Description

The KAU-BCMD dataset contains 1416 case images. Each case contains both right and left breast with MLO and CC view to provide sufficient 5386 images based on BIRADS classification for deep learning-based computer-aided diagnosis. All these images are reviewed and annotated thrice by expert radiologists and stored in DICOM and JPG format. Dataset grouped in 5 folders for 5 BIRADS case. In each of these folders 2 subfolders one for mammograms and the other for ultrasound images. Mammogram images are annotated for suspicious area segmentation via hand drawing. Images in JPG format are for tumor masks denoting benign and malignant cases [3]. In addition, it contains 405 ultrasound images to support hybrid imaging systems for deep-learning CAD. The summary of a number of mammograms for various BIRADS classes in the dataset is shown in Table 6.

4.7   OMI-DB Mammogram Dataset

The OPTIMAM Mammography Image Database (OMI-DB) was developed to study how different factors affect the diagnosis of breast cancer on mammograms.

Table 6. Summary of class distribution on KAU-BCMD

Density→ BIRADS ↓	A	B	C	D	Total
1	593	897	300	94	1884
2	880	1426	626	186	3118
3	61	140	61	28	290
4	20	28	20	4	72
5	8	10	4	0	22
Total	1562	2501	1011	312	5386

It is financed by Cancer Research United Kingdom for two projects OPTIMAM (2008–2013) and OPTIMAM2 (2013–2018). National Breast Screening System's is using to mark abnormalities in for each case. It is private data set ans consist of 3072878 mammogram image of 172282 women [8]. This is continue updated and the Table 7 shows the clients of various cases and more details are given in tabular form in [8].

Table 7. The summary of the number of clients in OMI-Databse

Classes	Totals
Normal	166694
Benign	4055
Malignant	7417
Interval Cancers	1160

4.8   The Chinese Mammography Database (CMMD)

The insufficiency of sample data, nondiversity in patient groups, absence of pathology confirmations, and unidentified molecular sub-types limit current publicly accessible mammography databases. So CCMD try to fill this gap it provides a large set of mammogram image in *.DCM format in Kaggle repository for research purposes. But it does not have information about the mass region, BIRADS classes, and Breast density. The database contains 3,728 mammographies from these 1,775 patients with biopsy-proven types of benign or malignant tumors. This data set has only abnormal cases for Calcification, benign/Malignant. Image data were collected on a GE Senographe DS mammography system [20]. The summary of various classes of each case is shown in Table 8.

The database contains 3,728 mammographies from these 1,775 patients with a biopsy-confirmed type of benign or malignant tumors. This dataset has only abnormal cases for Calcification, benign/Malignant. Image data were acquired on a GE Senographe DS mammography system [20]. The summary of various classes of each case is shown in Table 8.

Table 8. A summary of the CMMD

Number of Patients	1775	#Calcification	262
Number of Mammogram	3,728	# of Mass	1149
Images data Size (GB)	22.9 GB	# Benign	556
# of both	461	# Malignant	1316

5   Mammogram Mass Patch Dataset

The fully automatic and efficient CAD system designing is various difficult task for machine learning engineers because they have no knowledge about the suspicious region on mammogram images. An efficient mass classification algorithm can be designed with corrected mass images annotated with different classes. A correct suspicious region can be annotated by an expert radiologist which is provided in some full film DICOM images. In Kaggle dataset have 55890 mass images of size 299 × 299 extracted from DDSM and CBIS-DDSM data sets

5.1   IRMA Mammogram Patch

IRMA dataset is a collection of 2796 mass patches that cropped from full mammogram images of DDSM and MIAS dataset. This repository is created by Deserno et al. [5] for their experimentation and made available for research purposes. The mammogram patches is categorized into 20 classes i.e., five types of tissue density (A, B, C, D, G), and each BIRADS category is divided into three classes - normal, benign, and malignant (Table 9).

5.2   Mandaley Mammogram Patch Dataset

Huang et al. a breast mammogram data set which is a collection of 7632 breast mass images. These images are cropped from 106 in the breast dataset, 53 Mass based MIAS, and 2188 mass base DDSM. Database images are resized to 227*227 pixels. The whole dataset categorizes into 4 folders DDSM, breast, MIAS, and hybrid [9] Generated by CLAHE (Adaptive histogram equalization) and augmentation. The dataset is prepared by the following steps (Table 10).

1 As the density feature of the Mammogram affect the early detection of cancer, hence 106 original images are categorized for 4 breast density (A, B, C, D) (with benign/malignant class) making the dataset more powerful for research and diagnosis purpose.

Table 9. IRMA Databse Summary

BIRADS	Normal	Benign	Malignant	Total
DDSM Patches				
I	203	219	222	644
II	168	232	228	628
III	195	225	227	647
IV	207	224	226	657
Total	773	900	903	2576
MIAS Patches				
I	12	14	11	37
II	28	1	5	34
III	24	8	6	38
IV	26	9	6	41
Total	90	32	28	150

2 After filtration via CLAHE these 106 images are enhanced for mass detection to 212 images i.e. 106 original+106 CLAHE processed images.
3 In this each image Augmented via rotation at 30,60,90,120,150,180,210,240, 270,300 and 330 angle. Further, each image and original image flipped horizontally and vertically make a complete set of 7632 images in PNG format.

Table 10. Category Image After Data Augmentation

Classes	All	Density-C+Benign	936
Density-A+Benign	864	Density-C+Malignant	576
Density-B+Malignant	2160	Density-D+Benign	432
Density-B+Benign	288	Density-D+Malignant	72
Density-B+Malignant	2304	Total	7632

6 Conclusion

The selection of mammogram images depends on the domain of the classification problem. This review article helps the to researcher for selecting an appropriate dataset for implementing deep learning base diagnosis using Mammograms. The annotation of mammogram images of any dataset is a very challenging task, it requires a team of expert radiologists for correct annotation. From the study, we found that the mammogram images for separate breast densities for abnormal cases are not sufficient to develop a deep learning-driven CAD system that can be solved by augmentation or combining the mammogram images of various datasets. Table 11 compares the mammogram datasets for various problem domains.

Table 11. Analysis of Database Selection based on problem domain

Dataset	Size (GB)	Number mammo.	Breast Den.	BIRADS Clas.	Cal. det.	ROI Seg	Mass Clas.	DL Lrn.
MIAS	327(MB)	322	✓	x	✓	✓	✓	x
Inbreast	2(GB)	412	x	✓	✓	✓	✓	x
BCDR	2GB	3703	x	x	✓	✓	✓	x
CBIS-DDSM	162(GB)	6671	✓	✓		✓	✓	✓
VinDr	50GB	20,000	✓	✓	✓	✓	✓	✓
KAU-BCMD	607MB	5386	✓	✓	✓	✓✓	x	✓
OMI-DB	NA	3728	x	✓	✓	x	✓	✓
CMMD	5(GB)	2(lakh)	x	✓	✓	✓	✓	✓

Only VnDr Mammo, Kau, and OPTIMUM datasets have enough number mammogram images. Further, for future scope, these datasets are hybridized by separating the images from each public dataset with density and BIRADS classes marked images and combined in a single compatible database for density-wise assessment of breast cancer.

References

1. Agarwal, R., Diaz, O., Lladó, X., Yap, M.H., Martí, R.: Automatic mass detection in mammograms using deep convolutional neural networks. J. Med. Imaging **6**(3), 031409 (2019)
2. Alkhaleefah, M., et al.: Connected-segnets: a deep learning model for breast tumor segmentation from X-ray images. Cancers **14**(16), 4030 (2022)
3. Alsolami, A.S., Shalash, W., Alsaggaf, W., Ashoor, S., Refaat, H., Elmogy, M.: King Abdulaziz university breast cancer mammogram dataset (KAU-BCMD). Data **6**(11), 111 (2021)
4. Ciatto, S., et al.: Categorizing breast mammographic density: intra-and interobserver reproducibility of BI-RADS density categories. Breast **14**(4), 269–275 (2005)
5. Deserno, T.: Irma database (2012). https://irma-project.org
6. Dhungel, N., Carneiro, G., Bradley, A.P.: A deep learning approach for the analysis of masses in mammograms with minimal user intervention. Med. Image Anal. **37**, 114–128 (2017)
7. Duggento, A., et al.: An ad hoc random initialization deep neural network architecture for discriminating malignant breast cancer lesions in mammographic images. Contrast Media Mol. Imaging **2019** (2019)
8. Halling-Brown, M.D., et al.: Optimam mammography image database: a large-scale resource of mammography images and clinical data. Radiol. Artif. Intell. **3**(1), e200103 (2020)

9. Huang, M.L., Lin, T.Y.: Dataset of breast mammography images with masses. Data Brief **31**, 105928 (2020)
10. Kegelmeyer Jr, W., et al.: The digital database for screening mammography. University of South Florida, Department of Radiology, Massachusetts General Hospital, Department of Radiology, Massachusetts General Hospital (2001)
11. Lee, R.S., Gimenez, F., Hoogi, A., Miyake, K.K., Gorovoy, M., Rubin, D.L.: A curated mammography data set for use in computer-aided detection and diagnosis research. Sci. Data **4**(1), 1–9 (2017)
12. Lee, Z.Y., Goh, Y.L.E., Lai, C.: Classification of mammographic breast density and its correlation with BI-RADS in elder women using machine learning approach. J. Med. Imaging Radiat. Sci. **53**(1), 28–34 (2022). https://doi.org/10.1016/j.jmir.2021.10.004. https://www.sciencedirect.com/science/article/pii/S1939865421002368
13. Li, H., Zhang, S., Wang, Q., Zhu, R.: Clinical value of mammography in diagnosis and identification of breast mass. Pak. J. Med. Sci. **32**(4), 1020 (2016)
14. Lopez, M.G., et al.: BCDR: a breast cancer digital repository. In: 15th International Conference on Experimental Mechanics, vol. 1215 (2012)
15. Mohamed, A.A., Berg, W.A., Peng, H., Luo, Y., Jankowitz, R.C., Wu, S.: A deep learning method for classifying mammographic breast density categories. Med. Phys. **45**(1), 314–321 (2018)
16. Moreira, I.C., Amaral, I., Domingues, I., Cardoso, A., Cardoso, M.J., Cardoso, J.S.: Inbreast: toward a full-field digital mammographic database. Acad. Radiol. **19**(2), 236–248 (2012)
17. de Nazaré Silva, J., de Carvalho Filho, A.O., Corrêa Silva, A., Cardoso de Paiva, A., Gattass, M.: Automatic detection of masses in mammograms using quality threshold clustering, correlogram function, and SVM. J. Digit. Imaging **28**(3), 323–337 (2015)
18. Nguyen, H.T., et al.: Vindr-mammo: a large-scale benchmark dataset for computer-aided diagnosis in full-field digital mammography. medRxiv (2022). https://doi.org/10.1101/2022.03.07.22272009
19. Sharma, A.: DDSM utility (2015). https://github.com/trane293/DDSMUtility
20. Stadnick, B., et al.: Meta-repository of screening mammography classifiers. arXiv preprint arXiv:2108.04800 (2021)

A Hybrid Approach for Preserving Source Location Privacy for Wireless Sensor Networks

Nisha[✉] and S. Suresh

Department of Computer Science, Banaras Hindu University, Varanasi, India
nisha.singh17@bhu.ac.in

Abstract. The advancing development in wireless sensor networks (WSNs) has led to its immense applications in location-based services. However, along with various advantages it suffers from privacy concerns. The protection of location information of the sensor node that reports the occurrence of an event in its sensing range is very important. In this paper, we propose a hybrid approach of randomized and angular routing that is guaranteed at preserving the location of the source. The proposed scheme makes use of fake packets to create a time discrepancy in backtracking the packet flow that increases difficulty for the adversaries. Even though the use of fake packets to create traffic leads to a slight increase in energy consumption, it has also provided a higher level of security to source node.

Keywords: location privacy · confounding time-domain transmission · safety period · delay

1 Introduction

The use of wireless sensor network (WSNs) has grown substantially in recent times. Deployment of sensor nodes is done with the intent to monitor the presence or movement of the event of interest. The basic functionality of the sensor nodes is to collect the data and transmit the information to the base station [1, 2]. WSNs have found its application in fields like wildlife, agriculture, military, etc. [3, 4]. In application-oriented domains, the sensor nodes continuously send information informing about the location of the event to the base station. However, while this wireless transmission between sensor nodes take place, there are unethical entities who keep track of the information flow with the aim to locate the source node. The unethical entities are known as adversaries. The adversaries try to locate the source node to cause harm to the event for their personal greed. This leads to the need for designing mechanisms to preserve the location discovery of the source node.

Location privacy [5] is classified into two types: i. Source location privacy ii. Sink location privacy. Source location privacy [6, 7] suggests securing the location of the sensor node that transmits information about the asset against the adversaries who are continuously monitoring the network traffic to obtain its location. Adversaries use various methods such as blocking the route, compromising the sensor node, creating tremendous traffic, etc. to exploit the open nature of communication channels to reach the source.

© The Author(s), under exclusive license to Springer Nature Switzerland AG 2025
M. Gupta et al. (Eds.): MISS 2023, CCIS 1952, pp. 242–252, 2025.
https://doi.org/10.1007/978-3-031-69115-7_16

This can put the lives of the events at great risk and ultimately the failure of security promising wireless communication.

In this paper, we propose a source location privacy preservation scheme based on confounding domain and angular routing (CDSLP) and the steps of CDSLP are as below:

1. At first, the source node sends the data containing packet to another ring termed as interference ring or basic ring via a confounding time-domain transmission.
2. Then, the terminal node decides an angle 'α' and transmits the packet in a clockwise manner in the interference ring for 'α' angle.
3. After the event packet covers the 'α' angle, it travels towards the base station through shortest path routing.

In our approach, several rings with various radii are formed around the sink in the network region. Sink is placed at the exact center of the network region, and rings are formed around it. All the sensor nodes in the particular ring have the same number of hops towards the sink. Our main motive for designing this algorithm is that for each packet the route taken to reach to the sink through multiple rings must be random. The simulations have been performed in MATLAB and it shows that our algorithm provides longer attack time, more diversified routes along with a small increase in energy consumption and hop count.

The structured for the paper is given as follows. We provide a descriptive overview of various location protection mechanisms in Sect. 2. In Sect. 3, we provide information about different models i.e. network, adversary and energy consumption model. We discuss about the proposed mechanism in Sect. 4 and put forward the experimental results in Sect. 5. Conclusion and direction for future works are discussed in Sect. 6.

2   Related Work

Location privacy preservation is a very crucial area to be worked upon by the researchers. There have been various solutions proposed for preserving source location privacy. Solutions have been categorized into i. Random Walk ii. Phantom Routing iii. Ring loop mechanisms iv. Network Location Anonymization v. Cross-layer routing etc. Random Walk [8, 9] generally deals with transmitting the packets randomly to one of its neighbor nodes without following any pattern. The selection of the next node from among its neighbors can either be random or probabilistic reasoning. The routing path looks completely random to the adversary and makes it difficult to attack through backtracking and hop-by-hop analysis. However, variations in the mechanism of random routing have led to its even more efficient routing schemes such as directed random walk, probabilistic random walk, opportunistic random routing, routing through an intermediary selected node, etc.

Phantom routing was first discussed by Ozturk et al. [10] in which the mechanism was a 2-phased approach. The first phase was a random walk routing and then it was followed by the flooding of packets till it reaches the base station. An improved version was proposed by Kamat et al. [11] termed as Phantom Single Path Routing (PSPR) to reduce the high energy consumption that occurred in the Phantom routing. In PSPR, the packets are sent to phantom nodes through a random path, and then from phantom nodes using shortest path routing it is transmitted to the base station.

In the Ring Loop mechanism, several rings with variable radii are created around the sink node and the transmission of event packet or fake packet follows a certain route in the rings. Han et al. [12] proposed the Source Location Protection protocol based on Dynamic Routing (SLPDR) that worked on ring loop mechanisms. The event packet was transmitted in a cyclic route along with a fake packet transmitted in a different direction in the same ring itself. A node termed as forwarder is chosen from the outermost and transmits a dummy packet that follows a greedy path to the sink. When the dummy packet reaches the event ring, the contents get replaced with event data, and then through greedy routing the packet is transmitted to the sink. SLPRR [13], iHIDE [14], CT [15], CDR [16], etc. are some of the mechanisms that make use of ring loop mechanisms to successfully evade the adversaries' attacks and provide location security.

Network Location Anonymization schemes work towards anonymizing the geographic location information along with some other personal information that is usually fed into Location-Based Server (LBS). LBS is greatly helpful in providing location-based services and route planning methods but it also accumulates personal identifiable information along with the service requests. Location privacy preservation mechanisms focus on anonymizing the query with other requests of the users. K-anonymity [17–19], LocLok mechanism [21], PIR [21], Voronoi diagram [22], Combinatorial Optimization are some of the approaches that guarantee location anonymization and privacy in source location preservation.

Though most of the techniques explained above have helped in attaining a certain level of privacy in location preservation but each has come with a higher cost in energy consumption or lower network lifetime. Moreover, all the routes taken by the event packet till the sink is somewhat inclined towards major traffic near the event ultimately making it predictable to the adversaries. In our approach, we make use of fake source nodes to clone the real source node to increase the difficulty of the adversaries and increase the complexity of the algorithm.

3   System Model

3.1   Network Model

The network model that is being used in this work is the Panda-Hunter model which was first brought into light by Ozturk et al. [10]. As seen in Fig. 1, the network region is covered with multiple sensors that reports to the base station whenever they locate any panda within their sensing region. Data is transmitted to the base station after passing through multiple hops between sensor nodes. The network model has the characteristics given below:

i. Sensor nodes are deployed with a specific density in the network region. Their position once set is fixed for the entire transaction and all the sensor nodes are homogeneous in the operational capabilities like energy, computation capability and memory.

ii. Base station or sink is placed at the center of the network region.

iii. For ease of understanding the routing process, the network region is divided into several rings, each with multiple clusters. Information is exchanged between clusters through the cluster heads.

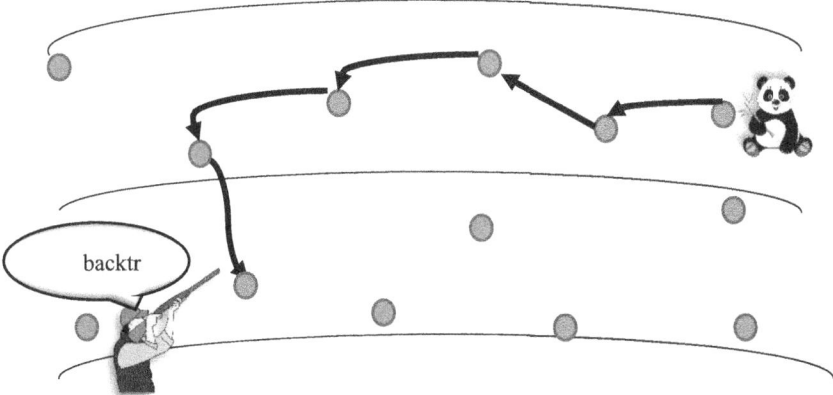

Fig. 1. Panda-Hunter model

3.2 Adversary Model

Adversaries have the ultimate motive of eavesdropping on the packet flow to trace and move to the start of the packet transmission leading to disclosure of source node. Following assumptions are made for the adversaries:

i. Adversaries are equipped with advanced equipment to track the communication taking place between sensor nodes.
ii. Adversaries are considered to be passive i.e. they can only eavesdrop and backtrack without manipulating the content of the packet.
iii. They have a local range of view and hence can not monitor the entire network range at a time.

3.3 Energy Consumption Model

Energy consumption is a very important factor to decide the efficacy of an algorithm. The model [16] that is used to measure the energy loss while transmitting a packet is calculated by the Eq. 1 and 2. Equation 1 provides the formula to calculate energy loss while transmitting a packet and Eq. 2 gives the formula for energy loss during receiving a packet.

$$E_t = l * E_{elec} + l * \varepsilon_{fs} * d^2, d \leq d_0 \tag{1a}$$

$$E_t = l * E_{elec} + l * \varepsilon_{amp} * d^4, d > d_0 \tag{1b}$$

$$E_r = l * E_{elec} \tag{2}$$

where l is the length of the packets in terms of bits and d is the distance between receiving and sending sensor node distance. The details of the system and network parameters are given in Table 1.

Table 1. Parameters

Parameters	Value
Initial energy	0.5 J
d0	87 m
E_{elec}	50 nJ/bit
ε_{fs}	10 pJ/bit/m^2
ε_{amp}	0.0013 pJ/bit/m^4
Network radius (R)	400 m
Network density (ρ)	0.0015
Communication radius (r)	40 m

4   The Proposed CDSLP Scheme

The proposed CDSLP consists of three phases: i. Confounding-domain routing ii. Angular routing iii. Shortest path routing. Before the start of the transmission, the sink floods the entire network with a beacon message that leads to neighbor node discovery for each node. Each sensor node then divides its neighbors into closer neighbor list, far neighbor list and equal distance list. The classification of neighbor nodes in three lists is done by determining their distance from the sink with respect to the particular sensor node. A detailed discussion of each step is provided in the subsection below.

a. Confounding time-domain routing

Source node is the sensor node that senses the presence of event and informs the sink by sending the information through multiple hops and the ring where the event is sensed is termed as the event ring. Before the transmission starts, the source node selects the basic ring from among those outer rings. Confounding time-domain routing [13] takes the event packet outside of the event ring and makes use of fake packets to create time discrepancies. The packet is received by a node in the basic ring which is termed an intermediary node.

b. Angular routing

Before this phase starts, the intermediary node selects two parameters: angle 'α' where $0 < α \leq π$ and direction of moving the packet. The intermediary node transmits the packet to the node from the equal set and this process continues till it covers the 'α' angle. The node at which α is attained is called as a terminal node.

c. Shortest path routing

In this phase, the terminal node selects one of its neighbor nodes from among the near set and routes the packet towards the sink. This same process continues till the packet reaches its destination.

The entire mechanism is shown in Fig. 2.

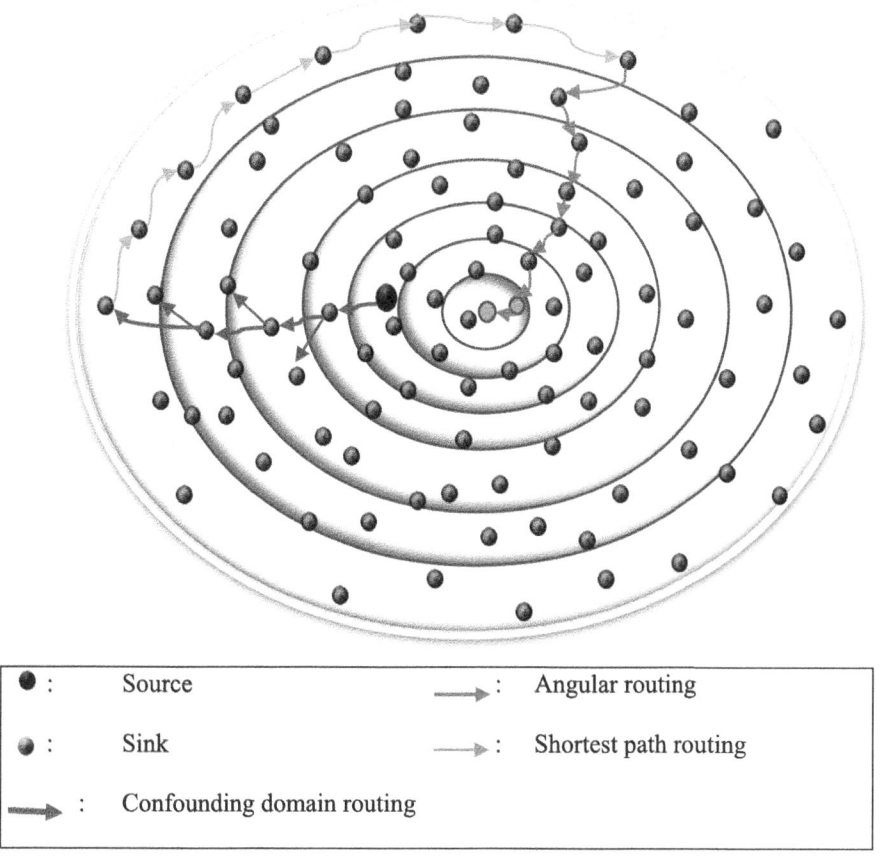

● :	Source	⟶ :	Angular routing
◑ :	Sink	⟶ :	Shortest path routing
⟶ :	Confounding domain routing		

Fig. 2. Network model of CDSLP mechanism

5   Performance Analysis

5.1   An Overview

In this section, we discuss about the parameters that will determine the efficiency of the proposed algorithm CDSLP. Four parameters are majorly taken into evaluation, namely, safety period, transmission delay, energy consumption, and network lifetime.

i. *Safety period:*

We define safety period as the calculated sum of packets transmitted by source to the sink before the adversary is successful in reaching to the source location. Higher safety period suggests greater security.

ii. *Transmission latency:*

It is defined as the average number of hops that the event packet takes to finally reach to the base station.

iii. *Energy consumption:*

It is the average amount of energy consumed during the entire process of packet transmission in a simulation run.

5.2 Simulation Setup and Results

P We consider a circular network region that is divided into multiple rings with the sink being placed at the center statistically. Sensor nodes are deployed randomly with a network density of ρ. We conducted the simulation in MATLAB 2018 and the parametric details are provided in Table 1. We have compared our proposed scheme CDSLP with SLP-R [8] mechanism.

i. Safety period

Figure 3 depicts the variation of the safety period measured in terms of hops for the various location of the source node from the sink. It is observed that SLP-R contribute lower to the privacy level as compared to CDSLP. The reason behind the higher privacy level of the proposed scheme is attributed to the presence of fake packets transmission in the routing process that increase the network traffic and create more difficulties for the adversaries. Greater safety period of CDSLP contributes to the objective of providing a higher privacy level.

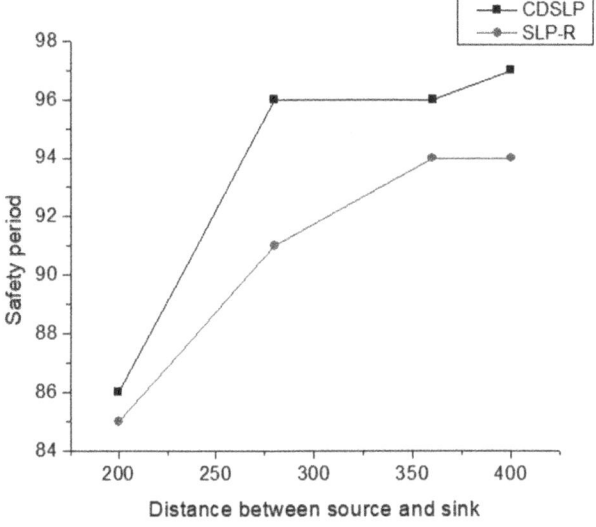

Fig. 3. Safety period

ii. Transmission latency

The average transmission delay taken by the packet to reach from source to sink for the differing location of the source is depicted in Fig. 4. Transmission delay is lesser for SLP-R and it is maximum for CDSLP. The increased transmission delay for our proposed mechanism is because randomization in choosing the next forwarding node leads to a longer routing length. Longer the routing length more is the transmission delay.

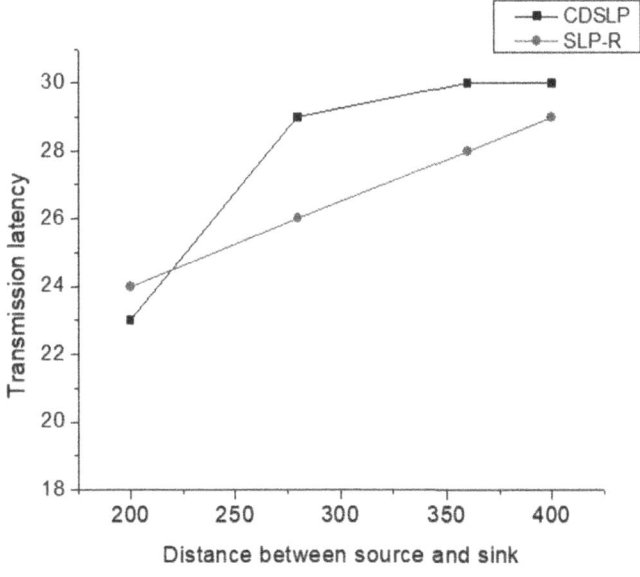

Fig. 4. Transmission latency

iii. Energy consumption

The average energy consumption per packet in each simulation for the differing source location is shown in Fig. 5. We observe that CDSLP consumes comparatively higher energy for each packet delivery as the transmission length is longer. The average of the total amount of energy spent by the sensor nodes for each simulation versus the location of source node is presented in Fig. 6. It is seen that our proposed mechanism CDSLP consumes higher energy as compared to SLP-R. As visible from Fig. 4, the transmission process for the proposed mechanism has longer routes that causes more energy loss as compared to shorter routes. This leads to greater energy loss for CDSLP.

iv. Entropy

Entropy is a metric that shows how random the transmission route is while the packet is being sent from source to destination. Entropy of the network for different location of the source node is presented in Fig. 7. When the source node is closer to the sink, we observe that SLP-R provides a good entropy level. It is because the packet constantly selects a different route for being moved to outer rings. But as the source node starts moving away, entropy level is better for our proposed CDSLP mechanism. Higher entropy level means better security which is achieved by CDSLP.

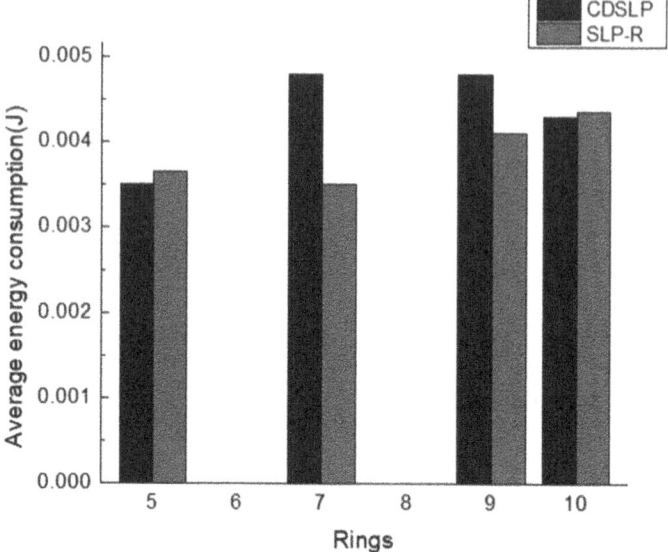

Fig. 5. Average energy consumption

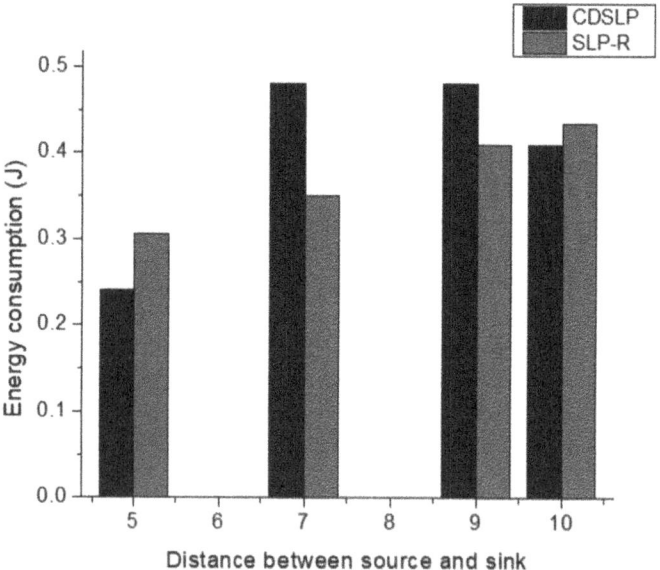

Fig. 6. Total energy consumption versus source location

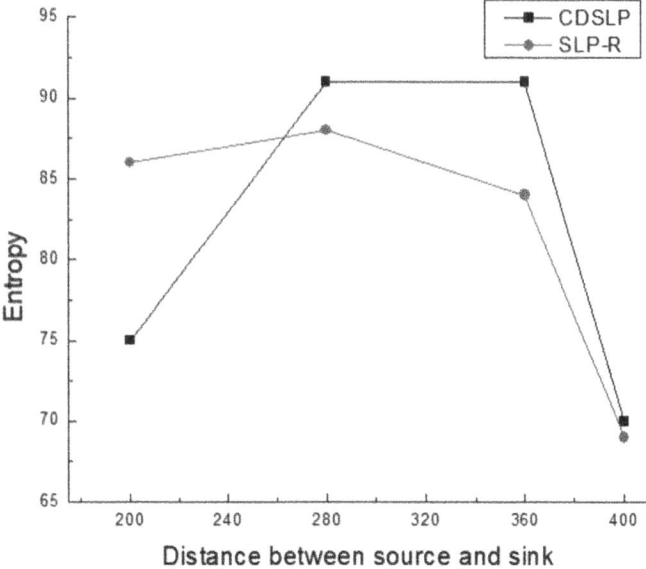

Fig. 7. Entropy versus source location

6 Conclusion

Location privacy preservation of the source node is a critical concern. To address this concern, we have proposed CDSLP routing scheme. The first phase exports the event packet out of the event ring to basic ring through Confounding-time domain transmission. This phase is followed by an angular routing of the event packet which after being attained is transmitted to the sink through greedy routing. We observed that while this mechanism provides a better privacy level as compared to SLP-R technique, it leads to an increased energy consumption. Privacy is the main concern and CDSP assures stronger privacy. Future research focus includes framing mechanisms that hold a good security level for multiple assets.

References

1. Kandris, D., Nakas, C., Vomvas, D., Koulouras, G.: Applications of wireless sensor networks: an up-to-date survey. Appl. Syst. Innov. **3**(1), 14 (2020)
2. Yick, J., Mukherjee, B., Ghosal, D.: Wireless sensor network survey. Comput. Netw. **52**(12), 2292–2330 (2008). ISSN 1389-1286
3. Akyildiz, I.F., Su, W., Sankarasubramaniam, Y., Cayirci, E.: A survey on sensor networks. IEEE Commun. Mag. **40**(8), 102–114 (2002)
4. Jiang, H., Li, J., Zhao, P., Zeng, F., Xiao, Z., Iyengar, A.: Location privacy-preserving mechanisms in location-based services: a comprehensive survey. ACM Comput. Surv. (CSUR) **54**(1), 1–36 (2021)
5. de Fuentes, J.M., González-Manzano, L., Mirzaei, O.: Privacy models in wireless sensor networks: a survey. J. Sens. **2016** (2016)

6. Conti, M., Willemsen, J., Crispo, B.: Providing source location privacy in wireless sensor networks: a survey. IEEE Commun. Surv. Tutor. **15**(3), 1238–1280 (2013)
7. Jiang, J., Han, G., Wang, Z., Guizani, M.: A survey on location privacy protection in wireless sensor networks. J. Netw. Comput. Appl. **125** (2018). https://doi.org/10.1016/j.jnca.2018.10.008
8. Raja, M., Datta, R.: An enhanced source location privacy protection technique for wireless sensor networks using randomized routes. IETE J. Res. **64**(6), 764–776 (2018). https://doi.org/10.1080/03772063.2017.1371652
9. Tian, H., Shen, H., Matsuzawa, T.: Random walk routing for wireless sensor networks. In: Sixth International Conference on Parallel and Distributed Computing Applications and Technologies (PDCAT 2005), pp. 196–200. IEEE (2005). https://doi.org/10.1109/PDCAT.2005.193
10. Ozturk, C., Zhang, Y., Trappe, W., Ott, M.: Source-location privacy for networks of energy-constrained sensors. In: 2004 Proceedings of the Second IEEE Workshop on Software Technologies for Future Embedded and Ubiquitous Systems, Vienna, Austria, pp. 68–72 (2004)
11. Kamat, P., Zhang, Y., Trappe, W., Ozturk, C.: Enhancing source-location privacy in sensor network routing. In: 25th IEEE international conference on distributed computing systems (ICDCS 2005), pp. 599–608. IEEE (2005)
12. Han, G., Zhou, L., Wang, H., Zhang, W., Chan, S.: A source location protection protocol based on dynamic routing in WSNs for the social internet of things. Future Gener. Comput. Syst. **82**, 689–697 (2018). ISSN 0167–739X
13. Wang, H., Han, G., Zhou, L., Ansere, J.A., Zhang, W.: A source location privacy protection scheme based on ring-loop routing for the IoT. Comput. Netw. **148**, 142–150 (2019). ISSN 1389-1286
14. Kazatzopoulos, L., Delakouridis, C., Marias, G.F., Georgiadis, P.: ihide: hiding sources of information in WSNS. In: Second International Workshop on Security, Privacy and Trust in Pervasive and Ubiquitous Computing (SecPerU 2006), pp. 8-p. IEEE (2006)
15. Wang, Y., Liu, L., Gao, W.: An efficient source location privacy protection algorithm based on circular trap for wireless sensor networks. Symmetry **11**, 632 (2019). https://doi.org/10.3390/sym11050632
16. Ren, J., Zhang, Y., Liu, K.: An energy-efficient cyclic diversionary routing strategy against global eavesdroppers in wireless sensor networks. Int. J. Distrib. Sens. Netw. (2013). https://doi.org/10.1155/2013/834245
17. Wang, H., Huang, H., Qin, Y., Wang, Y., Wu, M.: Efficient location privacy-preserving k-anonymity method based on the credible chain. ISPRS Int. J. Geo Inf. **6**, 163 (2017). https://doi.org/10.3390/ijgi6060163
18. Zhang, C., Huang, Y.: Cloaking locations for anonymous location based services: a hybrid approach. GeoInformatica **13**, 159–182 (2009). https://doi.org/10.1007/s10707-008-0047-2
19. Guo, X., Wang, W., Huang, H., Li, Q., Malekian, R.: Location privacy-preserving method based on historical proximity location. Wireless Commun. Mob. Comput. **2020** (2020)
20. Xiao, Y., Xiong, L., Zhang, S., Cao, Y.: Loclok: location cloaking with differential privacy via hidden Markov model. Proc. VLDB Endow. **10**(12), 1901–1904 (2017). https://doi.org/10.14778/3137765.3137804
21. Khoshgozaran, A., Shahabi, C., Shirani-Mehr, H.: Location privacy: going beyond K-anonymity, cloaking and anonymizers. Knowl. Inf. Syst. **26**, 435–465 (2011). https://doi.org/10.1007/s10115-010-0286-z
22. Wang, B., Guo, Y., Li, H., Li, Z.: k-anonymity based location privacy protection method for location-based services in internet of thing. Concurr. Comput.: Pract. Exp. e6760 (2021)

Multi-criteria Decision Making Based Optimal Clustering Method for WSN

Shivendra Kumar Pandey[✉] and Buddha Singh

School of Computer and System Sciences, Jawaharlal Nehru University,
New Delhi, India
shivaiert@gmail.com

Abstract. In Wireless Sensor Networks (WSNs), energy optimization is a very crucial phenomenon. Clustering is a prevalent method of energy conservation. Lots of clustering methods are proposed in the literature for energy optimization. In this paper, we have proposed a clustering and data transmission method for efficient data transmission based on the AHP method. We have compared our proposed method with some existing popular methods, and the outcomes demonstrate that our suggested approach performs better.

Keywords: Wireless Sensor Network · Clustering · Optimization · MCDM

1 Introduction

The WSNs includes sensors disseminated in a pre-defined area. The WSNs collect important sensing field data and transmit it to BS for further processing. The BS or sink nodes are the access able nodes with high computing power, high energy and internet connectivity. The data at BS can be accessed from any part of world through the internet. In WSN, the researcher's primary concern is to collect the important data at the BS efficiently. Many models are proposed for efficient data transmission from sensor nodes to BS. The clustering method is one of the most popular methods among researchers. Three categories of clustering methods are discernible: flat, hierarchical and location based. In this paper, we have proposed a hierarchical clustering method based on multi-Criteria Decision-Making (MCDM) to select the better Cluster Heads (CHs) to optimise energy consumption in WSN. The significant contribution of our proposed model is as follows:

– We have calculated five parameters of CHs. These parameters can select the optimal CHs with a balanced load among them.
– We have used the AHP method to generate the weights of each criterion. The AHP method is discussed in section two in detail.
– We have applied the MCDM method to rank the set of CHs. The nodes of the top set will be selected as CHs.

© The Author(s), under exclusive license to Springer Nature Switzerland AG 2025
M. Gupta et al. (Eds.): MISS 2023, CCIS 1952, pp. 253–265, 2025.
https://doi.org/10.1007/978-3-031-69115-7_17

– We have compared our proposed model with some existing popular models. The result of the comparison is shown in Sect. 4.

The remaining sections are arranged as follows: In Sect. 2, the state of the art is covered. In Sect. 3, the methodology of our proposed model is explained. We contrast our suggested model with two current models. The result of the comparison is shown in Sect. 4. References used in this paper are added in the last.

2   Related Work

Sensors are the most crucial part of WSN. In each WSN, sensor nodes are dispersed in a physical environment for monitoring purposes. The sensors have low computing power, so all the sensed data has to be collected at BS. The BS has internet connectivity, so the data at BS, can be transmitted anywhere in the world. Sensors operate on battery power and it is very challenging to replace the battery. So once a sensor's battery is dead, it is of no use. Once the 50–75% sensors of a WSN are dead, the WSN is discarded. To monitor an area for longer time, it is crucial to use the energy of sensors in a very energy-efficient way. Since BS has unlimited energy or is connected to an energy source, our main motive is to transmit the data sensed by sensors to BS very energy efficiently. Many methods are proposed in WSN for efficient data routing from WSN to BS. In literature, clustering seems to be the most energy-efficient model of data from sensors to BS. LEACH, proposed by Heinzelman et al. [1], was the first model to propose the clustering method of data routing. In LEACH, a few nodes are first chosen as CHs randomly. All the other nodes, known as normal nodes, find the closest CH and create clusters. Every normal node sends data to their BS. The BS accumulates all the received data and transmits it to BS directly. Since BS has to transmit a larger amount of data at a longer distance, they consume higher energy. The LEACH changes the CHs in each round to overcome this problem.

The problem with LEACH was that it selects the CHs only on a probability basis. Several updated versions of LEACH are developed to address the problems of LEACH. In TL-LEACH [2] two levels of hierarchy is proposed having level one primary CHs and level two subsidiary CHs. It uses CH rotation policy to improve lifetime of WSN. In LEACH-C [3], the nodes with higher Remaining Energy (RE) than the highest energy consumed by a CH in the last round will be selected as CHs. In HEED [4] the selection of CHs is done using the residual energy of the sensors. PEGASIS [5] creates a chain of the data stream from sensors to BS to optimize energy consumption. In [6], a residual energy-based CH selection model is proposed. Several nature-inspired methods are also proposed for the selection of optimal CHs. In [7–9] and [10], Particle Swarm Optimization (PSO) base clustering methods are proposed. In [11] Multi-Attribute Decision Making (MADM) method is proposed using fuzzy logic for CH selection, and shark smell optimization (SSO) with genetic algorithm (GA) is used for data routing. In [12], Energy, QoS and location are used with MCDM for optimal CH selection. In [13] delay, the average distance to CHs and distance to BS is used with the Sum of Product (SOP) method of MCDM for optimal CH selection. In [14] Energy based probabilistic model is proposed for optimal CH selection. In our proposed model, we have been attempting to minimize the intra-cluster and inter-cluster communication cost with load distribution to minimize the energy consumption of WSN. We have used the Analytic hierarchy process (AHP) in our proposed model to calculate optimal criteria weights. The AHP process is explained below:

2.1 Analytic Hierarchy Process (AHP)

Analytical Hierarchy Process proposed by Saati et al. [15] in 1990 is an organised mathematical method for analysing difficult decisions. It models the problem as a hierarchy in which the target is placed at the top of the hierarchy, followed by the criteria, and then the options. Each alternative has its own value of criteria. The steps involved in AHP are shown below:

Step 1: Create pairwise comparison matrix $(C_{i,j})_{n*n}$ using scale of relative importance. This matrix will indicate the relative weight of the criterion in relation to other criterion. The used scale of relative importance is shown in Table 1:

Table 1. Relative Importance Scale

S.No	Importance	Assigned Value
1	Same/Equal	1.0
2	Medium	3.0
3	Strong	5.0
4	Very Strong	7.0
5	Absolute	9.0
6	In-between	2.0, 4.0, 6.0, 8.0

Step 2: Generate the normalized pairwise comparison matrix $(C^n_{i,j})_{n*n}$ by using Eq. 1.

$$C^n_{i,j} = \frac{C_{i,j}}{\sum_{i=1}^{n}(C_{i,j})}.$$ (1)

Step 3: Calculate the criteria weights $(W_j)_{n*1}$ by using Eq. 2:

$$W_i = \frac{\sum_{j=1}^{n}(C_{i,j})}{n}.$$ (2)

Step 4: Check the consistency.

Step 4.1: Calculate weighted sum matrix $(M^1_{i,j})_{n*n}$ using Eq. 3

$$M^1_{i,j} = C_{i,j} * W_j. \tag{3}$$

Step 4.2: Generate matrix $M^2_{i,j}$ by using Eq. 4

$$M^2_{i,j} = \frac{\sum^n_{i=1} M^1_{i,j}}{W_j}. \tag{4}$$

Step 4.3: Calculate λ_{max}

$$\lambda_{max} = \frac{\sum^n_{j=1}(M^2_{i,j})}{n} \tag{5}$$

Step 4.4: Calculate consistency index CI using Eq. 6

$$CI = \frac{\lambda_{max} - n}{n - 1} \tag{6}$$

Step 4.5: Calculate Consistency Ratio CR using Eq. 7

$$CR = \frac{CI}{RI} \tag{7}$$

where RI is Random Index, the value of RI is taken from the random index table. For the five criteria, the value will be 1.12. If $CR < 0.10$, we can consider our matrix consistent and proceed with decision-making using AHP; otherwise, we have to update the relative importance until the condition is satisfied.

2.2 Energy Consumption Model

The sensors in a WSN require energy for data transmission, data reception, and data processing. We have used the first-order radio model proposed in [1] to compute the energy utilization by sensor nodes as given in [1].

Energy Consumption by Transmitter Node (E_{TN}). The energy needed to transport a packet of k bits at distance l is proportional to the l^α. If l is less than d_o, then the value of α will be 2 otherwise 4. The d_o is a constant whose value can be calculated by Eq. 9. The (E_{TN}) can be calculated as 8

$$E_{TN}(k,l) = \begin{cases} k * E_{elec} + k * \epsilon_{fs} * l^2, & \text{if } l < d_o \\ k * E_{elec} + k * \epsilon_{mp} * l^4, & \text{otherwise} \end{cases} \tag{8}$$

The d_o can be calculated as 9.

$$d_o = \sqrt{\frac{\epsilon_{fs}}{\epsilon_{mp}}} \tag{9}$$

Energy Consumption by Receiver Node (E_{RN}): If a sensor node receives K bit of data, then it will consume energy due to electronic circuitry, which can be calculated by Eq. 10.

$$E_{RN}(k) = k * E_{elec} \tag{10}$$

Energy Consumption Due to Data Aggregation (E_{agr}): If a node receives a data packet of k bits from n sensor nodes and E_{DA} is aggregation ratio, then the energy consumed in data aggregation E_{agr} can be calculated by Eq. 11.

$$E_{agr} = k * m * E_{DA} \tag{11}$$

where:

E_{TN}: Energy consumption in data transmission.
E_{RX}: Energy consumption in receiving the data.
E_{agr}: Energy consumption in aggregation of the data.
k: Size of the data packet.
E_{elec}: Energy consumed by electronic circuitry in processing 1-bit data.
ϵ_{fs}: Per bit energy consumption in RF amplification for distance less than d_o.
ϵ_{mp}: Per bit energy consumption in the RF amplification for distance greater than d_o. The value of d_o may be computed by Eq. 4:
E_{DA}: Data aggregation constant.

We have used five parameters with AHP for the choosing the CHs. These parameters are mentioned below:

2.3   Parameters Used

The value of all the parameters is shown in Table 2

Table 2. Parameters used in our proposed model

S.No	Name	Type	Abbreviation
1	$Average\_RE$	BN	$C1$
2	$Std\_Devi\_of\_RE$	BN	$C2$
3	$Avg\_Dist\_Node\_to\_CH$	NBN	$C3$
4	$Std\_Devi\_of\_Dist\_Node\_to\_CH$	NBN	$C4$
5	$Avg\_Dist\_CH\_to\_BS$	NBN	$C5$

Here BN represents beneficial criteria, and NBN represents non-beneficial criteria. A detailed description of our parameters is given below:

Average_RE. The parameter *Average_RE* represents the average residual energy of CHs. It is a beneficial parameter. We prefer the higher energy CHs, so they don't die sooner. The formula to calculate the *Average_RE* is given below:

$$Average\_RE = \frac{\sum_{i=1}^{n} RE_i}{n} \tag{12}$$

Std_Devi_of_RE. The criteria *Std_Devi_of_RE* represents the standard deviation of the remaining energy of CHs. If two sets of nodes have the same average remaining energy, then the set with minimum *Std_Devi_of_RE* will perform better because, in this set, nodes have similar remaining energy, so no node dies sooner.

$$Std\_Devi\_of\_RE = \sqrt{\frac{\sum_{i=1}^{n}(RE_i - Average\_RE)^2}{n}} \tag{13}$$

Avg_Dist_Node_to_CH. The criteria *Avg Dist Node to CH* denotes the average distance between sensors and CHs. The minimum value of *Avg_Dist_Node_to_CH* will result in lower intra_cluster communication cost. If there are n nodes in a cluster then the average distance between nodes and CH, *davg*, can be determined as:

$$d_{avg} = \frac{\sum_{i=1}^{n}(d_{i\_CH})}{n}; \tag{14}$$

Std_Devi_of_Dist_Node_to_CH. This criteria represents the standard deviation of the distance between normal nodes and CHs. It is represented by d_{N-CH}^{std}. This criterion will prioritise the set in which similar distances exist between CHs and nodes. so that the normal nodes consume similar energy during data transmission to their CHs.

$$d_{N-CH}^{std} = \sqrt{\frac{\sum_{i=1}^{n}(d_{i\_CH_i} - d_{avg})^2}{n}} \tag{15}$$

Avg_Dist_CH_to_BS. This criteria represents the average distance between CHs and BS. If the average distance between CHs and the BS is lower, then lower energy will be used during data transfer from CHs to BS. If there are n CHs in a WSN and d_{i_BS} represents the distance between CH i and BS, then *Avg_Dist_CH_to_BS* represented by d_{CH-BS}^{avg} can be calculated as:

$$d_{CH-BS}^{avg} = \frac{\sum_{i=1}^{n}(d_{i\_BS})}{n} \tag{16}$$

The methodology of our proposed model is shown in Sect. 3.

3 Methodology

We have calculated five parameters of sensor nodes which will be used as criteria of AHP for selecting CHs. These parameters are shown in Table 2. We assume that the location of all the nodes is known to BS, and the location of sensor nodes is fixed. All the computation work is performed by the BS because BS has unlimited energy. Initially, 4% of sets are created randomly with 10% nodes in each set. So if we have 100 sensor nodes, the BS randomly creates four sets of ten nodes. Now the criteria values of all the sets are calculated. We have used the AHP algorithm to find the criteria weight of each criterion. Once we get the criteria weights, we use the Multi-Criteria Decision Making model to rank the alternatives. The set with the best rank will be selected as the final CH. The methodology of our proposed work is shown in Fig. 1;

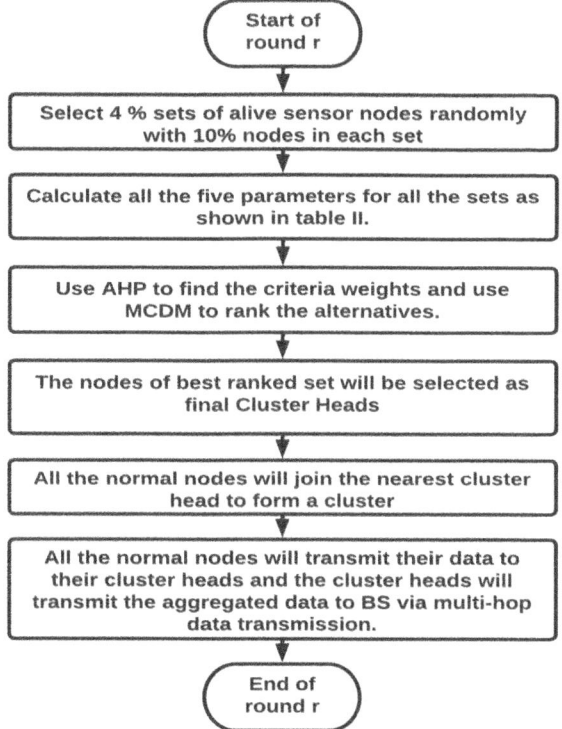

Fig. 1. Methodology of our proposed model

Our proposed work may be categorized into three steps:

Cluster Head Selection. Our first step is identifying the optimal CHs. Initially, some random sets of alive nodes are created to choose the final CHs. The number of nodes in each set is the same, and a node may participate in more than one set. We have constructed four percent sets, each of which contains ten percent of the network's alive nodes. The BS will calculate the value of all five sets' parameters. We have used AHP to find the weight of each criterion. After finding the criteria weights and the value of each criterion of each set, To get the fitness function (Fit_fun) of each set, we will utilise the Eq. 17 and 18.

$$Fit\_fun = \alpha * C1 + \beta * C2 + \gamma * C3 + \delta * C4 + \eta * C5; \tag{17}$$

where

$$\alpha + \beta + \gamma + \delta + \eta = 1 \tag{18}$$

The criteria scores for criteria C1, C2, C3, C4, and C5 are indicated here as *alpha*, *beta*, *gamma*, *delta*, and *eta*, respectively which will be calculated with the help of AHP. Once the Fit_fun of each set is calculated, the set with the best Fit_fun will be declared as CH, so the member nodes of this set will be CHs for the current round.

Cluster Formation. After selecting CHs, all the normal nodes calculate their distance from all the CHs of WSN and choose the closest CH to form the cluster. Once the clusters are formed, the data transmission begins for the current round.

Data Transmission. Each normal node will use the TDMA protocol to convey its data to its CH when clusters have been created. Equation 8 can be used to determine the normal nodes' energy usage. If $d_{i-BS} \leq TH_{MAX}$ is greater than TH_{MAX}, the CHs preprocess and combine all incoming data with their own data and transmit it to the BS; otherwise, the CH i will broadcast its data to the CH j, meeting the equation presented in Eq. 19

$$d_{i-BS}^2 > d_{i-j}^2 + d_{j-BS}^2 \tag{19}$$

Here d_{i-BS} represents the distance from node i to BS, and d_{i-j} from node i to node j. The CHs consume energy in receiving data, aggregating it and transferring the aggregated data to sink or nearby CHs. The energy used by CHs can be demonstrated by Eqs. 8, 10 and 11.

Criteria Weights Calculation Process. To determine the weights for each of the five criteria listed in Table 2, we have used the AHP method, as discussed in Sect. 2. The pairwise comparison matrix of our model is shown in Table 3:

After following steps 2 and 3 of the AHP process, we get the criteria weights. The calculated criteria weights are shown in Table 4. Now we have to check the consistency of these criteria weights. We have to calculate the consistency ratio using Eqs. 3, 4, 5, 6 and Eq. 7. We are using five criteria, so the value of the Random Index (RI) will be 1.12, which is taken from the random index table for five criteria. Our proposed model's calculated consistency ratio (CR) is 0.066. So the calculated criteria weights are consistent and can be used. Now in each

Table 3. Pairwise Comparison Matrix

Criteria	C1	C2	C3	C4	C5
C1	1	5	3	5	4
C2	0.2	1	1	1	0.25
C3	0.33	1	1	4	1
C4	0.2	1	0.25	1	0.33
C5	0.25	4	1	3	1

Table 4. Criteria Weights Calculated by AHP

Criteria	C1	C2	C3	C4	C5
Weights	0.48	0.09	0.17	0.07	0.20
Abbreviations	α	β	γ	δ	η

round the CHs will be elected using Eq. 17. Our suggested model is compared with two other existing popular models, and the comparison result is shown in Sect. 4.

4   Experiment and Results

We have implemented our proposed model using MATLAB R2015a. We have compared the proposed model with the two existing models named LEACH [1] and the model proposed in [13]. The details of the compared models and abbreviations used are shown in Table 5.

Table 5. Compared Models

Model Name	CH selection Criteria	Abbreviations
LEACH [1]	Probability Basis	Algo 1
Algo proposed in [13]	3 criteria with MCDM	Algo 2
Proposed Model	5 Criteria with AHP & MCDM	Proposed

We have considered the WSN with 100 nodes distributed in (0,0)m to (300, 300)m area and the BS is located at (150, 400). The same WSN is used for all three models. Some other parameters are shown in Table 6:

The comparison of the number of alive nodes in each round of each model and the remaining energy of WSN in each model is shown in Fig. 2. It can be observed that in Algo1, the nodes start dying earlier. The reason is that in Algo1, only the probability is used for the CHs selection. The random selection of CHs causes unbalanced energy consumption among the CHs, which can be observed

Table 6. Parameters

S. No.	$Parameter\_Used$	Values
1	WSN Area	$(0,0)$ m to $(300,300)$ m
2	Sensors	100
3	Position of Sink	$(150,400)$ m
3	E_{elec}	50 nj
4	ϵ_{fs}	10 $pj/bit/m^2$
5	ϵ_{mp}	0.0013 $pj/bit/m^4$
6	Packet Size	4000 $bits/packet$
7	E_{DA}	5 nJ/bit/signal
8	d_o	87.7058 m
9	Initial Energy (e)	0.25 j
10	p (LEACH)	0.1
11	R_{aggr}	10%
12	TH_{MAX}	85

from Fig. 2 also. The Algo tries to solve the problem of Algo1 by considering three parameters, 'latency', 'Average distance from node to CH' and 'distance to BS', but it does not consider other parameters for load balancing. In Algo2, the criteria weights are random, which may be inefficient. In our proposed model, we have tried to solve these problems by considering enough parameters for load balancing, and we have also used AHP to decide the criteria weights.

The comparison of energy consumption by each model in each round is shown in Fig. 3. The purpose of this comparison is to show the energy consumed in each round by all the sensor nodes. It is evident that our proposed model consumes the lowest power and does so more consistently than the other two models under comparison. It comes as a result of our efforts to balance the workload among the CHs to reduce the energy consumption.

The comparison of the First_Node_Dead (FND), Half_Node_Dead (HND) and Last_Node_Dead (LND) of all the compared algorithms are shown in Fig. 4. It can be seen that our proposed model performs best in the case of FND and HND, but Algo1 performs best in the case of LND. The reason for the better performance of Algo1 in the case of LND is that Algo1 uses direct data transmission from CH to BS, so there is a lower load and, thus, lower energy consumption among the CHs near BS. Because of this, certain nodes continue to exist for longer.

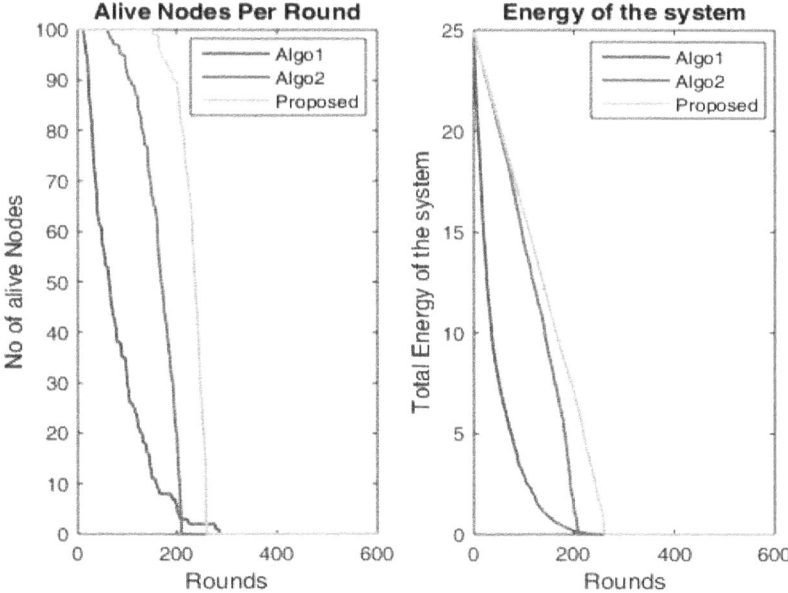

Fig. 2. Comparison of alive nodes and RE of WSN

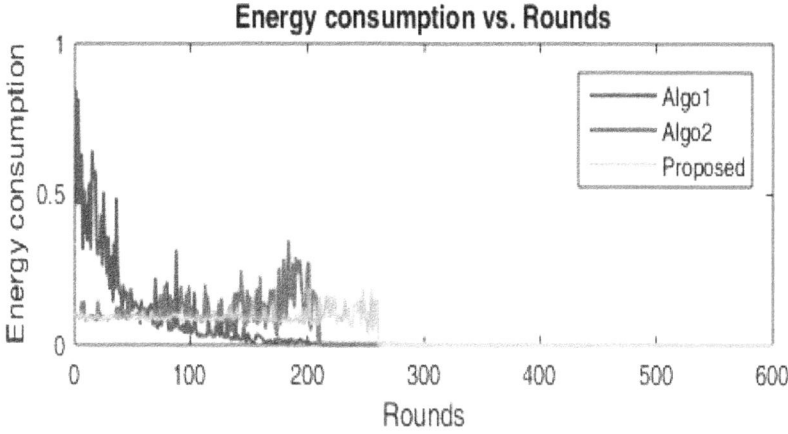

Fig. 3. Comparison of alive nodes and RE of WSN

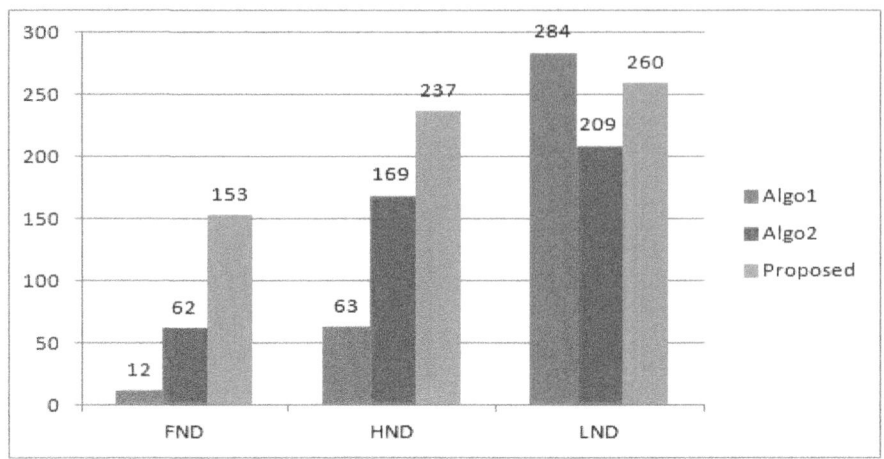

Fig. 4. Comparison of alive nodes and RE of WSN

5   Conclusion and Future Scope

In the proposed model, we have tried to select the optimal CHs and the load balancing among them. For the same purpose, we have defined five criteria shown in Table 2. Instead of considering the criteria weights randomly, we have used AHP to get optimal criteria weights. The outcome demonstrates that our suggested model optimises energy consumption and gives WSNs a longer lifespan. Due to load balancing, the energy consumption in our proposed model is consistent in each round which can be seen from Fig. 3. We have tried to cover most of the aspects of sensor nodes, but there are always chances for improvements. In future, the considered parameters may be optimized. In the proposed work, we have used the Some of Product (SOP) method to rank the alternatives. In future, some better meta-heuristic or MCDM approaches may be applied. The proposed model works for a 2-D area with one fixed BS, but in the future, it can be extended for more than one BS as well as moving BS.

References

1. Heinzelman, W.R., Chandrakasan, A., Balakrishnan, H.: Energy-efficient communication protocol for wireless microsensor networks. In: Proceedings of the 33rd Annual Hawaii International Conference on System Sciences, pp. 10–pp. IEEE (2000)
2. Loscri, V., Morabito, G., Marano, S.: A two-levels hierarchy for low-energy adaptive clustering hierarchy (TL-LEACH). In: IEEE Vehicular Technology Conference, vol. 62, no. 3, p. 1809. IEEE (2005)
3. Shi, S., Liu, X., Gu, X.: An energy-efficiency Optimized LEACH-C for wireless sensor networks. In: 7th International Conference on Communications and Networking in China, pp. 487–492. IEEE (2012)

4. Younis, O., Fahmy, S.: HEED: a hybrid, energy-efficient, distributed clustering approach for ad hoc sensor networks. IEEE Trans. Mob. Comput. **3**(4), 366–379 (2004)
5. Lindsey, S., Raghavendra, C.S.: PEGASIS: power-efficient gathering in sensor information systems. In: Proceedings of the IEEE Aerospace Conference, vol. 3, p. 3. IEEE (2002)
6. Behera, T.M., Mohapatra, S.K., Samal, U.C., Khan, M.S., Daneshmand, M., Gandomi, A.H.: Residual energy-based cluster-head selection in WSNs for IoT application. IEEE Internet Things J. **6**(3), 5132–5139 (2019)
7. Vijayalakshmi, K., Anandan, P.: A multi objective Tabu particle swarm optimization for effective cluster head selection in WSN. Clust. Comput. **22**(5), 12275–12282 (2019)
8. Pitchaimanickam, B., Murugaboopathi, G.: A hybrid firefly algorithm with particle swarm optimization for energy efficient optimal cluster head selection in wireless sensor networks. Neural Comput. Appl. **32**(12), 7709–7723 (2020)
9. Ali, H., Tariq, U.U., Hussain, M., Lu, L., Panneerselvam, J., Zhai, X.: ARSH-FATI: a novel metaheuristic for cluster head selection in wireless sensor networks. IEEE Syst. J. **15**(2), 2386–2397 (2020)
10. Preethiya, T., Muthukumar, A., Durairaj, S.: Double cluster head heterogeneous clustering for optimization in hybrid wireless sensor network. Wirel. Pers. Commun. **110**(4), 1751–1768 (2020)
11. Sreedharan, P.S., Pete, D.J.: A fuzzy multicriteria decision-making-based CH selection and hybrid routing protocol for WSN. Int. J. Commun. Syst. **33**(15), e4536 (2020)
12. Gao, T., Jin, R.C., Song, J.Y., Xu, T.B., Wang, L.D.: Energy-efficient cluster head selection scheme based on multiple criteria decision making for wireless sensor networks. Wirel. Pers. Commun. **63**(4), 871–894 (2012)
13. Rajpoot, P., Harsh Singh, S., Verma, R., Dubey, K., Kumar Pandey, S., Verma, S.: Multi-factor-based energy-efficient clustering and routing algorithm for WSN. In: Pant, M., Sharma, T.K., Verma, O.P., Singla, R., Sikander, A. (eds.) Soft Computing: Theories and Applications. AISC, vol. 1053, pp. 571–581. Springer, Singapore (2020). https://doi.org/10.1007/978-981-15-0751-9_53
14. Pandey, S.K., Patel, S.: Energy distance clustering algorithm (EDCA) for wireless sensor networks. In: 2019 9th International Conference on Cloud Computing, Data Science & Engineering (Confluence), pp. 287–292. IEEE (2019)
15. Saaty, T.L.: How to make a decision: the analytic hierarchy process. Eur. J. Oper. Res. **48**(1), 9–26 (1990)

Phish-EYE: A New Approach to Detect Homograph Domain Phishing Attack Using Domain Binary Visualization and TensorFlow

Pankaj Pandey[(✉)] [ID] and Nishchol Mishra [ID]

School of Information Technology, Rajiv Gandhi Proudyogiki Vishwavidyalya, Bhopal, India
pankaj82gh@gmail.com, nishchol@rgpv.ac.in

Abstract. A homograph phishing attack uses character script similarities to create and register fake domains of well-known brands to trick users into visiting these websites. This paper proposes a new visual inspection-based approach, Phish-EYE, for homograph phishing domain detection through binary visualization of domain names using TensorFlow. Phish-EYE can determine if a website domain is homograph or legitimate. Phish-EYE classifies homographs and non-homograph domains by converting the URL string into a binary string and using the byte class colour scheme to create an image. The TensorFlow machine learning framework is used to classify homographs and legitimate domains because it can learn image differences well. The Phish-EYE experiments have shown reliable and promising results. The average performance of Phish-EYE in identifying the homograph domain results in a 96.51% sensitivity (true positive rate) and 98.01% accuracy. The Phish-EYE performance demonstrates the feasibility and usefulness of leveraging the domain binary visuals to identify homograph phishing scams.

Keywords: Homograph phishing · Binary visualization · Phishing detection · Security · TensorFlow · Machine learning

1 Introduction

Internationalized Domain Names (IDNs) have expanded from English-only web services to those in non-English languages across the globe. IDNs let non-English users register their domains in their home languages. International domain names use Punycode. Punycode allows for the representation of Unicode characters using the constrained ASCII character set; for instance, domain **žugec.sk** is a domain xn–ugec-kbb.sk .

IDNs introduced security issues like the homograph attack [4]. Homograph attacks return with IDN, and a big Web security issue [2]. As the number of

© The Author(s), under exclusive license to Springer Nature Switzerland AG 2025
M. Gupta et al. (Eds.): MISS 2023, CCIS 1952, pp. 266–281, 2025.
https://doi.org/10.1007/978-3-031-69115-7_18

Unicode-based online pages and non-ASCII web addresses rises, researchers [2, 33] predicted that Unicode-based homograph attacks will become a larger threat to web security. Identity theft, fraud, and corruption will undoubtedly become more sophisticated.

In a homograph attack, attackers target their victims by using international characters that seem similar to English characters to spoof the spelling and appearance of legal domains but have different underlying Unicode values [40]. These international characters that seem similar to English could belong to non-English languages like Armenian, Latin, Hebrew etc. [13].

For example, Scammers can create homograph domains for legitimate domains: paypal.com and apple.com with matching Bulgarian Cyrillic characters that replace the English characters 'p', 'a' and 'y' in paypal.com and 'a', 'p' and 'e' in apple.com (shown in Figs. 2 and 4).

To launch a successful homograph domain attack, scammers first pick a well-known brand that they can use to trick the victim [9]. Then attackers make a fake URL by replacing a few Unicode non-Latin characters with Latin ones. Scammers then make the puny code for their phishing URL, which they buy and register with any domain name providers. Scammers send fake URLs to their victims via email, SMS, or Whatsapp. Now, even if the victim sees that the URL looks identical or almost identical, they may still give the scammer the information scammers want. IDN homograph attacks, like the Bitcoin exchange scam, are an increasing threat [5]. Matching characters prevent phishing in some solutions [15,43]. However, attackers use homograph domains with more Unicode-like characters. Although modern web browsers, email filters, and content filters can identify IDN homograph attacks, they do not eradicate the threat of IDN homograph attacks.

Previous homograph domain name analysis focused on the edit distance, also called Levenshtein distance, between the domain pairs, with the pair with the smallest distance being a homograph domain. However, methods that only use edit distance result in many false positives, specifically for domain names with minimal characters. In addition to edit distance, OCR [39,45] can find a domain name and online content similarities. These tactics increase processing costs when capturing and analysing many domain names or web pages, making it difficult to identify things quickly. Deep learning detection methods like the siamese CNN (Convolutional Neural Network) have been proposed [47]. With Euclidean distance, training takes longer, and detection is poorer.

This paper proposes a new visual inspection-based approach, Phish-EYE, for homograph phishing domain detection through binary visualization of domain names using TensorFlow. Figures 2 and 3 show an example of the legitimate and homograph domain names and their binary visualizations of the world's leading payment service organization(PayPal). Figures 4 and 5 illustrate binary visualizations of homograph and legitimate domains of the technology company(Apple), respectively. It is evident from both these figures that homograph domain names slightly possess different binary visualizations from their legitimate counterparts and are hence suitable for classification by the TensorFlow

Machine learning model. Therefore, the Phish-EYE framework is inspired by the idea that binary visualizations of domain names could be an intriguing way to examine homograph phishing attacks. The main contributions of the proposed work are:

- The Phish-EYE classifies homographs and non-homographs by converting the URL string into a binary string and then transforming this binary string into an image using the byte class colour scheme. Therefore, Phish-EYE is adaptable and resistant to code obfuscation techniques making crawling more difficult.
- The Phish-EYE works in a way that is similar to how a real person does when it comes to spotting phishing homograph domains by looking at them.
- The Phish-EYE project is an exciting way for anti-phishing researchers and developers to develop techniques or tools to detect homograph domain phishing scams using URL's binary visuals.

This article is structured as follows: Sect. 2 explores relevant research, and Sect. 3 provides a high-level summary of Phish-EYE. Section 4 discusses the implementation details of Phish-EYE. Section 5 discusses the performance evaluation of Phish-EYE, and Sect. 6 describes the benefits of Phish-EYE. Finally, Sect. 7 wraps up the article and discusses future work.

2   Literature Review

This section discusses existing techniques for detecting homograph-based phishing attacks and related works done by the previous researchers.

The early study by Holgers et al. [18] revealed that IDNs might spread homograph phishing scams, and attackers could use homograph domain names to lure people to bogus sites. Similarly, Quinkert et al. [34] discovered that homograph attackers aggressively targeted technological corporations and financial institutions. They found over 3,000 homograph domains aimed at technology and financial organizations.

Researchers have mainly used distance-based analysis and character or visual similarity analysis to differentiate homograph domain names from conventional ones. Traditional edit distance methods, on the other hand, overlook the visual confusion that can develop on characters. In addition to conventional methods, researchers now employ cutting-edge approaches like machine learning and deep learning to identify homograph domain names.

Accordingly, Liu et al. [37] introduced a an approach to estimate domain name visual similarity. Their method examines suspicious keywords, URLs, website layout, and appearance after finding them. However, the small data set used in phishing detection renders their proposed method ineffective. This technique finds the highest visual resemblance of a domain across a popular website set using generalised edit distance and distinctive visual features.

In the distance analysis method, researchers consider a pair of domain names with a minor edit distance to be homograph domain names. However, approaches merely relying on edit distance will have many false positives, particularly for domain names with fewer characters.

Woodbridge et al. [47] used a Siamese CNN for feature extraction from domain name images to examine homographic challenges. Their proposed method then uses Euclidean distance to detect similar and distinct strings.

Similarly, Ya et al. [48] used an LSTM-based siamese network to evaluate squatting domain name character sequences and calculate vector distances in Euclidean space.

Yu, Guangxi, et al. [50] designed a dual-channel CNN classifier that detects homograph domain names using minimal hash (MinHash) and locality-sensitive hash (LSH) retrieval techniques. CNN classifiers with dual channels evaluate pictures. MinHash and LSH reduce large amounts of data by locating domain names with similar characters. Their method for discovering homograph domain names is 98.5% accurate. MinHash and LSH are used to reduce large DNS actual log files.

Sawabe et al. [39] proposed OCR-based techniques for similarity analysis to detect homograph domains. This OCR technique builds matching mappings using ASCII characters based on input IDNs that contain non-ASCII characters. The relevant input IDN is a homograph domain name if these mappings contain legal domains.

In addition, K. Tian et al. [45] proposed a method for generating homographs that use cybersquatting. They extracted the most important visual elements from the page screenshots using optical character recognition (OCR) and visual analysis. However, the OCR system must execute preliminary image-processing steps before generating domain images. Meanwhile, this approach has a low detection rate.

Roshanbin et al. [36] used Normalized Information Distance and Kolmogorov Complexity theory to define visual similarity between Unicode glyphs.

Furthermore, some research [13,17] builds candidate homograph domain names from famous ones and investigates their WHOIS, DNS, and web page behaviour to find homograph attacks.

C. Daiki et al. [8] extracted 4.4 million IDNs using the commercial Whois database. They extracted 2,310 of the most popular brand domains from Alexa, Umbrella, and Majestic.

They calculated SSIM similarly to [42] but chose a threshold of 0.99 instead of 0.95. They classified IDN homographs with SSIMs of 0.99.

A. Pieter et al. [1] focused primarily on typosquatting. They did a Whois lookup on each domain and used visual metrics like page screenshot perceptual hash and locality-sensitive hash to create a cluster.

In order to identify DNS typo-squatting attacks with little complexity, Abdallah et al. [30] created an ensemble-based feature selection and classification methodology.

Ravi et al. [35] detected arbitrarily generated domain names and DNS homograph exploits without reverse engineering or DL model evaluation of the nonexistent domain.

T. Thao et al. [44] devised a system for classifying both IDNs and non-IDN homograph domains using machine learning and SSIM scores for each character in the domain string. Their proposed system improves accuracy and reduces the number of false positives. C. Daiki et al. [8] extracted 4.4 million IDNs using the commercial Whois database. They extracted 2,310 of the most popular brand domains from Alexa, Umbrella, and Majestic. They calculated SSIM similarly to [42] but chose a threshold of 0.99 instead of 0.95. They classified IDN homographs with SSIMs of 0.99.

L. Baojun et al. [42] proposed a system that uses Similarity(SSIM) to compare the image of each Internationalized domain name (IDN) to that of each brand domain among 1.4 million IDNs retrieved by them. They chose IDNs whose SSIM was more significant than or equal to 0.95 and classified them as homographs after setting a threshold for SSIM at that value. To enhance the classification, Zhu et al. [52] proposed a method that combines SSIM with Peak Signal-to-Noise Ratio (PSNR) and Means Squared Error (MSE).

Suzuki et al. [41] created ShamFinder to discover IDN homographs automatically. To analyze IDN homographs in the wild, ShamFinder was employed. Thus, the authors were able to suggest IDN homograph attack defences.

Y. Yazdani et al. [49] employed homoglyph tables on OpenINTEL data to create IDN ASCII counterparts. The authors extended homoglyph tables to find 2.97 times the number of homograph domains as prior homoglyph tables. The approach only fingerprints suspect domains before appearing on OpenINTEL's blacklists.

However, these methods [8,42] may need more scalability because many new domains are registered daily. Another disadvantage of these methods is that non-homographs can have SSIMs that are both higher and lower than these thresholds.

Almuhaideb et al. [3] suggested a homoglyph attack detection approach that uses hash functions and machine learning. Homoglyph attack detection was 99.8% accurate using Random Forest with the hash function.

Agten et al. [32] evaluated the Alexa top 500 typosquatting domains for seven months. They grouped content from probable typosquatting web pages. Typosquatters attacked 95% of the 500 domains examined, and most did not defend their domains.

Khan et al. [25] suggested an intent inference method for analyzing the effects of typosquatting cybercrime on victims. Compared to a browser error page, typosquatting costs the average user 1.3 s for each incidence. Unregistered typos cost legitimate websites 5% of their visitors.

Dam et al. [11] extensively studied typosquatting domain popup scams. This fraud uses a JavaScript warning popup to attract attention. The Alexa top 1M list revealed 9,857 popup notifications on 8,255 sites, most of which only appear to one HTTP user agent.

Abdallah et al. [30] proposed an ensemble-based feature selection and classification model to detect DNS typosquatting attacks with low complexity.

Using CSS similarity, Mao et al. [27] suggested a method that compares phishing and original websites. Identical CSS is needed to design an exact website. This method compares new CSS to previously viewed pages and cannot detect zero-hour attacks.

Afroz and Greenstadt [19] suggested PhishZoo, a website profiling system that compares new websites to database ones. PhishZoo first checks URL and SSL certifications. If the URL and SSL certificate match, PhishZoo checks saved profiles for the page text. If the logo changes, PhishZoo can't detect phishing.

Rosiello et al. [21] use DOM trees to detect phishing scams by comparing the DOM trees of two websites and looking for similarities. However, their system is ineffective if the attackers use a different DOM or phishing websites containing only images.

Huang et al. [26] proposed a web page-based signature system to compare a new page's signature to the database when a user opens it. Phishing if the domain name matches the signature. It can't detect zero-hour phishing frauds and requires extensive image collection.

Chen et al. [7] proposed a two-module Contrast Context Histogram system to record invariant information around vital points (CCH). CCH attributes determine page similarity in its second module. Passive methods need large image databases.

Utkarsh et al. [38] suggested several phishing tactics for detection, correction, and prevention. Statistic Database check, DNS and who- is check, URL Ascii content verification, and Visual similarity are the four approaches they use to detect phishing attempts.

Khan et al. [24] suggested a method to detect phishing websites using structural information from the standard dataset. They compared phishing website detection machine learning methods for false positives, true positives, false negatives, and accuracy.

3   Methodology

This section discusses the Phish-EYE architecture and machine learning algorithm TensorFlow employed. This section also provides brief information about binary visualization for domain image generation. Figure 1 depicts the proposed experimental setup model.

The Phish-EYE first converts the URL string into a binary string and then transforms this binary string into an image using the byte class colour scheme. The byte class colour scheme divides bytes into four groups: low bytes, ASCII text, high bytes, and unique colours for 0x00 and 0xff. It considers Tab (09), newline (0a) and carriage return (0d) as text.

3.1   Domain Binary Visualization

Phish-EYE first converts each URL string in the dataset into RGB images using a binary data visualization tool Binvis [6]. Phish-EYE uses these RGB images

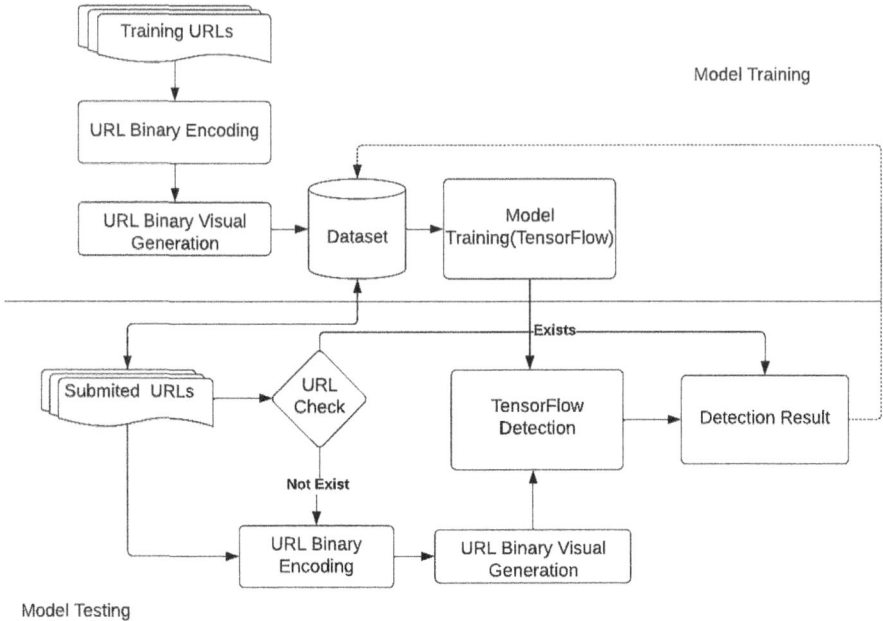

Fig. 1. Phish-EYE System Architecture

of each homograph and the legitimate domain classification using TensorFlow. The homograph domain pairs have different RGB values. Figure 2 and Fig. 3 illustrate binary visualization of PayPal's homograph and legitimate domains, respectively.

Similarly, Figs. 4 and 5 show binary visualization of Apple company's homograph and legitimate domains, respectively. It is evident from Figs. 2, 3, 4 and 5 that these binary visualizations of the homograph and legitimate domains have different RGB values and are hence suitable for classifying the phishing and legitimate domains with any machine learning algorithms.

3.2   TensorFlow

Phish-EYE uses TensorFlow, which uses artificial neural networks to analyze and process domain images to detect the homograph phishing domains. It has an excellent image recognition ability to learn the differences between the images. Therefore TensorFlow is best suitable for Phish-EYE applications to detect homograph domains for phishing attacks using MobileNet [28]convolutional neural network. TensorFlow analyses and classifies the domain visual images against its in-depth training.

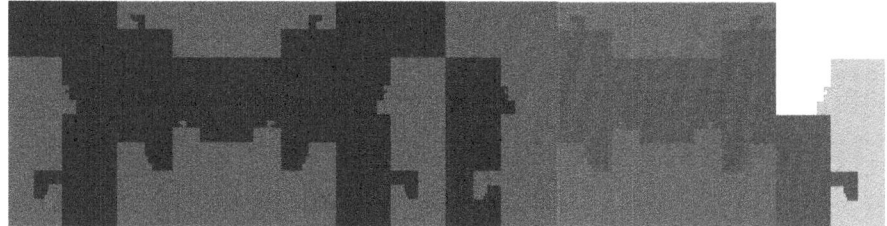

Fig. 2. Binary visualization of homographed PayPal domain.

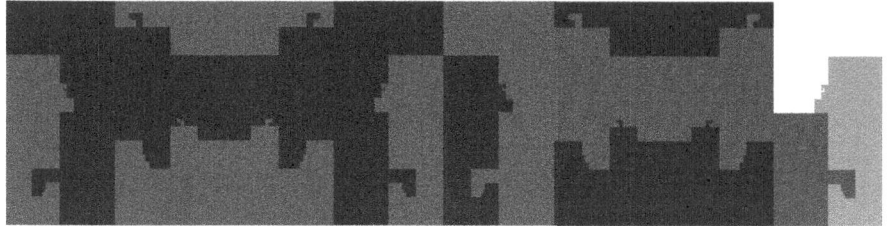

Fig. 3. Binary visualization of legitimate PayPal domain

Fig. 4. Binary visualization of homographed Apple company domain

Fig. 5. Binary visualization of legitimate Apple company domain

4   Implementation

Given a suspected domain name, Phish-EYE needs to predict whether it is a legitimate or homograph domain. Phish-EYE uses TensorFlow to classify if a domain is legit or homograph.

4.1   Tools Used in Implementation

Phish-EYE uses domain name binary visualizations to classify homograph and legitimate domains using TensorFlow. The proposed research uses the Binvis [6] tool to generate binary visualizations of domains. The Binvis uses the Hilbert space-filling curve clustering algorithm [20] and converts domain name strings to RGB values(128-pixel image) as its final output, as shown in Figs. 1, 2, 3 and 4. The Binvis divides the different ASCII characters of the domain name into the following classes of colours:

- Blue colour for print - ready ASCII characters.
- Green Colour for Control characters.
- Red colour for Extended ASCII characters.
- Black colour (0x00) for Null and white (0xFF) colour for spaces

4.2   Dataset

Phish-EYE uses dnstwist [46] service to collect 848 homograph domains of top finance, retail, technology and telecommunication brands. Table 3 shows the popular target brands under the categories mentioned above. Table 3 also illustrates the number of training and testing samples to train and test the TensorFlow machine learning model (Table 1).

Table 1. The top 15 most phished brands.

Sector	Brands
Finance	Intuit, American Express Inc, Paypa, Chase Bank, M&T Bank, DHL
Retail	Amazon, CVS, Sam's Club, Walmart
Technology	Microsoft, Netflix, Linkedin, Apple, facebook, DocuSign, Dropbox, Adobe, Google, Whatsapp
Telecommunications	Zoom, Ring Central, CamCast, Verizon

4.3   Performance Evaluation Measures

Metrics for performance evaluation of Phis-Eye is described as below:

- Recall sensitivity or True positive rate (TPR): It classifies homograph domains(h) as homograph phishing(Nh→h) among all homograph phishing domains (Nh).

$$TPR = \frac{N_{l \to h}}{N_l} * 100 \tag{1}$$

– False alarm or False positive rate (FPR): It classifies the legitimate domains (l) as homograph$((N_l{\rightarrow}h)$ among legitimate domains (N_l).

$$FPR = \frac{N_{h \to h}}{N_h} * 100 \tag{2}$$

– False negative rate (FNR):
It classifies homograph phishing domains (h) as legitimate $(N_h{\rightarrow}l)$ among legitimate domains (N_h).

$$FNR = \frac{N_{h \to l}}{N_h} * 100 \tag{3}$$

– Specificity or True negative rate (TNR):
It classifies legitimate domains (l) as legitimate $(N_l{\rightarrow}l)$ out of total legitimate domains(N_l).

$$TNR = \frac{N_{l \to l}}{N_l} * 100 \tag{4}$$

– Accuracy:
It predicts correctly all homograph phishing domain and genuine domains $(N_l{\rightarrow}l{+}N_p{\rightarrow}h)$ w.r.t. all the web pages $(N_l + N_h)$.

$$Accuracy = \frac{N_{l \to l} + N_{h \to h}}{N_l + N_h} \tag{5}$$

5   Evaluation

5.1   Result and Analysis

In the phishing attack detection process by homograph domains, Phish-EYE experimented on the dataset of 848 homograph domains of popular target brands shown in Table 2.

Table 2 describes the target brands with their training and testing samples used in the Phish-EYE. We conducted Phish-EYE experiments on the Colab environment to measure the efficiency of the trained TensorFlow model to predict the well-crafted phishing scam through homograph domains.

To prevent bias, we used the average value of the performance indicator specified in Sect. 4.3 as the expected value. Our experiment examined Phish-EYE with binary visuals of 848 homograph domains in its final dataset.

The Phish-EYE training stage trains the TensorFlow algorithm with 704 domain binary images(size of 128 pixels) per category (Legitimate and homograph domain pairs) for 2000 training steps. The learning rate was 0.005. The homograph domain names dataset contained several samples from popular target brands.

Table 3 compares how well Phish-EYE performed with well-known brands that meet the accuracy rate needed for practical use. The Phis-eYe achieved more than 99% accuracy for brands like Microsoft, LinkedIn, Facebook, Dropbox, Google, Zoom and CamCast. Phish-EYE performed well for other brands except for ChaseBank and DHL but with acceptable accuracy.

Table 2. Training and testing samples of popular target brands used in Phish-EYE

Target Brands	Training Samples	Testing Samples
Intuit	13	5
American Express	20	5
Paypal	85	5
Chase Bank	10	5
M&T Bank	22	5
DHL	25	5
Amazon	72	5
CVS	05	5
Sam's Club	13	5
Walmart	15	5
Microsoft	58	5
Netflix	27	5
Linkedin	50	5
Apple	27	5
Facebook	63	5
DocuSign	15	5
Dropbox	18	5
Adobe	15	5
Google	89	5
Whatsapp	14	5
Zoom	43	5
Ring Central	12	5
CamCast	08	5
Verizon	09	5

5.2 Comparison of Phish-EYE with Existing Anti-phishing Approaches

The proposed Phish-EYE method compared its average performance across all brands to other well-known anti-phishing approaches. The FPR, TPR, and accuracy (ACC) comparisons are shown in Table 4.

6 Advantage of Phish-EYE

6.1 Language Independent

To effectively target non-English websites, language-independent anti-phishing strategies are essential. The key benefit of Phish-EYE is that it is language-independent as it uses only URLs to generate their binary visual features for Tensorflow to detect phishing scams.

6.2 Third Party Independent

Several anti-phishing strategies, including those in [12,14,16,24,29,31,51], rely on third-party services for their feature extraction process, including DNS, whois

Table 3. Phish-EYE Test Results for Popular Brands.

Brand	FPR%	TPR%	Accuracy
Intuit	1.71	94.23	97.67
American Express Inc	1.34	95.89	97.93
Paypal	3.43	93.65	98.13
Chase Bank	4.89	91.41	95.18
M&T Bank	1.68	97.58	98.14
DHL	7.16	93.14	95.23
Amazon	1.13	98.65	98.83
CVS	3.78	95.23	96.17
Sam's Club	2.15	95.84	96.78
Walmart	2.46	96.47	97.95
Microsoft	1.23	98.73	99.01
Netflix	1.67	97.81	98.10
Linkedin	1.29	98.23	99.07
Apple	1.78	97.56	98.23
Facebook	1.17	98.45	99.24
DocuSign	2.89	96.71	98.25
Dropbox	3.17	94.57	99.21
Adobe	3.28	95.79	97.16
Google	1.61	98.69	99.29
Whatsapp	2.53	96.57	97.93
Zoom	1.15	98.57	99.24
Ring Central	3.67	96.27	97.37
CamCast	1.98	98.65	99.87
Verizon	2.65	97.63	98.15

Table 4. Comparison of Phish-EYE with other standard approaches.

Approach	FPR%	TPR%	ACC%
Chiew et al. [10]	1.30	99.80	93.40
El-Alfy et al. [12]	3.88	97.24	96.74
F. Feng et al. [14]	0.17	96.68	97.71
Ankit et al. [22]	1.25	99.39	99.09
Ankit Jain et al. [23]	1.51	98.30	98.42
Gotham et al. [29]	1.71	98.24	98.25
A.A. Orunsolu et al. [31]	0.04	99.60	99.60
Tan et al. [24]	7.48	99.68	96.10
Zhang et al. [51]	0.53	98.64	99.04
Proposed approach(Phish-EYE)	2.49	96.51	98.01

records, blocklist/allowlist, and others. A lengthy prediction time may result from third-party feature extraction. To extract the features for domain classification as homograph or legitimate domains, Phish-EYE only uses the domain names of well-known brands.

7   Conclusion

This study presented a new method for spotting domain homograph phishing scams based on the binary representation of domain names. The primary goal of this research is to investigate whether binary visuals of domain names can detect a phishing scam that is both flexible and resistant to code obfuscation. To prove the efficacy of Phish-EYE, we compiled a data set of specifically targeted well-known brands. With a true positive rate of over 97.63% and a false positive rate of only 2.65%, the binary visuals of domains are highly effective, as demonstrated by the experimental results of Phish-EYE on the dataset for various well-known brands. Comparatively, Phish-use eYe's domain binary visuals to detect phishing scams are just as effective as existing anti-phishing methods. The domains' binary visuals in the proposed work are determined solely by the URL. Therefore, our future work will focus on discovering homograph phishing attacks that use languages other than English.

Compliance with Ethical Standards

Data Availability. The corresponding author will provide datasets created and analysed during the current investigation upon reasonable request.

Conflict of Interest. The authors declare that they have no known financial or personal interests that might conflict with or impact the work described in this publication.

Ethical Approval. This article does not contain any studies with human participants or animals performed by any authors.

References

1. Agten, P., Joosen, W., Piessens, F., Nikiforakis, N.: Seven months' worth of mistakes: a longitudinal study of typosquatting abuse. In: Proceedings of the 22nd Network and Distributed System Security Symposium (NDSS 2015). Internet Society (2015)
2. Al Helou, J., Tilley, S.: Multilingual web sites: internationalized domain name homograph attacks. In: 2010 12th IEEE International Symposium on Web Systems Evolution (WSE), pp. 89–92. IEEE (2010)
3. Almuhaideb, A.M., et al.: Homoglyph attack detection model using machine learning and hash function. J. Sens. Actuator Netw. **11**(3), 54 (2022)
4. Baasanjav, U.B.: Linguistic diversity on the internet: Arabic, Chinese and Cyrillic script top-level domain names. Telecommun. Policy **38**(11), 961–969 (2014)

5. Binance: Summary of the phishing and attempted stealing incident on binance (2022). https://www.binance.com/en/support/announcement/summary-of-the-phishing-and-attempted-stealing-incident-on-binance-360001547431. Accessed 15 Dec 2022

6. Binvis: Binvis.io (2022). http://binvis.io/. Accessed 15 Nov 2022

7. Chen, K.T., Chen, J.Y., Huang, C.R., Chen, C.S.: Fighting phishing with discriminative keypoint features. IEEE Internet Comput. **13**(3), 56–63 (2009)

8. Chiba, D., Hasegawa, A.A., Koide, T., Sawabe, Y., Goto, S., Akiyama, M.: {DomainScouter}: understanding the risks of deceptive {IDNs}. In: 22nd International Symposium on Research in Attacks, Intrusions and Defenses (RAID 2019), pp. 413–426 (2019)

9. Chiba, D., Hasegawa, A.A., Koide, T., Sawabe, Y., Goto, S., Akiyama, M.: Domainscouter: analyzing the risks of deceptive internationalized domain names. IEICE Trans. Inf. Syst. **103**(7), 1493–1511 (2020)

10. Chiew, K.L., Chang, E.H., Tiong, W.K., et al.: Utilisation of website logo for phishing detection. Comput. Secur. **54**, 16–26 (2015)

11. Dam, T., Klausner, L.D., Buhov, D., Schrittwieser, S.: Large-scale analysis of pop-up scam on typosquatting URLs. In: Proceedings of the 14th International Conference on Availability, Reliability and Security, pp. 1–9 (2019)

12. El-Alfy, E.S.M.: Detection of phishing websites based on probabilistic neural networks and k-medoids clustering. Comput. J. **60**(12), 1745–1759 (2017)

13. Elsayed, Y., Shosha, A.: Large scale detection of IDN domain name masquerading. In: 2018 APWG Symposium on Electronic Crime Research (eCrime), pp. 1–11. IEEE (2018)

14. Feng, F., Zhou, Q., Shen, Z., Yang, X., Han, L., Wang, J.: The application of a novel neural network in the detection of phishing websites. J. Ambient Intell. Hum. Comput. 1–15 (2018)

15. Fouss, B., Ross, D.M., Wollaber, A.B., Gomez, S.R.: Punyvis: a visual analytics approach for identifying homograph phishing attacks. In: 2019 IEEE Symposium on Visualization for Cyber Security (VizSec), pp. 1–10. IEEE (2019)

16. Gandotra, E., Gupta, D.: Improving spoofed website detection using machine learning. Cybern. Syst. **52**(2), 169–190 (2021)

17. de Gouveia Dal Pino, E.M., Kowal, G.: Particle acceleration by magnetic reconnection. In: Magnetic Fields in Diffuse Media, pp. 373–398 (2015)

18. Holgers, T., Watson, D.E., Gribble, S.D.: Cutting through the confusion: a measurement study of homograph attacks. In: USENIX Annual Technical Conference, General Track, pp. 261–266 (2006)

19. Huang, C.Y., Ma, S.P., Yeh, W.L., Lin, C.Y., Liu, C.T.: Mitigate web phishing using site signatures. In: 2010 IEEE Region 10 Conference, TENCON 2010, pp. 803–808. IEEE (2010)

20. Jagadish, H.V.: Analysis of the hilbert curve for representing two-dimensional space. Inf. Process. Lett. **62**(1), 17–22 (1997)

21. Jain, A.K., Gupta, B.B.: Phishing detection: analysis of visual similarity based approaches. Secur. Commun. Netw. **2017** (2017)

22. Jain, A.K., Gupta, B.B.: Towards detection of phishing websites on client-side using machine learning based approach. Telecommun. Syst. **68**(4), 687–700 (2018)

23. Jain, A.K., Gupta, B.B.: A machine learning based approach for phishing detection using hyperlinks information. J. Ambient. Intell. Humaniz. Comput. **10**(5), 2015–2028 (2019)

24. Khan, M.F., Tiwari, R.K., Saroj, S.K., Tripathi, T.: A comparative study of machine learning techniques for phishing website detection. In: Role of Data-Intensive Distributed Computing Systems in Designing Data Solutions, pp. 97–109. Springer, Cham (2023)
25. Khan, M.T., Huo, X., Li, Z., Kanich, C.: Every second counts: quantifying the negative externalities of cybercrime via typosquatting. In: 2015 IEEE Symposium on Security and Privacy, pp. 135–150. IEEE (2015)
26. Liu, W., Deng, X., Huang, G., Fu, A.Y.: An antiphishing strategy based on visual similarity assessment. IEEE Internet Comput. **10**(2), 58–65 (2006)
27. Mao, J., Li, P., Li, K., Wei, T., Liang, Z.: Baitalarm: detecting phishing sites using similarity in fundamental visual features. In: 2013 5th International Conference on Intelligent Networking and Collaborative Systems, pp. 790–795. IEEE (2013)
28. MobileNet: Mobilenet (2022). https://ai.googleblog.com/2017/06/mobilenets-open-source-models-for.html. Accessed 18 Nov 2022
29. Montazer, G.A., ArabYarmohammadi, S.: Detection of phishing attacks in Iranian e-banking using a fuzzy-rough hybrid system. Appl. Soft Comput. **35**, 482–492 (2015)
30. Moubayed, A., Aqeeli, E., Shami, A.: Detecting DNS typo-squatting using ensemble-based feature selection & classification models. IEEE Can. J. Electr. Comput. Eng. **44**(4), 456–466 (2021)
31. Orunsolu, A., Sodiya, A., Akinwale, A.: A predictive model for phishing detection. J. King Saud Univ.-Comput. Inf. Sci. (2019)
32. Pour, M.S., Nader, C., Friday, K., Bou-Harb, E.: A comprehensive survey of recent internet measurement techniques for cyber security. Comput. Secur. 103123 (2023)
33. Qiu, B., Fang, N., Wenyin, L.: Detect visual spoofing in unicode-based text. In: 2010 20th International Conference on Pattern Recognition, pp. 1949–1952. IEEE (2010)
34. Quinkert, F., Lauinger, T., Robertson, W., Kirda, E., Holz, T.: It's not what it looks like: measuring attacks and defensive registrations of homograph domains. In: 2019 IEEE Conference on Communications and Network Security (CNS), pp. 259–267. IEEE (2019)
35. Ravi, V., Alazab, M., Srinivasan, S., Arunachalam, A., Soman, K.: Adversarial defense: DGA-based botnets and DNS homographs detection through integrated deep learning. IEEE Trans. Eng. Manag. **70**(1), 249–266 (2021)
36. Roshanbin, N., Miller, J.: Finding homoglyphs-a step towards detecting unicode-based visual spoofing attacks. In: Web Information System Engineering–WISE 2011: 12th International Conference, Sydney, Australia, 13–14 October 2011. Proceedings 12, pp. 1–14. Springer (2011)
37. Rosiello, A.P., Kirda, E., Ferrandi, F., et al.: A layout-similarity-based approach for detecting phishing pages. In: 2007 Third International Conference on Security and Privacy in Communications Networks and the Workshops-SecureComm 2007, pp. 454–463. IEEE (2007)
38. Sanjay, U., Ananad, P., Danthi, A.A., Akshay, G., Ravi, P.: Real-time phishing detection using statistic database check, DNS and who is check, verifying ASCII content of the URL and visual similarity. In: Cognition and Recognition: 8th International Conference, ICCR 2021, Mandya, India, 30–31 December 2021, Revised Selected Papers, pp. 332–339. Springer, Cham (2023)
39. Sawabe, Y., Chiba, D., Akiyama, M., Goto, S.: Detecting homograph IDNs using OCR. Proc. Asia-Pac. Adv. Netw. **46**, 56–64 (2018)

40. Simpson, G., Moore, T., Clayton, R.: Ten years of attacks on companies using visual impersonation of domain names. In: 2020 APWG Symposium on Electronic Crime Research (eCrime), pp. 1–12. IEEE (2020)

41. Suzuki, H., Chiba, D., Yoneya, Y., Mori, T., Goto, S.: Shamfinder: an automated framework for detecting IDN homographs. In: Proceedings of the Internet Measurement Conference, pp. 449–462 (2019)

42. Thao, T.P.: Improving homograph attack classification. arXiv preprint arXiv:2009.08006 (2020)

43. Thao, T.P., et al.: Human factors in homograph attack recognition. In: International Conference on Applied Cryptography and Network Security, pp. 408–435. Springer, Cham (2020)

44. Thao, T.P., Sawaya, Y., Nguyen-Son, H.Q., Yamada, A., Omote, K., Kubota, A.: Hunting brand domain forgery: a scalable classification for homograph attack. In: IFIP International Conference on ICT Systems Security and Privacy Protection, pp. 3–18. Springer, Cham (2019)

45. Tian, K., Jan, S.T., Hu, H., Yao, D., Wang, G.: Needle in a haystack: tracking down elite phishing domains in the wild. In: Proceedings of the Internet Measurement Conference, pp. 429–442 (2018)

46. Ulikowski, M.: dnstwist (2022). https://github.com/elceef/dnstwist. Accessed 15 Nov 2022

47. Woodbridge, J., Anderson, H.S., Ahuja, A., Grant, D.: Detecting homoglyph attacks with a Siamese neural network. In: 2018 IEEE Security and Privacy Workshops (SPW), pp. 22–28. IEEE (2018)

48. Ya, J., Liu, T., Li, Q., Lv, P., Shi, J., Guo, L.: Fast and accurate typosquatting domains evaluation with Siamese networks. In: 2018 IEEE Military Communications Conference (MILCOM), MILCOM 2018, pp. 58–63. IEEE (2018)

49. Yazdani, R., van der Toorn, O., Sperotto, A.: A case of identity: detection of suspicious IDN homograph domains using active DNS measurements. In: 2020 IEEE European Symposium on Security and Privacy Workshops (EuroS&PW), pp. 559–564. IEEE (2020)

50. Yu, G., Yang, X., Zhang, Y., Cui, H., Yang, H., Li, Y.: Towards homograph-confusable domain name detection using dual-channel CNN. In: Information and Communications Security: 21st International Conference, ICICS 2019, Beijing, China, 15–17 December 2019, Revised Selected Papers 21, pp. 555–568. Springer, Cham (2020)

51. Zhang, D., Yan, Z., Jiang, H., Kim, T.: A domain-feature enhanced classification model for the detection of Chinese phishing e-business websites. Inf. Manag. **51**(7), 845–853 (2014)

52. Zhu, Z., Thao, T.P., Nguyen-Son, H.Q., Yamaguchi, R.S., Nakata, T.: Enhancing a new classification for IDN homograph attack detection. In: 2020 IEEE International Conference on Dependable, Autonomic and Secure Computing, International Conference on Pervasive Intelligence and Computing, International Conference on Cloud and Big Data Computing, International Conference on Cyber Science and Technology Congress (DASC/PiCom/CBDCom/CyberSciTech), pp. 507–514. IEEE (2020)

Effect of Cosine Decay Restart Learning Rate Scheduler on Movie Recommender System

Sonu Airen[✉][ID] and Jitendra Agrawal

SOIT, RGPV, Bhopal, India
sonugoyal24@rediffmail.com, jitendra@rgpv.ac.in

Abstract. Recommender System is an information filtering tool to filter the relevant information from given information in the present era of big data. Movie Recommender System is machine learning based autonomous tool that filters the movies from big movie database like Netflix, Amazon etc. according to user preferences. Learning Rate Scheduler is optimization technique in Deep Neural Network. Learning rate is most important hyper-parameter for Deep Neural Network training. The main focus of this paper is to propose Learning Rate Scheduler for Movie Recommender System. The primary objective of this research article is to find effect of Cosine Decay Restart Learning Rate Scheduler on Movie Recommender System. Test results obtained from the Movie database show that the proposed method can bring more accurate personalized recommendations for the movie as compared to existing methods of the order of 2.224%.

Keywords: Deep Neural Network · Learning Rate Scheduler · Loss Function · Machine Learning · Recommender System · Stochastic Gradient Descent

1 Introduction

1.1 Recommender System

Digital platforms are a must for human life in the digital contemporary world. In order to conduct digital payments, communicate, and purchase online, this platform is mandatory. In some digital applications, Recommender System is required. Recommendation become a mainstream feature in nowadays e-commerce because of its significant contribution in promoting revenue and customer satisfaction. Suppose for example online movie application. For this application, thousands of movies are presented and to select an appropriate movie to watch, we have to choose a movie from this large movie database. Here to select a good movie to see, the person does not depend upon their friend's recommendation due to the large size of the database. To avoid searching in a large database Recommender System is used. A tool or programme for information

© The Author(s), under exclusive license to Springer Nature Switzerland AG 2025
M. Gupta et al. (Eds.): MISS 2023, CCIS 1952, pp. 282–295, 2025.
https://doi.org/10.1007/978-3-031-69115-7_19

filtering called a "Recommender System" [1, 7] suggests products based on the tastes or desires of the user. For example in the case of movies when we perform a search for the movie it will give the same result for 10 years old boy as well as for a 50-year-old man. But in the case of the Movie Recommender System, it provides results according to the user's age and choice.

For given user-item pair, Recommender System predicts the rating for this pair. Content-based Recommender System, Collaborative Filtering Recommender System, Hybrid Recommender System, Demographic Recommender System and Knowledge-based Recommender System are the main techniques of Recommender System [1, 7]. In Content-based Recommender System the target item profile is matched to the collection of items that the user has previously assessed. Collaborative Filtering Recommender System [14] finds similar users to the active user and called it the active user's neighborhood (usually referred to as "K Nearest Neighbors" or KNN). Active user has a positive correlation with this neighborhood. Active user recommendations are generated by combining ratings from neighborhoods. The demographic Recommender System classify user based on user's demographic like age, gender, city, state and religion etc. Knowledge-based Recommender System matches domain knowledge of users' requirements with the feature vector of items and finds how the items fulfill users' requirements. Content-based, Demographic, Utility-based, Collaborative, Knowledge based, context aware recommendation are the main recommendation techniques. For better performance, the hybrid recommendation combines two or more of these methods. They try to use the advantage of one technique to remove the disadvantage of the second technique.

1.2   Neural Network

An artificial intelligence technique called a neural network instructs computers to analyse data in a manner modelled after the human brain. Deep learning is a type of machine learning that employs linked neurons or nodes in a layered structure to learn hidden pattern present in data. Neural Network uses numerous hidden layers and non-linear activation functions to describe complicated patterns in data-sets. Input for the neural network is applied on the input layer in the form of vectors. These input vectors are processed via several hidden layers of the network and output is produced by the Neural Network. Neurons in layers are the main processing elements that apply nonlinear activation function to the weighted sum received from the previous layer neuron.

Gradient Descent [13] is one optimization approach used to iteratively train neural networks. The difference between the prediction and the target is used to create an error metric at the conclusion of each training cycle. Using a method known as back-propagation, the feedback signal is passed from the last layer of the Neural Network to the previous layer using the error metric's derivative. The coefficients (weights) of each neuron are then modified according to their contribution to the overall error. Iteratively repeating this strategy continues until the network error falls below a desirable level. The following objects are vital to neural network architecture [3].

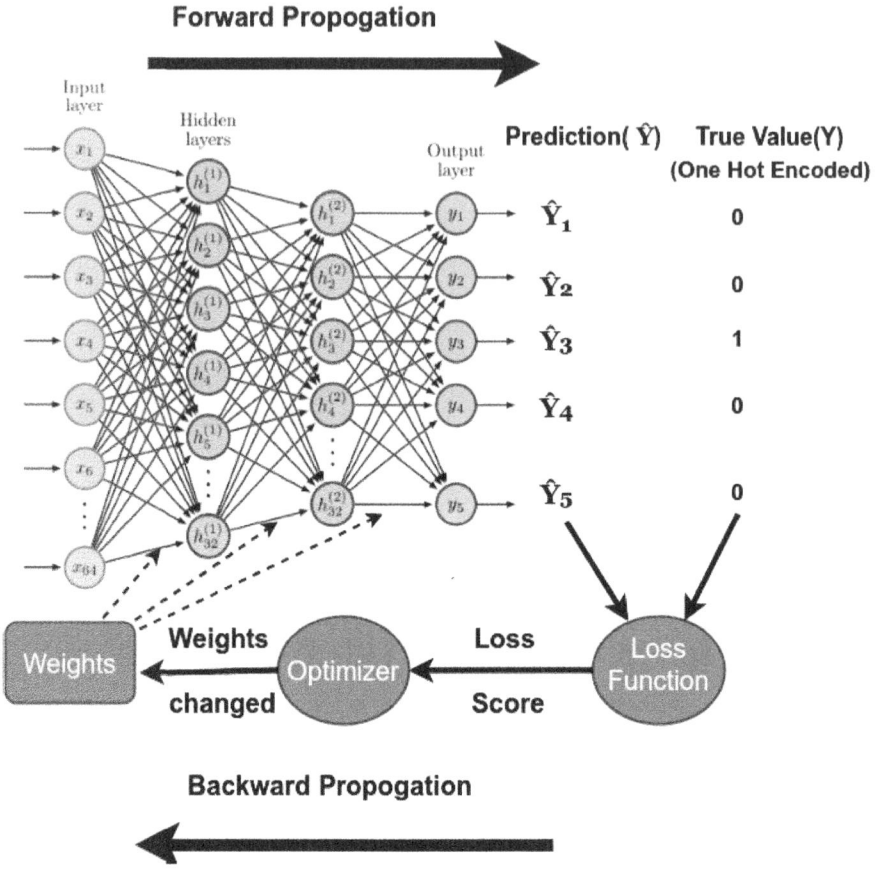

Fig. 1. Neural network architecture

Layers

The network, which is made up of layers connected together, transforms input data into predictions. A layer is a data processing module. The neural network's input layer is its initial layer. The data used to train the model is stored in the input layer. A distinct attribute from the data-set is represented by each neuron in the input layer (e.g. height, hair color, etc.). A hidden layer executes an activation function between the input and output layers and transmits the results afterward. In order to learn complex decision boundaries for categorization, deep learning neural networks stack many hidden layers. Typically, fully-connected dense layers make up hidden layers in conventional networks. Input to each neuron is the output produced by all the neurons in the previous layer and each neuron in the next layer receives the output of this neuron as input. The network's output layer is the last layer. It takes input from the preceding hidden layer, may or may not apply an activation function, and then returns an output that represents a model prediction.

Activation Functions
Complex non-linear interactions between features are modelled by activation functions. Binary Step Function, Sigmoid/Logistic, Tanh (Hyperbolic Tangent), ReLU and Softmax are the most prevalent neural network activation function.

Loss Functions
The feedback signal is determined by the loss function. The model's prediction is wrapped in a loss function, also known as a cost function, which indicates "how good" the model is at making predictions for a particular set of parameters. By comparing the guess made by the model to the desired outcomes loss values are calculated by the loss function. The model's parameters are updated using these loss values.

Optimizer
Optimizer controls the trajectory of learning. In response to the output of the loss function, Optimizer modifies the model parameters during training. Whether the optimizer is heading in the right direction or not is determined by the loss function.

Learning Rate
From examples in the training data-set, a Deep Neural Network optimally maps inputs to outputs to learn or approximate a function. The method used for training deep learning neural networks is Stochastic Gradient Descent.

With the guidance of the back-propagation method, neural networks are optimized. The data is fed forward into the model, the parameters are chosen at random, and the learning rate-dependent error of the result is propagated backwards into the model. The "learning rate" or "step size" is a hyper-parameter, that corresponds to the amount that the weights are updated during training. The lowercase Greek letter eta (η) is frequently used to denote the learning rate.

The relationship between these elements can be seen in the Fig. 1 Neural network architecture.

1.3   Neural Network Architecture

Four commonly used neural network architecture are as follows:

A neural network architecture called a multilayer perceptron, generally described as a conventional neural network, is developed to approximate arbitrary continuous functions and can address non-linear separable issues. A multilayer perceptron is composed of three primary layers: input, hidden, and output.

Computer vision problems, like image identification and classification are handled by Convolutional neural networks. Convolutional, pooling, and fully connected layers make up the three primary layers of this neural network architecture.

A neural network architecture known as a recurrent neural network excels at processing sequential information, like data from time series and natural language processing.

A type of neural network design known as a generative adversarial network can produce new data with the same properties as the training data. The generator and discriminator networks in this neural network architecture are positioned crosswise of each other. Every network plays a unique role.

2 Motivation

Given the training epochs number, nodes per layer number, and layers number the model will learn. Most of the time, a high learning rate makes it possible for the model to learn more quickly, but at the expense of producing a final set of weights that is less than ideal. Although training can take a lot longer, a model with a lower learning rate might be able to learn a more ideal or even globally optimal set of weights.

As a result, neither a large nor a small learning rate should be used. However, the model must be set up so that, on average, a "good enough" combination of weights is discovered to approximate the mapping issue as represented by the training data-set. Therefore, it is essential to determine the model's proper learning rate using the provided training data-set.

Variating the learning rate throughout the training process is an alternative to using a fixed learning rate. The learning rate schedule describes how the learning rate changes over time (training epochs). The learning rate can be decreased linearly from a high initial value to a low initial value, which is arguably the simplest learning rate schedule. As a result, it is possible to make substantial weight adjustments at the beginning of the learning phase and smaller weight changes or fine-tuning towards the conclusion of the learning phase.

Using the cosine function as the learning rate decay function is referred to as cosine decay. In actual applications, it has been demonstrated that the cosine function outperforms substitutes like straightforward linear decay. Cosine decay and warm restarts are the two components of the Cosine Decay Warm Restart learning schedule. Warm restarts refer to periodically restating the learning rate, i.e., raising it again. The scheduler for the cosine decay learning rate offers both high and low learning rates. The scheduler's high learning rates serve to keep the learner from becoming trapped in a local cost minima, while the scheduler's low learning rates enable it to converge to a nearly true optimal point inside the global minima it discovers.

Stochastic Gradient Descent with Warm Restarts [9] abbreviated SGDR at times, is the combination of Cosine Decay with Warm Restarts and Stochastic Gradient Descent.

When the model is trained, and if it is in a steep trough, in search of a more stable local minimum, it jumps from one local minimum to another. It is the restart phase of SGDR.

$$\eta_t = \eta_{min}^i + \frac{1}{2}(\eta_{max}^i - \eta_{min}^i)(1 + \cos(\frac{T_{cur}}{T_i}\pi)) \tag{1}$$

where:

η^i_{min} and η^i_{max} are the minimum and maximum LR for the ith run.

T_i is the number of epochs before we'll do a restart.

The contributions of this paper are:

1. We propose the methodology for setting the SGDR learning rate scheduler for the movie Recommender System.
2. In the best of our knowledge, no previous paper proposed SGDR learning rate scheduler for a movie Recommender System.
3. We significantly perform our proposed approach on representative data-set from a movie Recommender System, and empirical results found that prediction error is reduced by 2.224%.

3   Related Work

This research [10] employed a collaborative filtering system that predicts a user's movie rating based on ratings from other users on the MovieLens data set using a scalable deep learning technique based on auto-encoders. By using regularisation to further reduce the prediction errors, the authors claimed that this study performed better than the neighborhood-based baseline and improved at predicting movie ratings and recommending movies to consumers.

To enhance stochastic gradient descent's anytime performance when training deep neural networks, the authors of this research offer a straightforward warm restart method [9].

Slanted triangle learning rates (STLR), which initially linearly increase the learning rate and subsequently linearly decline it, are suggested by [6]. The concept of cyclical learning rates is introduced in the work [16], which effectively gets rid of the necessity to conduct experiments to determine the ideal schedule and values for the global learning rates. Instead of the learning rate monotonically declining, this technique enables the learning rate to fluctuate between acceptable cutoff values. When cyclical learning rates are used during training rather than constant values, classification accuracy may often be increased without the need for tuning and with fewer iterations.

In order to adapt learning rates as a function of mini-batch size, [5] employ a linear scaling strategy and create a new warm-up technique that overcomes optimization difficulties early in training. The regulation of step-size and progressive noise reduction on the network are the goals of the learning rate. [18] will use a set learning rate and a decaying loss approach in this research to regulate the update's size. To validate their approach, they employed image classification, semantic segmentation, and GANs.

In this study [17], authors presented a novel learning rate scheduler that switches between triangular mode with restarts and hyperbolic tangent decay. The suggested scheduler switches the learning rate from epoch to epoch. A new learning rate is computed as training progresses during each epoch using the difference between the gradient values of the previous two repetitions.

This paper [8] proposes an automated LR scheduler designed specifically for a quick training with huge batch sizes. Two phases are suggested for LR schedule consisting of a warm-up phase followed by a decay phase. During the warm-up phase, the LR is exponentially raised until it stops decreasing, after which it is dropped to zero until training is complete. By using GP regression, the online identification of the lowest loss has been effectively and reliably accomplished.

In this study, authors [20] offers a system based on reinforcement learning that can automatically develop an adaptive learning rate schedule using data from previous training histories. To do this, they present a powerful collection of features to describe the dynamic training process, a valuable reward function, action space, and a sample-efficient RL algorithm to dynamically adjust the learning rate. Authors [19] suggest a Graph-Network-based learning rate Scheduler using the concept of reinforcement learning. The proposed system incorporates the current dynamics of Graph-Network using a graph message-passing network.

In [12] authors investigate recent deep neural network models and various learning rate scheduling techniques and also propose a scheduling method by the combination of sigmoid with a warm-up.

The authors [15] suggest a new approach for learning rate scheduler called "AdaDec" that separates per-parameter learning rate variation from long-term learning-rate scheduling.

4   Methodology

Suppose $(X^{(1)}, y^{(1)}), (X^{(2)}, y^{(2)}), ..., (X^{(m)}, y^{(m)})$ is given data set of m examples. $X^{(i)}$ and $y^{(i)}$ is input data and output label respectively. A prediction function $h_w : X- > y$ from input space X to label space y is determined by the model parameter w that goes along with it. The disagreement between the true label $y^{(i)}$ and the prediction $h_w(x^{(i)})$ defines the differentiable loss function as shown in Eq.

$$Loss = Error(Y, \widehat{Y}) \tag{2}$$
$$f^{(i)}(w) = J(h_w(x^{(i)}), y^{(i)}) \tag{3}$$

Loss function (error function) are reduced by Optimization techniques. Gradient descent [13] minimizes a given function to its local minimum by iteratively modifying the parameters of the model. Gradient Descent moves in the direction opposite to the steepest climb to iteratively reduce a loss function. Finding minima is reliant on the derivatives of the loss function.

$$f(W) = \frac{1}{m} \sum_{i=1}^{m} (J(h_w(x^{(i)}), y^{(i)})) \tag{4}$$

$$= \frac{1}{m} \sum_{i=1}^{m} f^{(i)}(w) \tag{5}$$

$$w^* = argmin_w F(W) \tag{6}$$

$$w_{t+1} = w_t - \eta \frac{\partial L}{\partial w} \tag{7}$$

where f is associated with parameter $W = (w_1, w_2, ..., w_m)$. The objective of the optimizer is to find the optimal parameter w* by minimizing the empirical loss incurred on a given set of training data $(X^{(1)}, y^{(1)}), (X^{(2)}, y^{(2)}), ..., (X^{(m)}, y^{(m)})$. The stochastic gradient descent method, which updates the solution weights at each iteration t based on gradient information, is mostly used to optimize supervised learning frameworks, which often require a large number of training data.

$$w^{(t+1)} = w^{(t)} - \eta \Delta f(w^{(t)}) \tag{8}$$

suppose $b^{(t)}$ is batch

$$w^{(t+1)} = w^{(t)} - \eta(1/B \sum_{i \epsilon b^{(t)}} \Delta f_i(w^{(t)})) \tag{9}$$

Softmax
The probability distribution of an event across 'n' special events is calculated by the Softmax function. This function, to put it simply, will determine the odds of each target class among all potential target classes. The probabilities that have been generated will later be useful in identifying the target class for the supplied inputs.

For multi-class single-label classification.

$$\sigma(z_i) = \frac{e^{z_i}}{\sum_{j=1}^{K} e^{z_j}} \tag{10}$$

for $i = 1, 2, ..., K$ Categorical Cross-Entropy loss
It is a Cross-Entropy loss combined with a Softmax activation [2,4,11]. Multi-class classification tasks use the categorical cross-entropy loss function. For these tasks, the model must choose which of the many possible categories a loss can fall under. It is intended to quantitatively measure the variation between two probability distributions.

In multi-class classification, for each class label in each sample, distinct losses are determined and their results are added.

$$-\sum_{k=1}^{K} y_k \log(p_k) \tag{11}$$

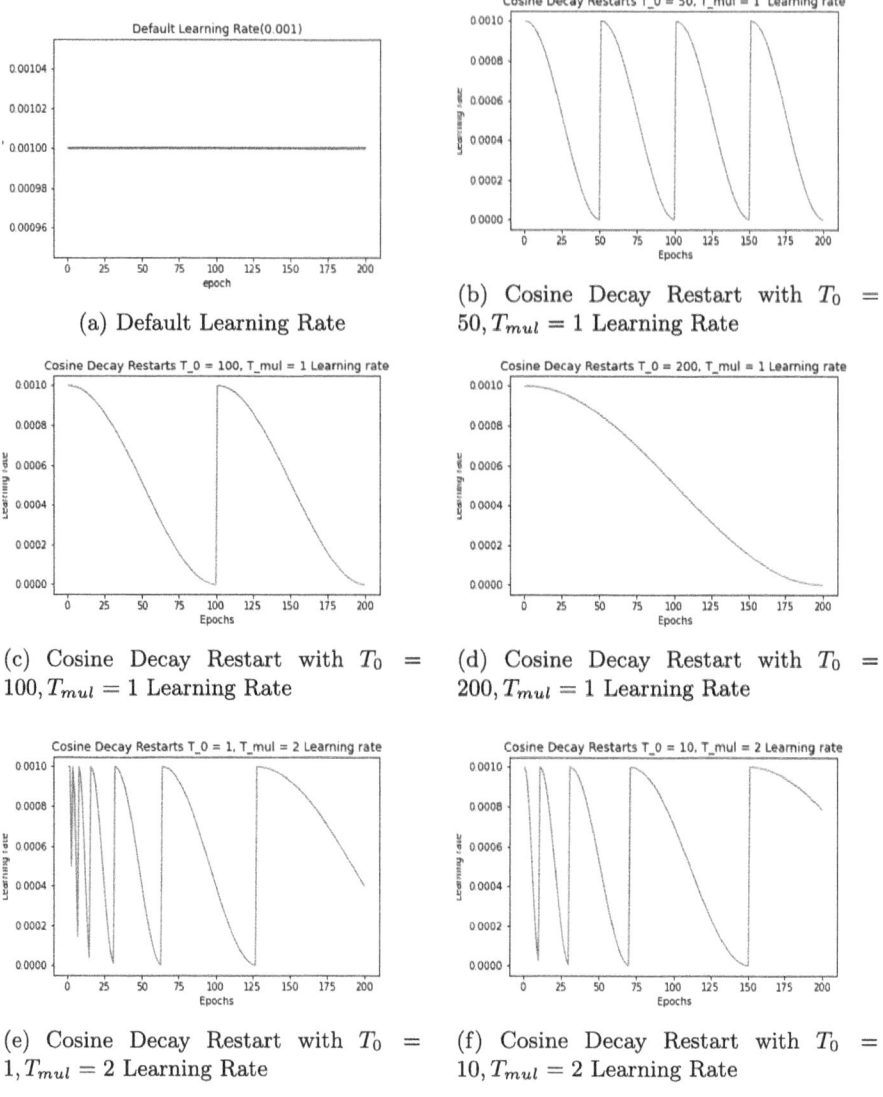

Fig. 2. Cosine Decay Restart Learning Rate

For entire data-sets and K classes, the categorical cross-entropy loss function calculates the loss:

$$-\sum_{i}^{m}\sum_{k}^{K} y_{true}^{(i)} \log\left(y_{predict}^{(i)}\right) \qquad (12)$$

Since one example can be assumed to belong to one category with a probability of 1, and to other categories with a probability of 0, the categorical cross-entropy

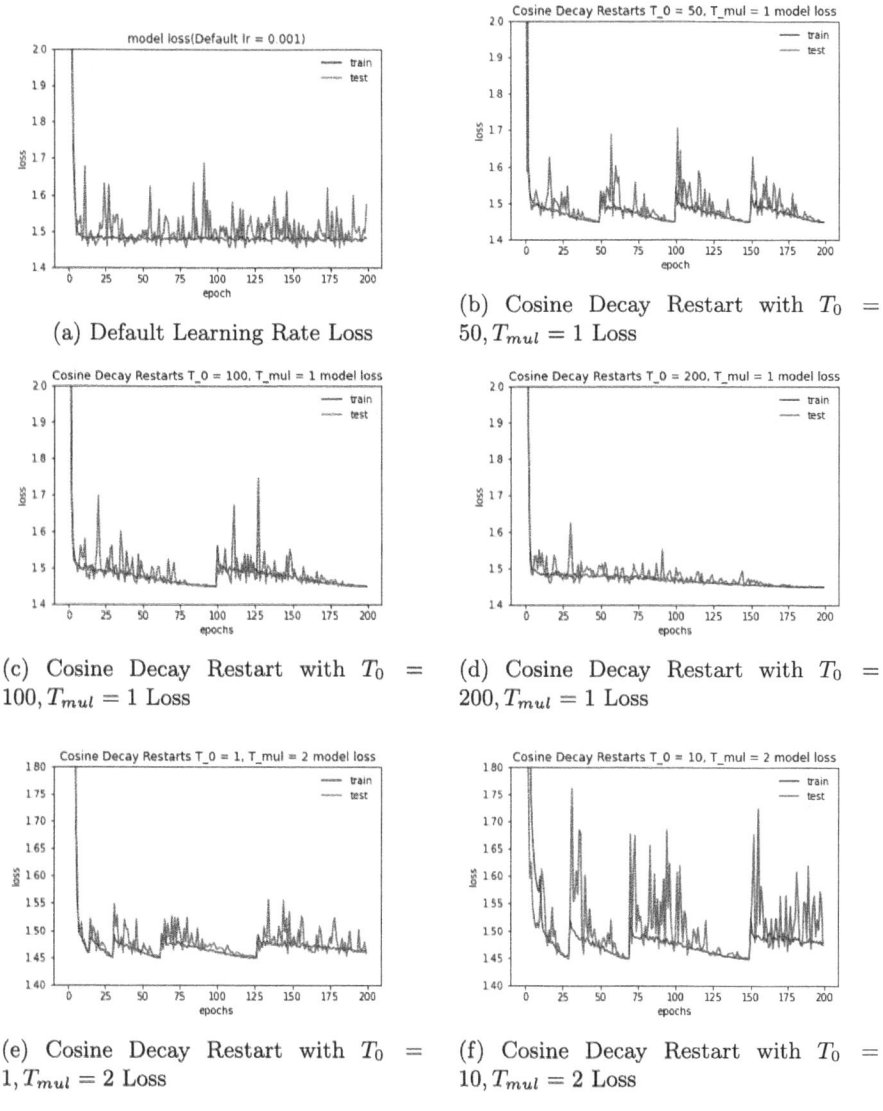

(a) Default Learning Rate Loss

(b) Cosine Decay Restart with $T_0 = 50, T_{mul} = 1$ Loss

(c) Cosine Decay Restart with $T_0 = 100, T_{mul} = 1$ Loss

(d) Cosine Decay Restart with $T_0 = 200, T_{mul} = 1$ Loss

(e) Cosine Decay Restart with $T_0 = 1, T_{mul} = 2$ Loss

(f) Cosine Decay Restart with $T_0 = 10, T_{mul} = 2$ Loss

Fig. 3. Cosine Decay Restart Loss

is ideal for classification tasks. One-of-many classification is a multi-class classification. One of the K classes can contain any one of the samples. K output neurons from the neural network can be aggregated into vector s (Scores). The ground truth (target) vector t will be a one-hot vector with C - 1 negative classes and a positive class. The assignment is viewed as a single sample classification problem involving K classes.

5   Our Proposed Recommender System

Our proposed model consists of a fully connected, dense, sequential deep neural network. A two-dimension vector of userID and movieID from the rating matrix is applied as input to the model and the true rating for the given (userID, movieID) pair is converted to categorical dummy variables using one hot encoding transformation. For hidden layers, the Relu activation function is used and the network ends with the dense layer of FIVE nodes with the softmax activation function. A stochastic gradient descent optimizer is used with categorical cross-entropy. In our proposed model batch size is 256 and the total number of epochs is 200. Given MovieLens data is divided into 80% training data and 20% validation data.

For each (userID, movieID) pair our proposed network produces a vector of 5 scores. These 5 output scores correspond to the probability that this particular (userID, movieID) pair belongs to this particular rating category. The sum of all 5 output scores is one. The class with the highest probability output score is declared as the rating label for the given (userID, movieID) pair. The purpose of model training is to predict the output label as close as possible to the true label by minimizing the categorical cross-entropy loss. By reducing the gap between the actual distribution of labels and the probability distribution of the network's output, this is accomplished.

In our proposed system the input (userID, movieID) pair is classified into only one category out of possible FIVE categories of movie ratings, so this problem is an instance of a single label, multiple class classification problem.

Table 1. Minimum Loss Value for Default (Learning Rate = 0.001) and Cosine Decay Restart Variation

LR Scheduler	Min Loss Value	Epoch No.
Default (Learning Rate = 0.001)	1.470708	121
Cosine Decay Restart with $T_0 = 50, T_{mul} = 1$	1.448819	199
Cosine Decay Restart with $T_0 = 100, T_{mul} = 1$	1.448773	199
Cosine Decay Restart with $T_0 = 200, T_{mul} = 1$	1.448461	199
Cosine Decay Restart with $T_0 = 1, T_{mul} = 2$	1.44934	126
Cosine Decay Restart with $T_0 = 10, T_{mul} = 2$	**1.448394**	149

Table 2. Same Loss Value for Default (Learning Rate = 0.001) and Cosine Decay Restart Variation at different number of epochs

LR Scheduler	Min Loss Value	Epoch No.
Default (Learning Rate = 0.001)	1.470708	121
Cosine Decay Restart with $T_0 = 50, T_{mul} = 1$	1.47972	17
Cosine Decay Restart with $T_0 = 100, T_{mul} = 1$	1.474009	50
Cosine Decay Restart with $T_0 = 200, T_{mul} = 1$	1.470196	79
Cosine Decay Restart with $T_0 = 1, T_{mul} = 2$	1.470728	86
Cosine Decay Restart with $T_0 = 10, T_{mul} = 2$	**1.470222**	20

6 Experimental and Results

6.1 Experimental Setup

For the proposed work experimental environment was set up on Ubuntu 20.04.4 LTS. Hardware used for the experiment is Intel R Core TM i7- 8557U CPU 4.50 GHz, 4 processors with 500 GB hard disk. Models are simulated on Python 3.9 package running on Anaconda Navigator - version 2.1.1.

6.2 Data Set

For the proposed research work, the data-set used is the MovieLens data-set[16]. We have used ML-100K for our experimental work. In ML-100K, 943 users rated 1682 movies, on a rating scale of 1 to 5 stars. In the MovieLens data set, The least sparse data-set is ML-100K, which has a sparsity level of 93.696%. Experiments exploring features of SGDR for categorical cross-entropy loss for different parameters setting of T_0 and T_{max} based on the MovieLens data-set ML100K.

6.3 Result and Discussion

In this experiment, we used SGDR for different parameters setting of T_0 and T_{max} (see Fig. 2). We used the categorical cross-entropy loss metric, and the outcomes are displayed in Fig. 3. The data-set was first divided into train and test sets at random, with 80% of the data serving as the training set and the remaining 20% serving as the test set. The train loss and test loss are shown in relation to the number of training epochs in Fig. 2(i). We choose a 128 batch size and run the model for 200 epochs.

As discussed in our proposed deep neural network architecture (userId, movieID) pair for the ML100K dataset is applied as input and ratings are converted to categorical data by using one hot encoder. Before predicting the rating value, input data is passed to fully connected dense hidden layers. The output layer applies the softmax function and produces probability distribution for each output category for a given sample. Lastly categorical cross-entropy loss is calculated and used for the calculation of back-propagation error.

Our empirical results suggest that SGDR requires 2x to 4x fewer epochs than the default learning rate scheduler. The lowest value of loss for default learning rate scheduler is 1.4707 and it is achieved at epoch number 121. The lowest value of loss for Cosine Decay Restart with $T_0 = 10, T_{mul} = 2$ learning rate scheduler is 1.4483 and it is achieved at epoch number 149 as shown in Table 1.

The lowest value of loss i.e. 1.4707 for default learning rate scheduler is achieved at epoch number 121 but the same value of loss is achieved by all variation of SGDR in much less number of epochs as shown in Table 2.

7   Conclusion and Future Research Work

In this digital era, a Recommender System helps users in finding their favorite movies according to their preferences saving their valuable time. In this paper, we find the effect of the Cosine Decay Restart Learning Rate Scheduler on the Movie Recommender System. Our proposed approach has been explained systematically and experimentally tested on the real-time datasets from MovieLens Dataset. We performed experiments on dataset ML100K and empirically found that the proposed approach provides better accuracy in term of loss when compared to previous work. The accuracy achieved by our proposed approach is 2.224% higher than the existing approach.

Some limitations of our proposed approach are: First, the accuracy has not been calculated with other datasets other than ML100K. The second limitation was with the experimentation due to the limitation of computation power on the hardware used. For future research work advances, deep learning techniques may be used.

References

1. Adomavicius, G., Tuzhilin, A.: Toward the next generation of recommender systems: a survey of the state-of-the-art and possible extensions. IEEE Trans. Knowl. Data Eng. **17**(6), 734–749 (2005)
2. Bishop, C.M.: Pattern Recognition and Machine Learning (Information Science and Statistics). Springer, Heidelberg (2006)
3. Goodfellow, I., Bengio, Y., Courville, A.: Deep Learning. MIT Press (2016)
4. Goodfellow, I.J., Bengio, Y., Courville, A.: Deep Learning. MIT Press, Cambridge (2016). http://www.deeplearningbook.org
5. Goyal, P., et al.: Accurate, large minibatch SGD: training imagenet in 1 hour. arXiv preprint arXiv:1706.02677 (2017)
6. Howard, J., Ruder, S.: Universal language model fine-tuning for text classification. arXiv preprint arXiv:1801.06146 (2018)
7. Isinkaye, F., Folajimi, Y., Ojokoh, B.: Recommendation systems: principles, methods and evaluation. Egypt. Inform. J. **16**(3), 261–273 (2015)
8. Kim, C., Kim, S., Kim, J., Lee, D., Kim, S.: Automated learning rate scheduler for large-batch training. arXiv preprint arXiv:2107.05855 (2021)
9. Loshchilov, I., Hutter, F.: SGDR: stochastic gradient descent with warm restarts. arXiv preprint arXiv:1608.03983 (2016)

10. Lund, J., Ng, Y.K.: Movie recommendations using the deep learning approach. In: 2018 IEEE International Conference on Information Reuse and Integration (IRI), pp. 47–54. IEEE (2018)
11. Murphy, K.P.: Machine Learning: A Probabilistic Perspective. MIT Press, Cambridge (2013)
12. Nakamura, K., Derbel, B., Won, K.J., Hong, B.W.: Learning-rate annealing methods for deep neural networks. Electronics **10**(16), 2029 (2021)
13. Ruder, S.: An overview of gradient descent optimization algorithms. arXiv preprint arXiv:1609.04747 (2016)
14. Sarwar, B.M., Karypis, G., Konstan, J., Riedl, J.: Recommender systems for large-scale e-commerce: Scalable neighborhood formation using clustering. In: Proceedings of the Fifth International Conference on Computer and Information Technology, vol. 1, pp. 291–324 (2002)
15. Senior, A., Heigold, G., Ranzato, M., Yang, K.: An empirical study of learning rates in deep neural networks for speech recognition. In: 2013 IEEE International Conference on Acoustics, Speech and Signal Processing, pp. 6724–6728. IEEE (2013)
16. Smith, L.N.: Cyclical learning rates for training neural networks. In: 2017 IEEE Winter Conference on Applications of Computer Vision (WACV), pp. 464–472. IEEE (2017)
17. Vidyabharathi, D., Mohanraj, V., Kumar, J.S., Suresh, Y.: Achieving generalization of deep learning models in a quick way by adapting T-HTR learning rate scheduler. Pers. Ubiquit. Comput. 1–19 (2021)
18. Wei, J.: Forget the learning rate, decay loss. arXiv preprint arXiv:1905.00094 (2019)
19. Xiong, Y., Lan, L.C., Chen, X., Wang, R., Hsieh, C.J.: Learning to schedule learning rate with graph neural networks. In: International Conference on Learning Representations (2021)
20. Xu, Z., Dai, A.M., Kemp, J., Metz, L.: Learning an adaptive learning rate schedule. arXiv preprint arXiv:1909.09712 (2019)

How Optimization Will Influence a Software Quality Characteristics Recommendation Model

Kamal Borana[1]([✉]), Meena Sharma[1], and Deepak Abhyankar[2]

[1] IET, DAVV, Indore, India
kamalborana2006@gmail.com
[2] SCSIT, DAVV, Indore, India

Abstract. The software is engineered to solve real-world problems in a specific area. In order to deliver good quality software quality evaluation is an essential task. During the quality evaluation, we test the software on various quality attributes by using some predefined models, but model-based software quality assessment is expensive in terms of cost and time. Therefore, recently we introduced a software quality matrix recommendation system using Particle Swarm Optimization (PSO). In this paper, we are investigating the influence of different optimization techniques over the previously proposed software quality attribute recommendation model. This paper involves an overview of the recommendation system, and different optimization techniques i.e., Fuzzy C Means (FCM), Genetic Algorithm, PSO, and Ant Colony Optimization (ACO) algorithm. Finally, a simulation is implemented, and the influence of the different optimization algorithms is measured. The findings of the simulation demonstrate the cost of software testing can be reduced by selecting appropriate quality attributes. Additionally, the FCM-based approach reduces the time of recommendation more effectively as compared to other optimization techniques, but the ACO and PSO deliver higher accurate results as compared to others.

Keywords: Optimization · Software Quality Assessment · Quality Attribute Selection · Comparison · Implementation · Model Recommendation

1 Introduction

Optimization techniques are employed in real-world applications to find optimal solutions for complex problems. It is a type of search process, which will help in improving and finding best matched solution for complex problems [1]. Basically, it is different from the classical Machine Learning (ML) algorithms, which are used for prediction and classification. The optimization of solution is performed by generating the solutions in phases, and a fitness function is used to discover the most optimal solution. However, a number of optimization techniques are available. Additionally, the working and their objective convergence rate is different from each other. Therefore, in this paper, we study the influence of optimization algorithms in order to select optimal software quality evaluation attributes.

© The Author(s), under exclusive license to Springer Nature Switzerland AG 2025
M. Gupta et al. (Eds.): MISS 2023, CCIS 1952, pp. 296–309, 2025.
https://doi.org/10.1007/978-3-031-69115-7_20

However, in the literature, a number of software quality-measuring models are available with different quality-measuring characteristics. These quality characteristics could be used to express the quality of the software [2]. In recent years, different software engineers are contributing different software quality models for the assessment of the software product. These models are defined by a set of quality characteristics or attributes. Sometimes, the assessment of software quality using these models is becoming expensive and time-consuming because when we follow a software quality model then all the attributes of the model need to be verified during the evaluation of the software product. Therefore, the selection and employment of the appropriate quality matrix are required. In this presented work we formulate this problem as the recommendation problem for suggesting the appropriate set of attributes for the evaluation of software quality. The aim of recommendation systems is to work as a filter to reduce the less relevant attributes and rank the most appropriate set of attributes. In this paper, we highlight the following work:

1. **Background:** In this section, we discuss different software quality models and previously own contributed recommendation models for suggesting the software quality characteristics.
2. **Proposed work:** Overview of different optimization techniques and impact on the software quality characteristic recommendation model.
3. **Results analysis:** Discuss the simulation conducted with the optimization techniques and compare them to conclude the consequences.

2   Literature Review

This section involves the study of existing software quality models and also provides an overview of the previously proposed software quality attribute recommendation models. That section will help to understand the proposed work carried out in this presented paper.

2.1   Software Quality Models

In the literature, we have two types of software quality estimation models which are discussed in this section.

2.1.1   Basic Software Quality Models

There are many software quality models with a number of characteristics. This section discusses some essential models:

McCall's Quality Model: It includes three main tasks first is Product operation to understand, operate, and provide the results. It includes the properties of reliability, correctness, efficiency, usability, and integrity. Product revision includes changes, errors, and adaptation. It covers flexibility, testability, and maintainability. Product deployment considers the new environments, and changes in hardware [3]. The accuracy is the main concern of the model.

Boehm's Quality Model: It adds maintainability to McCall's model [4]. It is easy to use, reliable and efficient. And has the ability to modify, testable, and understand.

Dromey's Quality Model: It consists of three models, i.e. Requirement quality model, the Design quality model, and the Implementation quality model. It evaluates if some principles are violated, with functionality and reliability. It is dedicated to evaluating how perfectly a component has been employed based on its use, maintainability, efficiency, and reliability. Additionally, it also evaluates a component's descriptiveness. But architectural integrity is not fully addressed. Testability is implicitly embedded. Domain-specific attributes are not addressed [4, 5].

FURPS Quality Model: The FURPS model [4, 5] includes Functionality like security, a set of attributes, and capabilities. Usability considers human-centric attributes such as user interface, helping documents, wizards, user manual, and training process. Similarly, reliability describes the chances and complexity of failure, recovery, prediction, accuracy, and time between failures. Next, the performance describes the need for execution speed, availability, throughput, response time, and resource usage. Finally, supportability demonstrates testing, extension, adaptation, maintenance, and compatibility.

ISO 9126 Quality Model: It [6–8] is an international standard based on McCall and Boehm's models. It has four parts Quality Model, Internal Attribute, External Attribute, and Quality in use. It focuses on attributes like Functionality, Reliability, Usability, Efficiency, Maintainability, and Portability, where Internal Attribute is referred to the properties that can be evaluated without including, external references observed during execution [9, 10].

ISO 2510: An improved version of ISO 9126 is ISO 2510. This model divides quality into 8 properties. The model includes security and compatibility as new property.

Summary of Basic Quality Models: Here, we found that models work on a total of 28 Characteristics among 24 existing ISO-2510 Models. This model has the highest number of Quality factors. This basically resulted from applying effort from [11–14], and [15] depending on the more important.

2.1.2   Tailored Software Quality Models

These models were built from basic quality models. There are many tailored models but we are considering four models:

Bertoa Model: This Model [16] is based on the Model [6]. It gives the definition of the quality attributes for the evaluation of Commercial Off-The-Shelf Components.

GEQUAMO Model: This model is called GEQUAMO [16] and consists of the sub-layers, and features and encapsulates user requirements.

ALVERO Model: This model is used for the certification of software components. It is consisting of four parts, quality components, Framework for technical certification, Certification Process, and Metrics.

Rawashdeh Model: This Model [17] was offered by combining the Dromey model and ISO 9126. It covers the need of users in four steps. First identification of a group of attributes, then a top-down technique is utilized for segmenting it to a set of characteristics. Next, internal and external metrics are used for differentiating between them.

This model considers the behavior, specification, external dependencies, and line of code during operations. The model is also identifying relevant users for each characteristic.

Summary of Models: according to the studied models, we can define the models from three different perspectives. First is product-centric [18], domain-centric [16], and user-centric [17].

2.1.3 Recent Quality Assurance Models

Quality models for AI have been proposed, but there is a lack of Systematic Mapping Studies (SMS) for quality models in AI systems, software, and components. The goal of M. A. Ali et al. [18] is to understand, classify, and evaluate existing quality models. This study conducts an SMS to investigate quality models proposed by various authors. The SMS has revealed no work on AI software component quality models. Finally, the limitations of the quality models and the implications for future research and development efforts are discussed.

In order to ensure the behavior of a software system, there are standards that define necessary qualities of the system. Due to the nature of ML, we have to re-interpret existing qualities for ML systems or add one more like trustworthiness. J. Siebert et al. [19] have to be very precise about which quality property is relevant for which entity, and how to evaluate quality requirements. Authors present how to construct quality models for ML systems. This quality model enables to specify and assess qualities. In addition, overall construction process described, the outcomes include a meta-model for specifying quality models, reference relevant views, entities, quality properties, and measures for ML systems, an example of a quality model for industrial use case, and lessons learned. They found that it is crucial to follow a systematic process to come up with quality properties. They want to learn how the term quality differs between different types.

The opportunities of ML and AI in healthcare are promising, for the growth of data-driven prediction models, which require quality and applicability assessment before that are applied in practice. A. A. H. de Hond et al. [20] identify actionable guidance involved AI-based prediction model (AIPM), evaluation and implementation and to identify gaps. The review is providing guidance or quality criteria of AIPMs using a multi-stage screening. Topics were extracted and summarized in six phases: (1) data preparation, (2) AIPM development, (3) AIPM validation, (4) software development, (5) AIPM impact assessment, and (6) AIPM implementation. It was found for data preparation, AIPM development and AIPM validation, later phases have received less attention. The phases of the AIPM cycle provide a framework for responsible of AI-based prediction. Additional domain and technology specific research may be necessary and practical to support.

COVID-19 has accelerated the growth of E-commerce, causing the quality of E-commerce systems. Y. H. Chung et al. [21] attempts to maximize feasibility and confidence of the development of e-commerce systems. They present different types of development methods, including traditional and agile. Also presents related works on development methods, comprising Extreme Programming, Scrum, and Kanban, and a software quality model, ISO/IEC 25010. The sub-characteristics of ISO/IEC 25010 are mapped to the features of systems. As a result, Scrum with frequent and effective meetings can minimize technical debt, design failures, stress, miscommunication, ambiguity, shared vision, feedback for verification, productivity, team morale, delivery predictability, project visibility, risk reduction, and engineering.

Different techniques have been adopted by the software industry to discover defects and improve software reliability. Businesses provide beta versions and other releases accessible to the general public. These releases keep track of the various bugs that users report. N. Kumar et al. [22] employ ML methods like decision tree models, KNN, linear regression models, ensemble learning with both boosting and bagging. The SPSS 14.0 dataset will be used, and the multidimensional result analysis will be used to determine which model is best for predicting software reliability.

ML can help to construct data-driven software. The ML-constructed software is created without direct control, quality management is a significant challenge, requiring a better understanding of the ML software models and data. C. Pahl [23] explores open issues toward a quality-oriented management of ML software. Various functions such as predictors, classifiers or self-adaptive systems are built using ML techniques to create respective ML models. Author explores the challenges to maintain the quality of these systems. And argue for a need of a DevOps approach for ML-constructed software that links construction and dynamic quality management.

2.1.4 Previously Introduced Model

This model identifies appropriate characteristics of software quality estimation. The aim is to reduce the quality estimation time and cost. First, we categorize the software quality attribute of both kinds of quality estimation models i.e. tailored and basic. Both kinds of matrices involve some common characteristics which are given in Table 1. Thus, the proposed quality attributes consider these parameters mandatory.

Table 1. Characteristics Followed by both models

S. No.	Properties
Tailored models	
1	Functionality
2	Maturity
3	Resource Utiliasation
4	Testability
5	Compliance
6	Understability
7	Usability
8	Learnability
Basic models	
9	Reliability

Based on the software requirements the attributes can be categorized to decide the additional characteristics. Table 2 demonstrates the additional category of attributes:

The proposed model for suitable attribute selection for software product quality estimation is demonstrated in Fig. 1. In this diagram, project requirements and a set of attributes are needed to be used during the optimization process. The project requirements as defined in Table 2 are selected first then attributes are selected for initializing the population of the particle swarm optimization (PSO). The PSO is a search-based optimization algorithm. This returns at least one optimal solution. The PSO algorithm requires a number of epochs to minimize possible errors and enhance the solution quality. There are two stopping criteria to stop the algorithm process are:

1. Setting limited number of epoch
2. Objective satisfied

In order to set objective function, let the attributes of Table 1 is denoted as T_1 and project requirements are defined as T_2. Then the total required attributes R is given by Eq. (1):

$$R = T_1 \cup T_2 \tag{1}$$

The distance between the required attribute R and the model's attribute M is calculated using the Euclidean distance. In this context when the distance between R and M becomes near zero then the objective is satisfied. Thus the distance function is considered an objective function. This equation is trying to find out the best model which contains the required project assessment characteristics.

3   Proposed Work

In previous work, we utilized the PSO algorithm for performing the search process. Therefore, we are proposing to utilize different optimization techniques to measure the influence of the optimization techniques for performing the search process. In this section, we are discussing some of the popular optimization techniques for utilizing the previously proposed model for getting the influence.

3.1   Fuzzy C Means (FCM)

Fuzzy c means (FCM) is a clustering algorithm, which works on membership calculations. In this algorithm, a single instance can belong to multiple clusters. The FCM clustering involves an objective function:

$$J_m = \sum_{i=1}^{M} \sum_{j=1}^{C} u_{ij}^m \left\| x_i - c_j \right\|^2 \tag{2}$$

where, $m > 1$ is a real number, u_{ij}^m is degree of membership, x_i is i^{th} element of data, and c_j is centroid.

Table 2. Demonstrate Requirement based attributes

Requirements	Attributes
calculations	Accuracy
	Correctness
	Efficiency
security, privacy, and communication modules	Integrity
	Fault Tolrence
	Time Behaviour
data collection and analysis	Human Engg
	Analyzability
Utilized by non-technical person	Recoverability
	Suitability
	Attractiveness
	Operability
Needed to change, modify, or scaled	Adaptability
	Changeability
	Flexibility
	Modifiability
	Reusability
	Operability
	Suitability
Deployed in multiple places/multiple machines/ multiple clients	Flexibility
	Installability
	Maintainability
	Portability
	Transferability
	Configurability
	Compatibility
	Reusability
	Interoperability
Deployed in resource constrained scenarios	Stability
	Resource Utilisation
	Self Contained
	Replicability

(*continued*)

Table 2. (*continued*)

Requirements	Attributes
	Manageability
Large number of modules	Supportability

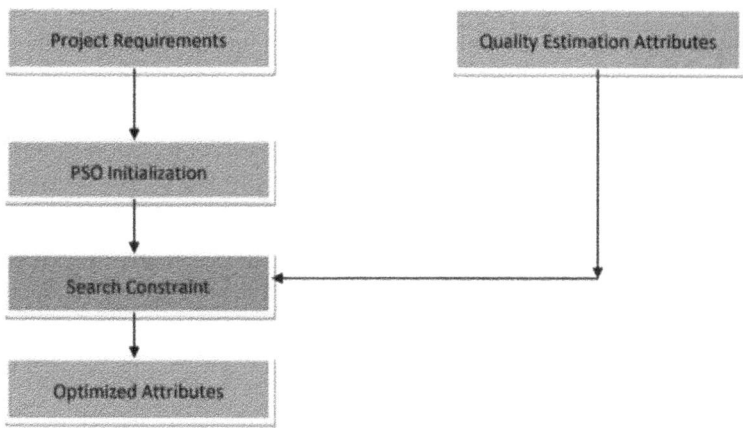

Fig. 1. Optimization Model for Software Quality Attribute

In order to categorize the data, we need to select similar number of data instances randomly as centroid as c_j, where $j = 1, 2, ..., n$. Next need to compute degree of membership for partitioning the data, the degree of membership u_{ij}^O is computed as:

$$u_{ij} = \frac{1}{\sum_{k=1}^{C} \left(\frac{\|x_i - c_j\|}{\|x_i - c_k\|} \right)^{\frac{2}{o-1}}} \tag{3}$$

After first phase of clustering that is essential for update the previous centroid. The new centroids are calculated using:

$$c_j = \frac{\sum_{i=1}^{o} u_{ij}^o * x_i}{\sum_{i=1}^{o} u_{ij}^O} \tag{4}$$

The optimization process can take a significant time thus a termination condition is required. The algorithm stops working when the degree of membership in the step k and $k + 1$ remains constant.

$$\left\{ \left\| u_{ij}^{(k+1)} - u_{ij}^{k} \right\| \right\} < \varepsilon \tag{5}$$

3.2 Genetic Algorithm (GA)

GA is a genetically inspired search process, to find the optimal solution. The solutions are genetically treated to find the fittest solution among available solutions. The iterative optimization process guarantees finding the best solution. But intermediate solutions can be used as a solution. The GA considers reproduction, selection, and diversity for new solution generation [15]. GA uses pair of solutions. The solutions are produced using random sequences of symbols that are unique in solution space. The steps of the GA are given as:

Initial Population – GA is initiated with the random sequences, with the allowed symbols. The generated population has the same number of symbols in each sequence.

Termination Condition – To stops the process and to produce results the following conditions are considered:

1. The fitness function starts providing same results.
2. Reaches maximal number of iterations
3. By limiting number of generation

Selection – It is a process of selecting individuals as a new population. Additionally using two-operator crossover and mutation new solutions are generated. The new and best solutions are creating a new generation. The elitism guarantees, the quality of the solution.

Crossover – Crossover is a process of recombining the individuals chosen by selection with each other. Using this new sequence is obtained. The aim is to get a new population, which inherits the best characteristics of their parents.

Mutation – the random change in the genes guarantees that even if none of the individuals contain the required solution genes, it is still possible to generate them using the mutation. The genetic algorithm can be described using Table 3.

3.3 Ant Colony Optimization (ACO)

The ant colony optimization algorithm is a heuristic search algorithm based on the behavior of ants. Initially the ants start searching for their food in random directions. Once the ant finds a path to the food source, it leaves a chemical substance, called pheromone, while returning back to its nest. Similarly, different ants would discover different paths to the food source. The density of the pheromone will evaporate if the path to the food source is very long. If the density of the pheromone is very high, it implies that many ants have used this path and hence the next Ant can follow this particular path. On the other hand, if the pheromone density is very low and starts to evaporate, it means that the path to the food source is very long and hence the next ant to will not follow this path. Based on the path length the pheromone concentration varies. Finally, the remaining ants will follow the path with the highest pheromone concentration.

4 Results Analysis

This section evaluated and investigates the performance influence of the implemented optimization techniques for finding appropriate software quality characteristics. Additionally, the comparison with the previously proposed PSO-based model is also discussed

Table 3. Genetic algorithm

Input: instance Π, size α of population, rate β of elitism, rate γ of mutation, number δ of iterations
Output: solution X

Process:
1. Randomly create α solutions;
2. Store solution as population P;
3. for $i = 1$ to δ do
    a. Calculate elitism $ne = \alpha \cdot \beta$;
    b. Select best ne solutions from P and store in P_1;
    c. Crossover $nc = (\alpha - ne)/2$;
    d. For $j = 1$ to nc do
        i. Select two solutions X and Y from P;
        ii. Using crossover of X and Y create new solution X_1 and Y_1;
        iii. Save X_1 and Y_1 to P_2 ;
    e. End for
    f. For $j = 1$ to nc do
        i. Select solution X_j from P_2;
        ii. Mutate X_j and generate a new solution X_j' ;
        iii. if X_j' is unfeasible
            1. update X_j' by repairing;
        iv. end if
        v. update X_j with X_j' in P_2;
    g. end for
    h. update $P = P_1 + P_2$;
4. End for
5. Return the best X solution from P;

---

in this section. In order to evaluate and compare the optimization process we considered two parameters i.e., time and accuracy.

4.1 Time Consumption

The time consumption is the amount of time required to achieve required objective during the optimization process. The time difference between training start time T_s and training finish time T_f is calculated as time consumed T_c:

$$T_c = T_f - T_s \tag{6}$$

The comparative performance of the PSO-based software quality recommendation model and other techniques are given in Fig. 2 and Table 4 in terms of milliseconds (MS). In this diagram, the X axis demonstrates the number of optimization cycles and the Y axis shows the time taken. According to experimental results as the number of

Table 4. Comparative time consumption

Epoch	PSO	FCM	GA	ACO
1	4.3	2.1	5.7	4.1
10	37.2	22.7	43.9	33.4
50	180.9	89.3	201.8	156.2
100	320.4	178.5	366.1	299.7

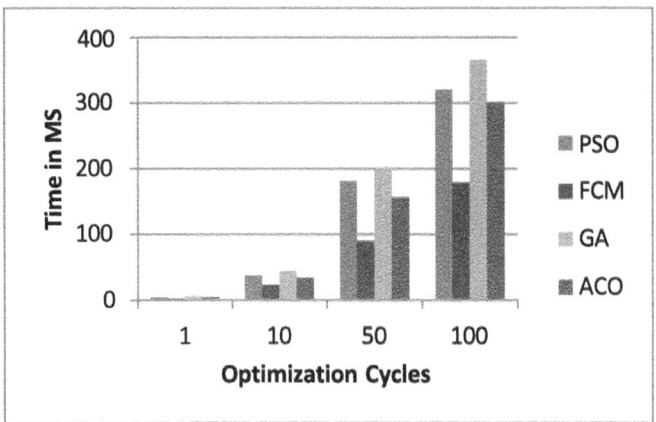

Fig. 2. Required Time to Recommend Software Quality Model using different optimization techniques

epoch cycles is increasing the amount of time for finding the best solution is increasing for all the algorithms. Based on the results we found the FCM is more time efficient as compared to others. And ACO algorithm is more time efficient than the GA and PSO.

4.2 Accuracy

The next parameter of the performance analysis and comparison is accuracy in terms of percentage (%). Here the accuracy of the recommendation model is estimated by using the total fit solutions which have a lower distance than 50%, among the total solutions generated by the different optimization algorithms. The below-given formula is used for calculating the accuracy:

$$accuracy(\%) = \frac{total\ fit\ solution}{total\ solutions\ genrated} \times 100 \tag{7}$$

Figure 3 and Table 5 contain the accuracy of the recommendation model with the different optimization techniques. The performance is measured by the increasing amount of epoch cycles.

In this diagram, the X axis contains the optimization cycles, and the Y axis shows the accuracy. According to the obtained accuracy, findings show that the less number

Table 5. Comparative Accuracy (%)

Epoch	PSO	FCM	GA	ACO
1	27	21	23	26
10	39	26	31	40
50	43	34	37	44
100	74	65	69	73

of epochs is producing less accurate results as compared to higher number of epochs. However, in the initial two experiments when the number of epoch cycles is less the accuracy of the recommendation is below than expectations but as the epoch cycles are increasing the model provides accurate results. Finally, in comparison of algorithms, the PSO and ACO-based techniques provides more accurate results as compared to GA and FCM. Additionally, ACO is effective for utilizing the recommendation system as compared to other algorithms in terms of both accuracy and time. Therefore, if we replace the PSO with the ACO algorithm then the software quality recommendation system becomes more accurate and time efficient.

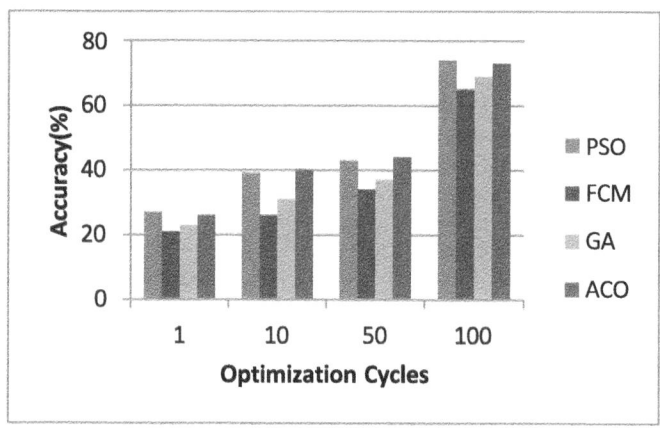

Fig. 3. Comparative Accuracy (%) of Recommendation Model

5 Conclusion and Future Work

The aim of the proposed work is to explore the different optimization-based techniques in order to enhance the recently introduced model for software quality characteristics recommendation based on the project requirements. This recommendation problem is formulated here as the optimization problem and solved using different optimization algorithms like particle swarm optimization (PSO), genetic algorithm (GA), Fuzzy c means (FCM), and ant colony optimization (ACO). These algorithms are employed with the previously introduced software quality characteristic recommendation model. Additionally, their influence on the recommendation yield is measured. According to the obtained experimental results we found that FCM and ACO are the most time-efficient optimization techniques as compared to PSO and GA. Additionally, the PSO and ACO-based model is more accurate than the FCM and GA. Therefore, we recommend using the ACO algorithm in place of the PSO algorithm for making the enhanced software quality characteristic recommendation model.

References

1. Sircar, A., Yadav, K., Rayavarapu, K., Bist, N., Oza, H.: Application of machine learning and artificial intelligence in oil and gas industry. Pet. Res. **6**, 379–391 (2021)
2. Miguel, J.P., Mauricio, D., Rodríguez, G.: A review of software quality models for the evaluation of software products. Int. J. Softw. Eng. Appl. **5**(6) (2014)
3. Li, H.F., Cheung, W.K.: An empirical study of software metrics. IEEE Trans. Softw. Eng. **SE-13**(6), 697–670 (1987)
4. IEEE STD 610.12-1990: IEEE Standard Glossary of Software Engineering Terminology (1990). http://web.ecs.baylor.edu/faculty/grabow/Fall2013/csi3374/secure/Standards/IEEE610.12.pdf
5. Bassam, A.B.A.: Software quality evaluation: user's view. Int. J. Appl. Math. Inf. **5**(3), 200–207 (2011)
6. ISO/IEC 9126-1: Software Engineering - Product Quality- Part 1: Quality Model, International Organization for Standardization, Switzerland (2001)
7. ISO/IEC 9126-2: Software Engineering - Product Quality- Part 2: External Metrics International Organization for Standardization, Switzerland (2002)
8. ISO/IEC 9126-3: Software Engineering - Product Quality- Part 3: Internal Metrics, International Organization for Standardization, Switzerland (2003)
9. Alvaro, A., de Almeida, E.S., de Lemos Meira, S.R.: A software component quality framework. ACM SIGSOFT Sens. **35** (2010)
10. Boehm, B.W., Brown, J.R., Lipow, M.: Characteristics of Software Quality. North Holland (1978)
11. Waghmode, M.L., Jamsandekar, P.P.: Software quality models: a comparative study. ASM's Int. e-J. Res. Manag. INCON-viii-2013
12. Rawat, M.S., Mittal, A., Dubey, S.K.: Survey on impact of software metrics on software quality. Int. J. Adv. Comput. Sci. Appl. **3**(1) (2012)
13. Kitchenham, B.: What's up with software metrics? A preliminary mapping study. J. Syst. Softw. ACM **83**(1) (2010)
14. Frakes, W., Terry, C.: Software reuse: metrics and models. ACM Comput. Surv. **28**(2) (1996)
15. Aparna, et al.: Software quality improvement-clean room approach and implementation. Int. J. Adv. Res. Comput. Sci. Soft. Eng. **4**(7) (2014)
16. Jones, C.: Strengths and Weaknesses of Software Metrics, Version 5 (2006)
17. McCall, J.A., et al.: Factors in Software Quality, Griffiths Air Force Base. Rome Air Development Center, Air Force Systems Command, New York (1977)
18. Ali, M.A., Yap, N.K., Ghani, A.A.A., Zulzalil, H., Admodisastro, N.I., Najafabadi, A.A.: A systematic mapping of quality models for AI systems, software and components. Appl. Sci. **12**, 8700 (2022)
19. Siebert, J., et al.: Construction of a quality model for machine learning systems. Softw. Qual. J. **30**, 307–335 (2022)

20. de Hond, A.A.H., et al.: Guidelines and quality criteria for artificial intelligence-based prediction models in healthcare: a scoping review. Npj Digit. Med. **5**(1), 2 (2022)
21. Chung, Y.H., Thinakaran, R., Batumalay, M., Ismail, N.H.A.: Software quality assurance perspective on e-commerce system development. J. Theor. Appl. Inf. Technol. **100**(15) (2022)
22. Kumar, N., Joshi, D.: Enhancing software quality using machine learning on open source projects. In: International Conference on Intelligent Technologies & Science (2021)
23. Pahl, C.: Research challenges for machine learning-constructed software. In: Service Oriented Computing and Applications. Springer (2022)
24. Parnas, D.L., Lawford, M.: The role of inspection in software quality assurance. IEEE Trans. Soft. Eng. **29**(8) (2003)

N-Gram approach to prepare Crime-related Legal DataSet: A roadmap to classify Legal Text

Souraneel Mandal[1]([✉]) and Tanaya Das[2]

[1] Ghani Khan Choudhury Institute of Engineering and Technology, Malda, India
souranee197@gmail.com
[2] JIS University, Kolkata, India
tanayadas.das23@gmail.com

Abstract. While analyzing legal documents, legal professionals usually explore vital information in any legal case documents. Mostly, they manually extract major information from legal documents, which is time-consuming. Artificial Intelligence-based techniques can reduce this time-consuming nature. Dataset plays a major role in preparing any Artificial Intelligence based system in any application domain. Developing an Artificial Intelligence-based system for legal professionals requires knowledge of features that can be provided as a dataset. Hence in this paper, the authors have proposed an approach using NLP-based techniques like N-gram, Wordnet, Lemmatization, etc., on legal documents to prepare a legal dataset. As a case study, this research focused on dowry death cases, one of our society's alarming crimes. The proposed methodology to prepare datasets has been determined to capture the major concepts of dowry death cases with more than 95% precision. In the future, a good number of legal datasets can be prepared on other woman-centric crimes using the proposed methodology.

Keywords: Natural Language Processing · Bi-Gram tokens · Text analytics · Indian Penal Code Section 304B · Legal Judgments · Legal DataSet

1 Introduction

Intelligent systems can be developed using Artificial Intelligence and its allied subdomains like Natural Language Processing, Machine Learning, Computer Vision, etc. In practice, it has been observed that the datasets are required as knowledge to construct Artificial Intelligence based systems. Incorporating knowledge across any domain is made possible with the assistance of datasets. The datasets help represent certain features present in any domain. The dataset can assist in solving different problems like prediction [1], analysis [2], classification [3], information retrieval [4], etc. Datasets need some preparation to provide actionable insights. Some datasets lack values, have invalid values, or have other issues that make it difficult for an algorithm to handle them. If there are gaps in the data, the algorithm won't be able to use it. If the data are inaccurate, the algorithm will provide less accurate results or may even be misleading. Effective data preparation results in clean and well-curated data, which in turn leads to model findings that are more applicable and accurate.

© The Author(s), under exclusive license to Springer Nature Switzerland AG 2025
M. Gupta et al. (Eds.): MISS 2023, CCIS 1952, pp. 310–322, 2025.
https://doi.org/10.1007/978-3-031-69115-7_21

Nowadays, any Machine learning-based work is strongly reliant on learning data. It is the most important factor that has made it feasible to train algorithms in recent years. Data preparation refers to a collection of processes that, when completed, make your dataset more appropriate for machine learning. In a larger sense, the data preparation also includes building the appropriate framework for data collecting.

In recent years, the legal sector in India has been subject to various innovations in terms of technological advancement. Applying artificial intelligence-based systems could revolutionize the conventional approach that legal practitioners take to their work. The Artificial Intelligence committee of India's Supreme Court has produced a program called SUPACE [5]. Its purpose is to manage massive amounts of legal data. Some novel research work [6] discussed the results of their research on a legal knowledge base that utilized Artificial Intelligence and Legal Ontology [7] to deliver pertinent information to legal practitioners. Artificial Intelligence based applications can assist legal professionals can easily access a plethora of legal information that is vital to analyze the decision of legal cases in less time. The Artificial Intelligence-based Software Kira by Cyril Amarchand Mangaldas [8] analyses, separates and provides refined insights on the necessary information with substantial accuracy. Legal interpretation is one of the more difficult aspects of interpreting that involves the semantics of the language [9]. The majority of legal documents are often made available in a semi-structured format. The inference must be drawn from legal documents, which necessitates the use of legal practitioners or other domain specialists. The primary role of a domain expert is to aid in interpreting the meaning of the text following the semantics of the various legal parts. The authors of this paper have attempted to prepare datasets based on judgments related to a second heinous crime against women in India. Implementing Artificial Intelligence (A.I.) in the field of law will help to quicken the provision of legal services.

Section 2 expressed the literature survey for the multiple text analytics approach. Section 3 outlines the proposed methodology along with the algorithm. Section 4 discusses the results of the proposed methodology to create Legal DataSet using major parameters and various judgments with I.P.C. 304B. Finally, Sect. 5 concludes the paper and its future use in Indian Legal System.

2 Literature Survey

In recent years, various business sectors have used Artificial Intelligence in various use cases. From robot-based hotel attendants to entertainment and amusement, Artificial Intelligence has a high impact on improving business [10]. N.L.P. is utilized in different areas for text pre-handling for unstructured information. Indian Legal System has many spaces where N.L.P. can help in different ways, as most authoritative records are accessible in an unstructured organization. Today, legal research requires time-consuming hours where legal professionals manually read legal case documents to extract major information. Artificial Intelligence can potentially level up the legal profession to increase beneficiary legal services [11]. In this section, the authors reviewed artificial intelligence-based work in the Indian legal sector and looked for any research gaps from the past pieces of literature. With the help of Artificial Intelligence techniques like text analytics and processing, legal professionals frequently search the key information in legal cases.

In 2022 authors proposed their work on the Indian Legal System's Dowry death domain, where they extract dominant keywords on multiple Dowry death documents after cleaning outliers with the help of Natural Language Processing Tools. They visualized the highest frequent words or tokens across all legal documents [12]. In the same year, these authors upgraded their work, mainly visualizing major parameters of dowry death cases and trying to capture the specific parameters in dowry death legal rule, i.e., Indian Penal Code 304B. The authors have used a conventional Term Frequency – Inverse Document Frequency (TF-IDF) based vector to show the dispersion of words within the case records. This strategy displays relationships among words or concepts representing co-occurrence, semantic closeness, etc., present in dowry death cases [13]. The Pima Indian Diabetes dataset from the University of California Irvine (U.C.I.) repository is a tough dataset with a higher number of missing values (48%); therefore, authors in 2019 provided a data preparation approach for Pima Indian Diabetes for that they developed a dataset where they considered Data preprocessing for increasing the precision of the classification model, several data preparation techniques are carried out on the Pima Diabetes dataset. The suggested approach has three stages: stage one is outlier removal and imputation, stage two is normalization, and stage three is dataset balance. After each preprocessing stage, three classifiers are used to assess the model: Support Vector Machine, Random Forest, and K-nearest neighbor. The suggested approach was implemented for incremental situations using SMOTE to manage the skewed dataset [14]. This paper presents a study that aims to contribute to the development of software that enables law firms to more quickly and effectively gather information, allowing them to concentrate their efforts on activities that require a higher level of intellectual capacities, such as strategically determining how to best present the claim to the Court. In addition to lessening the number of times law firms check their databases, this software also helps higher courts keep tighter control over how subordinate courts apply jurisprudence. Categorizing texts via supervised machine learning methods was employed; such a prediction algorithm achieved more than 90% of ranking judgment correctness scores, providing reliable knowledge about each judge's tendencies [15].

From the abovementioned literature, the authors have observed that most research works focused on data processing and prediction approaches. However, few literatures discussed the approach to preparing the dataset for implementation. The freely available datasets for Artificial Intelligence-based work are from various domains worldwide. There are only a few Indian legal datasets where Artificial Intelligence-based work can be applied to assist legal professionals. There isn't any Text Classification oriented DataSet on the legal judgment that expresses the context of the corpus in the Indian legal System. Hence, the author of this paper created Dowry Death-related DataSet after extracting Bi-Gram words from various IPC 304B Judgement and extracted corpus linked with it with the help of N.L.P., which helped not only legal professionals but also data analysts.

3   Work Methodology

Natural Language Processing is one of the captivating areas of Artificial Intelligence to fathom human language. It has been used to preprocess unstructured information and examine definitive records to help in a genuine direction. Most of the information found in the Indian Legal System is unstructured. Legal professionals' conventional approach to breaking down and pursuing choices is tedious. In this paper, the authors have created DataSet after extracting N-Gram words based on Indian Penal Code (I.P.C.) Section 304B, which explained Sects. 3.1 to 3.3.

3.1 Origin of Work

At the origin [16] of this exploration work, authors had utilized Dowry Death Legal Documents containing unstructured information as the input data. The authors utilized NLTK and RE modules for pre-handling. The authors had taken out max size 2 letter tokens, punctuation, tokens containing any digits, complete name, HTML tag, Header and Footer, Criminal Record Number (C.R.N), digits, month and year name. Then the corpus was cleaned for applying the stemming and lemmatization method (where the parameter consists of a verb, adverb, and adjective). Then, at that point, put away the tokens of each archive in discrete records. Then outer join was applied to convert all the list series to a DataFrame. Then, at that point, the DataFrame was cleaned with NaN to 0. The authors applied the CountVectorizer method on each list of tokens and tracked down the recurrence of every token in a list. Yet, the CountVectorizer strategy doesn't decide each word's dominancy, so the authors applied the Term Frequency-Inverse Document Frequency (TF-IDF) technique to figure out the most dominant words. After that, we applied N-gram analysis to extricate the most dominant Uni-Gram and Bi-Gram words.

The major **Uni-Gram** words **are 'person', 'accused', 'evidence', 'husband', 'dowry', 'death', 'cash', 'court', 'victim', 'cruelty', 'trial'** etc.

AND

The major **Bi-Gram** words **are 'trial court', 'demand dowry', 'offense section', 'dowry death', 'cruelty harassment', 'section evidence', 'subject cruelty', 'cause death', 'matrimonial home'** etc (Fig. 1).

	Word	Case1	Case2	Case3	Case4	Case5	Case6	Case7	Case8
0	result	0.125884	0.126882	0.130130	0.851599	0.247439	0.134602	0.000000	
1	victim	0.877264	0.175799	0.557286	0.397618	0.354589	0.077917	0.372991	0.156518
2	cannot	0.298649	0.398085	0.791938	0.146075	0.233610	0.225878	0.000000	0.084026
3	court	0.429807	0.213367	0.218829	0.392113	0.355859	0.346749	0.207488	0.309575
4	cash	0.927403	0.056420	0.057894	0.108193	0.170691	0.000000	0.179559	0.245580
5	shal	0.692942	0.238684	0.061250	0.629880	0.180678	0.116465	0.128710	0.000000
6	send	0.707850	0.446165	0.114596	0.641686	0.337452	0.198761	0.473312	0.121376
7	occur	0.280432	0.580055	0.148726	0.278085	0.584980	0.141400	0.193838	0.315602
8	cruelty	0.879134	0.129888	0.155415	0.228322	0.109155	0.295519	0.160737	0.094228
9	suffer	0.764032	0.254467	0.000000	0.243989	0.171079	0.165417	0.809654	0.369208
10	burn	0.100844	0.052147	0.000000	0.549906	0.315329	0.254238	0.719164	0.056746
11	trial	0.983524	0.204911	0.377728	0.158946	0.330146	0.478828	0.309886	0.080062
12	cross	0.364466	0.269240	0.553266	0.464676	0.217214	0.472555	0.657125	0.000000
13	facts	0.304907	0.076835	0.481120	0.680300	0.318007	0.307482	0.083632	0.000000
14	find	0.106388	0.623480	0.300913	0.140860	0.369854	0.464897	0.311256	0.199547
15	default	0.287851	0.000000	0.152681	0.142721	0.450327	0.145141	0.789539	0.161976
16	come	0.180884	0.000000	0.383725	0.296951	0.314426	0.484631	0.529119	0.339283
17	submit	0.342252	0.351983	0.363024	0.254540	0.634673	0.172571	0.375501	0.000000
18	police	0.655315	0.338069	0.280458	0.487373	0.256301	0.000000	0.069872	0.276563
19	record	0.517913	0.238060	0.181115	0.741834	0.240073	0.116065	0.126273	0.064763
20	conduct	0.737714	0.305163	0.312995	0.428024	0.153882	0.148738	0.161876	0.000000

	Word	Case1	Case2	Case3	Case4	Case5	Case6	Case7	Case8
0	cruelty harassment	0.868146	0.046053	0.085240	0.255719	0.170479	0.355151	0.000000	0.127859
1	dowry death	0.694956	0.000000	0.057737	0.692845	0.057737	0.106916	0.124777	0.057737
2	subject cruelty	0.866996	0.183969	0.170253	0.340506	0.170253	0.157636	0.091985	0.085127
3	demand dowry	0.344917	0.281776	0.372525	0.447030	0.335272	0.351867	0.201268	0.074505
4	connection demand	0.629533	0.081633	0.075547	0.302188	0.302188	0.629533	0.081633	0.075547
5	trial court	0.519313	0.151517	0.630990	0.070110	0.350550	0.194742	0.378791	0.000000
6	offence section	0.353987	0.206561	0.000000	0.095580	0.095580	0.530981	0.722964	0.095580
7	matrimonial home	0.307078	0.119459	0.773867	0.000000	0.331657	0.204719	0.358377	0.110552
8	section evidence	0.228594	0.133391	0.370338	0.664121	0.000000	0.114297	0.133391	0.123446
9	information solutions	0.568191	0.331555	0.306835	0.306835	0.306835	0.284096	0.331555	0.306835
10	certify copy	0.352937	0.411896	0.381187	0.381187	0.381187	0.352937	0.000000	0.381187
11	cause death	0.099960	0.000000	0.323885	0.539808	0.647770	0.199921	0.349978	0.107962

Fig. 1. Tf-Idf or Weighted mean Matrix with N-Gram tokens[16]

3.2 Bi-gram Analysis on IPC 304B

It is observable that learned bigram language models [17]:

1. Perform better test set perplexity than unigram models constructed on similar Bag-of-Words (BoW) archives,
2. Assign greater probability to plausible bigram word pairings,
3. Understand the bibliomaniac to understand the context of a corpus, better ordered-document recovery from a BoW, and
4. Recover Bi-Gram **Language Modeling (L.M.)** that outperforms naïve unigram L.M.s on a similar B.O.W. corpus.

Due to this reason, the authors have chosen **Bi-Gram** over **Uni-Gram** *tokens* to construct DataSet.

In this research work, the authors have used N-Gram-based modeling to display certain Bi-Gram parameters in the form of classes and their associated tokens. The authors have focused on Indian Penal Code 304B Section [18], i.e., Dowry Death, which has been the second highest crime against women. According to the National Crime Records Bureau (NCRB) [19] study, more than 7,000 women in India succumbed to dowry death between 2020 and 2022. Legal professionals may benefit from using cutting-edge technology like Artificial Intelligence and its subfields like Natural Language Processing (N.L.P.) to analyze cases and organize their drafts with specific data [20].

Thus, the author of this paper extracted the most dominant keywords[7], which are *'demand dowry'*, *'dowry death'*, *'cruelty harassment'*, *'subject cruelty'*, *'cause death'*, and *'matrimonial home'*.

3.3 Proposed Approach

To construct a **MultiLabel Text Classification** based **DataSet** for **Dowry Death**, authors have collected various dowry death judgments from legal databases like indiankanoon [21] used as user input or corpus.

The proposed DataSet construction methodology is explained in Fig. 2 below.

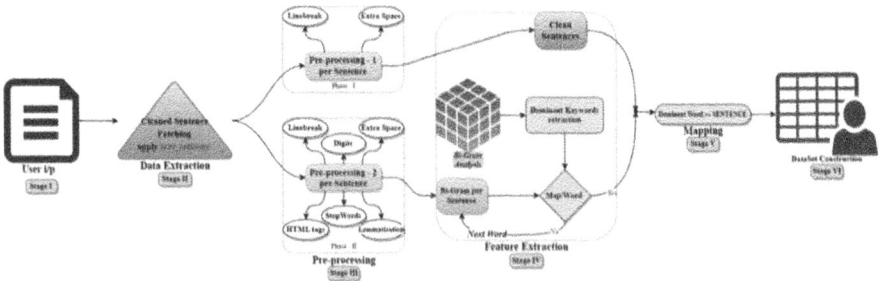

Fig. 2. Stagewise construction through Bi-Gram tokens to prepare the dataset

The proposed diagram consists of two phases for preprocessing, where user input is inputted twice for text preprocessing. The proposed *algorithm* and *pseudocode* contain all the steps with explanations and experimented output.

ALGORITHM: Crime-based DataSet Construction.

Start

Import-Module : → | *nltk, stopwords, sent_tokenize, WordNetLemmatization, panda, and matplotlib*

most dominant keyword : → | *'demand dowry', 'dowry death', 'cruelty harassment', 'subject cruelty', 'cause death', and 'matrimonial home'.*

N-Gram : → | 2 | CONTEXT: Bi-Gram – MULTILABEL TEXT CLASSIFICATION

Step 1: Initialize *Legal_Paragraph* as **input** where inputs are gathered from various **professionals** and the **indiankanoon** website regarding Dowry Death Judgment. *Legal_Paragraph* should be considered a corpus.

Preprocessing **Phase – I** *explained in* **Steps 2 – 4**
Step 2: *Legal_Paragraph* will be preprocessed with the *removal of unwanted space*.
Step 3: Instantiate a list named *sent_para* where each sentence will be stored after applying *the sent_tokenize* method on *Legal_Paragraph*.
Step 4: *sent_para* is a **Text** that will be used for **mapping**.

Preprocessing **Phase – II** *explained in* **Steps 5 – 13**
Step 5: *Legal_Paragraph* will be inputted again for preprocessing.
Step 6: Apply *LowerCase on Legal_Paragraph*.
Step 7: Apply the *sent_tokenize* method in *Legal_Paragraph*.
Step 8: For each sentence in *Legal_Paragraph,* remove escape sequence, numbers, and text with numbers.
Step 9: For each sentence in *Legal_Paragraph,* remove the letter with length two.
Step 10: For each sentence in *Legal_Paragraph,* remove unwanted spaces.
Step 11: For each sentence in *Legal_Paragraph,* apply *WordNetLemmatization* after removing the stopword technique.
Step 12: For each sentence in *Legal_Paragraph,* apply **Bi-Gram** tokens stored in the *n_gram* instance.
Step 13: *n_gram* is a **Bi-Gram token,** or two consecutive words, that will be used for **mapping**.

Mapping **Phase – I & II** *construction of* **DataSet** *explained in* **Steps 17 – 19**
Step 14: *zip sent_para* with *n_gram* via python *dict* function.
Step 15: Compare *most_dominant_keyword* with *n_gram*; if matched, extract matched **sentences** and pair with matched *n_gram* words.
Step 16: Export paired **sentences** and *n_gram* words to panda's DataFrame to create DataSet.

End

PSEUDOCODE: Crime-based DataSet Construction.

Start

Import-Module : →	*nltk, stopwords, sent_tokenize, WordNetLemmatization, panda, and matplotlib*
most dominant keyword : →	*'demand dowry', 'dowry death', 'cruelty harassment', 'subject cruelty', 'cause death', and 'matrimonial home'.*
Input : →	LEGAL_PARAGRAPH as corpus or docString ; WNL ← WORDNETLEMMATIZER() ; *n_g* ←2
Output : →	DOWRYDEATHLEGALDATASET.csv

sent_para ← replace(LineBreak with ' ' and stored in ***Legal_Paragraph***) using *re* **module**
sent_para ← replace(unwanted space with ' ' and stored in ***text***) using *re* **module**
sent_para ← sent_tokenize(sent_para)

Define ***generate_N_grams_map*** with Legal_Paragraph & n_g
    sent ← apply LowerCase on Legal_Paragraph
    sent ← replace(LineBreak with ' ' and stored in ***Legal_Paragraph***) using *re* **module**
    sent ← replace((number | text with number) with ' ' and stored in ***sent***) using *re* **module**
    sent ← replace(numbers written in word with ' ' and stored in ***sent***) using *re* **module**
    sent ← replace(length(word) ≤ 2 with ' ' and stored in ***sent***) using *re* **module**
    sent ← replace(unwanted space with ' ' and stored in ***sent***) using *re* **module**
    sent ← replace(stopwords.words('English') with ' ' and stored in ***sent***) using *re* **module**
    sent ← apply (*WordNetLemmatizer*) stored in ***sent***

    n_gram ← (pd.Series(nltk.ngrams(***sent***, n_g)).
    return **n_gram**
n_gram ← ***generate_N_grams_map***(Legal_Paragraph, n_g)

map_inner ← zip(***sent_para***, ***n_gram***)
for k, v in map_inner.items() **do**
    for i, j in map(sent_para, n_gram) **do**
        if most_dominant_keywords⟦i⟧ == v⟦j⟧ **do**
            sent_keys ← ⟦{v⟦j⟧, k}⟧

pd.**DataFrame**(sent_keys, columns ← ['CONTEXT_IPC304B',
'LEGAL_EXTRACTED_CORPUS']).to_csv('***DowryDeathLegalDataSet.csv***', header=True)

End

Based on the proposed methodology, results are articulated with proper figures and diagrams in the result and discussion segment.

4 Results and Discussion

This section describes the results obtained by the authors when generating the dataset using pseudocode. Sample input is shown in Fig. 3 as the ***Legal_Paragraph*** instance regarding **D**owry **D**eath is fetched from judgments available at various courts, and legal professionals explained in the proposed **algorithm** in step **1**.

After removing unwanted spaces and escape sequences, the authors have applied the sent_tokenize method. A sentence typically concludes with a full stop as a separator to break sentences as strings within a list or a list of strings. The ***sent_tokenize*** method is

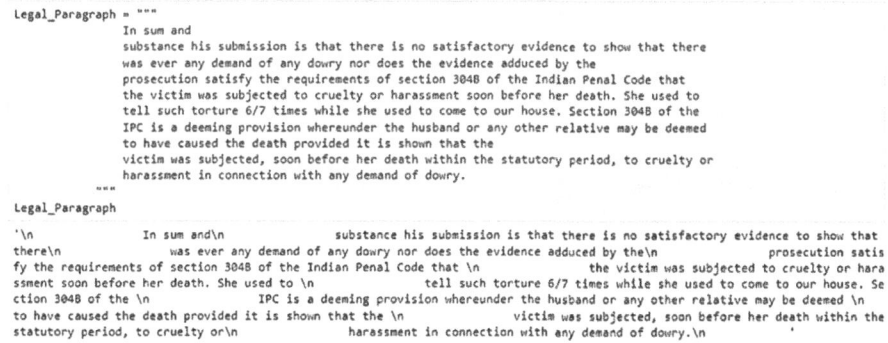

Fig. 3. User input related to dowry death.

applied to separate sentences from paragraphs after a full stop, as shown in Fig. 4 and is explained in the proposed **algorithm** in steps **2–4**.

```
sent_para

[' In sum and substance his submission is that there is no satisfactory evidence to show that there was ever any demand of any
dowry nor does the evidence adduced by the prosecution satisfy the requirements of section 304B of the Indian Penal Code that t
he victim was subjected to cruelty or harassment soon before her death.',
 'She used to tell such torture 6/7 times while she used to come to our house.',
 'Section 304B of the IPC is a deeming provision whereunder the husband or any other relative may be deemed to have caused the
death provided it is shown that the victim was subjected, soon before her death within the statutory period, to cruelty or hara
ssment in connection with any demand of dowry.']
```

Fig. 4. Snapshot of output after *sent_tokenize*

Legal_Paragraph instance is **re-inputted**, where the authors have applied the *sent_tokenize* method and removed tokens with max size 2, including punctuation, tokens containing any digits, date, month, year, unwanted spaces, and stopwords from each sentence. Then, the *WordNetLemmatization* method is applied to clean the corpus. In each sentence, Bi-gram is applied as shown in Fig. 5 and explained in the proposed **algorithm** through steps **5–13**.

```
for i in n_gram:
    print(i)

['sum substance', 'substance submission', 'submission satisfactory', 'satisfactory evidence', 'evidence show', 'show ever', 'ever demand', 'd
emand dowry', 'dowry evidence', 'evidence adduce', 'adduce prosecution', 'prosecution satisfy', 'satisfy requirements', 'requirements sectio
n', 'section indian', 'indian penal', 'penal code', 'code victim', 'victim subject', 'subject cruelty', 'cruelty harassment', 'harassment soo
n', 'soon death']
['use tell', 'tell torture', 'torture time', 'time use', 'use come', 'come house']
['section ipc', 'ipc deem', 'deem provision', 'provision whereunder', 'whereunder husband', 'husband relative', 'relative may', 'may deem',
'deem cause', 'cause death', 'death provide', 'provide show', 'show victim', 'victim subject', 'subject soon', 'soon death', 'death within',
'within statutory', 'statutory period', 'period cruelty', 'cruelty harassment', 'harassment connection', 'connection demand', 'demand dowry']
```

Fig. 5. Extracted Bi-gram

Output received from Fig. 5 and Fig. 4 will be *zipped* together as a python dictionary function where each sentence will be mapped with that sentence's Bi-gram token, shown in Fig. 6 and explained in the proposed **algorithm** in step **14**.

The most **dominant context, as keywords *demand dowry, cause death*, etc.,** are traversed with all **Bi-Gram tokens** and matched tokens along with the sentence are

```
map_inner_ = dict(zip(sent_para, n_gram))

for i in map_inner_:
    print(i, '->', map_inner_[i])
    print('-------')
```

In sum and substance his submission is that there is no satisfactory evidence to show that there was ever any demand of any dowry nor does the evidence a
dduced by the prosecution satisfy the requirements of section 304B of the Indian Penal Code that the victim was subjected to cruelty or harassment soon be
fore her death. -> ['sum substance', 'substance submission', 'submission satisfactory', 'satisfactory evidence', 'evidence show', 'show ever', 'ever deman
d', 'demand dowry', 'dowry evidence', 'evidence adduce', 'adduce prosecution', 'prosecution satisfy', 'satisfy requirements', 'requirements section', 'sec
tion indian', 'indian penal', 'penal code', 'code victim', 'victim subject', 'subject cruelty', 'cruelty harassment', 'harassment soon', 'soon death']

She used to tell such torture 6/7 times while she used to come to our house. -> ['use tell', 'tell torture', 'torture time', 'time use', 'use come', 'come
house']

Section 304B of the IPC is a deeming provision whereunder the husband or any other relative may be deemed to have caused the death provided it is shown th
at the victim was subjected, soon before her death within the statutory period, to cruelty or harassment in connection with any demand of dowry. -> ['sect
ion ipc', 'ipc deem', 'deem provision', 'provision whereunder', 'whereunder husband', 'husband relative', 'relative may', 'may deem', 'deem cause', 'cause
death', 'death provide', 'provide show', 'show victim', 'victim subject', 'subject soon', 'soon death', 'death within', 'within statutory', 'statutory per
iod', 'period cruelty', 'cruelty harassment', 'harassment connection', 'connection demand', 'demand dowry']

Fig. 6. Mapped Bi-grams with each sentence

fetched. Python's dictionary data structure is used where each sentence in the ***sent_para***
instance is **key,** and the ***n_gram*** word is the **value** shown in Fig. 7 and is explained in
the proposed **algorithm** in steps **14–15.**

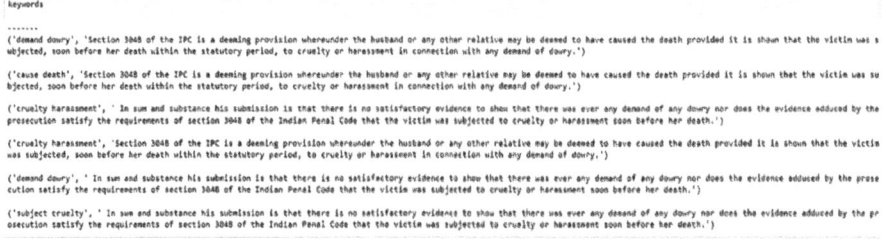

Fig. 7. Legal sentences mapped with major keywords

As shown in Fig. 8, python's **dictionary is** exported to panda's **DataFrame.**
One dataset has been created for Bi-Gram words, which is implemented concerning the proposed pseudocode. The DataSet consists of columns named
"**LEGAL_EXTRACTED_CORPUS**" and "**CONTEXT_IPC304B**" with *100* **instances.**
The input and output column is ***Text*** and ***Bi-Gram*** token, respectively. Since both the
columns' type is category, in future work, the proposed dataset will be applied as **MultiLabel Text Classification** based on **dowry death cases** is explained in the proposed
algorithm in step **16.**

From the proposed DataSet, as shown in Fig. 8, the authors observed that the proposed dataset is based on the majority of the instances of dowry death cases extracted
from the case documents considered for this research work. These are the instances for
which legal professionals concentrate while analyzing any judicial documents on dowry
death. Hence, the authors have done some numerical analysis on the percentage of occurring major instances out of the total extracted instances. Table 1 depicts instances like
'demand dowry', 'dowry death', 'cruelty harassment', 'subject cruelty', etc.

The essential concepts of dowry death are presented in the dataset from Fig. 8 and
Table 1; the authors have analyzed and visualized the major context of dowry death
cases as Pie Chart depicted in Fig. 9. In the proposed dataset, crucial information such

Fig. 8. Proposed DataSet for Dowry death

Table 1. Numerical analysis of Proposed DataSet.

Context_IPC304B	Proposed DataSet	Instances	Frequency of Instance (%)
Cause Death	100	14	15.38
Cruelty Harassment		20	18.68
Demand Dowry		18	18.68
Dowry Death		19	15.38
Matrimonial Home		14	16.48
Subject Cruelty		15	15.38

as **demand dowry (18.68%), cruelty harassment (18.68%), cause death (15.38%)**, etc., as stated in Indian Penal Code 304B [15] achieving an overall precision of more than 95%.

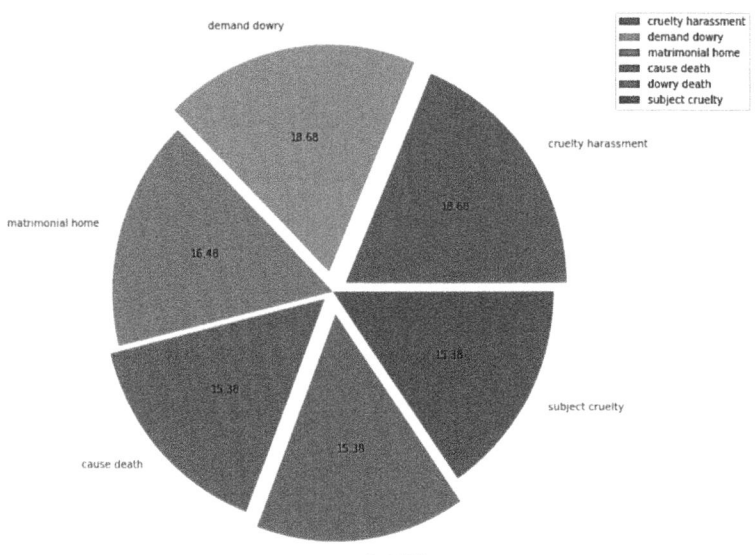

Fig. 9. Percentage of the major concepts in dowry death cases captured by dataset

This proposed pseudocode can be applied to other criminal domains to prepare the dataset for training models for Artificial Intelligence based applications to assist legal professionals. This proposed methodology helps to create a dataset in other woman-centric crimes like Domestic Violence (**S. 498a I.P.C.**), Femicide, Sexual Violence (**S. 354 A IPC**), Women Trafficking (**S. 370 & 370A I.P.C.**), Stalking (**S. 354D I.P.C.**), Assault to Outrage Modesty (**S. 354 & 364B I.P.C.**), Acid Attack (**S. 326A I.P.C.**), etc.

5   Conclusion and Future Work

In this paper, authors have centered NLP-based different perception techniques to extricate the significant boundaries present in authoritative reports. The lawful experts can comprehend the prevailing words in authoritative reports utilizing the measurable methodology, i.e., N-gram analysis. The weighted mean matrix plainly shows the critical boundaries present regardless, and these are the very boundaries that lawful experts chase while breaking down cases physically. Later, this proposed dataset can be utilized for different Artificial intelligence-based work. This work will assist experts with expecting legitimate and analyzed decisions per the law in an authoritative report like case draft, judgment etc. Recognizing the expository motivations behind terms in a legitimate case report can assist with semantic pursuit synopsis and case regulation examination. It would be truly useful for legal professionals to comprehend the expected outcome. Our proposed work can be utilized in different areas of criminal cases, which will help legitimate experts to examine cases. This study looks at the possibility of utilizing Natural Language Processing and Neural Networks to improve statistical research in the legal domain.

References

1. Zareapoor, M., Seeja, K.R.: Feature extraction or feature selection for text classification: a case study on phishing email detection. In. J. Inf. Eng. Electron. Bus. **7**(2), 60–65 (2015). https://doi.org/10.5815/ijieeb.2015.02.08
2. Sowmyayani, S.: Machine learning. In: Prediction and Analysis for Knowledge Representation and Machine Learning, pp. 1–31 (2022). https://doi.org/10.1201/9781003126898-1
3. Aggarwal, C.C.: Text classification: basic models. In: Machine Learning for Text, pp. 113–157 (2018). https://doi.org/10.1007/978-3-319-73531-3_5
4. Croft, W.B.: Machine learning and information retrieval. In: Machine Learning Proceedings, vol. 587 (1995). https://doi.org/10.1016/b978-1-55860-377-6.50078-5
5. https://analyticsindiamag.com/behind-supace-the-ai-portal-of-the-supreme-court-of-india/. Accessed 25 Sept 2022
6. Das, T., Roy, A., Majumdar, A.K.: A study on legal knowledge base creation using artificial Intelligence and ontology. In: Raj, J.S., Abul Bashar, S.R., Ramson, J. (eds.) ICIDCA 2019. LNDECT, vol. 46, pp. 647–653. Springer, Cham (2019). https://doi.org/10.1007/978-3-030-38040-3_73
7. Das, T., Moulick, J., Majumdar, A.K., Sarkar, P.K., Roy, A.: An ontology based interpretation of legal section for dowry death cases in India. Sebold Rep. **17**(7), 1433–1450 (2022). https://doi.org/10.5281/zenodo.6973807
8. Cyril Amarchand Mangaldas is India's first law firm to embrace artificial intelligence technology as part of Legal Innovation. Kira Systems, 2 February 2023. https://kirasystems.com/company-announcements/cyril-amarchand-mangaldas-is-indias-first-law-firm-to-embrace-artificial-intelligence-technology-as-part-of-legal-innovation/. Accessed 9 Feb 2023
9. Das, T., Roy, A., Majumdar, A.K.: Ontology based knowledge visualization for domestic violence cases. In: Innovations in Bio-inspired Computing and Applications, pp. 579–589 (2022). https://doi.org/10.1007/978-3-030-96299-9_55
10. Jain, S., Kumar, P.: Adoption of artificial Intelligence in industrial sectors and its impact. In: Artificial Intelligence and Global Society, pp. 129–134 (2021). https://doi.org/10.1201/9781003006602-12

11. Vijippriya, R.A.: Critical study on artificial intelligence (AI) in Indian legal sectors. Shanlax Int. J. Arts Sci. Hum. **9**(4), 58–64 (2022). https://doi.org/10.34293/sijash.v9i4.4871
12. Mandal, S., Saha, S., Das, T.: An approach to extract major parameters of legal documents using text analytics. In: Fong, S., Dey, N., Joshi, A. (eds.) ICT Analysis and Applications: Proceedings of ICT4SD 2022, pp. 331–338. Springer, Singapore (2023). https://doi.org/10.1007/978-981-19-5224-1_33
13. Mandal, S., Saha, S., Das, T.: A Text Analytics Approach of Exploratory Visualization of Legal Parameters of Dowry Death Cases. In: Smys, S., Kamel, K.A., Palanisamy, R. (eds.) Inventive Computation and Information Technologies. Lecture Notes in Networks and Systems, vol 563. Springer, Singapore (2023). https://doi.org/10.1007/978-981-19-7402-1_7
14. Padmavathi, M.S., Sumathi, C.P.: A new method of data preparation for classifying diabetes dataset. Indian J. Sci. Technol. **12**(22), 1–9 (2019). https://doi.org/10.17485/ijst/2019/v12i22/144929
15. Barros, R., et al.: Case law analysis with machine learning in Brazilian court. In: Mouhoub, M., Sadaoui, S., Mohamed, O.A., Ali, M. (eds.) Recent Trends and Future Technology in Applied Intelligence IEA/AIE 2018, pp. 857–868. Springer, Cham (2018). https://doi.org/10.1007/978-3-319-92058-0_82
16. Mandal, S., Das, T.: N-gram-based legal parameters retrieval: the state-of-the-art and future research trends of Indian Judiciary. In: Kaiser, M.S., Xie, J., Rathore, V.S. (eds.) Information and Communication Technology for Competitive Strategies (ICTCS 2022). Lecture Notes in Networks and Systems, vol 615. Springer, Singapore (2023). https://doi.org/10.1007/978-981-19-9304-6_63
17. Zhu, X., Goldberg, A.B., Rabbat, M., Nowak, R.: Learning bigrams from Unigrams. A.C.L. Anthology (n.d.). https://aclanthology.org/P08-1075/. Accessed 3 June 2022
18. Khastgir, J.: Criminal Manual, vol. 469 (2012)
19. Crimes in India 2018: National Crime Records Bureau, vol. 207 (2018)
20. Das, T., Sil, R., Roy, A., Majumdar, A.K.: UML-based modelling for legal rule using natural language processing. In: Sanyal, G., et al. (eds.) International Conference on Artificial Intelligence and Sustainable Engineering: Select Proceedings of AISE 2020, Volume 1, pp. 481–492. Springer, Singapore (2022). https://doi.org/10.1007/978-981-16-8542-2_39
21. Indian kanoon - search engine for Indian law. (n.d.). https://indiankanoon.org/. Accessed 3 Jan 2023

Handling Mouse Events Using Finger's Landmarks

Bhushan Yelure[1(✉)] , Niranjan Deokule[1] , Siddheshwar Patil[2] ,
and Shabnam Mujawar[1]

[1] Government College of Engineering, Karad, Karad, India
bhushanyelure2008@gmail.com
[2] D. Y. Patil College of Engineering & Technology, Kolhapur, India

Abstract. As the Computer technology is continuously developing a smaller and smaller electronic device to reduce the human efforts and increase the efficiency of using that device. Due to day by day changes in the computer technology it is very essential to find new and alternate way for Human Computer Interaction. As in today's world mostly of the device's system is made of touch screen technology, which is not affordable for everyone and for every devices, almost every device is changed but what has not changes is the way to provide input. So to increase the HCI developed system provides a new way of input with the help of Hand Gesture Recognition. As HGR plays very important role in the HCI, user can control most of the mouse events with the help of HGR, webcam and some built in libraries like medipipe, opencv, autopy etc. The proposed system is a "Handling Mouse Events using Finger's landmarks", which uses finger's landmarks that are captured from a camera with the help of hands module of the mediapipe library developed by the Google. Proposed system allows the user to perform mouse events like single click, double click, right click, scroll up and down, volume up and down by using web-camera or external attached camera with the computer without using a traditional mouse. Designed system have improved accuracy as compare to the existing system.

Keywords: Gesture Reorganization · Image processing · Human Computer Interaction · Mouse Events · Mediapipe

1 Introduction

Human begins use hand gestures for communication from long generations, they know "Action speaks louder than words", and gestures are one of the easiest ways to interact with anyone. So it is better way to replace action event of any hardware input device with hand gesture to increase human computer interaction. As everyone knows that mouse is an input device and there are different variations in the mouse from rubber ball mouse to wireless mouse but the type of mouse is same, which is hardware. This variation increases the accuracy of

© The Author(s), under exclusive license to Springer Nature Switzerland AG 2025
M. Gupta et al. (Eds.): MISS 2023, CCIS 1952, pp. 323–337, 2025.
https://doi.org/10.1007/978-3-031-69115-7_22

the mouse but there are some limitations like durability, click not functioning properly, extra space for external attached mouse etc.

Image processing consists of videos or images as input and output is also in the form of image or a corresponding feature of that image. An image is matrix as a 2D function f(x, y) where x and y are the co-ordinates where x, y represents intensity of the image at that point. As image is an array where pixels contains information about intensity and color. Image processing helps in detection, tracking and recognition. In recent years, number of image processing techniques has been proposed. It help in hand tracking, which is used in several applications like HMI, motion capture, Human behavior analysis, etc. for this different types of sensors, detection gloves are used. Instead of using these expensive extra devices user can simply use webcam of their system for detection, tracking and reorganization in image processing. Main objective is to find the ways to replace the hardware mouse with virtual mouse with the help of image processing in finger tracking, as control of cursor of a computer is still working in physically, as there are some limitations like durability, click not functioning properly, extra space for external attached mouse, etc. So, user can make use of camera of their system, by using some built in libraries and algorithms of python user can perform mouse events without touching the physical mouse by using only movements of his fingers.

"Handling Mouse Events using Finger's landmarks", system detect and track the finger's with the help of mediapipe as finger's are involved in making different gestures which can be used to replace the mouse events in real time. Here autopy and pyautogui library help us to generate mouse events. Basically finger tracking is done by using 'hands' module which is mediapipe's solution. It helps to detect hand and draw the landmarks. As shown in Fig. 1 there are 21 landmarks for single hands ranges from 0 to 20, each finger have 4 landmarks. With help of fingertip landmark we can track the finger's position like which finger is up or down and also distance between up fingers. This system performs mouse events like left click, right click, double click, scroll up, scroll down, volume up, volume down, mouse move with the help of specified finger's position. Objective of the developed system is to create cost-free mouse for laptops and PC's with camera attached to it, external camera also work.

1.1   Objectives

The purpose of the designed system is to automate the mouse where user can execute mouse operation or provide mouse input to the system by using his finger's movements.

– Mouse events should have quick action and reaction with better accuracy and system should provide user friendly gestures.
– To automate the mouse need of camera to capture the image, image processing library is applicable to process the image and to execute mouse operation pyautgui and autopy python libraries is required. Need to use hands module for gesture recognition. System attached camera is required to capture the

finger's movements. As shown in Fig. 1. System stores landmarks position to identify the finger's position with the help of hands module and then map with the predefined gestures if matches then mouse operation will executed.
- System is designed to work for some mouse operation which are mouse move, left click, right click, double click, scroll up, scroll down, volume up, volume down with good accuracy.

1.2 Motivation

In today's world of virtualization, everything becomes virtual where virtualization creates an abstraction layer on hardware device, which help us reduce cost, downtime and help to increase efficiency, productivity, enhance resiliency in disaster recovery situations. Overall virtualization is best solution to replace any possible hardware device.

Gesture are more user friendly for literate and illiterate peoples, so they easily handle designed system.

As a result when we combine virtualization and gestures, it became best alternate solution for any existing hardware input system.

1.3 Organization

Introduction section explain why this system is designed, where we can use and how to use this system in deep. Literature survey section describe already developed system to handle such mouse events virtually without using physical hardware mouse. There is methodology section which describe system work flow in some steps with the help of workflow diagram and algorithm of the system. Implementation part describes all mouse operation in-detail that how to use the system. Result shows the accuracy of the designed system with the help of confusion matrix using some formula's. At the end conclusion and future scope has described.

2 Literature Survey

Many researchers have developed different technologies that convert physical mouse into virtual mouse. As shown in Table 1 different researcher tried to replace the existing hardware mouse to any other user-friendly technique or device. Vantukala VishnuTeja Reddy and Thumma Dhyanchand [1] has developed a system which replaces hardware mouse with hand. They have used some color which is applied on tip of the finger's to identify the gesture and then detect the motion of the color with the help of background subtraction. Madhur Rajadhyaksha and Manav Ranawat [5] has developed a system to solve physical mouse problem by using gesture recognition where dataset for gesture recognition is used based on gesture prediction, mouse events trigger where need of dataset for gesture recognition. Prof. Monali Shetty and Christina A. Daniel

[2] tried to develop a solution for this problem where they have used object tracking techniques to replace mouse operation with object movements. Finger should contains some object to track and based on the that object movement cursor action taken place, but accuracy is based on the background for plain background it is better and for background with external noise it very poor. Kollipara Sai Varun and Puneeth [3] try to solve this problem with the help of color detection using HSV method where different system defined color has been applied on finger's and moved in air, in front of camera.

As per literacy survey there are several such existing system are designed but the way of working, methodologies used, type of mouse operation performed, ease of use is very different from the designed system. As compared to the previously developed system the designed system have extra mouse events and better accuracy of mouse operations.

3   Methodology

"Handling Mouse Events using Finger's landmarks", system detect and track the finger's with the help of mediapipe as finger's are involved in making different gestures which can be used to replace the mouse events in real time. As shown in Fig. 1 each finger have 4 landmarks and each landmarks contains specific name associated with it. With the help of fingertip landmark it is possible to track the finger's position like which finger is up or down and also distance between up fingers to track the position of finger's have to compare central x and y position of the finger's landmarks. This system has been implemented in four major stages namely, convert extracted image to RGB from BGR, detect and track the hand, draw and track the landmarks.

- Convert extracted image to RGB from BGR: Hands module of the mediapipe require image in RGB which is true image. So image is converted into RGB from image that is already in BGR form. It helps in hand detection phase of system flow as shown in Fig. 2
- Detect and track the hand: Converted image provided to hands module which detect the hand and then draw the landmarks and based on landmarks position mouse events executed as shown in Fig. 2. Hands solution provided by mediapipe library developed by google. Based on flag's value Hands module works. Hands are different flag's like static image mode, max number of hands, model complexity, min detection confidence, min tracking confidence. These flag's helps to find accurate result.
- Draw and tack the landmarks: Every hand contains 21 landmarks as shown in Fig. 1. System tracks that landmarks and store their position which help in gesture recognition.

Table 1. Literature Survey

reference	Objective	Methodology	Used Parameters	Remark
[1]	Replace mouse operation with hand gesture	Color detecting	Tracking the fingertip on which system defined color is applied. where color is applied on fingertip	Solution for the Finger tracking in the real world & control the cursor of a computer which is still performed physically
[2]	Replacement of the mouse with any object	Object Tracking	Collecting Frame	Developed Systems help to perform mouse operation based on object movement where background causes effect on detection
[3]	Replace mouse operation with movement of specified color	Color Tracking	Color detection and mouse movement based on highlighted color which is given from the user for movement of the mouse	System work as mouse where movement recognized based on the color movement
[4]	Replace mouse operation with hand gesture	Tracking, detecting hand	Tracking, detecting hand and recognize the movement by processing dataset	System performs mouse event based on recognized hand gesture
[5]	Replace mouse operation with Gesture	Gesture Reorganization	Image Capturing, processing, detection gesture recognition,event triggering	Developed a system which control the position of cursor with the hands and eliminate use of Physical mouse
[6]	Replace mouse operation with hand gesture	Fingertip tracking	Tracking the fingertip on which system defined color is applied. where color is applied on fingertip	Designed framework developed a solution for the finger tracking in the real world and control the cursor of a computer. which is still performed physically
[7]	Automate mouse operation with hand gesture	Fingertip tracking	Tracking the fingertip where hand have data-gloves on on it	A solution for the finger tracking in the real world to control the cursor of a computer

Algorithm 1: Mouse Events Execution

Data: Declare *tipId* & *Cap* as finger's tip & camera object
Result: System defined mouse event performed.
while *True* **do**
 Capture & captured image to RGB store to *img*.
 result ← hands.process(imgRGB)
 myhand ← result.multi$_h$and$_l$andmarks[0]
 for *id and lm in myHand.landmark* **do**
 end
 length(llist)!=0 Find which finger is up.
 if *llist[tipId[0]][2]≤ llist[llist[0]-1][1]* **then**
 | finger[0]←1
 end
 else
 | finger[0]←0
 end
 for *i=1* **to** *to 5* **do**
  if *llist[tipId[i]][2]≤ llist[tipId[i-2]][2]* **then**
  | finger[i]←1
  end
  else
  | finger[i]←0
  end
 end
 if *Index,middle,ring, little finger is up* **then**
 | *Scroll up mouse event executed.*
 end
 if *Middle,ring, little finger is up* **then**
 | *Scroll down mouse event executed.*
 end
 if *Index Thumb finger is up* **then**
  if *length less than 100 pxls* **then**
  | *volume up mouse event executed.*
  end
  else
  | *volume down mouse event executed.*
  end
 end
 if *Index finger is up* **then**
 | *Mouse move mouse event executed.*
 end
 if *Index,middle,ring finger is up* **then**
 | *Left click mouse event executed.*
 end
 if *Index,middle finger is up* **then**
  if *length ≤ than 30 pxls* **then**
  | *Double click mouse event executed.*
  end
  else
  | *Right click mouse event executed.*
  end
 end
 end
end

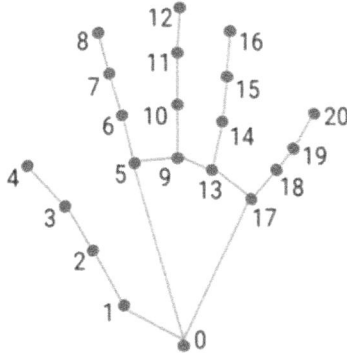

Fig. 1. Landmarks

3.1 Mouse Events Execution Algorithm

Algorithm 1 shows system flow that how system is defined in such a way that it handle mouse events using finger's landmarks as hand is moved in front of camera and Camera captured live movements. System has implemented this algorithm in object and classes format where class contains all system required methods in predefined from and they are used whenever system needs by achieving re-usability. In detail explanation of all such methods are given in next module.

4 System Implementation

4.1 Event Generation

Based on specified finger's gestures appropriate mouse events occurred. This is an infinite loop, until we press exit key(ESC). Table 2 shows all mouse events and associated finger's movements which system uses to execute corresponding mouse events.

- Case "Mouse Move": Check is there any hand in front of camera and if only index finger is up. Find the fingertip position and scale it to screen position, mouse moved from current position to fingertip scaled position. Finger's position must be like as shown in Fig. 3.
- Case "Right click": Check is only index and middle finger is up with at least 60 pixels distance in between. Get current cursor position press right mouse key down. Finger's position must be like as shown in Fig. 3.
- Case "Left click": Check is only index, middle and ring finger is up. Get current cursor position press left mouse key down and wait for very short period. Finger's position must be like as shown in Fig. 3.
- Case "Double click": Check is only index and middle finger is up with no distance in between. Get current cursor position and press left mouse key. Finger's position must be like as shown in Fig. 3.

- Case "Scrolling UP": Check is Index, Middle, Ring, little Finger up. If yes scrolling up cursor event generated. Finger's position must be like as shown in Fig. 3.
- Case "Scrolling Down": Check is Middle, Ring and Little finger is up. If yes scrolling down cursor event generated. Finger's position must be like as shown in Fig. 3.
- Case "Volume UP": Check is Thumb and Index Finger with less than 100 pixels distance. If yes volume up cursor event generated. Finger's position must be like as shown in Fig. 3.
- Case "Volume Down": Thumb and Index Finger with greater than 100 pixels distance. If yes volume down cursor event generated. Finger's position must be like as shown in Fig. 3.

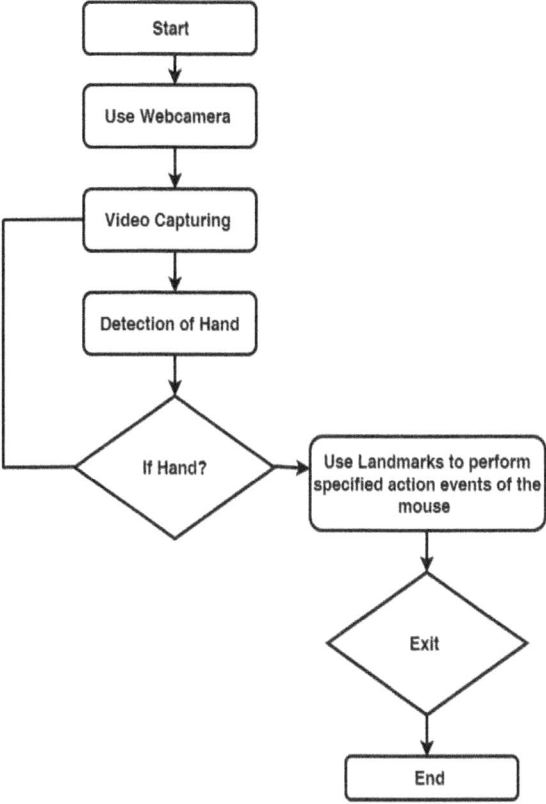

Fig. 2. Work flow of proposed system

Table 2. Hand gestures and their related mouse actions.

Finger's Gesture	*Mouse action*
Index finger up	Move Mouse
Index and middle finger up with some distance	Right click Mouse
Index, middle and ring fingers up	Left click Mouse
Index, middle finger up with no distance between	Double Click Mouse
Index, middle, ring and little finger up	scrolling up
Middle, ring and little finger up	scrolling down
Thumb, index up with less than 100 pixels distance	Volume up
Thumb, index up with max distance 100 pixels	Volume down

5   Results and Discussion

This system is designed to just make mouse operation more easy and user-friendly. Here user need only active webcam so that it can capture the image then system detects and track the hand and draw the finger's landmarks which help to track finger's different gesture so that mouse cursor work accordingly. Almost all mouse operation performed by this designed system like single click, double click, right click, scroll-up, scroll-down, volume up, volume down and mouse moves etc. To detect, track the hand and then draw landmarks mediapipe library is used. This system is working very good for 720p = 1280×724 pixels 0.9MP (16:9) and provide very clear and accurate result in plain background and background with some noise with at least Core-3 processor with 4 GB RAM required.

For accuracy of all mouse operation, user feedback for every mouse event is collected. Table 3 shows success and failure rate of all mouse operations where sample of every 20 events is taken. While performing mouse operations correct distance from webcam is maintain.

Table 4 shown Accuracy(%), Precision and Recall of every Mouse events, where accuracy of mouse move, left click, right click, double click, scroll up is very good but for rest it is very less mainly for volume up and down.

5.1   Accuracy

Accuracy [8–10] is defined as fraction of sum of all true positive and true negative predicted with number of samples. It's value ranges from 0% to 100%. With the help of Confusion matrix, accuracy of every mouse event is calculated by using formula. As per the observation in result represented in Table 4 and shown in Fig. 4, accuracy of all events is 100% except volume up and down events.

$$Accuracy = \frac{TP + TN}{No. of samples} \tag{1}$$

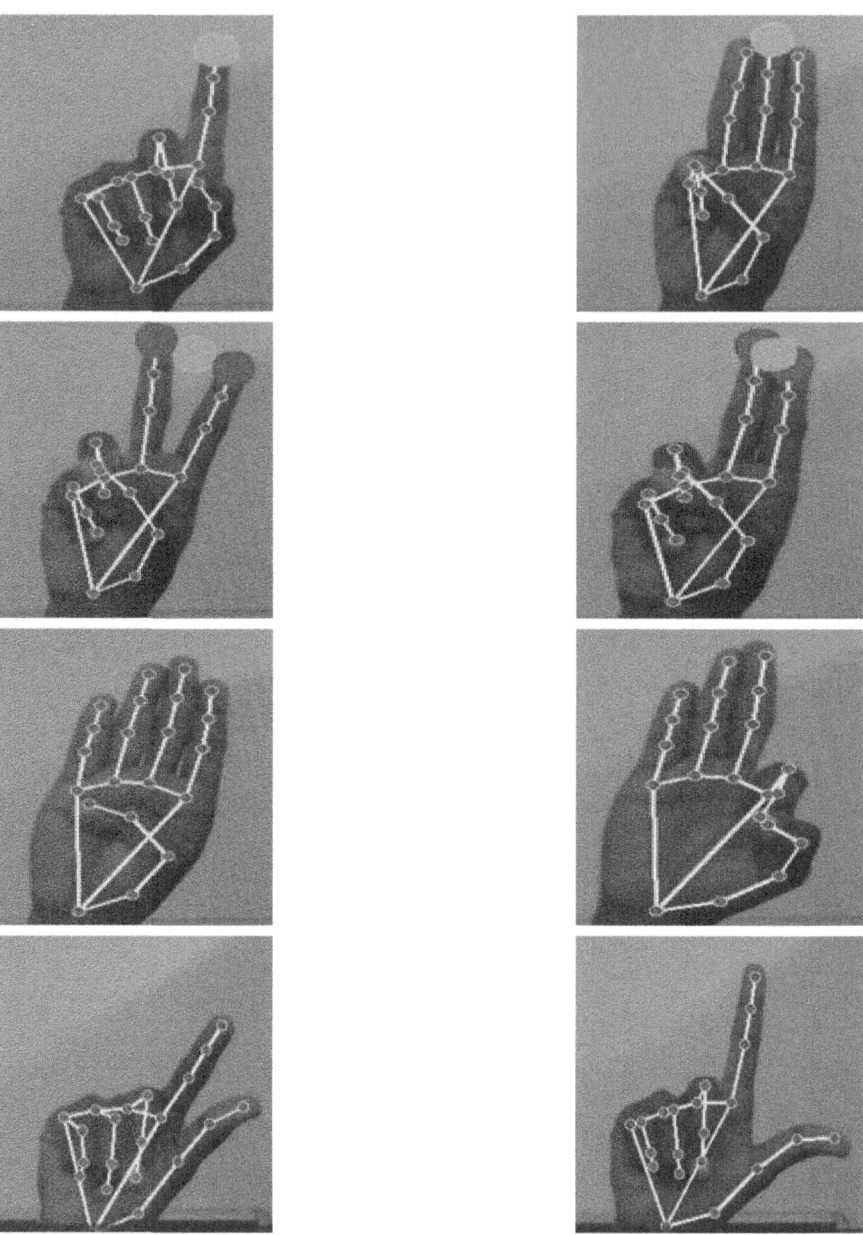

Fig. 3. All Hand gestures required to perform mouse operation

Table 3. Success rate(%) of Every mouse operation.

Mouse action	Success	Failure	Success-rate (%)
Move Mouse	20	0	100%
Left click Mouse	20	0	100%
Right click Mouse	20	0	100%
Double Click Mouse	20	0	100%
Scroll UP	20	0	100%
Scroll Down	20	0	100%
Volume UP	10	10	50%
Volume Down	14	06	70%

Table 4. Mouse Events Accuracy(%), Precision and Recall.

Mouse action	TP	TN	FP	FN	Accuracy (%)	Precision	Recall
Move Mouse	20	20	0	0	100%	1	1
Left click	20	20	0	0	100%	1	1
Right click	20	20	0	0	100%	1	1
Double Click	20	20	0	0	100%	1	1
Scroll UP	20	20	0	0	100%	1	1
Scroll Down	20	20	0	0	100%	1	1
Volume UP	11	12	09	08	57.5%	0.55	0.57
Volume Down	10	18	10	02	70%	0.5	0.83

5.2 Precision

Precision [8–10] is defined as fraction of all true positive predicted with all true positive and false positive predicted value. Value for precision always lies in between 0 and 1 and it is depicted in Fig. 5.

$$Precision = \frac{TP}{TP + FP} \qquad (2)$$

If we see precision value of all mouse events in Table 4. Precision of all events is 1 except volume up and down events, 0.5 for volume up and down.

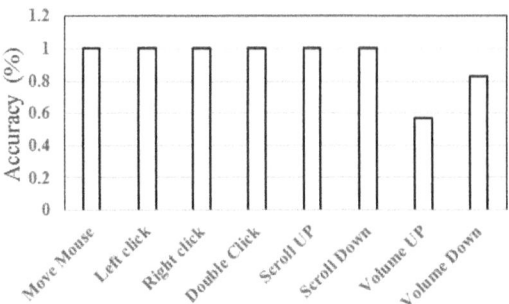

Fig. 4. Mouse Event's Accuracy Graph

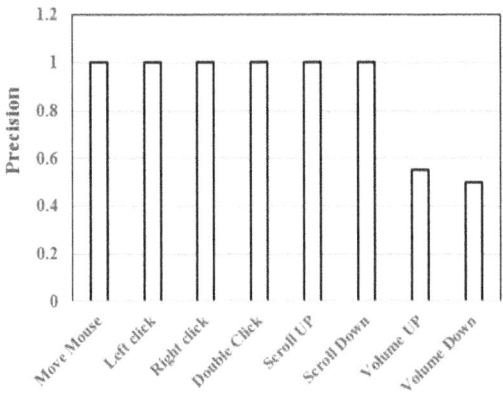

Fig. 5. Mouse Event's Precision Graph

5.3 Recall

Recall is defined as fraction of true positive with true positive and false negative.

$$Recall = \frac{TP}{TP + FN} \tag{3}$$

Recall [8–10] value of all mouse events are shown in Table 4. Recall of all events is 1 except volume up and down events, 0.5 for volume up 0.7 for volume down and results are shown in Fig. 6.

Time Complexity of proposed Algorithm 1. is $T(n) = O(n + m)$, where n is number of fingers and m is number of landmarks.

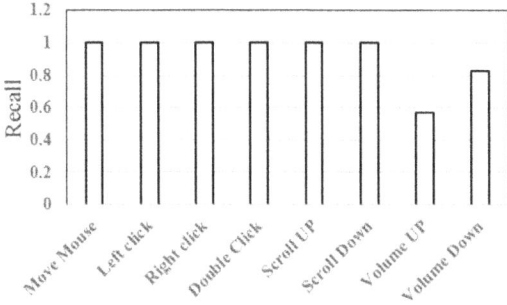

Fig. 6. Mouse Event's Recall Graph

6   Computational Comparison of Existing System

Designed system is compared with existing system to find accuracy difference, user-friendly nature and performance of the designed system. As shown in comparison Table 5, designed system has better accuracy with user-friendly Gestures and multiple mouse event options. Existing system designed by Manav Ranawat, Madhur Rajadhyaksha [5] is close to the designed system but less mouse oper-

Table 5. Comparison of Existing system

Reference	Accuracy (%)	Designed System Accuracy (%)	Reference Technique	Designed System Technique	Additional Features
[2]	82.4%	90.94%	Object Tracking	Finger's	Volume up, down and user friendly gesture
[3]	82%		Color Tracking	Tracking	Left/Right/Double click. new operations are there
[1]	81.98%		Color detection applied on fingertip	Using Hand's module	Volume up, down and user friendly gesture
[4]	80%		Gesture Recognition using processed and Trained dataset		Double click, Scroll up and down, Volume up and down
[5]	85%		Gesture reorganization		Volume up, down and user friendly gesture

ation options are there. Also we don't need any external predefined object with predefined color i.e. there no need to use any color in designed system to track the finger's gestures as hands module is used to track the hand and draw the land marks.

7   Conclusion

This system is proposed to recognize finger's movements and replace mouse events with that movement. This system illustrates that how to use computer's operation with the help of hand gestures i.e. finger's movements. Basically Finger tracking is done by using 'hands' which is mediapipe's solution, and it helps to detect hand and draw the landmarks. There are 21 landmarks for single hands ranges from 0 to 20. With the help of fingertip landmark we can track the finger's position like which finger is up or down and also distance between up fingers. This System performs mouse events like left click, right click, double click, Scroll up, scroll down, volume up, volume down, mouse move with the help of specified finger's position. Designed system creates user friendly mouse for laptops and PC's with camera.

The usability of the developed system depend on features such as the distance from camera, camera resolution, lighting condition. This system helps to increase the Human Computer Interaction and also useful in current pandemic, as it reduces direct human touch with machine up to some level.

8   Future Scope

Designed System works for only few of mouse operation like left click, right click, double click, scroll-up, scroll-down, volume up and volume down, mouse move with the help of specified finger's position. Some other operation need to be developed like drag, drop etc. Also need to increase the accuracy of the system as volume up and volume down function have very less accuracy. Sometimes camera is unable to track the hand as it goes beyond the given distance range. Thus it needs to provide such option to maintain distance between camera and human hand.

References

1. Reddy, V.V.T., Dhyanchand, T., Krishna, G.V.: Virtual mouse control using colored fingertips and hand gesture recognition. In: 2020 IEEE-HYDCON, Hyderabad, India, pp. 1–5 (2020). https://doi.org/10.1109/HYDCON48903.2020.9242677
2. Shetty, M., Daniel, C.A., Bhatkar, M.K., Lopes, O.P.: Virtual mouse using object tracking. In: Fifth International Conference on Communication and Electronics Systems (ICCES 2020) (2020)
3. Varun, K.S., Puneeth, I., Prem Jacob, T.: Virtual mouse implementation using open CV. In: Third International Conference on Trends in Electronics and Informatics (ICOEI 2019) (2019)

4. Matlan, R., Dadlani, R., Dumbre, S., Mishra, S.: Virtual mouse using hand gestures. In: 2021 International Conference on Technological Advancements and Innovations (ICTAI) (2021)
5. Ranawat, M., Rajadhyaksha, M., Shankarmani, R., Lakhani, N.: Hand gesture recognition based virtual mouse events. In: 2021 $2^n dx$ International Conference for Emerging Technology (INCET) (2021)
6. Tran, D.S., Ho, N.H., Yang, H.J., et al.: Real-time virtual mouse system using RGB-D images and fingertip detection. Multimed. Tools Appl. **80**, 10473–10490 (2021)
7. Ganzeboom, M.: How hand gestures are recognized using a data- glove. University of Twente, The Netherlands (2010)
8. Shibly, K.H., Dey, S.K., Islam, M.A., Showrav, S.I.: Design and development of hand gesture based virtual mouse. In: 1st International Conference on Advances in Science, Engineering and Robotics Technology (ICASERT), Dhaka, Bangladesh (2019)
9. Vuyyuru , G.M., Shirke, M.R.: Performing basic tasks on computer using hand gestures & ultrasonic sensors. Int. J. Eng. Res. Technol. (IJERT) **10**(05) (2021)
10. Juhong, A., Treebupachatsakul, T., et al.: Smart eye-tracking system. In: 2018 International Workshop on Advanced Image Technology (IWAIT) (2018)

Analysis on Stock Market Stream Data Using Kafka, AWS and PowerBI

K. K. Baseer[1], B. Siva Siddartha Reddy[2], D. Vishnuvardhan[2(✉)], K. Chandravathi[2], L. Abhishek[2], and M. Karthikaa[2]

[1] Department of Information Technology, Mohan Babu University (Erstwhile Sree Vidyanikethan Engineering College), Tirupati, India
[2] Department of Information Technology, Sree Vidyanikethan Engineering College, Tirupati, India
vishnud2535@gmail.com

Abstract. Real-time data analysis is becoming increasingly important, as it allows businesses to act quickly and with greater decision accuracy. Traditionally, data analysis occurs after data has been collected and saved, but real-time data analysis provides an opportunity for businesses to extract valuable insights from data as it's generated. A streaming platform like Kafka is necessary to handle the constant intake of data in a systematic and efficient way, and to build data pipelines that reliably integrate and transfer information between platforms. Cloud computing services like AWS S3, AWS Crawler, AWS Glue Data Catalog, and AWS Athena can be used in conjunction with Kafka to process and analyze the data. This framework is useful for processing real-time data in various sectors, such as the Stock market, e-commerce applications, gaming, and more, and extracting useful features and hidden patterns. Machine learning models can also be trained on the data for better analysis and decision-making.To enhance this framework, it's important to integrate various cloud computing services with Kafka more seamlessly to better handle large-scale data. Additionally, real-time data visualization tools could help generate insights and provide interactive feedback on data streams. The incorporation of predictive analytics capabilities and intelligent, data-driven decision-making processes could help businesses make more informed decisions. By integrating real-time data analysis with intelligent automation tools, human error can be reduced, and overall efficiency and effectiveness in the decision-making process can be improved.

Keywords: Real-Time data · kafka · AWS S3 · stock market data · big data analytics · visualization · analysis · zookeeper

1 Introduction

Real-time big data analytics is a technique that allows for the analysis of massive amounts of incoming data as it is being created or stored by the IT infrastructure. Due to the influx of data and the requirement to respond quickly to data triggers, real-time analytics is

© The Author(s), under exclusive license to Springer Nature Switzerland AG 2025
M. Gupta et al. (Eds.): MISS 2023, CCIS 1952, pp. 338–353, 2025.
https://doi.org/10.1007/978-3-031-69115-7_23

becoming more and more popular in commercial and social applications. Big Data analytics expanded their applications in various sectors such as Stock Market, HealthCare, and Manufacturing, E-commerce, Media & Entertainment, IoT, Gaming, Government, etc. [2]. A financial market is a space where people can trade securities and financial derivatives for minimal or no money. Some examples of securities are stocks, bonds, and raw resources. And precious metals, which are referred to as necessities in the financial system. The advantages of the financial sector include Economic Development, Infrastructural Development, Promotes Regional Balances, Employment Generation, Provide Funds, and Credit Creation. The disadvantages of the financial sector are Restrictions on the Borrower, a System of Collateral Securities, and Complex processes. Information about financial sector in the past, present, and future values of stocks is known as stock market data [3]. In order to make investments, investors need share market data for analysis of stocks. With advancements in technology, the introduction of new products, diversification, and of course lessons learned from various frauds and crises among other factors that went unnoticed over time, this journey of stock markets began way back in the 1600s, or even earlier (in a different form), and has led to what we perceive as today's global equities, which seem to be equitable today compared to any other period of time [3]. The stock market involves many issues such as Current events, Natural calamities, Exchange rates, Real-time analysis, etc. Our current research focusses on analyzing the stock market data in real-time [15]. A streaming platform must therefore be able to manage this continuous inflow of data and analyse it progressively and sequentially. Kafka is used to generate real-time streaming data pipelines and applications.Along with Kafka cloud computing services such as AWS S3 (simple storage service), AWS Crawler, AWS Glue Data Catalog, and AWS Athena are used to process and analyze the data.

2 Literature Survey

In [9], the authors Tapashi Gosswami, Sanjit Kumar Saha, Mahmudul Hasan tell about how most of the researches on analyzing stock market data has been done using common ways such as fundamental analysis, statistical analysis and machine learning.In this research work the authors have used neural network, as it is the most effective tool to predict and analyze data. After specification of various experimental parameters the authors started the implementation of Backpropagation neural network (BPNN). In a gradient descent approach, it decreases a discrepancy between the intended output and the actual output. That is why the authors chose BPNN.

In [2], the authors Prit Modi, Shaival Shah, Himani Shah described about how big data can be used for accurate prediction and analysis of huge amounts of data. The process involved in their research includes data collection, feature selection and applying machine learning algorithms like logistic regression with regularization and SVM with different kernels.The computed results support the authors' conclusion that support vector machines with regularization and a Gaussian kernel perform better than logistic regression and SVM with other kernels.

In [3], the authors predicted the stock market behaviour by compiling past historical data on shares or any other traded securities in the market and attempting to predict the value of the company's stock or securities in the future. The authors experimented on various methods that could become the best fit for prediction of stock market like Simple moving Average model, ARIMA, Holt-Winters method. The authors concluded that the Simple Moving Average model gives low prediction error which makes it the best solution whereas the ARIMA model provided unreliable results.The only limitation in this model is it can be used only on linear trends.

In [5], The authors assess distributed stream processing frameworks' present acceptability and maturity among IoT applications and look at whether they may be used in the data processing layer of smart cities. Their experiments compare Apache Storm, Apache Spark Streaming, and Apache Flink's performance the most among the three DSPFs. Their findings indicate that selecting a suitable framework for a Smart City's data analytics layer necessitates thorough knowledge of the traits of the intended applications. Finally, they draw the conclusion that each of the frameworks discussed here has benefits and drawbacks. Their tests reveal that Storm and Flink perform relatively similarly, but Spark Streaming has significantly higher latency while offering more throughput.

In [3], the authors presented a method that forecasts the indices of the Indian Stock Market uses a Differential Evolution based (FLANN). For performance evaluation, the (MAPE) and (RMSE) are calculated.

In [37], Big data analytics is a crucial component of many business and service sectors' success. The military, disaster preparedness, logistics, energy, and finance are a few examples of these fields.Numerous big data techniques in manifold fields rely on rapid analyses based on the available data to make reliable conclusions. The technical challenges provided by real-time big data analytics systems are examined in this paper.

In [21], To predict changes in stock values and help investors make wise decisions, a number of researchers and analysts have devised techniques and systems. Researchers can predict the market using novel textual social media data utilising advanced trading techniques.Text data analytics and ensemble methods are few modern machine learning techniques which has significantly improved prediction accuracy. The volatility, unpredictability, and chaos of the data continue to make stock market forecasting and analysis one of the most difficult academic fields.In order to understand the systematics of machine learning-based systems for stock market prediction, this study implements a broad framework.

Many researchers and analysts has always been concerned about predicting Stock Market. Many of them used various Machine Learning techniques to perform the prediction and fore casting the data. One experimental tool called TinTo seeks to explain the value and viability of using traditional SQL queries for evaluating a variety of data streams. Multiple technical indicators, numeric values which have been generated from previous stock data can be used to analyze the stock market using TinTo tool [1]. Few authors mentioned about how the stock market is predicted using two different menthods namely technical and fundamental analysis [3]. Technical analysis is carried out using sentiment analysis on social media data, and fundamental analysis is carried out using machine learning algorithms on past stock prices [6, 7, 15, 18, 30].

As technology advances in the twenty-first century, one of the best technologies is Apache Kafka. Kafka's attributes, such as its scalability, ability to enable distributed stream processing, fault-tolerant messaging system, and speed, have made it a better option for data handling and analysis [4, 11]. Kafka is designed in such a way that a central data schema of a large firm can be served by a single cluster [12, 29, 31].

The Research objectives of the work are as below:

- Setup the Kafka framework on AWS EC2 instance, to configure Kafka broker and Zookeeper server.
- Configure Python API and send the real-time stock data to Kafka server (Producer).
- The stream of data from Kafka server stores in AWS S3 bucket using Python API (Consumer).
- To configure AWS crawler for scanning the stock data of S3 bucket which automatically defines relational schema and stores data in AWS Glue data catalog.
- To generate views on AWS Glue data catalog for analysis of stock exchange data through AWS Athena.
- Extract the data from the AWS Athena results and import the relational data which can be visualized through Power BI.

3   Architecture

In this section we stated the whole process that is involved in building the architecture (Fig. 1).

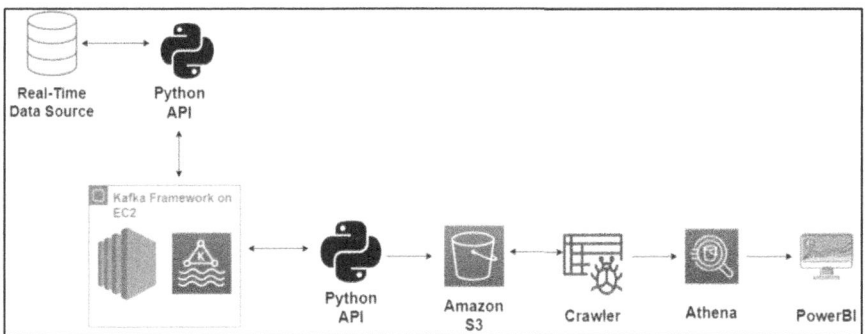

Fig. 1. Data flow diagram

- Setup the Kafka framework on AWS EC2 instance, to configure Kafka broker and Zookeeper server.
- Configure Python API and send the real-time stock data to Kafka server (Producer).
- The stream of data from Kafka server stores in AWS S3 bucket using Python API (Consumer).
- To configure AWS crawler for scanning the stock data of S3 bucket which automatically defines relational schema and stores data in AWS Glue data catalog.
- To generate views on AWS Glue data catalog for analysis of stock exchange data through AWS Athena.
- Extract the data from the AWS Athena results and import the relational data which can be visualized through Power BI.

Let K be the set of Kafka topics, P be the set of producers, and C be the set of consumers.

The equation for describing the basic operation of Kafka can be represented as:

$$K = P + C \tag{1}$$

where K is the collection of all topics, P is the collection of all producers that produce messages to the topics, and C is the collection of all consumers that consume messages from the topics.

In Eq. (1) represents the fundamental relationship between producers, topics, and consumers in Kafka. Consumers receive messages from themes that producers have written. The set of topics is the union of the sets of messages produced and consumed by the producers and consumers.

In addition, the equation can be extended to represent the concept of partitioning in Kafka:

$$K = P + C + P(x)C(y) \tag{2}$$

where $P(x)$ represents the collection of x partitions for each producer, and $C(y)$ represents the collection of y consumers for each partition. This equation represents the concept of partitioning in Kafka, where each topic is divided into multiple partitions, and each partition is consumed by a specific group of consumers.

Overall, the Eq. (1) represents the basic operation of Kafka, and the Eq. (2) represents the concept of partitioning in Kafka. Together, these equations provide a mathematical representation of the Kafka architecture and its key features.

Let S_1 be the set of data sources(S3 Bucket), $D_{(f)}$ be the set of extracted data by crawler(Data Catalog), and C be the set of AWS Crawlers.

The equation for describing the basic operation of an AWS Crawler can be represented as:

$$D_{(f)} = S_1 x C \tag{3}$$

where $D_{(f)}$ is the collection of all data that is extracted from the websites, S is the collection of all data source that are being crawled, and C is the collection of all AWS Crawlers that are responsible for extracting the data.

This equation represents the fundamental relationship between data sources, and AWS Crawlers.

In addition, the equation can be extended to represent the concept of filtering and categorizing the data:

$$s_2 = D_{(f)}(F(X)) \tag{4}$$

where $F(X)$ is collection of queries applied on data(athena),X is number of queries $D_{(f)}$ is AWS glue Data Catalog(provides data to perform queries.S_2 id AWS S3 bucket (2) and the query results are stored in S2

Overall, the Eq. (3) represents the basic operation of AWS Crawler, and the Eq. (4) represents the concept of filtering and categorizing the data. Together, these equations provide a mathematical representation of the AWS Crawler's architecture and its key features.

Visualization of AWS Athena results can be performed using the following mathematical equation:

$$V(R) = f(A, T)$$

where:

V = Visualization
R = AWS Athena Results
A = PowerBI tool (Data Analysis)
T = Data Transformation tool (e.g. SQL, Python)

The function f(A, T) describes the process of using the data analysis tool A and the data transformation tool T to transform and analyze the AWS Athena results R, and create a visual representation (V) of the data.

This equation highlights the importance of utilizing both data analysis and data transformation tools in order to effectively visualize the results obtained from AWS Athena queries.

4 Implementation

4.1 Setup Kafka Framework Through AWS

Kafka. In order to process flowing data in real time, Kafka was created as a decentralized system for storing data. Data that is continuously produced by a variety of sources, the majority of which provide data records simultaneously, is referred to as "streaming data." It is vital to have a platform that can handle the continuous inflow of data gradually and progressively.Kafka blends publish-subscribe and queuing messaging technologies to give users the key advantages of each. Queuing is very scalable because it enables the distribution of data processing across numerous consumer instances. Multiple separate applications can read from data streams independently and at their own pace because to Kafka's model's reusability.

Amazon Elastic Compute Cloud (EC2)
A safe, scalable cloud computing web service Amazon EC2 offers compute capacity. You can design and launch apps more quickly using Amazon EC2 since there is no initial hardware expenditure. Using Amazon EC2, you may launch quite so many or even as few vms as you require, specify networking and security parameters, and control storage. Without having to anticipate traffic, Amazon EC2 enables you to move up or down to handle shifts in demand or popular spikes.Instances are created in AWS EC2 service and kafka is connected with SSH client also setting up of JDK, running Zookeeper and Kafka servers are performed in this module.

In Fig. 2, the flow diagram shows the creation of AWS EC2 instance that setup the environment to download and extract Kafka. Start the Kafka and Zookeeper servers under this module.

Steps involved in this module are:

1. Creating an EC2 instance in AWS Cloud
2. Connect to EC2 Instance
3. Install Java
4. Installing Kafka on an AWS EC2 Instance
5. File Directory of Kafka
6. Start running Zookeeper
7. Running Kafka Broker

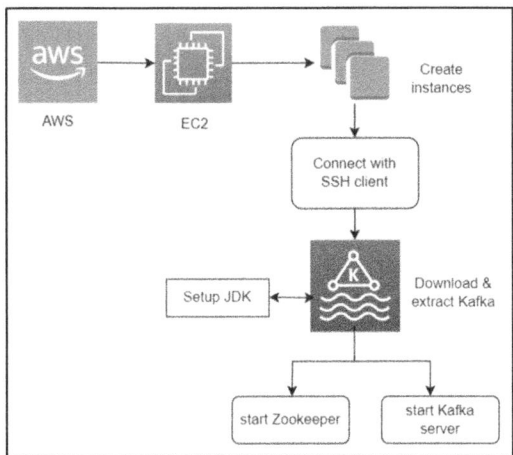

Fig. 2. Flow diagram to setup kafka framework

Your EC2 instance of Kafka has now been successfully launched. By making the necessary adjustments to the security groups, you can use the Kafka-Python API or Kafka-Java API to access your Kafka-server.

4.2 Data Produced to Kafka Node Python API(Producer)

Real time data source is shared to Kafka node which then performs monitoring.

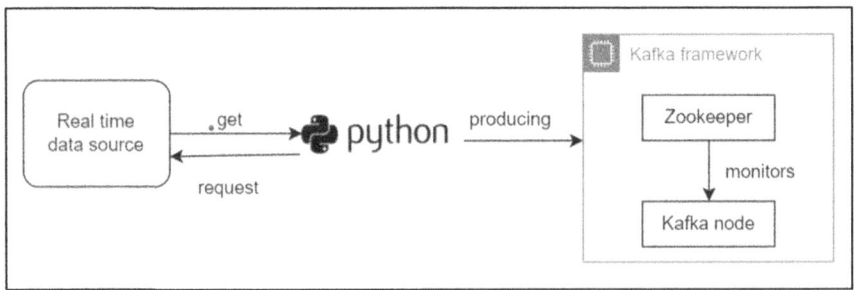

Fig. 3. Flow diagram for producer

In Fig. 3, the data produced to Kafka node by using Python API under this module. The real-time stream data is produced to kafka node.

The following steps are involved in this module:

1. Python libraries
2. Data produced to kafka node

Algorithm 1: Producer

Function: $\mu(e, T) \rightarrow KB(T, Ack)$

begin

//μ - Producer

//e - event

//T – Topic

//KB – Kafka Broker

//Ack – Acknowledgement

if T.append(e) then

Ack(d, a)

else:

Ack(d, r)

Ack(d, a, r)->Z(0|1)

//Z: Zookeeper

// d: acknowledgement data

// a/r: approve/reject
//Ack returns 0 if record is successfully written on

//Topic otherwise returns 1 to the Zookeeper node.

end

The algorithm above describes how a producer node works in Kafka. The function μ takes in three parameters: e, T, and Ack. μ is a producer that sends an event e to a Kafka broker T, and then waits for an acknowledgement from Ack. If the event is successfully appended to the topic T, Ack returns a success acknowledgement with data d and approval a. If the event cannot be appended to the topic T, Ack returns a failure acknowledgement with data d and rejection r.

Ack then sends the acknowledgement data d, and approval/rejection status a/r to the Zookeeper node Z. If the record is successfully written to the topic, Ack returns a 0 to the Zookeeper node. If the record cannot be written to the topic, Ack returns a 1 to the Zookeeper node.

4.3 Data Consumed to AWS S3 from Kafka Node Through Python API(Consumer)

The data is consumed in AWS S3 buckets that are being generated from Kafka node through Python API.

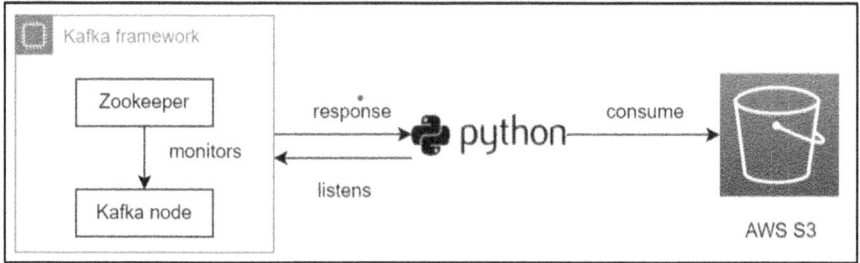

Fig. 4. Flow diagram for consumer

In Fig. 4, data consumed to AWS S3 from Kafka node through Python API under this module. The data to be seen from Zookeeper server.

Various steps involved in performing this module involves:

1. Create bucket in S3
2. Data consumed to S3

Algorithm 2: Consumer

Function: KB(T, Ack) ->Ω(T, r)

begin

        //KB – Kafka Broker

        //T - Topic

        //Ack – Acknowledgement

        //r – record

        //Ω - Consumer

    if Ω(T, r) then

        Ack(d, a)

    else:

        Ack(d, r)

Ack(d, a, r)->Z(0|1)

        // Z: Zookeeper

        // d: acknowledgement data

        // a/r: approve/reject

        //Ack returns 0 if record is successfully written on

//Topic otherwise returns 1 to the Zookeeper node.

end

The above algorithm illustrates the work of a consumer node in Kafka. A function KB (kafka Broker) is created with a topic and acknowledgement as its parameters. If the consumer inputs a topic and a record, an acknowledgement is generated with approval. Otherwise, a rejection acknowledgement is raised.

4.4 Data Analysis Through AWS Crawler, Glue Data Catalog and Athena

The data stored in Amazon S3 bucket is scanned by AWS crawler which builds a schema by itself and the queries are run on the data using AWS Athena to generate results.

In Fig. 5, the flow diagram is related to analyze the straming data through AWS Crawler, Glue data catalog and AWS Athena.

The following steps are included in this module:

1. Data from S3 is scanned using AWS Crawler
2. Queries are performed using AWS Athena

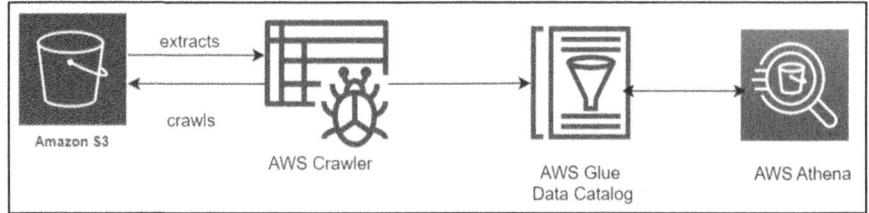

Fig. 5. Flow diagram to analyze the data

4.5   Comparing Results and Analyse the Visualizations in Power BI Dashboard

The results generated in AWS Athena are then gathered in power BI to generate visualization.

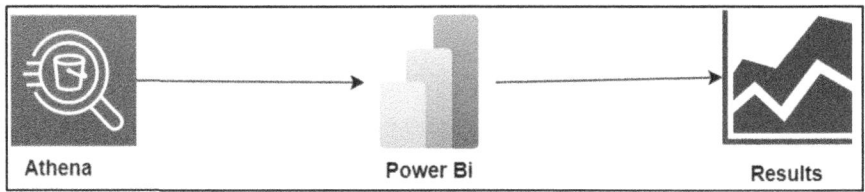

Fig. 6. Flow diagram to compare results

In Fig. 6, the flow diagram is related to compare the results and analyze the different visualizations for case studies on stock-market data in Power BI desktop.

Firstly, the data from athena is exported in a.CSV format and then imported in PowerBI to visualize the generated results.

Power BI Desktop.   Power BI desktop is the primary composing and documenting tool for Power BI. It is used by developers and advanced users to build entirely new models and reports from beginning. Costs: Null.

Power BI Service.   Power BI's data models, reports, and dashboards are hosted using an online cloud service (SaaS).In the cloud, management, sharing, and collaboration take place.

Steps involved in this module are:

1. Exporting results from Athena result to Power BI
2. Analyze visualizations

The results generated from running queries on the stock market data in AWS Athena are imported into PowerBI for further analysis and visualization. This is achieved by utilizing the "Get Data" feature in PowerBI, which allows you to connect to various data sources and retrieve the information you need for your analysis.

The comparison of the stock market data in PowerBI is done by using line charts and stacked column charts. These visualizations allow you to see trends and patterns over time, as well as compare the relative performance of different stocks or market segments. The line chart displays data points connected by a line, making it easy to see how the data changes over time. The stacked column chart, on the other hand, displays data in columns, with each column representing a different category. By stacking the columns on top of each other, the chart allows you to see how the categories contribute to the overall total. By using these chart types, you can get a clear and effective visual representation of the stock market data and compare the performance of different stocks or segments.

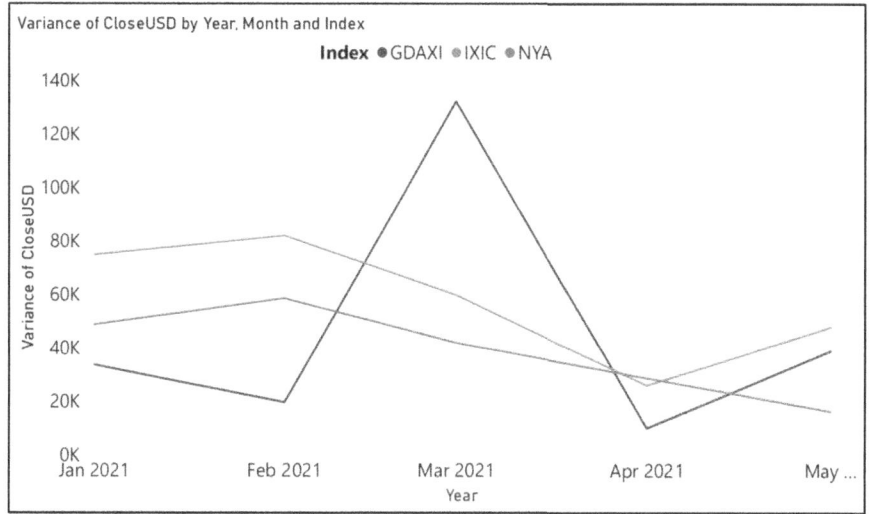

Fig. 7. Line chart

The Fig. 7 is a line chart which describes the variance of closing stocks of three topics in USD from the period January 2021 to May 2021. A drastic change has been observed in the GDAXI. It rose dramatically from February and March 2021 and at the same time the value fell in the same way. The NYA and IXIC stocks followed a similar pattern. They experienced a slight increase until February 2021 and reached a certain point and then decreased gradually. Although, the IXIC had seen a growth at the end of the period (Fig. 8).

The above figure is a clustered column chart that illustrates the variance of closed stocks in particular to the month of May 2021. It compares the stocks on a daily basis where GDAXI remained on top with around 19k as its peak. The least value stock was IXIC which closed around 13–14k every day. And the NYA stock was closed around 16–17k.

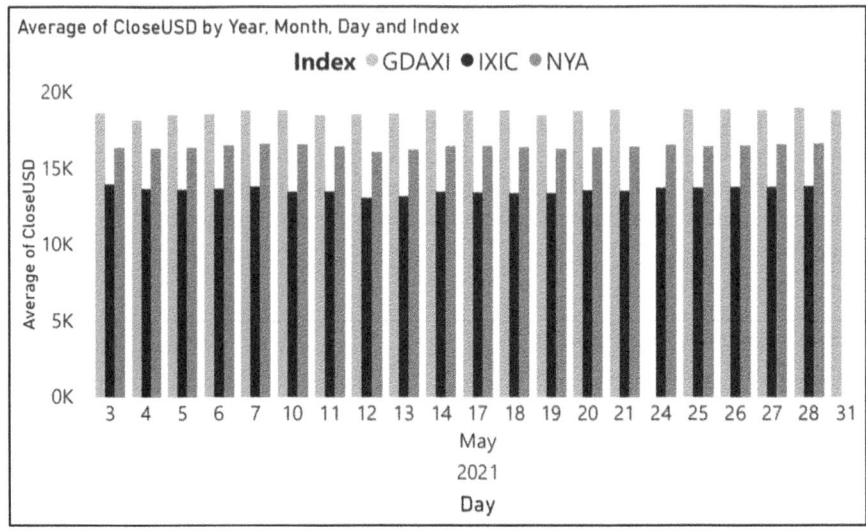

Fig. 8. Clustered Column chart

5 Conclusion

The process of analyzing real-time stock market data is not limited to just one sector, and is utilized in several industries such as financial, gaming, and e-commerce. The first step involves extracting the data using Python where the Kafka framework helps to capture it in real-time. The data is then transferred from the Kafka framework to an Amazon S3 bucket for storage and further processing. The AWS Glue data catalog is used to scan the data stored in the S3 bucket, building a schema for the data to be stored in a structured format. AWS Athena is utilized for querying the data stored in the AWS Glue data catalog, making it fast and efficient to gain insights into the stock market. Finally, the data is visualized in PowerBI to provide a clear and effective representation of the data, making it easier to understand and analyze trends and patterns in the stock market. This process demonstrates the use of technology tools to effectively analyze real-time stock market data in various industries, providing organizations with meaningful insights to make informed decisions. Further extensions of this project can be made by incorporating machine learning algorithms to make predictions and automate the analysis process.

References

1. Behrend, A., Dorau, C., Manthey, R.: TinTO: a tool for the view-based analysis of streams of stock market data. In: Kotagiri, R., Krishna, P.R., Mohania, M., Nantajeewarawat, E. (eds.) Advances in Databases: Concepts, Systems and Applications. DASFAA 2007. Lecture Notes in Computer Science, vol. 4443. Springer, Heidelberg (2007). https://doi.org/10.1007/978-3-540-71703-4_111
2. Modi, P., Shah, S., Shah, H.: Big data analysis in stock market prediction. Int. J. Eng. Res. Technol. (IJERT) **8**(10) (2019). http://www.ijert.orgISSN. 2278-0181 IJERTV8IS100224. www.ijert.org

3. Joshi, P., Afli, H.: Affective analytics and visualization for ensemble event-driven stock market forecasting. In: Reis, T., Bornschlegl, M.X., Angelini, M., Hemmje, M.L. (eds.) Advanced Visual Interfaces. Supporting Artificial Intelligence and Big Data Applications. AVI-BDA ITAVIS 2020. Lecture Notes in Computer Science, vol. 12585. Springer, Cham (2021). https://doi.org/10.1007/978-3-030-68007-7_3

4. Belacel, N., Richard, R., Rangavajjala, D.P., Adhaduk, R.: Online anomaly detection for streaming data implemented on top of Kafka, scikit-multiflow and river. In: Arai, K. (ed.) Proceedings of the Future Technologies Conference (FTC) 2021, Volume 3. FTC 2021. Lecture Notes in Networks and Systems, vol. 360. Springer, Cham (2022). https://doi.org/10.1007/978-3-030-89912-7_63

5. Deshpande, K., Rao, M.: An open-source framework unifying stream and batch processing. In: Smys, S., Balas, V.E., Palanisamy, R. (eds.) Inventive Computation and Information Technologies. Lecture Notes in Networks and Systems, vol. 336. Springer, Singapore (2022). https://doi.org/10.1007/978-981-16-6723-7_45

6. Kulkarni, M., Jadha, A., Dhingra, D.: Time series data analysis for stock market prediction. In: Proceedings of the International Conference on Innovative Computing & Communications (ICICC) (2020). SSRN https://ssrn.com/abstract=3563111 or https://doi.org/10.2139/ssrn.3563111

7. Attigeri, G.V., Manohara Pai, M.M., Pai, R.M., Nayak, A.: Stock market prediction: a big data approach. In: TENCON 2015 - 2015 IEEE Region 10 Conference, pp. 1–5 (2015). https://doi.org/10.1109/TENCON.2015.7373006

8. Mishra, A., Kumar, G.: Big data analytics on AWS cloud. Int. J. Eng. Res. Technol. (IJERT) **10**(04) (2021)

9. Gosswami, T., Saha, S., Hasan, M.: Stock market data analysis and future stock prediction using neural network. Int. J. Comput. Sci. Inf. Secur. **16**, 6 (2018)

10. Lima, L., Portela, F., Santos, M.F., Abelha, A., Machado, J.: New Contributions in Information Systems and Technologies, vol. 353 (2015). ISBN 978-3-319-16485-4

11. Mobility based on Apache Kafka. In: Obaidat, M.S., Oren, T., Rango, F.D. (eds.) Simulation and Modeling Methodologies, Technologies and Applications. SIMULTECH 2020. Lecture Notes in Networks and Systems, vol. 306. Springer, Cham (2020). https://doi.org/10.1007/978-3-030-84811-8_2

12. Srivastava, A., Johri, P., Kumar, A., Gaur, N., Jain, R.: Real-time data pipeline in confluent Kafka and Mule4 ESB with ActiveMQ. In: Prateek, M., Singh, T.P., Choudhury, T., Pandey, H.M., Gia Nhu, N. (eds.) Proceedings of International Conference on Machine Intelligence and Data Science Applications. Algorithms for Intelligent Systems. Springer, Singapore (2021). https://doi.org/10.1007/978-981-33-4087-9_54s

13. Cañibano, R., Rozas, C., Orlandi, C., Balladini, J.: Data management optimization in a real-time big data analysis system for intensive care. In: Rucci, E., Naiouf, M., Chichizola, F., De Giusti, L. (eds) Cloud Computing, Big Data & Emerging Topics. JCC-BD&ET 2020. Communications in Computer and Information Science, vol. 1291. Springer, Cham (2020). https://doi.org/10.1007/978-3-030-61218-4_7

14. Su, F., et al.: A survey on big data analytics technologies. In: Long, K., Leung, V., Zhang, H., Feng, Z., Li, Y., Zhang, Z. (eds.) 5G for Future Wireless Networks. 5GWN 2017. Lecture Notes of the Institute for Computer Sciences, Social Informatics and Telecommunications Engineering, vol. 211. Springer, Cham (2018). https://doi.org/10.1007/978-3-319-72823-0_34

15. Mary Auxilia, P.A., Alvarado-Tolentino, J., Gonzales-Yanac, T., Huaman-Osorio, A., Durga, S., Hasan, M.F.: The dynamic role of big data analytics in learning and development and its impact on risk analysis in stock market. In: Yadav, S., Haleem, A., Arora, P.K., Kumar, H. (eds.) Proceedings of Second International Conference in Mechanical and Energy Technology.

Smart Innovation, Systems and Technologies, vol. 290. Springer, Singapore (2023). https://doi.org/10.1007/978-981-19-0108-9_17

16. Yasmin, A., Kamalakkannan, S.: Analytical performance in data lake storage of big data analytics by databricks delta lake for stock market analysis. In: Smys, S., Lafata, P., Palanisamy, R., Kamel, K.A. (eds.) Computer Networks and Inventive Communication Technologies. Lecture Notes on Data Engineering and Communications Technologies, vol. 141. Springer, Singapore (2023). https://doi.org/10.1007/978-981-19-3035-5_17

17. Kalra, S., Gupta, S., Prasad, J.S.: Predicting trends of stock market using SVM: a big data analytics approach. In: Batra, U., Roy, N., Panda, B. (eds.) Data Science and Analytics. REDSET 2019. Communications in Computer and Information Science, vol. 1229. Springer, Singapore (2020). https://doi.org/10.1007/978-981-15-5827-6_4

18. Lima, L., Portela, F., Santos, M.F., Abelha, A., Machado, J.: Big data for stock market by means of mining techniques. In: Rocha, A., Correia, A., Costanzo, S., Reis, L. (eds.) New Contributions in Information Systems and Technologies. Advances in Intelligent Systems and Computing, vol. 353. Springer, Cham (2015). https://doi.org/10.1007/978-3-319-16486-1_67

19. Bellini, V., Guidolin, M., Pedio, M.: Can big data help to predict conditional stock market volatility? An application to brexit. In: Machine Learning, Optimization, and Data Science. LOD 2020. Lecture Notes in Computer Science, vol. 12565. Springer, Cham (2020). https://doi.org/10.1007/978-3-030-64583-0_36

20. Chen, X.: The application value of big data analysis technology in financial forecasting. In: Macintyre, J., Zhao, J., Ma, X. (eds.) The 2021 International Conference on Machine Learning and Big Data Analytics for IoT Security and Privacy. SPIoT 2021. Lecture Notes on Data Engineering and Communications Technologies, vol. 97. Springer, Cham (2022). https://doi.org/10.1007/978-3-030-89508-2_112

21. Sharma, K., Bhalla, R.: Stock market prediction techniques: a review paper. In: Luhach, A.K., Poonia, R.C., Gao, XZ., Singh Jat, D. (eds.) Second International Conference on Sustainable Technologies for Computational Intelligence. Advances in Intelligent Systems and Computing, vol. 1235. Springer, Singapore (2022).. https://doi.org/10.1007/978-981-16-4641-6_15

22. Canaj, E., Xhuvani, A.: Big data in cloud computing: a review of key technologies and open issues. In: Barolli, L., Xhafa, F., Javaid, N., Spaho, E., Kolici, V. (eds.) Advances in Internet, Data & Web Technologies. EIDWT 2018. Lecture Notes on Data Engineering and Communications Technologies, vol. 17. Springer, Cham (2018). https://doi.org/10.1007/978-3-319-75928-9_45

23. Wu, J., et al.: Stock cloud computing platform: architecture and prototype systems. In: Web-Age Information Management. WAIM 2013. Lecture Notes in Computer Science, vol. 7901. Springer, Heidelberg (2013). https://doi.org/10.1007/978-3-642-39527-7_35

24. Das, M., Dash, R.: Role of cloud computing for big data: a review. In: Mishra, D., Buyya, R., Mohapatra, P., Patnaik, S. (eds.) Intelligent and Cloud Computing. Smart Innovation, Systems and Technologies, vol. 153. Springer, Singapore (2021). https://doi.org/10.1007/978-981-15-6202-0_18

25. Soderi, M., Kamath, V., Breslin, J.G.: Toward an API-driven infinite cyber-screen for custom real-time display of big data streams. In: 2022 IEEE International Conference on Smart Computing (SMARTCOMP), pp. 153–155 (2022). https://doi.org/10.1109/SMARTCOMP55677.2022.00036

26. Shree, R., Choudhury, T., Gupta, S.C., Kumar, P.: KAFKA: the modern platform for data management and analysis in big data domain. In: 2017 2nd International Conference on Telecommunication and Networks (TEL-NET), pp. 1–5 (2017). https://doi.org/10.1109/TEL-NET.2017.8343593

27. Hiraman, B.R., Viresh, C.M., Abhijeet K.C.: A study of Apache Kafka in big data stream processing. In: 2018 International Conference on Information, Communication, Engineering and Technology (ICICET), pp. 1–3 (2018). https://doi.org/10.1109/ICICET.2018.8533771
28. Elgendy, N., Elragal, A.: Big data analytics: a literature review paper. Lecture Notes in Computer Science, vol.8557, pp. 214–227 (2014). https://doi.org/10.1007/978-3-319-08976-8_16
29. Kato, K., Takefusa, A., Nakada, H., Oguchi, M.: A study of a scalable distributed stream processing infrastructure using ray and Apache Kafka. In: 2018 IEEE International Conference on Big Data (Big Data), pp. 5351–5353 (2018). https://doi.org/10.1109/BigData.2018.8622415
30. Umadevi, K.S., Gaonka, A., Kulkarni, R., Kannan, R.J.: Analysis of stock market using streaming data framework. In: 2018 International Conference on Advances in Computing, Communications and Informatics (ICACCI), pp. 1388–1390 (2018). https://doi.org/10.1109/ICACCI.2018.8554561
31. Wu, H., Shang, Z., Wolter, K.: Learning to reliably deliver streaming data with Apache Kafka. In: 2020 50th Annual IEEE/IFIP international Conference on Dependable Systems and Networks (DSN), pp. 564–571 (2020). https://doi.org/10.1109/DSN48063.2020.00068
32. Drohobytskiy, Y., Brevus, V., Skorenkyy, Y.: Spark structured streaming: customizing kafka stream processing. In: 2020 IEEE Third International Conference on Data Stream Mining & Processing (DSMP), pp. 296–299 (2020). https://doi.org/10.1109/DSMP47368.2020.9204304
33. Apache Kafka by Nishant Garg. Packt Publishing (2013). https://dl.acm.org/doi/10.5555/2588385#cited-by-sec
34. Narkhede, N., Shapira, G., Palino, T.: Kafka: The Definitive Guide Real-Time Data and Stream Processing at Scale, 1st. edn. O'Reilly Media, Inc. (2017)
35. Rao, P.S., Srinivas, K., Mohan, A.K.: A survey on stock market prediction using machine learning techniques (2020). https://doi.org/10.1007/978-981-15-1420-3_101
36. Mohamed, N., Al-Jaroodi, J.: Real-time big data analytics: applications and challenges. In: Proceedings of the 2014 International Conference on High Performance Computing and Simulation, HPCS 2014 (2014). https://doi.org/10.1109/HPCSim.2014.6903700
37. Strader, T.J., Rozycki, J.J., Root, T.H., Huang, Y.-H.: Machine learning stock market prediction studies: review and research directions. J. Int. Technol. Inf. Manag. **28**(4), Article no. 3 (2020)

An Algorithm for Estimating Corrected-QT Interval in the Electrocardiogram

Amar Bahadur Biswakarma$^{(\boxtimes)}$, Jagdeep Rahul, and Kurmendra

Department of Electronics and Communication Engineering, Rajiv Gandhi University,
Arunachal Pradesh, India
amar.biswakarma2023@gmail.com

Abstract. The interpretation and estimation of the QT interval in the ECG can help to detect the risky malignant ventricular arrhythmia and prevent the sudden death caused by the unusual QT segment. In an ECG signal, a QT interval in one cardiac cycle is defined as a segment before the beginning of Q-wave to the end of T-wave. The proposed technique consists of a pre-processing stage followed by baseline identification for better R-peak detection. Thereafter, QRS is detected by implementing a decision based rule on the root mean square value of the signal. A RR interval was used to detect T-wave and T-wave offset points after the S-wave. Similarly, P-onset points were detected prior to the Q-wave. After that, the corrected QTc and QT interval were calculated between Q and T-waves. The proposed method is applied to the standard QT-database, and examines the short and prolonged QT intervals using Bazett, Fridericia, Framingham, and Hodges formulae to explore QT-RR co-variability for the stratification of cardiac diseases. Among the four different formulas, the Bazett method yields an overall average QT interval of 517 ms, which is consistent and accurate with different heart rates.

Keywords: ECG · QRS complex · QT interval and QTc (corrected Q-T interval)

1 Introduction

ECG is a widely used diagnostic electronic device that depicts the overall electrical function of a self-rhythmic heart. Various cardiovascular diseases (CVDs) can be identified by interpreting the ECG. The waveforms of an ECG signal are associated with the physiological events in the heart, which are denoted by P, Q, R, S, and T waves [1]. The P wave represents the depolarization of the atrium, the QRS represents ventricular depolarization, and the T wave illustrates the repolarization of the ventricles, respectively [2]. Most cardiac problems are diagnosed by analyzing the characteristics of these waves. Any changes in the morphology of these waves may lead to a cardiac disorder. Many physiological segments that are derived from these waves play a very important role in cardiac arrhythmia detection. These segments are the PR interval, RR interval, QR interval, QT interval, and ST segments. In these physiological segments, the QT interval is the most widely used segment for cardiac arrhythmia detection in the early stages [19–21]. The QT interval, which comprises the QRS complex and T wave

© The Author(s), under exclusive license to Springer Nature Switzerland AG 2025
M. Gupta et al. (Eds.): MISS 2023, CCIS 1952, pp. 354–364, 2025.
https://doi.org/10.1007/978-3-031-69115-7_24

period between the beginning of ventricular depolarization and the ending of ventricular repolarization, is, measured [3]. The QT interval duration differ in accordance with the rate of pulse. The QT interval duration depends on age, sex, and heart rate, ranging from 0.36 to 0.44 s, and must not be greater than half the distance between consecutive R waves. Prolonged QT intervals show that the relative refractory time is longer, which may increase the risk of a serious cardiac disorder. An extended QTc interval in the ECG is always a danger sign for torsade de pointes [4]. Torsade determines the presence of ventricular tachycardia, which is characterized by QRS complex fluctuation (twisting) around the ECG isoelectric line. A short QT interval could result from hyperkalemia in the ECG. Too long QT/QTc intervals create an electrophysiological environment that favors torsade de pointes, which can result in polymorphic ventricular tachyarrhythmia. If the heart rate is 70 beats per minute, then QT can be expected to be less than or equal to 0.40 s. Similarly, for 80 bpm pulse rate, QT is assumed to be lower than equal to 0.38 s [5, 17, 18]. The prolongation in the QTc may cause a premature action potential at the last phases of depolarization, which can increase the possibility of originating ventricular arrhythmia or fatal fibrillation. The excessively prolonged QTc can be observed in females, those with high systolic blood pressure, and in old patients. There could be various factors of prolonged QT intervals, such as genetic agencies associated to gene variation, adverse drug reactions like haloperidol, ziprasidone, and methadone, and past experience of cardiac attacks. In addition to this, a higher concentration of alcohol in the blood can prolong the QT interval. Antibiotic use is also linked to QT prolongation, and pathological conditions such as low thyroid gland function can cause QT prolongation [6]. Acute hypocalcemia causes prolong QT interval that leads to ventricular dysrhythmias. Typically, there is no such effective evidence that can prolong the QTc interval, which has been linked to a rise in mortality [27, 28]. The proposed method was used to detect P, QRS, and T waves in the ECG signal, followed by preprocessing. Thereafter, the offset and onset of these waves were identified to estimate the corrected-QT interval.

2 Literature Review

Many studies have been conducted to determine whether a prolonged QT interval can predict cardiovascular disease and prevent premature death. The study on rheumatoid arthritis and inflammatory arthritis shows that increase in 50-ms duration of QTc interval can lead to death [7]. Furthermore, patients with a longer QTc interval (>424 ms) were found to have a high mortality rate than the patient's having shorter QTc interval. The patient's suffering from diabetes may also increase the risk of cardiovascular diseases. In general, a QT interval below 400 ms is considered normal; the interval between 440 ms and 460 ms in men and 440 ms and 470 ms in women is assumed to be marginal. Since, QT interval is reciprocally proportional to heart rate, so different heart rate correction formulae are employed to turn the measured QT interval into a variable known as corrected QT interval (QTc), which is assumed to be independent of heart rate [7]. To diagnose a prolonged QT interval, it is important to determine whether the T wave extends beyond the midpoint of the RR interval or not. If the T-wave offset crosses the midpoint of the RR interval, then it is referred to as prolonged QT. The techniques based on machine learning, wavelet transform, mathematical modeling, and Gibbs sampling for the estimation of the QT interval are computationally complex and time consuming [8]. [12] developed a highly effective automated method using a continuous wavelet transform to detect QT intervals in different patterns of ECG signal. ECG delineation for QT interval analysis using an unsupervised learning method was employed in [16], where statistical parameters such as mean, SD, and TRP gave the most accurate estimation of the QT interval. Using a 10-s ECG signal, the QT interval variability index analysis was performed to assess the risk factor associated with the QT interval [24]. For conclusive remarks, a comparison of the variability index and standard deviation (SD) of the QT interval were performed in this study. The SD of the QT interval shows a better and more effective way to assess the risk factor [24]. The connection between the QT interval and the variability of heart rate (HRV) using a window of 30 s on ECG data was proposed in [25]. A systematic review was done in [26] to identify the sources of QT interval variability that can be implemented in the monitoring of ECG in clinical practices. QTc measurement by machine using Bazett formula exceeds the length of QTc interval and number of patients with prolonged QTc as compared to various formulas and QT-nomogram. Therefore, manually QTc measurement and correction by Fridericia formula or QT-Nomogram is precedent to improve antipsychotic therapies [27]. In [28] Framingham's formula yields higher proportion in the judgement of long or short QT interval, while Fridercia's formula tends to minimize them. Bazett's formula permits intermediate fraction of abnormal QT which can be used for ECG evaluation for strong dynamic football athletes. In [29] the method of Bland-Altman intimates better agreement between Hodge equation and simple rule as compared to other formulas. In [30] Bazett's formula is not effective for the evaluation of drug influenced QTc and ΔQTc at 60 bpm heart rate distant. While, combination of Hodges formula and Nomogram method found to be more relevant in estimating ΔQTc. Therefore, a simple and efficient approach for estimating QT intervals must be created. The approach for estimating the QT interval utilizing simple pre-processing and a root mean square (rms) based threshold for fiducial point detection is proposed in this paper. To analyse and validate the proposed approach, it was applied to a standard QT database and four different techniques were used to calculate the corrected QT interval. Therefore, a simple and efficient approach for estimating QT intervals must be created. This paper proposes an approach for estimating the QT interval using simple pre-processing and a root mean square (rms)-based threshold for fiducial point detection. To analyse and validate the proposed approach, it was applied to a standard QT database, and four different techniques were used to calculate the corrected QT interval.

3   Materials and Method

The Q-T database is compiled from subjects with various cardiac arrhythmias who are available on physionet.org [22, 23]. Each record is 15 min in duration with a 12-bit resolution and a normal input range of 20 mV, which is sampled at 250 samples per second [9]. The header file consists of detailed information like patient history, type and variety of electrocardiogram leads, clinical findings, and recording equipment details. The QT database is downloaded from physionet.org and converted into a readable format (.mat file). The signal is clearly readable in each lead, and the signal of lead-II is employed in this technique for the analysis. The methodology for QT-interval estimation and analysis was discussed in this section. The QT interval is inversely proportional to the heart rate, and the methodology's goal is to establish a system for detecting prolonged and shortened QT intervals that is reliable and quick. First, the ECG signal is pre-processed by a 2-stage median filter, and a threshold for R-peak detection was computed in relation to the baseline. Then, using present windows W1 and W2, peaks of P and T waves were recognized. The differential equations were then used to determine the Q-point and T-wave offset points. Finally, QT interval was calculated to estimate the QT-corrected interval for further segment analysis. Figure 1 depicts the flow diagram of the proposed technique.

3.1 Preprocessing

It is important to remove the noise from the raw ECG signal before performing the analysis. During the preprocessing stage, this method used a 2-stage median filter and smoothing filter to get rid of noise in the ECG [10].

(a) The first median filter was used to pass the raw ECG signal. The output of the filter was processed using a window length equal to the sampling frequency.
(b) The output of the first median filter was applied to the next median filter. The size of the window used to estimate the output of the next-step median filter using half the sampling frequency.
(c) The drift-free raw ECG signal is produced after eliminating the output of the next-step median filter from the raw ECG signal.

The drift-free raw ECG signal was smoothed using the least-squares smoothing filter. The least-squares filter output is superior to the average FIR filter because it retains the ECG signal's peak amplitude, and shape. While smoothing the signal, the FIR filter removes the major characteristics of the ECG signal as well as noise. The Savitzky-Golay filter is another name for the least-squares smoothening filter. This filter employs the notion of polynomial fitting to estimate single data points inside a preset window across the input data set. The order of the filter coefficients and window length were taken from [3].

3.2 Threshold Estimation and R-peak Detection

The R-peak threshold was computed using the baseline and RMS of the ECG signal. The significance of identifying the ECG baseline cannot be emphasized. The baseline, which is given in [11], serves as a point of reference for identifying both upward and descending QRS complex. The threshold value for identifying upward QRS complexes is the sum of the baseline and RMS values. Similarly, to determine the threshold for the descending QRS complex, the RMS value is subtracted from the baseline value. When the R-peak exceeds the higher threshold, it is called upward R-peak. If the R-peak is less than or equal to the lower threshold, it is considered a descending R-peak.

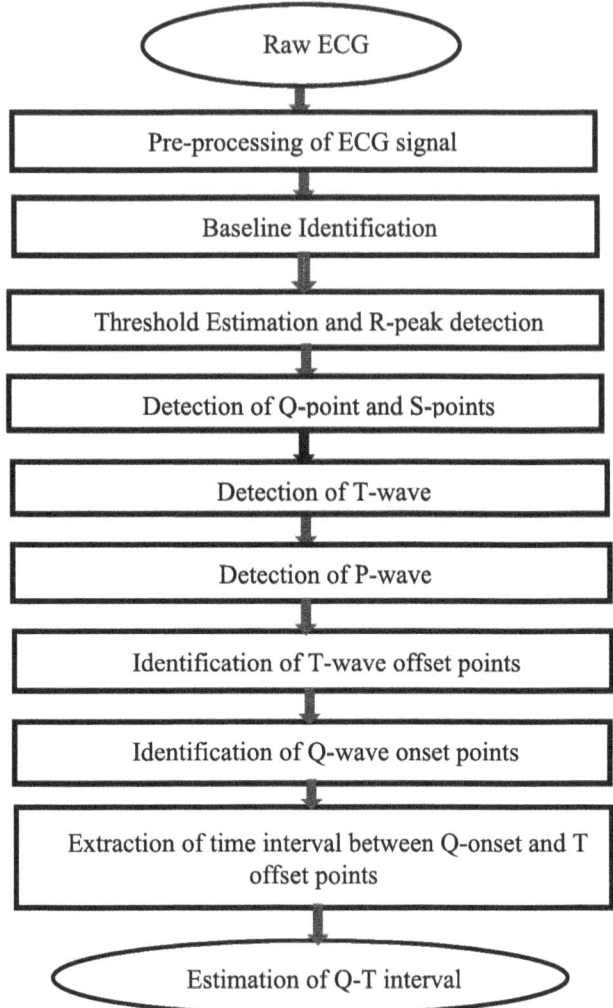

Fig. 1. Flow chart of Q-T interval estimation

3.3   Identification of Q and S Points Using Window (W1)

R-peaks were detected in the last section. After identifying the R-peaks, a window based on the RR interval must be assumed for the detection of Q and S-wave locations. In this section, the QRS complex was calculated using a window size of one-eighth of the RR interval. The Q-wave point appears just before the R-peak and the S-wave point arrives immediate after the R-peak within a specified range of an ECG cardiac cycle. The minima function is used to detect Q-waves before the R-peak within a predefined window (W1). The minima function is used to detect S-waves in the preset window (W1) immediately following the R-peak. The minima function was used to detect downward QRS complexes in the ECG signal.

3.4 Detection of P-peak Using Window (W2)

After the detection of QRS complex, P wave needs to be detected. A window (W2) of size one-fourth of the RR interval is considered for searching the P-peaks within the predetermined window. P-waves appear before the present cardiac cycle of Q-wave and after the preceding cardiac cycle of T-wave in a typical ECG record. For P-peak detection, predefined window (W2) is used to search the maxima in that region. The maxima point in that region is considered as a P-peak.

3.5 Detection of T-peak Using Window (W3)

The T-peak detect is a very important step while estimation of the corrected-QT interval. A window (W3) of size one-half of the RR intervals is considered for T-peak detection in the ECG signal. In this technique, we have detected both upright and inverted T-peak by utilizing the concept of baseline as discussed in [11]. While searching in the chosen window (W3), both maxima and minima functions were utilized to find the T-peak. Both upward and inverted T-peaks were detected in this step.

3.6 Identification of T-onset and T-offset Points

After detecting the fiducial points P, QRS, and T peaks in the previous steps, the boundaries of the T-wave need to be detected for QT interval estimation. For detecting the T-wave offset, the first inflection point was considered as T-onset point using differential equation before the T-peaks. Similarly, the T-offset is identified from the first inflection point after the T-peak using differential equation.

3.7 Extraction of Interval Between Q-onset and T-offset Points

Q-onset and T-offset point are used to calculate QT interval in ECG signal. It is the difference in time between the Q-onset point and T-offset point locations (milliseconds). Bazett, Fridericia, Framingham, and Hodges formulas are used to determine the corrected-QT interval. For the diagnosis of cardiac problems, the corrected-QT interval is divided into two categories: long QT interval and short QT interval.

4 Results and Discussion

The proposed method is used to estimate the corrected QT interval in an ECG signal. This technique is used on QT database to demonstrate the performance validation. A two-stage median filter and a SG filter were used to pre-process the raw ECG data. The method was implemented in two steps; in the first step, P and T-peaks detection were done followed by QRS complex detection. In the second step, other fiducial points such as offset and onset of P and T-waves were detected using differential equation. The raw ECG signal is depicted in Fig. 2 which contains the baseline drift and power-line interference. The output from first and second median filters is shown in Fig. 3 which reflects baseline and drift free ECG signals.

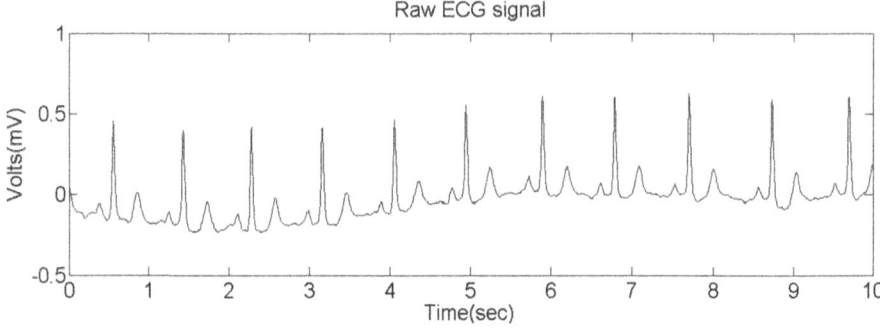

Fig. 2. Raw ECG signal

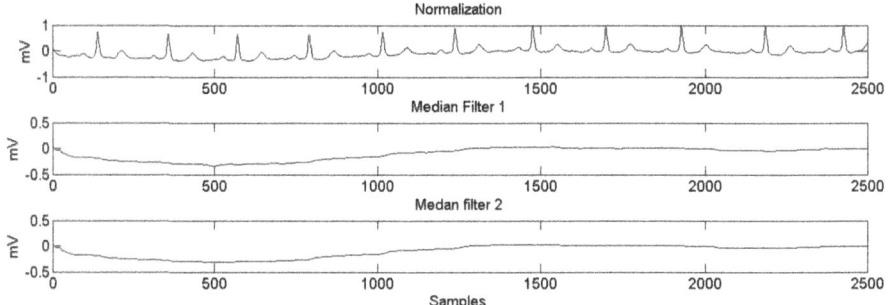

Fig. 3. Two stage median filtering process

To get the drift-free ECG signal, the changing baseline is retrieved from the output of second median filter and the subtraction from the raw ECG signal. The drift-free ECG data is then smoothed using least-squares to eliminate any signal outliers. Figure 4 shows the result of the least-squares filter. Following pre-processing, the resulting signal is squared to enhance the R-peaks while decreasing P and T-waves. Thereafter, the threshold value for R-peak detection was calculated using RMS and baseline value respectively. The denoised and squared ECG signal after the pre-processing is illustrated in Fig. 4.

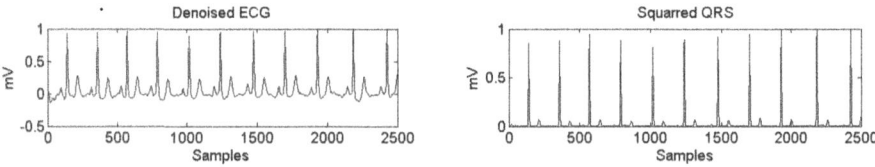

Fig. 4. De-noised and squared ECG signal

The positive and negative QRS complexes were identified using a threshold value and a simple decision procedure after the preprocessing of the ECG signal. If the peak value exceeds the threshold value, R-peak is detected. After that, minimum values were searched before the R-peak for Q-wave detection and after the R-wave for S-wave detection. P and T peaks were identified using the pre-determined window length and maxima/minima functions, The boundaries of a T-wave are determined using the differential equation before and after the T-peak. The detected fiducial points P, QRS, T, and T-offset are shown in Fig. 5.

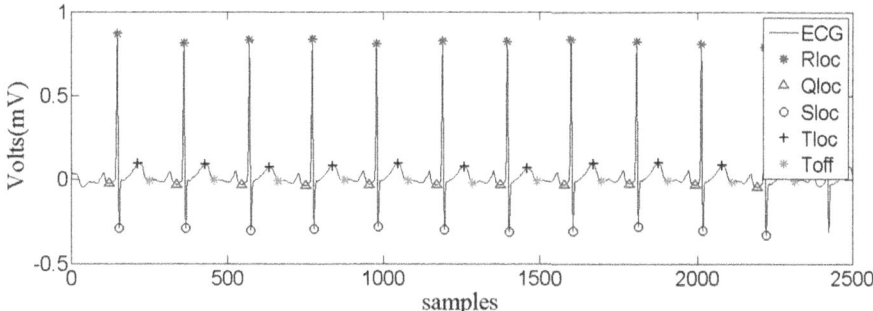

Fig. 5. Location of Q, R, S, T peaks and T-offset points

After Q-peak and T-offset detection, QT-interval is easily estimated by simply subtracting Q-location from T-offset point. The QT-interval is shown in Fig. 6.

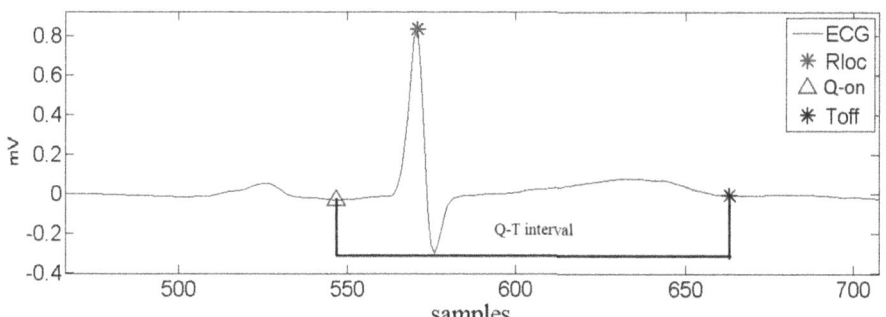

Fig. 6. Q-T interval

The detected QT interval shown in Fig. 6 was further used for estimating the corrected QT interval with the help of four different formulae, namely Bazett, Fridericia, Framingham, and Hodge. These formulas use the QT interval and RR interval to estimate the correct QT interval. The corrected QT interval was estimated using long-term ECG records and is helpful for early cardiac disorder detection.

QT-interval is a crucial parameter for the diagnosis of cardiac arrhythmia, and a change in its duration might incline patients to cardiac sudden arrest. Because the corrected QT interval is inversely proportional to heart rate and is generally rectifying relative to heart rate, its interpretation becomes contradictory to heart rate variability

Table 1. Performance evaluation of corrected QT (QTc) interval

Sl. No	Data	HR (bpm)	Mean (ms)	Median (ms)	Mode (ms)	Var (ms)	SD (ms)	QTcB (ms)	QTcFri (ms)	QTcFram (ms)	Hodge (ms)	Result
1	Sel30	66	530	528	528	157	17.54	559	549	530	540	Prolonged
2	Sel33	70	545	532	534	160	18.23	560	545	535	542	Prolonged
3	Sel34	64	493	500	504	152	12.32	509	504	493	500	Prolonged
4	Sel36	62	488	495	500	148	13.35	504	501	490	489	Prolonged
5	Sel37	75	550	561	489	248	21.22	550	552	551	520	Prolonged
6	Sel39	87	395	370	472	3615	60.13	430	448	395	442	Prolonged
7	Sel42	55	640	645	631	198	13.98	578	591	630	600	Prolonged
8	Sel44	104	431	434	460	810	28.47	335	340	376	339	Shortened
9	Sel46	60	555	564	484	2257	47.51	556	556	555	555	Prolonged
10	Sel47	62	525	532	526	157	17.50	552	545	525	542	Prolonged
11	Sel48	86	394	372	368	1854	43.06	474	446	394	440	Normal
12	Sel49	55	646	664	528	2937	54.19	624	632	646	638	Prolonged
13	Sel50	45	704	694	608	7893	88.84	611	641	704	678	Prolonged
14	Sel51	50	635	636	628	194	13.93	581	598	635	617	Prolonged
15	Sel100	110	376	386	388	473	21.76	338	411	342	336	Shortened
Overall average								517	523	520	518	

[12, 13]. Different strategies for correction of QT-intervals have been proposed in clinical practice, but the commonly used Bazett formula is found to be the most consistent in producing the result. The four different QTc calculation methods that have been considered are Bazett [14], Fridericia [15], Framingham [16], and Hodges [15]. Table 1 represents all four methods of QT correction and the statistics associated with them. The Bazett formula demonstrated the narrow range of values across the dataset. In all methods, the minimum corrected QTcB interval is 335 ms, the maximum corrected QTcB interval is 624 ms, and the difference is achieved to be 289 ms. The relative differences found for QTcFri, QTcFra, and QTcH are 301, 362, and 368 ms, respectively. We found that, among the four evaluated methods, the Bazett method best fits the relation between the uncorrected QT-interval and the heart rate. We found that using the Bazett method gave better accuracy and simplicity. Therefore, using the Bazett method as a standard correction method makes the QTc value more consistent. Our analysis shows that its use is appropriate as the current best option. The mean is an arithmetic average of all Q-T intervals. The median is the midpoint of the Q-T interval when all Q-T intervals are sorted. Mode represents the Q-T interval that occurs with the greatest frequency. Variance is the sum of the squares of the difference between each Q-T interval and the mean divided by the number of Q-T intervals, and standard deviation is known as the square root of the variance. The overall average of the corrected Q-T interval from all methods was calculated, and it was found that the Bazett method produced the best results as compared with the other methods considered in this study. Other statistical parameters were also calculated from the QT interval to demonstrate the validity of the proposed method.

5   Conclusion

In this paper, we propose a method for determining the time interval between a Q-wave and a T-wave offset using an approach for Q-T interval estimation. The proposed method utilised a two-stage median filter for baseline wander removal and a least-square smoothing filter for signal smoothing. Thereafter, P, QRS, and T-waves were detected using thresholding and a simple decision-based rule. After that, Q and T-wave offset points were identified with the help of a differential equation and a predefined window. We estimated the corrected Q-T interval using four distinct methods: Bazett, Fridericia, Framingham, and Hodges for predicting extended or short Q-T intervals in this methodology. This method was tested on a QT database, and we found that Bazett techniques for calculating the corrected QT interval are more accurate and appropriate as compared to other commonly used techniques.

References

1. Rahul, J., Marpe, S., Lakhan, D.S.: An overview on biomedical signal analysis. Int. J. Recent. Technol. Eng. **7**, 206–209 (2019)
2. Rahul, J., Sora, M., Dev Sharma, L.: Dynamic thresholding based efficient QRS complex detection with low computational overhead. Biomed. Signal Process. Control **67**, 102519 (2021)
3. Rahul, J., Sora, M., Dev Sharma, L.: A novel and lightweight P, QRS, and T peaks detector using adaptive thresholding and template waveform. Comput. Biol. Med.. Biol. Med. **132**, 104307 (2021)
4. Rahul, J., Marpe, S.: A novel adaptive window based technique for T wave detection and delineation in the ECG. Bio-Algorithms Med-Syst. 16(1) (2020)
5. Goldenberg, I., Moss, A.J., Zareba, W.: QT interval: how to measure it and what is "normal." J. Cardiovasc. Electrophysiol.Cardiovasc. Electrophysiol. **17**(3), 333–336 (2006). https://doi.org/10.1111/j.1540-8167.2006.00408.x
6. Vandael, E., et al.: Risk factors for QTc-prolongation: systematic review of the evidence. Int. J. Clin. Pharm. **39**(1), 16–25 (2017). https://doi.org/10.1007/s11096-016-0414-2
7. Darpo, B., Nebout, T., Sager, P.T.: Clinical evaluation of QT/QTc prolongation and proarrhythmic potential for nonantiarrhythmic drugs: the International Conference on Harmonization of Technical Requirements for Registration of Pharmaceuticals for Human Use E14 guideline. J. Clin. Pharmacol.Clin. Pharmacol. **46**(5), 498–507 (2006)
8. Dev Sharma, L., Sunkaria, R.K.: Novel T-wave detection technique with minimal processing and RR-interval based enhanced efficiency. Cardiovasc. Eng. Technol.. Eng. Technol. **10**(2), 367–379 (2019). https://doi.org/10.1007/s13239-019-00415-4
9. Goldberger, A.L., et al.: PhysioBank, PhysioToolkit, and PhysioNet: components of a new research resource for complex physiologic signals. Circulation **101**(23), e215–e220 (2000)
10. Dev Sharma, L., Sunkaria, R.K.: A robust QRS detection using novel pre-processing techniques and kurtosis based enhanced efficiency. Measurement **87**, 194–204 (2016)
11. Rahul, J., Sora, M., Dev Sharma, L.: Exploratory data analysis based efficient QRS-complex detection technique with minimal computational load. Phys. Eng. Sci. Med. **43**(3), 1049–1067 (2020)
12. Işcan, M., et al: Multilead QT interval analysis algorithm based on continuous wavelet transform. In: 2017 4th International Conference on Electrical and Electronic Engineering (ICEEE). IEEE (2017

13. Hajimolahoseini, H., Hashemi, J., Redfearn, D.: ECG delineation for QT interval analysis using an unsupervised learning method. In: 2018 IEEE International Conference on Acoustics, Speech and Signal Processing (ICASSP). IEEE (2018)
14. Ahnve, S.: Correction of the QT interval for heart rate: review of different formulas and the use of Bazett's formula in myocardial infarction. Am. Heart J. **109**(3), 568–574 (1985)
15. Indik, J.H., et al.: Bazett and Fridericia QT correction formulas interfere with measurement of drug-induced changes in QT interval. Heart Rhythm **3**(9), 1003–1007 (2006)
16. Sagie, A., et al.: An improved method for adjusting the QT interval for heart rate (the Framingham Heart Study). Am. J. Cardiol.Cardiol. **70**(7), 797–801 (1992)
17. Rahul, J., et al.: An improved cardiac arrhythmia classification using an RR interval-based approach. Biocybern. Biomed. Eng. **41**(2), 656–666 (2021). https://doi.org/10.1016/j.bbe.2021.04.004
18. Rahul, J., Dev Sharma, L., Bohat, V.K.: Short duration Vectorcardiogram based inferior myocardial infarction detection: class and subject-oriented approach. Biomed. Eng./Biomedizinische Technik **66**(5), 489–501 (2021). https://doi.org/10.1515/bmt-2020-0329
19. Manlong, N.A., Rahul, J., Sora, M.: ST segment analysis for early detection of myocardial infarction. Int. J. Comput. Sci. Eng.Comput. Sci. Eng. **6**(6), 1500–1504 (2018). https://doi.org/10.26438/ijcse/v6i6.15001504
20. Rahul, J., Sharma, L.D.: An enhanced T-wave delineation method using phasor transform in the electrocardiogram. Biomed. Phys. Eng. Express **7**(4), 045015 (2021)
21. Rahul, J., Marpe, S., Lakhan, D.S.: Baseline correction of ECG using regression estimation method. In: 2019 4th International Conference on Internet of Things: Smart Innovation and Usages (IoT-SIU). IEEE (2019)
22. Rahul, J., Sharma, L.D.: Artificial intelligence-based approach for atrial fibrillation detection using normalised and short-duration time-frequency ECG. Biomed. Signal Process. Control **71**, 103270 (2022)
23. Rahul, J., Sora, M.: Premature ventricular contractions classification using machine learning approach. In: 020 International Conference on Smart Electronics and Communication (ICOSEC). IEEE (2020)
24. Andršová, I., et al.: Heart rate influence on the QT variability risk factors. Diagnostics **10**(12), 1096 (2020). https://doi.org/10.3390/diagnostics10121096
25. Singh, I., Rabkin, S.W.: Circadian variation of the QT interval and heart rate variability and their interrelationship. J. Electrocardiol.Electrocardiol. **65**, 18–27 (2021)
26. Hnatkova, K., Malik, M.: Sources of QTc variability: implications for effective ECG monitoring in clinical practice. Ann. Noninvasive Electrocardiol.Noninvasive Electrocardiol. **25**(2), e12730 (2020)
27. Scott, J.L., Walls, R.M.: QT interval prolongation. J. Emerg. Med.Emerg. Med. **3**(3), 221–225 (1985)
28. Rentoukas, E., et al.: Extreme but not life-threatening QT interval prolongation? Take a closer look at the neck! J. Electrocardiol.Electrocardiol. **46**(2), 128–130 (2013). https://doi.org/10.1016/j.jelectrocard.2012.10.007
29. De Oliveira Neto, N.R., et al.: A practical method for QTc interval measurement. Cureus **12**(12) (2020).
30. Chiladakis, J., et al.: Preferred QT correction formula for the assessment of drug-induced QT interval prolongation. J. Cardiovasc. Electrophysiol.Cardiovasc. Electrophysiol. **21**(8), 905–913 (2010)

Enterprise of Fusion Cryptography-Steganographic Method for Cloud Loading Refuge with Social Spider Optimization Algorithm

Rahul Mishra$^{(\boxtimes)}$ and Saket Mishra

Centre for Interdisciplinary Research in Business and Technology, Chitkara University Institute of Engineering and Technology, Chitkara University, Rajpura, Punjab, India
`{rahul.mishra.orp,saket.mishra.orp}@chitkara.edu.in`

Abstract. Customers and businesses can take use of a wide range of options for storing and migrating data in remote data centres with the help of cloud computing and storage solutions. These options may be found in the cloud. It is similar to a utility, such as the electrical grid, in that it functions most effectively and efficiently when the people who use it combine their resources to achieve a common objective.

In this line of research, steganography is utilised in order to disguise the genuine significance of encrypted data. Additionally, Biorthogonal Wavelet Transforms (BWT) and social spider optimization algorithms are utilised in order to embed and extract the data, respectively. Two of the performance indicators that can be detected in experimental data are referred to as normalised correlation (NC) and peak signal-to-noise ratio (PSNR). Enhanced Social Spider Optimization (ESSO), Social Spider Optimization (SSO), and Cuckoo Search are all aspects that are taken into account, regardless of whether or not attacks (such as Noise, Blurring, and Filtering) are present (CS). The results of the experiments demonstrate that the proposed method possesses excellent performance as well as high levels of security.

Keywords: Normalized correlation · PSNR · BWT · ESSO

1 Introduction

Business challenges are always difficult for customers. By incorporating ideas from Service-oriented Architecture, cloud computing enables the user to transform these challenges into services that, when aggregated, can provide a cost-effective solution. All of the cloud's hardware, software, and network capacity are exposed as services, and cloud computing makes use of the tried-and-true procedures and best-practice discoveries made in the field of service-oriented architecture (SOA) to facilitate easy, unrestricted, and universal access to cloud resources [1].

Public cloud pay-as-you-go approaches revolve upon the metrics. In addition, standard services are a crucial component of autonomic computing's feedback loops, enabling cooperative scalability on demand and automatic crash recovery.

© The Author(s), under exclusive license to Springer Nature Switzerland AG 2025
M. Gupta et al. (Eds.): MISS 2023, CCIS 1952, pp. 365–375, 2025.
https://doi.org/10.1007/978-3-031-69115-7_25

When the Quos (quality of service) and security issues with grid computing arose, cloud computing emerged as a viable solution. Cloud computing provides the infrastructure necessary to develop data/compute intensive parallel applications at a fraction of the cost of more conventional parallel computing approaches. Even while SOA promotes "anything as a service," the leading cloud computing providers still segment their "services" into the three NIST-recognized models: Saas; Iaas; Paas [2].

These models provide a higher level of abstraction, and are thus represented as if they were independent stack layers, such as infrastructure-, platform-, and software-as-a-service. For instance, it is possible to execute a programme on IaaS and immediately access it without encapsulating it as SaaS, and it is also possible to provide SaaS done on physical machines (bare metal) without first using either the PaaS or IaaS layers. Multiple levels of service models for cloud computing.

1.1 Cloud Calculating Individualities

1.1.1 Provider-Independent, On-Demand Assistance

Automated resource provisioning in the cloud occurs in the absence of any user engagement with the cloud service provider.

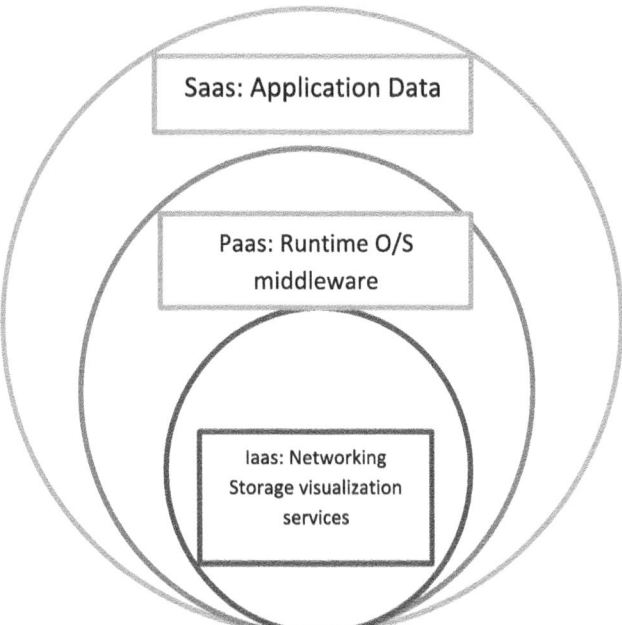

Fig. 1. Architecture of Service Model.

1.1.2 Allows for Access to a Large Network

Resources in the cloud can be accessed across the network and utilised by a wide variety of thin and thick client devices including mobile phones, workstations, laptops, tablets, and Smartphones (Fig. 1).

1.1.3 Sharing of Assets

The cloud's multi-tenancy features allow for multiple tenants to share the same set of servers. Virtual resources, which are layered above the underlying physical ones, are allocated to users based on their class. A sense of freedom of location is present despite the fact that the client has little to no say in or knowledge of the precise location of the provided resources. However, the client may be able to express location at a higher level of thought [3].

1.1.4 Speedy Elasticity

Fast and flexible provisioning of resources is possible. With cloud computing, it is easy to instantly increase or decrease the amount of resources used. Horizontal and vertical scaling options simplify the process by letting them specify parameters. Users can increase or decrease the amount of resources their apps require based on metrics unique to the programme itself, such as the number of transactions per second, the number of users at once, the latency of requests, and so on [3].

1.1.5 Sizing up the Horizon

Horizontal scalability refers to the capability of integrating different hardware or software components into a cohesive whole, hence increasing capacity.

1.1.6 Vertical Scaling

However, with vertical scalability, extra hardware components, such RAM or a second CPU, can be added to an existing system to boost its overall processing power [3]. Virtualization is the most important technology that can be used to boost cloud computing. Using software called virtualization, a physical computing device can be partitioned into one or more "virtual" device machines. These virtual machines can each be utilised and maintained independently to carry out computing activities. Virtualization at the level of the operating system enables the creation of a separate container that exists independently of any existing computing devices. This allows for the efficient allocation and utilisation of unused computer resources. Not only does virtualization give the agility necessary to speed up IT operations, but it also reduces expenses by increasing the amount of infrastructure that is put to use. Autonomic computing makes it easier for users to generate resources on demand by streamlining the method by which they do so. Automation speeds up the process, lowers labour costs, and reduces the likelihood of errors by requiring the user to participate in it as little as possible.

Customers often deal with difficult challenges in their businesses. The utilisation of concepts from service-oriented architecture is incorporated into cloud computing, and these concepts can assist users in transforming challenges into services that can be joined to form a workable solution. Cloud computing provides all of its resources as services and makes use of the well-established measures and best practises gained in the field of SOA to enable global and informal access to cloud services in a standardised manner. This is accomplished by providing all of its resources in the form of cloud computing services.

Cloud computing also makes use of ideas derived from utility computing in order to provide metrics regarding the employed services. Metrics are at the centre of pay-per-use pricing strategies for public cloud services. In addition to this, regular services are a crucial component of the feedback loops in autonomic computing. These loops make it possible for cooperation to grow on demand and for autonomous crash recovery to take place.

The term "cloud computing" refers to a type of grid computing that has developed in response to the challenges of ensuring high quality of service and data safety. Cloud computing makes it possible to construct data and computation-intensive parallel applications at prices that are far lower than those associated with traditional parallel computing methods, thanks to the availability of the necessary tools and technology.

2   Existing Work Done

If the attack succeeds, regular users can request the compromised service instance, which will then execute the malicious code [4]. In another use of this force, the attacker tries to infect the cloud with malware.

According to researchers a cloud service provider includes ensuring data security at every stage of its lifecycle. Data storage typically makes use of some storage technology, such as a network attached storage (NAS), a storage area network (SAN), object storage (S3 or S4), etc. All of these storage methods rely on a large number of users sharing a relatively small amount of storage space [5]. The safe isolation of consumer information is crucial in such a setup and must be ensured.

Some Chinese-affiliated web programmers stole company data from Google and other innovative companies. Since then, nation-state-sponsored attacks have grown in frequency and sophistication, from the Stuxnet attack on Iran's nuclear preparing capabilities to the Syrian Electronic Army's activist crusade and the Chinese government's almost endless supply of licenced technology. According to him, companies and government agencies need to protect data from data-stealing malware while personnel continue working, as stated by authors. When cloud storage is combined with robust encryption, it becomes possible to have a system that is both secure and reliable across the entire Internet [6]. According to Billy Lau, an examination researcher at the Georgia Tech Information Security Centre, each record is encoded and filed separately to facilitate fast retrieval.

Researchers provide measures to safeguard against flood attacks. Here, a hypervisor is employed to plan the workload of the cluster's servers. The Hypervisor will perform the viability analysis, and the system will be able to inspect the example to determine if any unapproved code is preventing the usual 40 calculation in the cloud system. The flooding problem may be alleviated to some degree if this is done. In the event that the designated hypervisor is compromised on a local level, additional research and diligence are required to restore security [7]. In addition, a PID can be included in the information to verify the genuine nature of the request. The Hypervisor can use the task of providing samples to the fleet of servers to verify the PID. This PID can be encrypted using a variety of techniques, including, for instance, a hash function performed using the RSA.

Investigator devised an algorithm based on elliptic curve cryptography to ensure the safety of cloud storage and data transmission. Here, the author focuses on an Online Alert methodology that, in comparison to current systems, offers multi-level security by alerting the data owner if a third-party attacker attempts to modify the data or any malpractice occurs during that period [8].

A strong unified character administration design and strategy inside the firm is the fundamental success factor in managing identities of the cloud service provider. Some aspects of character administration could be outsourced and coupled with cloud providers through the use of "Identity as a Service" providers. Authors offer a reputation system for protecting data-centre access at a coarse-grained level and secure data access at the fine-grained file level by means of software watermarking via data colouring and trust negotiation. While some security guidelines are broader in scope, others focus on more niche concerns. With HIPAA, patients' personal information and medical records are protected [9]. According to researchers, PCI DSS V2.0's purpose is to "enable and update cardholder information security and support the broad reception of predictable information safety activities globally".

The SEC2 solution, introduced by authors, allows users to tailor their security policy settings in the same way that they control their on-premise network. CloudSec, proposed by authors, is a virtualization-aware security solution designed to identify and prevent rootkits in the kernel data [10].

Attempts at data security were proposed by investigator. Instances of data abuse are heightened when multiple businesses use the same resources. Therefore, it is crucial to safeguard data repositories and, more generally, the data that comprises storage, transmission, or processing, in order to avoid random chance. Information security is one of the biggest challenges for cloud computing. Authentication, permission, and access control for data stored in the cloud are crucial for improving cloud computing security [11].

Data security and privacy must be given top priority as information is increasingly stored and accessed via the internet in "cloud" environments [12]. A company's profitability, reputation, and customer trust can take a serious hit in the event of a data breach. Similar to the difficulty in protecting sensitive information, the segregation of data stood at 92%. Authors offer a solution to the security problems associated with cloud computing by emphasising the need to keep sensitive information hidden. Encryption and tokenization of cloud-stored data are two key approaches to addressing these issues [13].

Encrypting the data stored in the cloud is a significant part of the solution to the problem of cloud security. As a result, encryption makes it more difficult to search for information by going through records, and it makes the process of coordinating efforts more difficult to adapt. Miller, in 1985, was the one who initially established the open key Crypto architecture, and since then, it has developed into a vital component of modern cryptography [14]. ECC's security strength is based on the difficulty of solving the Elliptic Curve logarithm problem (ECDLP), and it provides the same level of security that is obtained from RSA with a smaller key size. This is because ECDLP is difficult to solve. In addition to having applications in cryptography, prime testing, and expanding whole number factorization, an elliptic curve is a factor in all of these [15].

Authors put out the plan to reduce the risks associated with information security and a variety of cryptographic algorithms for protecting secret information. Researchers have recommended that the elliptic curve cryptography technique be used to produce cryptographic keys that are more compact, more efficient, and faster [16]. This would be beneficial for the development of secured information and the secure deployment or storage of data in a cloud environment.

In order to protect users' personal information, the authors suggested using the ECC algorithm and the RSA algorithm. They suggested methods for improving information security in the cloud by leveraging elliptic curve cryptography in conjunction with digital signature and encryption technologies. They have suggested cryptographic systems that place an emphasis on the protection of data as well as its privacy, thereby increasing the level of confidentiality [17].

The problem of authentication that the authors researched resulted in the proposal of a useful technology, which is exquisitely advanced for outfitting extraordinarily better confidential data storage to the cloud system. In the beginning, the customer is required to register his files in the cloud, produce his login and secret key, and in this manner get the general public and private keys [14]. This is a prerequisite for using the service. If the client's information is already stored in the cloud server, the user will be required to enter his username and password at this point. After that point, the cloud server will begin looking into the client's verification. In the event that the validation is successful, the client will be connected to the server. If the validation is unsuccessful, however, the server will refuse the request made by the client. With the presence of an intrusion recognition system, an evaluation is carried out to determine whether or not the information is regular or intrusive.

In order to protect the sensitive information contained within a patient's medical record, researchers suggest storing medicinal information in an encrypted format on a server and having another physician decrypt the data prior to downloading it from the server. Before 2009, they advocated using homomorphic encryption in cloud computing and voting systems, which helps with expansion homomorphism and augmentation homomorphism. This was done in order to circumvent security concerns [18].

Investigators have developed a secure circular cloud-user framework, which has been implemented for the purpose of maintaining information security. Within this framework, AES and RC6 mechanisms have been utilised. Using this strategy, the information can be sent on any of the available servers, which amounts to a total of three servers. A single server is utilised by the cloud server in order to back up data and perform preliminary processing on that data. Therefore, there are a total of four servers available [19]. There were a total of three servers involved in the transfer of information, and one of those servers is the cloud server.

3   The Proposed Work

This method utilises three distinct layers of cloud security to keep data safe. The first is the standard login/password setup, the second is cryptography using signcryption, and the third is steganography using various optimization methods. User authentication often entails the use of a user ID and password. Some hackers may be able to get private information such as login credentials. As a secondary layer of protection, the signcryption algorithm is utilised to encrypt the data. Band decay is achieved through the application of steganography, wherein encrypted data is implanted in the image using Biorthogonal wavelet transformations (BWT). Enhanced Social Spider Optimization (ESSO) techniques are used to incorporate the images once wavelet decomposition has been performed.

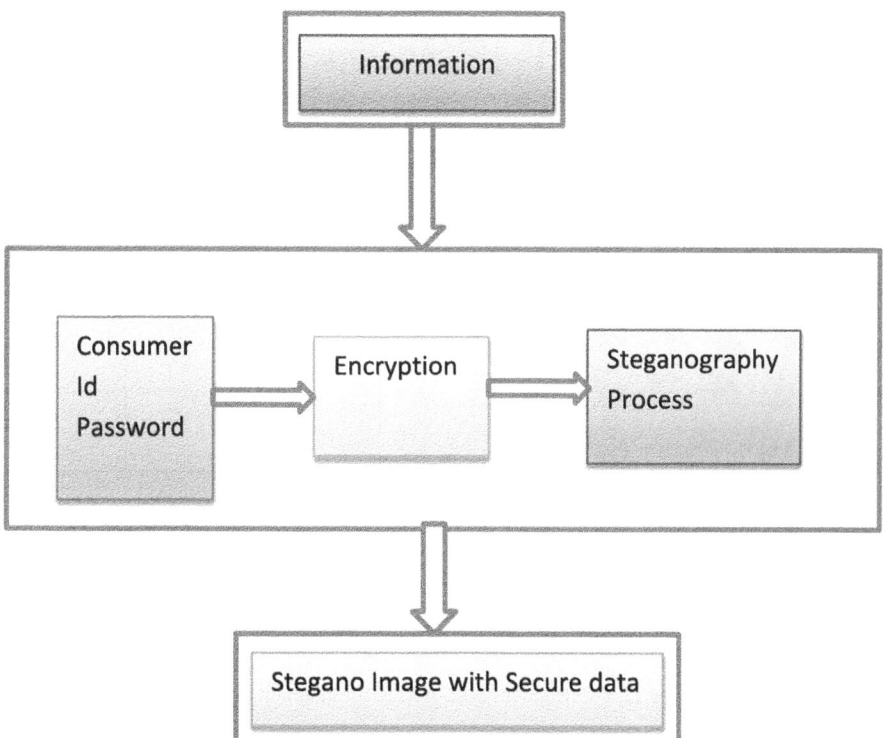

Fig. 2. The Proposed Architecture of the block diagram.

Figure 2 shows the proposed crypto-stegno architecture for this thesis, which consists of numerous parts, including user id and password, information, cryptography, and steganography. The signcryption technique is utilised in this cryptography to ensure the safety of the data being transmitted. The public-key encryption method known as "signcryption" serves both the numerical-naming and encryption functions concurrently.

From the preceding explanation, the information has been secured by diverse techniques, and assessment metrics have been applied on a stegno image and normal image. The various evaluation metrics specifically Peak signal-to-noise ratio (PSNR) and Normalized Correlation (NC) value is determined.

PSNR (Peak Signal to Noise Ratio): It is calculated by taking the square root of the Mean Square Error (MSE) between the noisy and the original image and dividing it by the maximum potential intensity value of the image (10log10). The mathematical definition of PSNR is:

$$PSNR = 10.log_{10}\frac{Max_i}{\sqrt{MSE}} \tag{1}$$

Image (Elephant.jpg)	Attack	PSNR	NC
	Noise	19.9618	0.8781
	Blurring	18.6758	0.6899

A variety of PSNR values are displayed in Table 1 after an optimization technique is applied to an elephant image and subjected to a noise assault, blurring attack, and filter attack. The enhanced social spider optimization method has a high PSNR value in comparison to the other algorithms (Fig. 3).

Table 1. PSNR values with Attack for elephant image.

Attacks	PSNR Value	
	ESSO	CS
Noise	19.9618	17.9812
Blurring	18.6754	16.6875
Filtered	33.4489	31.5781

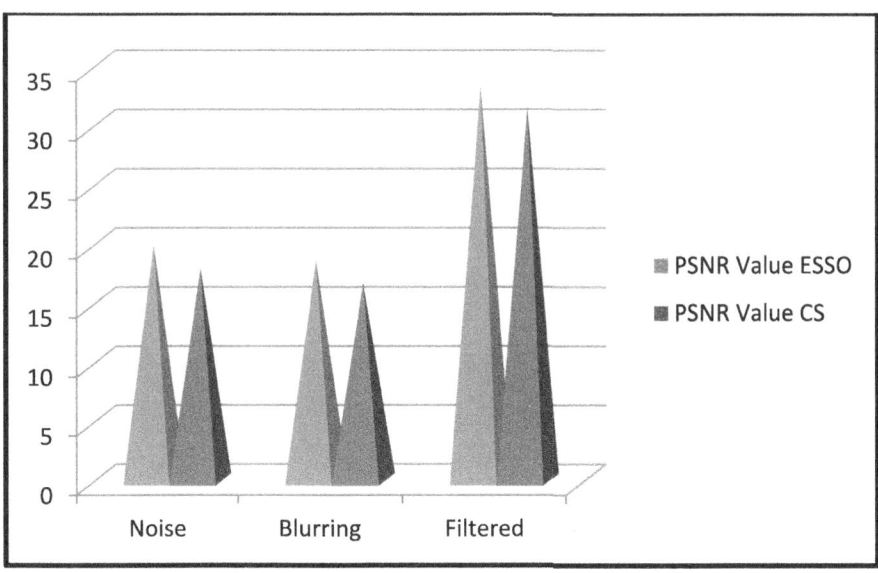

Fig. 3. PSNR values with Attack for elephant image.

The results of employing a variety of optimization algorithms, as well as a noise assault, blurring attack, and filter attack, are shown in Table 2. The table demonstrates that the enhanced social spider optimization method outperforms the alternatives.

Table 2. NC values with Attack for elephant image.

Attacks	NC Value	
	ESSO	CS
Noise	0.8782	0.8431
Blurring	0.6912	0.6214
Filtered	0.9541	0.9319

Figure 4 displays the results of applying an optimization algorithm to an image of an elephant, showing that the enhanced social spider optimization algorithm for a filter attack yielded a higher normalisation value (between a range of 0.02 and 0.03) and performed better than the other attacks.

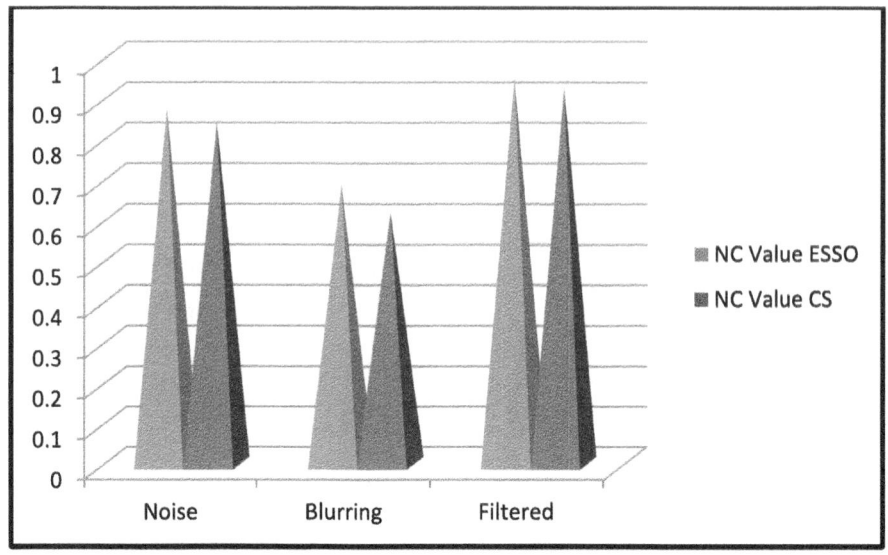

Fig. 4. NC values with Attack for elephant image.

4   Conclusion

The purpose of this work is to explore methods for ensuring the safety of cloud-based information. To ensure the safety of the data, a slew of different security measures were tried out and put into place. Additional goals include safeguarding data from hackers and preventing unwanted access. The data processing stage of a cryptographic algorithm typically takes the most time. There is little difficulty in incorporating this algorithm into pre-existing algorithms. An improved social spider optimization technique is proposed, together with a signcryption algorithm, to shorten the time it takes to transmit data and retrieve it safely from the cloud storage server. Finding the blue pixel is how information is hidden in this image. Since it is difficult to find data in pixel format, this reduces the likelihood of data being stolen by hackers. Errors are minimized, data is kept private, and the integrity of the cloud storage server is protected thanks to an improved optimization method. So, the suggested technique is a significant improvement over the prior art when it comes to cloud data security. This method can be used by any stakeholder to protect their data across any domain.

References

1. Amounas, F., El Kinani, E.H.: Ecc encryption and decryption with a data sequence. Appl. Math. Sci. **6**(101), 5039–5047 (2012)
2. Anitha, J., Sirmathi, H., Meenakshi, A.: Steganography based secure data storage and intrusion detection for cloud computing using signcryption and artificial neural network. Res. J. Appl. Sci. Eng. Technol. **13**(5), 354–364 (2016)
3. Chang, L.: Top-down leveled multi-replica merkle hash tree based secure public auditing for dynamic big data storage on cloud-. IEEE Trans. Comput. **64**(9), 2609–2622 (2015)

4. Chang, L., Zhang, X., Yang, C., Chena, J.: CCBKE Session key negotiation for fast and secure scheduling of scientific applications in cloud computing. Future Gener. Comput. Syst. **29**, 1300–1308 (2013)
5. Zhao, F., Li, C., Liu, C.F.: A cloud computing security solution based on fully homomorphic encryption. In: ICACT (2014)
6. Haiyong, X., Li, Z., Laxmi, B.: Architectural Analysis of Cryptographic Applications for Network Processors. Department of Computer Science & Engineering, University of California, Riverside (2014)
7. Loai, T., Darwazeh, N.S., Al-Qassas, R.S., AlDasari, F.: A Secure cloud computing model based on data classification. Procdia Comput. Sci. **52**, 1153–1158 (2015)
8. Manpreet, K., Navtej, G.: Security in Cloud Computing: A Review. ICRTEDC, vol. 1, no. 2 (2014)
9. Rao, R.V., Selvamani, K.: Data security challenges and its solutions in cloud computing. In: Proceedings of conference, International Conference on Intelligent Computing, Communication and Convergence, vol. 48, pp. 204–209 (2015).
10. Ravichandiran, C.: Data security challenges and its solutions for cloud computing. Int. J. Adv. Comput. Eng. Appl. **3**(6), 234–241 (2014)
11. Hanan, S., et al.: Cloud computing virtualization of resources allocation for distributed systems. J. Appl. Sci. Technol. Trends (JASTT) **1**(3), 98–105 (2020)
12. Chatura Chinthana, G.: Cloud computing: determinants of software as a service (SAAS) model adoption. Int. J. Sci. Technol. Res. **9**(10), 109–117 (2020)
13. Cho, J., Kim, Y.: A Design of server less computing service for edge clouds. In: International Conference on Information and Communication Technology Convergence (ICTC), Jeju Island, Korea, Republic of Korea (2021)
14. Cao, K., Liu, Y., Meng, G., Sun, Q.: An overview on edge computing research. IEEE Access **8**, 85714–85728 (2020). https://doi.org/10.1109/ACCESS.2020.2991734
15. Sukhpal, Singh G., et al.: AI for next generation computing: emerging trends and future directions. Internet of Things **19**, 100514 (2022). https://doi.org/10.1016/j.iot.2022.100514
16. Cho, J., Kim, Y.: A Design of server less computing service for edge clouds. In: International Conference on Information and Communication Technology Convergence (ICTC), Jeju Island, Korea, Republic of Korea (2021)
17. Ju, J., Wang, Y., Fu, J., Wu, J., Lin, Z.: Research on key technology in SaaS. In: 2010 International Conference on Intelligent Computing and Cognitive Informatics, Kuala Lumpur, Malaysia (2010)
18. Mansaf, A., Kashish Ara, S.: Recent Developments in Cloud Based Systems: State of Art. arXiv, New Delhi (2015)
19. Joaquim Molo, M., et al.: A review of evolutionary trends in cloud computing and applications to the healthcare ecosystem. Appl. Comput. Intell. Soft Comput. **2021**, 1–16 (2021). https://doi.org/10.1155/2021/1843671

FOG Grounded Observing Scheme for E-Healthcare Based Stroke Estimation and Alert Message in IoT Environment

Sakshi Pandey[(✉)] and Rahul Mishra

Centre for Interdisciplinary Research in Business and Technology, Chitkara University,
Rajpura 140401, Punjab, India
{sakshi.pandey.orp,rahul.mishra.orp}@chitkara.edu.in

Abstract. In the current day, appropriate characteristics are dogged from medical datasets by effectively allowing for meta-heuristic and machine learning systems. In addition, these methods can enable an automated, rather than manual, feature selection procedure. It is also observed that by combining the feature selection procedure with machine learning classifiers, prediction accuracy can be greatly enhanced. Several systems for providing healthcare remotely are discussed in the literature in an effort to spread healthcare more widely. Technology like the Internet of Things (IoT), fog computing, sensors, wearable devices, cloud computing, and so on all play a role in these kinds of systems. The scientific community is currently paying a lot of attention to cloud computing, fog computing, and the Internet of Things for their potential to improve the efficiency with which a variety of diseases may be diagnosed and monitored. Numerous healthcare and diagnostic systems have been built using cloud, fog, and IoT technologies. In this study, we investigate the potential of the Internet of Things, cloud computing, and fog computing for developing a reliable and efficient stroke prediction monitoring system. An ensemble classifier is used in the proposed system to make predictions. Patients are also notified of an impending stroke infection by alerts and warnings generated by the fog layer. The simulation results demonstrated the superior accuracy of the proposed system compared to the state-of-the-art models and classifiers. Users receive the stroke infection warning message from the proposed system. The suggested system achieves a lower latency rate than cloud computing and systems without cloud and fog computing, which is used to measure the effectiveness of the alert generation process.

Keywords: IoT · FOG Computing · SVM · Cloud Computing · Disease

1 Introduction

Personal healthcare technology, such as smartphones, wearables, and healthcare apps, has completely transformed the healthcare industry in a matter of years. In other words, these gadgets are claimed to transform institutionalised healthcare systems into person-centred care systems [1]. Until now, it has been considered that, with the exception of the

© The Author(s), under exclusive license to Springer Nature Switzerland AG 2025
M. Gupta et al. (Eds.): MISS 2023, CCIS 1952, pp. 376–386, 2025.
https://doi.org/10.1007/978-3-031-69115-7_26

iPhone and Fitbit, an active internet connection is a major necessity for internet-based connected applications. Only because of the internet of things is this even conceivable; the internet of things is a network that links together all electronic devices in digital form and allows them to exchange data and have conversations with each other and with humans. In addition, IoT can link existing healthcare facilities like hospitals and clinics to the many healthcare apps and wearables. Apps and wearables in the medical field are made possible by the network's connectivity. Wearable technology is not only more user-friendly, but it also provides users with suggestions on how to better their health statistics [2]. From this we get the term "Internet of Health Things," which describes the inter-connection of medical devices through the web (IoHT). The Internet of Health Things (IoHT) is the convergence of healthcare infrastructure, wearables, applications, classi-fiers, and emerging technologies such as cloud and fog computing, etc. The invisibility of IoHT's features—including their data-transfer capacities, contextual understanding, and the execution of predetermined actions in response to gathered information—is a major selling point for the technology. In addition, the Internet of Things's three main goals are to I lower healthcare costs, (ii) address the health-related concerns of the vast majority of people, and (iii) enhance healthcare quality [3].

Additionally, it is shown that IoT offers real time heterogeneous applications because of the utilisation of various sensors, wearable devices, and medical services. Internet of Things (IoT) provides healthcare services based on a wide range of sensors, communi-cation, and interaction options. Communication between medical devices and between humans and machines is commonplace in the healthcare industry. Wearable gadgets, such as wrist watches and Fit mattresses, can be used to characterise the end user in an Internet of Things setting. The primary function of these gadgets is to track the user's health data. In addition, these gadgets gather data for monitoring purposes, which can then be sent to a centralised or decentralised system for analysis. Those that have access to the data can check in on the user's health at regular intervals [4]. It is also noted that sen-sors and medical devices can introduce some variability into the obtained data. It is also highlighted that the transmission, processing, and storage of medical data can introduce noise due to infrastructure and health care architecture. However, the aforementioned problems cannot change the IoT's vision, which includes things like intended service response, medical support, and medical emergent. So, it's crucial to build a system that takes into account medical data analysis and other potentially life-saving measures [5].

Integrating computing models and technologies, the IoT platform can also perform healthcare data analysis. We discuss the many analytical and visual methods available for dealing with confidential data. Furthermore, machine learning classifier and intelligent data analytic approaches enhance data integrity and dependability [6]. These algorithms are also able to discriminate between data with high and low potential value in the health sector. Health data processing is claimed to be both sensitive and time-consuming, necessitating the use of cutting-edge computer and visualisation methods. It has also been noted that ML and DS systems are widely used in the IoT context to locate an accurate answer. Instantaneous, error-free analysis is used to determine how well these methods work [7]. These systems can also effectively compute the relevant attribute for the prediction task, and they can handle both static and dynamic data. Thus, it is difficult to create a reliable healthcare system that incorporates an IoT platform, means of communication, and data management. More so, specialist devices are needed for dealing with heterogeneity, as these devices/algorithms can manage the data reliance. Thus, it is concluded that in the near future, IoT-based systems are a viable alternative for remote patient monitoring.

In addition to this, the IoT platform does an analysis of the healthcare data using the computational models and technologies that are built into it. The presentation discusses the numerous methods of analytical and visual data processing that can be used to handle sensitive material. In addition to this, the use of machine learning classifiers and other intelligent data analysis techniques helps to increase the data's integrity as well as its reliability. It can also be observed that these algorithms are able to differentiate between the variability of health data and its potential utility. In addition to this, it has been noted that the processing of health data is sensitive, yet it is still necessary to use effective computation and visualisation techniques in order to handle the data. It has also been found that ML and DS systems are frequently used in IoT environments in order to locate precise solutions. The efficiency of these methods is judged based on the results of an analysis that is both immediate and free of mistakes. In addition to that, these systems are able to deal with both static and dynamic data, and they are also able to effectively compute the relevant characteristic for prediction tasks. Therefore, developing an efficient and reliable healthcare system that incorporates Internet of Things platforms, communication, and data management is a tough undertaking. In addition, specialist devices are necessary for dealing with heterogeneity because these devices and algorithms need to be able to cope with data dependencies. Therefore, it can be concluded that systems based on the Internet of Things are a workable solution for remotely monitoring the health of patients in the not too distant future.

2   Related Work Done

For missing value computation in the context of cerebral stroke prediction, authors [8] evaluated the random forest based regression method. To prepare the stroke dataset for the subsequent classification phase, we will use the missing value technique. With the goal of missing value computation for the cerebral stroke dataset in mind, the random forest regression technique can be used as a pre-processing tool.

For missing data imputation, particularly in medical datasets, researchers [9] developed a normalised mean technique. Mean and cube root of cubic mean make up the proposed method. It is also observed that the missing value can be determined using the maximum values from the aforementioned approaches. The authors assert that using the imputation method improves the accuracy of the predictions made.

Datasets with missing values can skew prediction outcomes. A new threshold-distance-based weighted imputation approach was developed by investigator [10]. In order to fill in the blanks of incomplete data instances, the suggested technique calculates the k-number of missing values, after which an optimised value may be used. The proposed method gives a higher imputation rate compared to other examined strategies when missing values are considered. Existing research has shown that the majority of imputation methods use a multivariate normal distribution approach; nonetheless, this can lead to unexpected findings in some cases. To address the problem of missing data, authors [11] devised a secure region-based data imputation method. The safe region technique comprises of k-NN and various imputation techniques to tackle incomplete attribute.

Data imputation is studied by researchers [12] to see if it can reliably identify diabetes risk variables. The median and group median were used to replace missing data by the authors. The results demonstrated that median-valued RF yields the best outcomes. Thus, it is concluded that the data imputation method enhances the accuracy of diabetes risk predictions.

To efficiently deal with missing data, authors [13] combined the two GRNN approach with the SGTM approach. Air conditioning systems are monitored using the aforementioned methods. There was evidence that using missing value imputation enhanced the final result.

According to researchers [14], current data imputation techniques are inadequate for dealing with high-dimensional data. Therefore, the authors came up with a new class-based clustering-based imputation method (CBCC-IM) to fix the aforementioned issue. Ten other imputation methods are evaluated alongside CBCC-output. IM's It was demonstrated that CBCC-IM outperforms the other methods in terms of accuracy.

Poor performance and inaccurate predictions might result from imprecise data imputation. Researchers [15] devised a hybrid strategy that takes into account both single and multiple data imputation strategies to address the problem of improper data imputation in medical datasets.

In the hybrid method, the multivariate imputation process is implemented as a chain of equations. It can handle both classified and numerical information. The proposed hybrid method was shown to increase the F-measure rate by up to 20% in experiments.

Imputation of missing data is a significant challenge in the area of medical informatics. To calculate missing values in partial data instances, authors [16] used the k-NN algorithm with a heterogeneous distance function. Improving the overall imputation outcome of biological datasets was shown to be possible with the use of a heterogeneous distance function. To deal with the missing value in EHR data [17] created an imputation technique. The multivariate similarities between patients form the basis of the suggested imputation method. In addition, patients are ranked based on how they compare to others with the same condition. It is also possible to deal with missing values by taking a weighted average of known values. The findings demonstrated the effectiveness of the aforementioned method.

Authors [18] created an auto encoder neural network-based method for missing value imputation. There are two distinct phases to the training in this method: The first part of the process involves training an automatic neural network without any missing data, and the second part involves using this model to forecast missing data. The eight existing imputation techniques are selected for comparing the results of auto encoder based data imputation technique. The results indicated that the auto encoder data imputation method outperformed the other methods.

Researchers [19] investigated selection of important features issue of stroke disease and proposed an integrated environment for selecting the relevant aspects and also diagnosis of ischemic stroke patient. In this work, a combination of shapiro-wilk and Pearson correlations are utilised for determine the significant characteristics for stroke dataset. Multiple ML methods, including RF, VC, MLP, and AdaBoost, are used to complete the prediction task. The effectiveness of the aforementioned algorithms is measured in terms of their accuracy. It has been observed that the accuracy of the stroke dataset is greatly improved by combining the aforementioned ML methods with a feature selection technique.

Combining greedy step wise with DT has been shown to increase stroke diagnosis accuracy. The purpose of greedy stepwise is to uncover more significant aspects of stroke dataset and create the processed dataset. While the analysed data is utilised to train DT to predict strokes. This paper considers the Japanese stroke dataset for measuring the performance of aforementioned combination [20]. According to the simulation findings, the proposed combination outperforms the competition in terms of accuracy.

Stroke illness detection prototyping using text mining and ML approaches [21]. To begin designing the prototype, we first establish a semantic and syntactic link between the features of the stroke dataset. There are 500 patients' records in the stroke dataset. The patient's medical record is reviewed for signs of stroke disease. In addition, the tagging and entropy techniques are used to mine these symptoms. More specifically, classifiers based on random forest, support vector machine, bagging, artificial neural network, and boosting techniques are utilised to make the stroke prediction. Results from several classifiers are compared, and it is shown that ANN achieves the most encouraging outcomes.

Several machine learning techniques are investigated in [22] for the purpose of diagnosing stroke-related pneumonia in Chinese AIS patients. LR with regulation, SVM, extreme GB, and fully connected DNN are the ML algorithms that are being discussed here. The performance of the aforementioned machine learning algorithms is evaluated using a number of different performance indicators. According to the results of the simulation, extreme GB produces more accurate results than other algorithms. When determining the eventual outcome of an ischemic stroke, numerous machine learning techniques are taken into consideration for successful treatment of stroke patients. In this work, the capabilities of three well-known techniques, namely DNN, RF, and LR, for accomplishing the same goal are investigated. In addition, the dataset on stroke patients includes information on 243 individuals. The accuracy parameter is utilised in order to assess the results of the simulation. It has been shown that the DNN technique possesses a greater accuracy rate compared to the RF and LR procedures.

The capability of a multi-neural network is investigated with the intention of preventing and predicting stroke disease. In addition, CNN is utilised to extract attributes that are more pertinent to the stroke prediction and prevention process. For the purpose of stroke dataset classification, a number of different network models, including DenseNet, VGG16, VGG19, and ResNet, were utilised. The findings of the simulation demonstrated that when compared to other models, VGG16 produces more accurate results.

In addition, authors [23] took into consideration the problem of predicting the accuracy of the stroke dataset and proposed a solution to this problem using ML algorithms. A number of machine learning techniques, including SVM, DT, LR, and NB, are utilised in an effort to improve the accuracy of the prediction. It has been observed that the accuracy of the aforementioned Ml algorithms is only slightly better than that of the DRAGON, ASTRAL, and THRIVE tools respectively. It should also be noted that the prediction accuracy of ML systems can be greatly enhanced by examining more new features. This point is brought up throughout the article. These new characteristics also have the potential to be helpful in providing effective treatment to patients while they are receiving treatment.

3   Objective of Research Paper

This paper's goal is to create a framework for a health monitoring system based on the Internet of Things and Fog computing. Some of the key features of the proposed system are described below.

- With the help of the Internet of Things, cloud, and fog computing, we can create a reliable prediction system for stroke data.
- The early implementation of suggestions for treating stroke patients.
- Patients experiencing a stroke can be monitored in real time, and alert messages can be sent if necessary.
- Proposed system has a method for warning message generation and suggestions based on record sharing.

4   The Projected Work

Improved stroke prognosis prediction with the use of a novel monitoring system is offered. Figure 1 displays the proposed system's three-tier architecture. In this section, we will focus on the levels of the proposed system, which are highlighted as

1. Patient Layer
2. Fog Computing Layer
3. Cloud Layer

The patient layer's job is to compile all relevant data about the patient, including demographics, whereabouts, medical history, routine activities, and environmental and behavioural factors. Different sensors on the body are used to get this data.

The data collected at the patient layer is sent to the fog computing layer for analysis and diagnosis. Fog layer is used as an ensemble classifier for the prediction task. As soon as the prediction task is finished, the patient will receive an alert message informing them of the outcome and suggesting steps they can take to lower their risk of having a stroke.

Information is stored on the cloud layer, and it may be efficiently processed. In addition, the doctor, the patient, and the hospital can all access the data for more efficient care. Using this data, we may also estimate how severely a stroke will affect a patient's health. Additionally, a warning message is delivered to the patient informing them of their current health state and how to take preventative actions against a stroke.

There are three distinct manifestations of the stroke condition. These categories are I negative, (ii) positive and (iii) recover. According to what has been claimed, real-time data on strokes are obtained from the patient layer by employing a variety of different sensors and wearable devices. The fog layer is responsible for the processing of the collected data and also manages a variety of data-related characteristics, such as missing value, among other things. It has also been brought to our attention that the final processed dataset is of a diverse character. This disparate set of stroke-related data is processed by the fog layer, after which it is converted into a single format that can be used to predict the stroke infection categories. The random forest boosting classifier is used to provide predictions about the different types of stroke infections.

The RFB algorithm is as follows:

Stroke dataset, number of features (D), and class label are input variables.
Output: Accuracy, Specificity, and Sensitivity of Predictions.

1. It is possible to pick "P" features out of a total of "D" features by setting P D.
2. Determine the node (N) using the best split measure (S) from the features (P).
3. The best split measure is used to produce child nodes.
4. Iterating steps (1–3) until no "l" nodes can be computed is done until the process fails.
5. The design of a forest tree relies on the first four procedures. Iterations can be performed up to some maximum number "n," and a total of "k" decision trees will be produced.
6. Model training using labelled data.
7. Each tree's error rate is calculated using a weighted error rate function ().
8. In the event that (k > 0), the weight of the training data is adjusted.
9. Otherwise, the steps 7–10 are repeated and the weight is not changed.

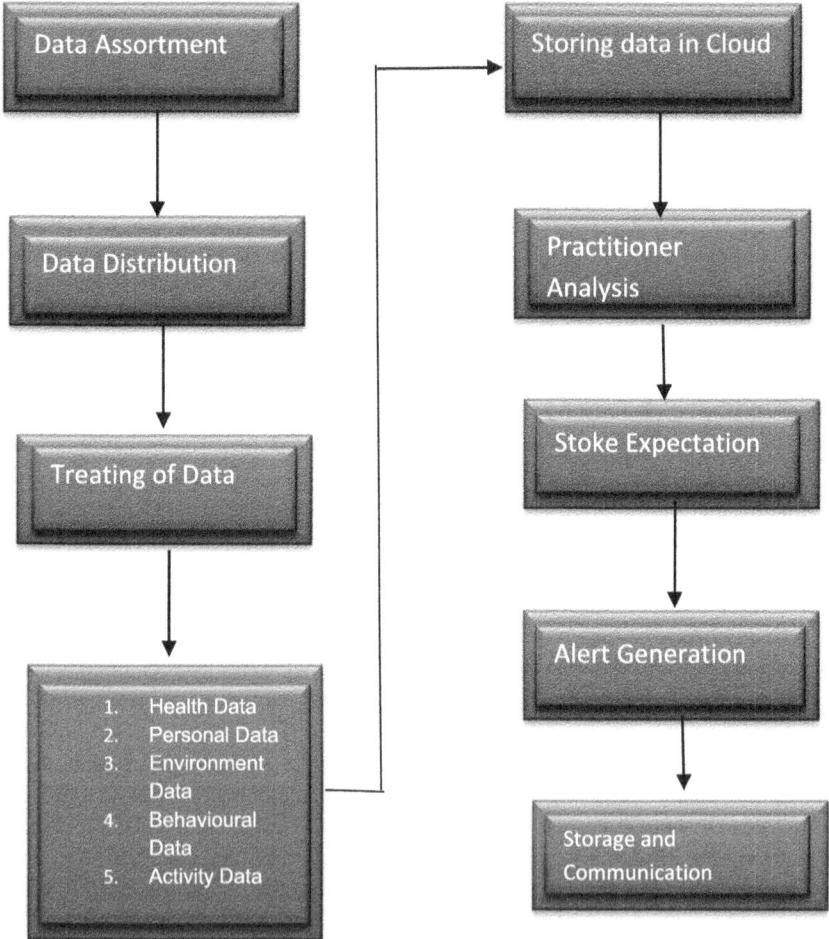

Fig. 1. Operational Arrangement of Projected Fog Computing Based Monitoring System

Whether or not the patient has made a full recovery, constant monitoring is a vital part of caring for a stroke victim. Periodic checks can be performed as needed. Initial checks will be performed every three hours, and if the patient is positive, additional checks can be performed every ten hours. Patients in the recovery phase have no set timeline but should instead visit their doctor. When a fog layer makes the diagnosis of a stroke infection, an alert message including the PI range can be sent to the patient. When the PI is within the normal range, the patient has not suffered from a stroke and is classified as having a negative PI. If the PI is out of the usual range, the patient receives a warning that specifies the signs and symptoms of a stroke. It's a huge boon to medical science because it allows for earlier stroke infection diagnosis, at which point preventative measures may be implemented to mitigate the disease's repercussions. Additionally, it is stated that the suggested system is able to assess stroke conditions and then produce an alarm message.

A primary purpose of the cloud layer is to supply a large enough data storage capacity. The information can be accessed whenever and wherever it is needed. As an added precaution, we make sure cloud data storage supports two distinct sharing modes: (i) public and (ii) private. Information like name, age, sex, qualifications, etc. is stored in shareable modes so that it can be accessed by others. Patients' private information is contained in the non-shareable mode, which means that it cannot be disclosed to outside parties. In general, information of this type includes things like a person's health status, test results, payment details, treatment history, and social contact information. In this way, the information is safe from prying eyes and can be archived indefinitely without fear of disclosure. The cloud layer has a dual-pronged permission approach to control who can access what. First, information about the patient's health and the patient's response to that report, as well as information about the patient's therapy, are all accessible to the patient and the patient's loved ones. Second, after reviewing a patient's history, medical professionals can provide informed recommendations.

It has been suggested that an ensemble RFB classifier can accurately predict the presence of infection in patients who have suffered a stroke. In order to assess how well RFB classifiers do their job, many different performance criteria are taken into account. F-value, precision, accuracy, sensitivity, specificity, error rate, and recall are some examples (Table 1).

Table 1. Simulation results of proposed system and other classifiers using different performance measures.

Parameter	Classifier			
	SVM	ANN	RF	Proposed Work
F-value	47.40	75.26	76.54	87.45
Recall	51.50	78.12	79.89	88.75
Precision	45.97	74.48	72.08	84.67
Accuracy	75.43	88.98	88.35	94.48
Sensitivity	51.00	58.54	69.16	88.47
Specificity	83.75	90.17	93.27	96.32

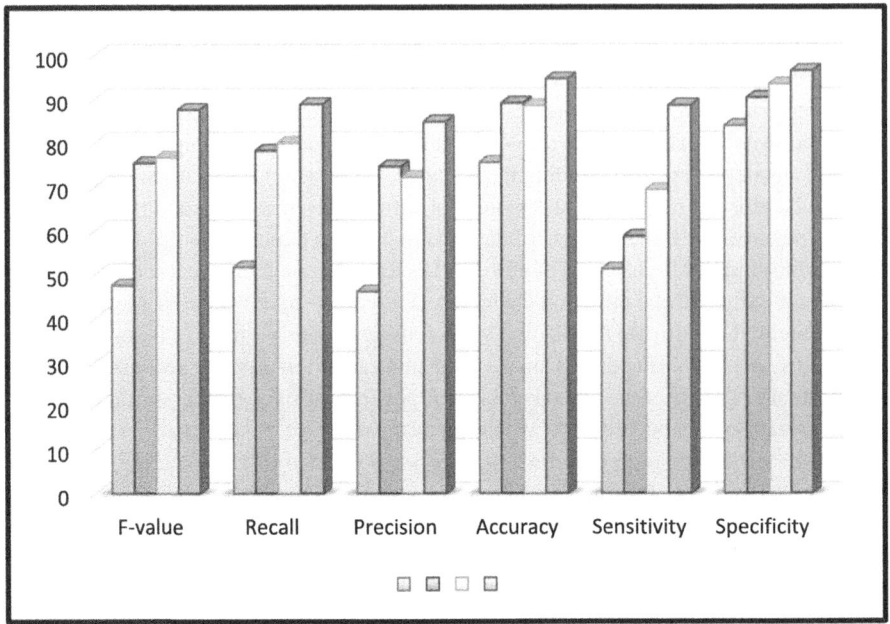

Fig. 2. Results of proposed system and other classifiers using different performance measures.

Figure 2 displays the simulation results of the proposed fog-based system with respect to accuracy, precision, f-measure, and recall. The proposed approach is clearly superior in terms of accuracy, precision, recall, and f-measure. The analysis of accuracy parameters reveals that ANN is the second most effective technique after other classifiers, however its performance is lagging behind the times in terms of precision, f-value, and recall. Fog-based systems have better accuracy, precision, and sensitivity. In addition, it is determined that the RF classifier is tied for second place with all other classifiers in terms of specificity and sensitivity. Therefore, it is claimed that the suggested approach is superior to other classifiers in identifying stroke patients. An additional use for the suggested fog-based system is in the creation of alerts and the sending of warning messages. The effectiveness of alert production and warning message is also analysed in this work.

5   Conclusion

There is a lot of interest in the scientific community right now in using cloud computing, fog computing, and the Internet of Things to better diagnose and monitor a variety of diseases. Cloud, fog, and the Internet of Things have inspired a plethora of new healthcare and diagnostic solutions. In this study, we investigate the potential of the Internet of Things, cloud computing, and fog computing for developing a reliable and efficient stroke prediction monitoring system. In this study, an Internet of Things (IoT) and Fog computing-based monitoring system for stroke patients' recovery was introduced. The design consists of three distinct layers: a "patient layer," "fog layer," and a "cloud layer

In the patient layer, various sensors and wearable technology are utilised to track vital signs and collect data in real time from stroke patients. In addition, RFB, an ensemble classifier, is taken into account to identify the stroke infection. The RFB classifier is incorporated into the fog layer to improve stroke infection prediction and decrease cloud layer congestion. The cloud layer, in turn, is in charge of tasks like data storage and transmission. Patients are also provided a warning notice that the fog layer has detected an infection related to stroke. The simulation results demonstrated the superior accuracy of the proposed system compared to state-of-the-art models and classifiers. The proposed method is able to warn patients of a stroke infection by sending them an alert or warning message. The suggested system achieves a lower latency rate than cloud computing and systems without cloud and fog computing, which is used to measure the efficiency of alert production.

References

1. Kavakiotis, I., Tsave, O., Salifoglou, A., Maglaveras, N., Vlahavas, I., Chouvarda, I.: Machine learning and data mining methods in diabetes research. Comput. Struct. Biotechnol. J. **15**, 104–116 (2017). https://doi.org/10.1016/j.csbj.2016.12.005
2. Yadav, G., Kumar, Y., Sahoo, G.: Predication of Parkinson's disease using data mining methods: A comparative analysis of tree, statistical and support vector machine classifiers. In: Computing and Communication Systems (NCCCS), 2012 National Conference on IEEE, pp. 1–8 (2012)
3. Kumar, Y., Sahoo, G.: Prediction of different types of liver diseases using rule based classification model. Technol. Health Care **21**(5), 417–432 (2013)
4. Kadhim, K.T., Alsahlany, A.M., Wadi, S.M., Kadhum, H.T.: An overview of patient's health status monitoring system based on internet of things (IoT). Wirel. Pers. Commun. **114**(3), 2235–2262 (2020)
5. Zgheib, R., Kristiansen, S., Conchon, E., Plageman, T., Goebel, V., Bastide, R.: A scalable semantic framework for IoT healthcare applications. J. Ambient Intell. Humanized Comput. 1–19 (2020)
6. Adi, E., Anwar, A., Baig, Z., Zeadally, S.: Machine learning and data analytics for the IoT. Neural Comput. Applic. **32**(20), 16205–16233 (2020)
7. Yadav, S.S., Jadhav, S.M.: Detection of common risk factors for diagnosis of cardiac arrhythmia using machine learning algorithm. Expert Syst. Appl. **163**, 113807 (2020)
8. Masood, A., et al.: Computer-assisted decision support system in pulmonary cancer detection and stage classification on CT images. J. Biomed. Inform. **79**, 117–128 (2018)
9. Devi, R.D.H., Bai, A., Nagarajan, N.: A novel hybrid approach for diagnosing diabetes mellitus using farthest first and support vector machine algorithms. Obes. Med. **17**, 100152 (2020)
10. Maniruzzaman, M., Rahman, M.J., Ahammed, B., Abedin, M.M.: Classification and prediction of diabetes disease using machine learning paradigm. Health Inform. Sci. Syst. **8**(1), 7 (2020)
11. Liu, T., Fan, W., Wu, C.: A hybrid machine learning approach to cerebral stroke prediction based on imbalanced medical dataset. Artif. Intell. Med. **101**, 101723 (2019)
12. Madhu, G., Lalith Bharadwaj, B., Sai Vardhan, K., Naga Chandrika, G.: A normalized mean algorithm for imputation of missing data values in medical databases. In: Saini, H.S., Singh, R.K., Beg, Mirza Tariq, Sahambi, J.S. (eds.) Innovations in Electronics and Communication Engineering. LNNS, vol. 107, pp. 773–781. Springer, Singapore (2020). https://doi.org/10.1007/978-981-15-3172-9_72

13. Santos, M.S., Abreu, P.H., Wilk, S., Santos, J.: How distance metrics influence missing data imputation with k-nearest neighbours. Pattern Recogn. Lett. **136**, 111–119 (2020)
14. Jazayeri, A., Liang, O.S., Yang, C.C.: Imputation of missing data in electronic health records based on patients' similarities. J. Healthcare Inform. Res. **4**, 295–307 (2020)
15. Li, X., et al.: Using machine learning to predict strokeǦ associated pneumonia in Chinese acute ischaemic stroke patients. Eur. J. Neurol. **27**(8), 1656–1663 (2020)
16. Heo, J., Yoon, J.G., Park, H., Kim, Y.D., Nam, H.S., Heo, J.H.: Machine learning–based model for prediction of outcomes in acute stroke. Stroke **50**(5), 1263–1265 (2019)
17. Ghanavati, S., Abawajy, J.H., Izadi, D., Alelaiwi, A.A.: Cloud-assisted IoT-based health status monitoring framework. Cluster Comput. **20**(2), 1843–1853 (2017)
18. Malathi, D., Logesh, R., Subramaniyaswamy, V., Vijayakumar, V., Sangaiah, A.K.: Hybrid reasoningbased privacy-aware disease prediction support system. Comput. Electr. Eng. **73**, 114–127 (2019)
19. Dastjerdi, A.V., Buyya, R.: Fog computing: Helping the Internet of Things realize its potential. Computer **49**(8), 112–116 (2016)
20. Malathi, D., Logesh, R.: Intelligent Cyber-physical System for an Efficient Detection of Parkinson Disease using Fog Computing. Multimedia Tools and Applications (2019)
21. Wu, W., Pirbhulal, S., Sangaiah, A.K., Mukhopadhyay, S.C., Li, G.: Optimization of signal quality over comfortability of textile electrodes for ECG monitoring in fog computing based medical applications. Futur. Gener. Comput. Syst. **86**, 515–526 (2018)
22. Jalali, F., Hinton, K., Ayre, R., Alpcan, T., Tucker, R.S.: Fog computing may help to save energy in cloud computing. IEEE J. Sel. Areas Commun. **34**(5), 1728–1739 (2016)
23. Sodhro, A.H., Luo, Z., Sangaiah, A.K., Baik, S.W.: Mobile edge computing based QoS optimization in medical healthcare applications. Int. J. Inform. Manag. **45**, 308–318 (2019)
24. Dev, A., Malik, S.: K: IoT and Fog Computing Based Prediction and monitoring System for Stroke Disease. Turk. J. Comput. Math. Educ. **12**(12), 3211–3223 (2021)

A System for Automatically Classifying Social Network Posts into Smart Cities Dimensions

C. Silpa[1], Avula Gayathri[2], Prem Chand Balu[2], P. Bharath Kumar Reddy[2],
Kaluvai Niranjan Reddy[2], and M. Pranay Kumar[2(✉)]

[1] School of Computing, Mohan Babu University (Erstwhile Sree Vidyanikethan Engineering College), Tirupati, India
[2] Department of Information Techology, Sree Vidyanikethan Engineering College, Tirupati, India
pranay30933@gmail.com

Abstract. Citizens concerns are addressed by the various departments that govern each province. The complaints are brought to light by the media which only focuses on the major problems and media is handled by private organizations. Real complaints are raised by people that are experiencing the problem. Specific departments of the government handling these issues can be notified. Data about complaints in a city such as water, fire accidents, police, education, economy, disaster, transport, health, telecommunication, electricity is collected from Twitter and classified into labels. The classification is done by eight different algorithms and the most efficient among them is chosen for classification. An automated system pulls data from twitter, categorizes it into labels, and then sends messages to the intended departments. This classified information is mailed to concerned departments so that the information reaches the government before any mainstream media or any manual complaints are reached. Citizens concerns are thus conveyed in a transparent manner to the respective departments.

Keywords: Term frequency-inverse document frequency (TFIDF) Vectorizer · Count Vectorizer · Text Classification · Real time Twitter Data

1 Introduction

In order to address challenges, public and private services are combined, accessible, and sustainable in smart cities. City services are expected to be continually improved in order to impact the economic and social well-being of citizens. Globally, urbanization has accelerated dramatically in recent decades leading to the development of sustainable and intelligent cities.

The setting can be created where the physical and digital worlds are constantly in conversation with one another thanks to technology and city systems, which opens up new opportunities for knowledge discovery but also presents a number of difficulties in the mining, manipulation and visualization of this city data. Urban big data is a growing field of multidisciplinary study. Data might be referred to as the "power and energy"

© The Author(s), under exclusive license to Springer Nature Switzerland AG 2025
M. Gupta et al. (Eds.): MISS 2023, CCIS 1952, pp. 387–401, 2025.
https://doi.org/10.1007/978-3-031-69115-7_27

of cities because it is the starting point of the transformation process that improves the quality of life.

Urban statistics include a substantial portion of online social networks (OSN). OSN data are increasingly being used for research and trading in both business and academics. As far as, OSNs have altered how people communicate and distribute information, and OSN analysis has almost entirely supplanted all conventional social scientific instruments. OSN-related problems have been addressed using a variety of machine learning (ML) applications.

In addition to providing information about citizens opinions, this model helps to monitor and manage various government bodies effectively. As part of this research, different keywords were used to extract the data. The extracted data was then filtered by removing hyperlinks and emoticons, converting the texts into lowercase for easy processing, and sorted by dimensions. To build a validation base for training and testing models, need to collect and categorize messages. In fact, there is a need to do a lot of work before we can normalize the messages. The use of trained models allows categorizing tweets in real time and sending relevant posts to corresponding authorities.

OSN posts are generally classified by Event detection, Topic detection, and Topic classification.

2   Literature Survey

Y. Liu, et al. [1] To solve the difficulty of detecting unidentified cities, an unsupervised framework dubbed NELPTW that uses of the enormous geographic location data present both in twitter and the Web is proposed. Using a linear function ranking model and named entity in data produces a variety of ranks for the candidate location. An unsupervised rank aggregation method is developed to aggregate the ranks and generate a more accurate ranking. An effective EM learning method can automatically learn the linear function ranking model specifications without the need for training labels.

Karami, et al. [2] proposes a methodical framework to comprehend studies related to Twitter and their hot and cold topics. Results demonstrate the significance of text mining and trend analysis in identifying semantic patterns and examining the evolution of research themes over time, as well as the potential of this research to comprehend large research corpora. The majority of the issues with a significant trend (P 0:05) had an ascending trend (Slope > 0), which is taken into account. Identifying dominant research topics on Twitter, summarizing the topics, and interpreting the evolution of topics over the past 10 years are the objectives of this study. Researchers, educators, and publishers can all benefit from the contributions made in this work, which can be applied to the expanding field of Twitter-based research.

Bencke, et al. [3]. When a city offers a variety of problem-solving tools, it qualifies as a smart city or high-tech city. Online social networks (OSN) have changed how individuals connect and share knowledge, and they are now a crucial part of urban data. One of the OSNs that is becoming increasingly popular and important for current news is Twitter. The Indian Standard Organization's ISO 37120, which focuses on city services, is broken down into city dimensions. A city has various dimensions like economy, transport, education etc. This study focused on categorizing OSN communications according to ISO 37120 dimensions. It collects data from the twitter and clean it for further processing. It applies 8 classifiers to compare the results. Based on that the effective prediction will be plotted and this could help the authorities to solve the problems in the city.

Alotaibi, et al. [4]. Startups who use social media sites, notably Twitter, for the majority of their business operations have found them to be helpful tools. It is necessary to evaluate the impact of particular business activities utilizing public opinion, or, to be more precise, the dissemination of such initiatives using user-generated material on Twitter. This study provides Startup Initiative Response Analysis (SIRA), an analytics-based methodology for assessing the effectiveness of startup initiatives using text categorization, sentiment analysis, and statistical analytical techniques.

Z. Shah, et al. [5]. Major applications look for spatiotemporal patterns to analyze high-volume streaming data from social media. Traditional methods, on the other hand, do not account for the localization of events in cities and nations, as well as within hours, days, and weeks. By keeping an eye out for changes in the language that is being used, events that indicate subjects that arise and degrade over time can be identified. An innovative method for classifying and finding events that may be used to Twitter data streams is created and tested in this work. To implement and validate the approach, 11.7 million tweets from users in 100 cities were released throughout the course of the 203-day observation period. The events represented by these tweets include suicide, shootings, elections, sports, and sentiment, and they were discovered across five example domains.

C. H. Mendhe, et al. [6]. Due to the restrictive API limits, it is very expensive for researchers in academia to get only the most recent data. Users can easily access huge amounts of Twitter data by focusing on their search criteria. Specific Domain Data is available (Health Care) and further extendable based on needs.

The data is collected by keyword search from the Twitter API and stored in MongoDB in JSON format. Data is classified and then machine learning algorithms are applied to it as time-based multiple systems linked by a network.

Amit Agarwal, et al. [7]. Social media tweets have been analyzed to identify complaints about traffic, accidents, and potholes in road transportation. Previously, In order to find and classify tweets about various topics, keyword-based algorithms have been utilised. However, these techniques simply rely on manually provided seed keywords, which are insufficient to handle all tweets. This provides a hybrid method for extracting geographic information from tweet text content utilising Regular Expression, Named Entity Recognition and parts of speech.

P. Vyas, et al. [8]. The rapid expansion of information exchange on social media platforms has ushered in a new era of information for humanity. During the COVID-19 epidemic, Twitter and other micro blogging services experienced rapid growth. Developed an automated algorithm to pull out tweets neutral, negative, and positive sentiments. Large numbers of tweets could be automatically categorised using the hybrid approach. The location-rich tweets are taken into account, and other tweets are analysed using regular expression and parts of speech. A variety of metrics, including precision, accuracy, recall, and F1 score, are used to assess this hybrid architecture.

Jain Yang, et al. [9]. One of the most widely used platforms for getting up-to-the-minute news on events happening across the globe is Twitter. Topic derivation is crucial due to the huge volume of information that is constantly pouring through the Twitter ecosystem. Researchers have noticed a the relation which real world has with people's responses, including sentiments, and behaviours, in a virtual setting by analysing social media data. In this study, we propose a text classification and feature extraction technique for textual data from Twitter related to an unusual event, such as Hurricane Sandy. This research suggests a topic derivation approach that takes interactions and tweet text similarity into account to solve this issue. The technique considers multiple forms of interactions between tweets, including retweets, replies, and tweets that reference the same person. Two-step matrix factorization is used for topic derivation. On several Twitter datasets, we ran a variety of experiments to demonstrate the individual and combined effects of the different variables under consideration. Our test results demonstrate that the suggested strategy outperforms more complex topic derivation techniques.

Xiaoyu Sean Lu, et al. [10]. Natural disasters are unusual events that don't happen often but may have disastrous effects on people and their surroundings when they do. People have a quick and accessible means to express their ideas thanks to social media. Thus, it may be used by researchers to look at how people behave, feel, and think during certain unusual events. In order to extract usable features from each text message, a new feature extraction approach is first suggested. The separation of event-related and irrelevant communications is then proposed using a fuzzy logic-based classification algorithm. Finally, the development of a rare occurrence is studied using our suggested technique and the frequently utilized keyword search method. The outcome shows that the suggested method can distinguish between text messages connected to related to unrelated events.

Q. Wang, J. Bhandal, et al. [11]. We have seen a lot of potentially private or sensitive communications being published to OSNs accidentally or voluntarily as a result of Twitter's rising popularity and tweet explosion. Many times, the people who posted these messages may have regretted doing so. Therefore, it is crucial for both users and service providers to be able to recognise tweets that reveal sensitive or private information. Users' topic choices based on individual tweet histories were indeed taken into account in this model's tweet semantics and term distribution features.

J.G. Fiscus, et al. [12]. Extract information from hundreds of public Twitter posts about locations in NYC. The extraction is carried out using the Filter Real-time Tweets Twitter Streaming API functionality. The captured tweets are put into a text file for further NLP processing, like regular expressions. The pertinent components of the tweets, such as their text and hyperlinks, are kept. These are kept in a clean tweets database. The SCC mapping is then carried out using the relevant aspect of our technique as shown in the Algorithm on the cleaned tweets.

3   Methodology

At present, all grievances should be raised in the grievance cell, where a separate grievance needs to be allocated for every department in the city. There is a need for a lot of resources to maintain them, and it is very hard to listen to and take one after another.

Proposed System
It endeavours to categorize postings using the city classification framework shown in Fig. 1 into Smart City service dimensions.

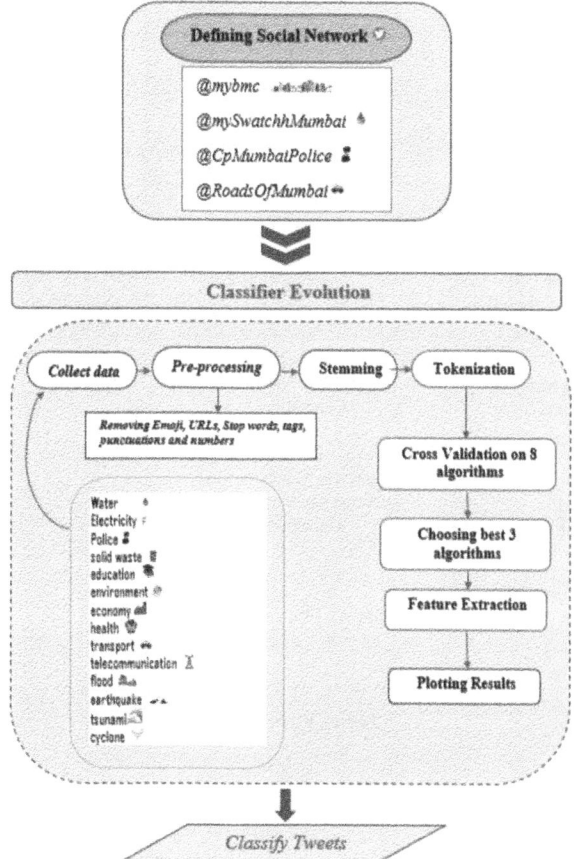

Fig. 1. Classification Framework.

It consists of three main components. First, we collect relevant messages [1] from citizens about different aspects of the city through public and official accounts related to those dimensions. Once collected, the message is stored [6] in the database and can be sent to the classification service, where it is pre-processed and classifier models are applied to find the best fit.

The third component represents the streaming of posts in order to classify [7] them into respective dimensions and notify the concerned authorities.

It is required to define the supervised learning model (SLM) and train and test it in order to determine the optimum SLM for sorting citizen messages from OSN into the appropriate dimensions. Figure 2 provides a summary of the steps involved in determining the SLM.

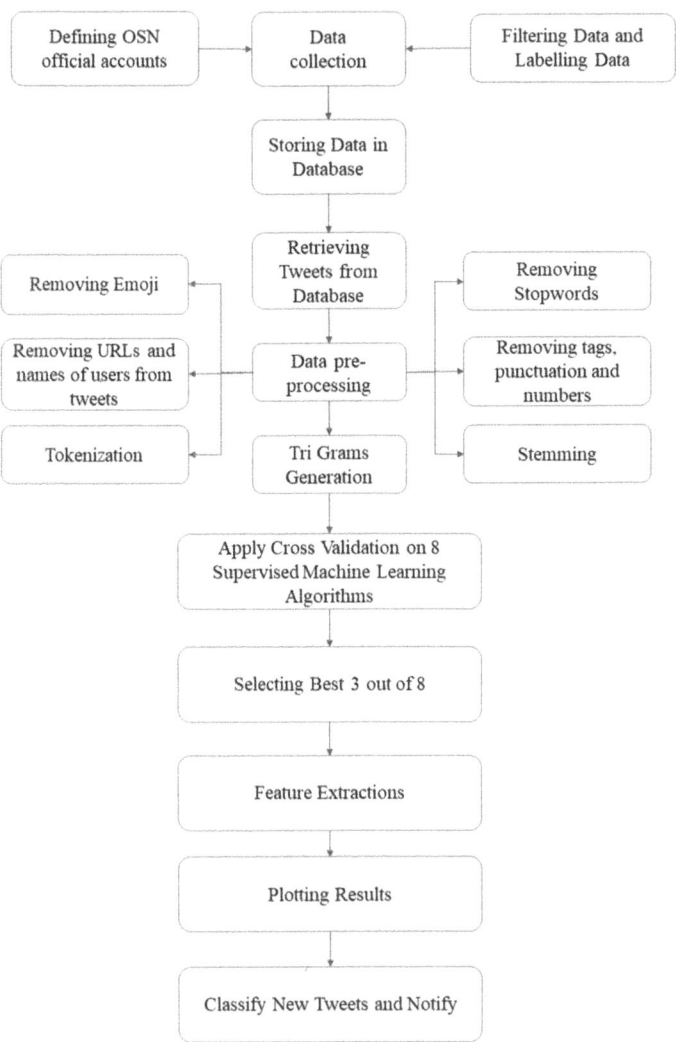

Fig. 2. Tweet Classification Workflow.

The true selection techniques and algorithms used in the classification method are combined into a wrapper strategy. F1 macros were used to evaluate a group of algorithms.

In order to improve classification results, the three most effective algorithms were extracted and selected.

Effective SLM is determined based on measures of F1 macro and internal category performances.

Based on the most effective SLM selected in the previous step, it can be used for classifying new tweets extracted using Streaming Client.

Algorithm 1: Finding Best SLM

Posts ← preProcessing(data)

// data collected from twitter api using official osn accounts

macroScore ← []

for all the algo in SLM do

// where slm contains 8 supervised machine learning algorithms

    i ← 0

    Mean_macro=0

    while i<10 do

        Mean_macro=mean_ macro +cross_validate(algo,posts)

        i ← i+1

    end while

    AddmacroScore (algo,Mean_macro)

end for

bestalgos ← findBest3Algorithms(macroScore)

features ← FeatureExtraction(bestalgos,data)

bestalgo ← max(features)

Data collection is done using the Twitter streaming API and the tweepy library, with the help of Official Online Social Network Accounts. As part of data pre-processing, tweets should be converted to lowercase, removing account names, links, punctuation, numbers, and symbols. Stop words, which contain articles, prepositions, conjunctions, and pronouns, should also be removed. Tokenization and Stemming are applied to the data.

A 3-g tokenization process was used to extract features, which resulted in word matrices, where columns represent features and rows represent documents (tweets). With Count Vectorizer, each cell of the matrix represents the number of occurrences of the corresponding feature in the respective document, while with TFIDF Vectorizer, it represents the weights of the respective TF-IDF feature. Cross-validation was applied to eight different supervised machine learning algorithms in order to find the best algorithm for the data. Ten cross-validations were used to find the top three algorithms out of the eight. The eight supervised learning algorithms are Random Forest Classifier, Linear SVC, SVC, Multinomial Naïve Bayes, Logistic Regression, K-Neighbours Classifier, Complement Naïve Bayes, and Decision Tree Classifier. Since we are using supervised learning, training must be provided for the model. The best model is determined by plotting graphs and using F1-macro and F1-micro, and it is used for tweets classification.

Algorithm 2: Classifying Posts using Streaming Listener

```
DEFINE CLASS StreamListener(StreamingClient):

  DEFINE FUNCTION on_tweet(self, tweet):

    label ← Model.predict(tweet)

    Notify(email,label)

SET stream_listener TO StreamListener()

SET rule TO StreamRule

stream_listener.add_rules(rule)

stream_listener.filter()
```

Finally, a streaming client is developed to listen for relevant tweets that can be pulled from Twitter by placing stream rules on it, helping us to find specific tweets instead of all the tweets posted by users.

Based on the best model determined, it can be used to classify new tweets and notify the concerned authorities efficiently.

4 Results

A. **Dataset Distribution**
    The data extracted consists of citizen complaints organized based on the dimensions of smart cities services. Below is a distribution of the results.
B. **F1 Macro Mean of Models**
    The mean of the 8 supervised machine learning models using 10 cross validation.

Using Count Vectorizer

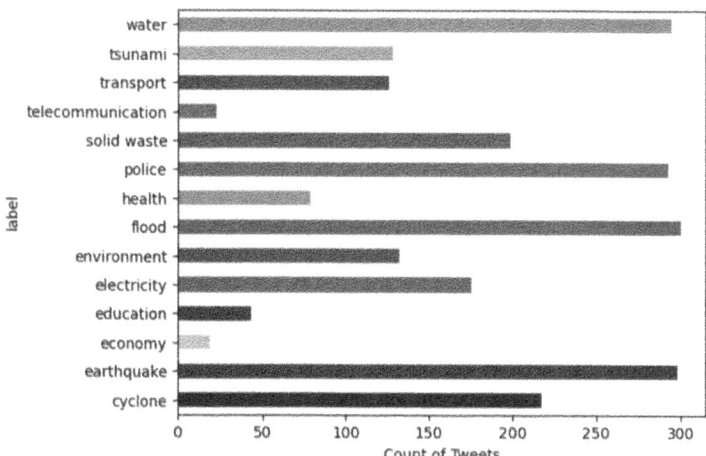

Fig. 3. Distribution of Dataset.

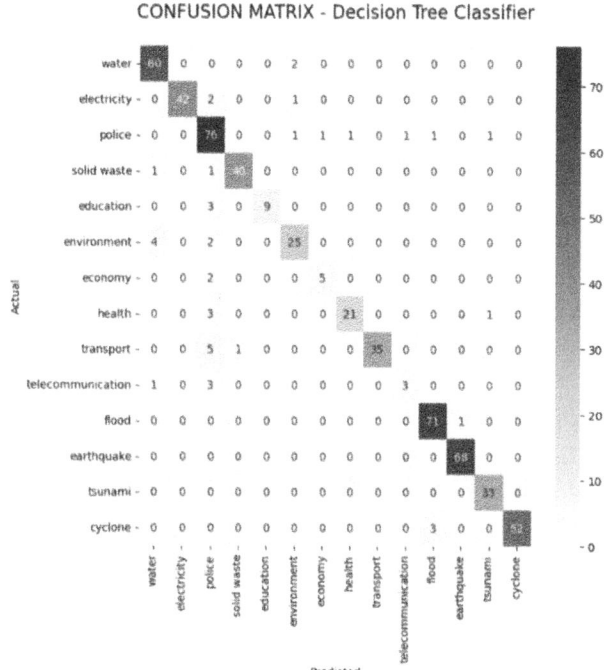

Fig. 4. Confusion Matrix Using TFIDF Vectorizer.

Table 1. Cross Validation Macro Score (Count Vectorizer) Using TFIDF Vectorizer

S. No	Model Name	F1 Macro Mean
1	DecisionTreeClassifier	0.886255
2	LinearSVC	0.872128
3	LogisticRegression	0.870617
4	ComplementNB	0.814195
5	SVC	0.750554
6	MultinomialNB	0.727694
7	KNeighborsClassifier	0.641448
8	RandomForestClassifier	0.532968

Table 2. Cross Validation Macro Score (TFIDF Vectorizer):

S. No	Model Name	F1 Macro Mean
1	DecisionTreeClassifier	0.882951
2	LinearSVC	0.870025
3	LogisticRegression	0.814330
4	ComplementNB	0.765371
5	SVC	0.741961
6	MultinomialNB	0.553676
7	KNeighborsClassifier	0.529558
8	RandomForestClassifier	0.402475

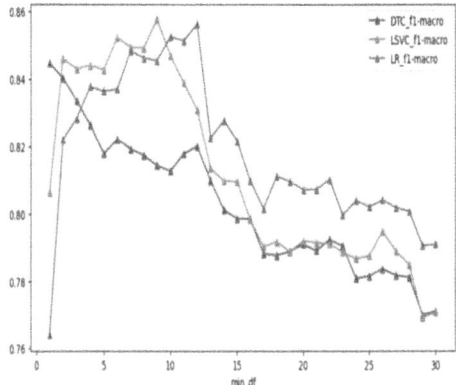

Fig. 5. Min_df plot using Coun Vectoizer.

Based on the F1 Macro Score, the three best algorithms chosen for further analysis were the Decision Tree Classifier, Linear SVC, and Logistic Regression. For Count Vectorizer and TFIDF Vectorizer, the best three algorithms were the same (Fig. 3 and Tables 1 and 2).

C. **Confusion Matrix**

From the Fig. 4 Decision Tree Classifier with TF-IDF vectorizer performed better than any other model when the MNF value was set to 10000.

D. **Minimum Document Frequency**

From the Fig. 5 the plot shows the relationship between the minimum document frequency value (ranging from 1 to 30) and the corresponding F1 macro mean for the top three models.

Using Count Vectorizer

For certain values, Linear SVC works well in the range of 1 to 10, while Logistic Regression suits best from 11 to 20.

Using TFIDF Vectorizer

Linear SVC generally yields positive results when combined with a TFIDF Vectorizer (Fig. 6).

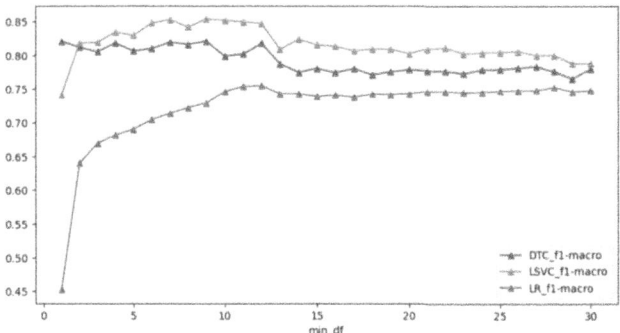

Fig. 6. Min_df plot using TFIDF Vectorizer.

E. **Classification Metrics**
  See Fig. 7.

```
                              CLASSIFICATIION METRICS

                        precision    recall  f1-score   support

              water        0.91      0.97      0.94        62
        electricity        1.00      0.93      0.97        45
             police        0.78      0.93      0.85        82
        solid waste        0.98      0.95      0.96        42
          education        1.00      0.75      0.86        12
        environment        0.86      0.81      0.83        31
            economy        0.83      0.71      0.77         7
             health        0.95      0.84      0.89        25
          transport        1.00      0.85      0.92        41
   telecommunication       0.75      0.43      0.55         7
              flood        0.95      0.99      0.97        72
         earthquake        0.99      1.00      0.99        68
            tsunami        0.94      1.00      0.97        33
            cyclone        1.00      0.95      0.97        55

           accuracy                            0.93       582
          macro avg        0.92      0.86      0.89       582
       weighted avg        0.93      0.93      0.93       582
```

Fig. 7. Classification Metrics for Decision Tree Classifier Using TF-IDF Vectorizer.

F. Maximum Number of Features

Using Count Vectorizer
Linear SVC generally yields positive results when combined with a Count Vectorizer for Maximum Number of features (Fig. 8).

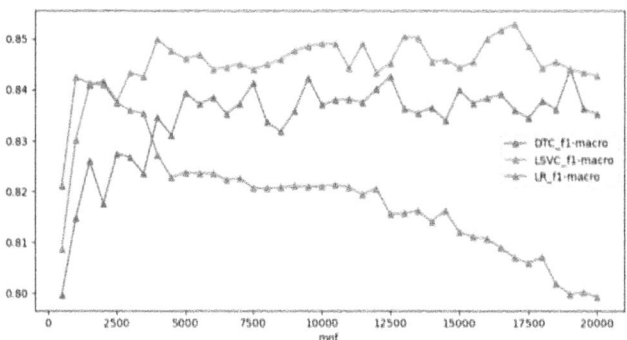

Fig. 8. MNF plot using count vectorizer.

Using TFIDF Vectorizer
For certain values, Linear SVC works well in the range of 0 to 5000, while Decision Tree Classifier suits best from 5000 to 20000 (Fig. 9).

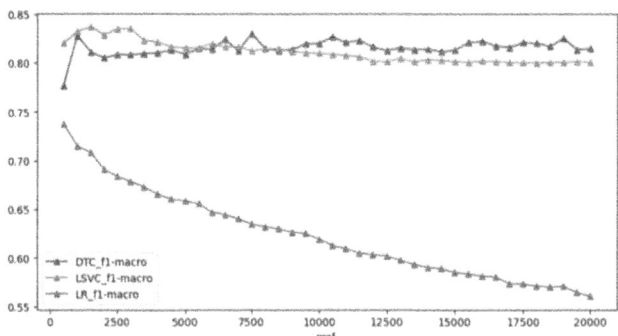

Fig. 9. TFIDF plot using count vectorizer.

During the streaming, we will receive some tweets concerning city dimensions, and after classifying them, these will be sent to relevant email addresses such as those of the police, environment, education, and so on. For example, the majority of public water supply problems can be resolved based on tweets received.

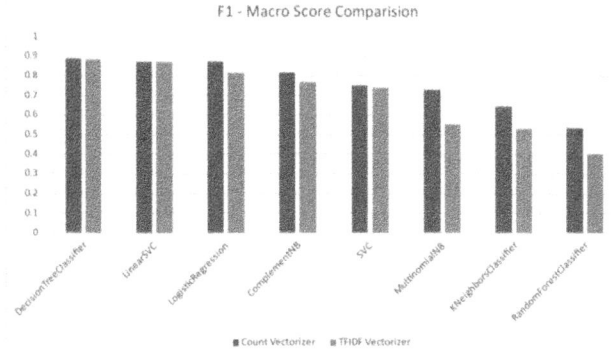

Fig. 10. Count Vectorizer and TFIDF comparison.

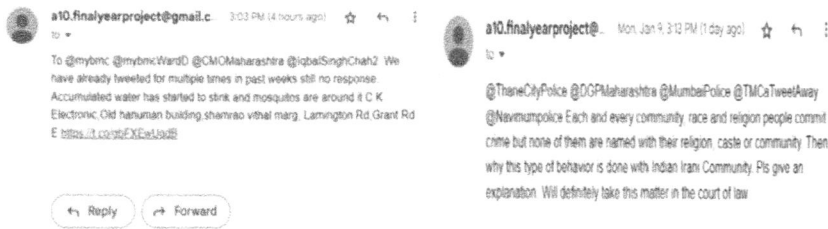

Fig. 11. Messages sent to relevant dimensions.

5   Conclusion and Future Work

The two vectorizers are applied with their default min df and MNF values, and the best three for each vectorizer are identified using the F1 macro score. From the Fig. 10, we can observe that F1 macro scores for eight supervised machine learning algorithms using Count Vectorizer and TF-IDF Vectorizer. We can conclude that the Decision Tree Classifier, Linear SVC, and Logistic Regression models performed well with both vectorizers. Therefore, these three models, maximum features and minimum document frequency are determined for the vectorizer. Based on the graphs plotted, the best-suited supervised machine learning is selected for classifying new tweets i.e., Decision Tree Classifier using TFIDF vectorizer (Fig. 11).

New tweets can be obtained from the Twitter API by continuously listening to the stream. By adding the necessary rules, relevant tweets can be identified and extracted from the stream. These tweets can then be used to classify them and notify the relevant authorities by sending an email.

Currently, this system is specific to a city, but it can be extended to multiple metropolitan cities, which will help concerned authorities quickly resolve the issues that the public addresses. We can prioritize problems, so that the most important ones can be taken care of as soon as possible.

References

1. Liu, Y., Shen, W., Yao, Z., Wang, J., Yang, Z., Yuan, X.: Named entity location prediction combining twitter and web. IEEE Trans. Knowl. Data Eng. **33**(11), 3618–3633 (2021). https://doi.org/10.1109/TKDE.2020.2973261

2. Karami, A., Lundy, M., Webb, F., Dwivedi, Y.K.: Twitter and research: a systematic literature review through text mining. IEEE Access **8**, 67698–67717 (2020). https://doi.org/10.1109/access.2020.2983656

3. Bencke, L., Cechinel, C., Munoz, R.: Automated classification of social network messages into smart cities dimensions. Futur. Gener. Comput. Syst. **109**, 218–237 (2020). https://doi.org/10.1016/j.future.2020.03.057

4. Alotaibi, B., Abbasi, R.A., Aslam, M.A., Saeedi, K., Alahmadi, D.: Startup initiative response analysis (SIRA) framework for analyzing startup initiatives on Twitter. IEEE Access (2020). https://doi.org/10.1109/access.2020.2965181

5. Shah, Z., Dunn, A.G.: Event detection on Twitter by mapping unexpected changes in streaming data into a spatiotemporal lattice. IEEE Trans. Big Data **8**(2), 508–522 (2022). https://doi.org/10.1109/TBDATA.2019.2948594

6. Mendhe, C.H., Henderson, N., Srivastava, G., Mago, V.: A scalable platform to collect, store, visualize, and analyze big data in real time. IEEE Trans. Comput. Soc. Syst. **8**(1), 260–269 (2021). https://doi.org/10.1109/TCSS.2020.2995497

7. Agarwal, A., Toshniwal, D.: Face off: travel habits, road conditions and traffic city characteristics bared using Twitter. Received April 3, 2019, accepted April 14, 2019, date of publication May 15, 2019, date of current version June 4, 2019. https://doi.org/10.1109/ACCESS.2019.2917159

8. Vyas, P., Reisslein, M., Rimal, B.P., Vyas, G., Basyal, G.P., Muzumdar, P.: Automated classification of societal sentiments on Twitter with machine learning. IEEE Trans. Technol. Soc. **3**(2), 100–110 (2022). https://doi.org/10.1109/TTS.2021.3108963

9. Yang, J., Zhao, W.: What and with whom? Identifying topics in twitter through both interactions and text. IEEE Trans. Serv. Comput. **13**(3), 584–596 (2020). https://doi.org/10.1109/TSC.2017.2696531

10. Lu, X.S., Zhou, M., Wu, K.: A novel fuzzy logic-based text classification method for tracking rare events on twitter. IEEE Trans. Syst. **51**(7), 4324–4333 (2021). https://doi.org/10.1109/TSMC.2019.2932436

11. Wang, Q., Bhandal, J., Huang, S., Luo, B.: Classification of private tweets using tweet content. In: 2017 IEEE 11th international conference on semantic computing (ICSC), pp. 65–68 (2017). https://doi.org/10.1109/ICSC.2017.36

12. Fiscus, J.G., Doddington, G.R.: Topic detection and tracking evaluation overview. In: Allan, J. (ed.) Topic detection and tracking, pp. 17–31. Springer, Boston, MA (2002)

13. Gupta, B.B., Sangaiah, A.K., Nedjah, N., Yamaguchi, S., Zhang, Z., Sheng, M.: Recent research in computational intelligence paradigms into security and privacy for online social networks (OSNs). Future Gener. Comput. Syst. **86**, 851–854 (2018). https://doi.org/10.1016/j.future.2018.05.017

14. Sebastiani, F.: Machine learning in automated text categorization. ACM Comput. Surv. **34**(1), 1–47 (2002). https://doi.org/10.1145/505282.505283

15. Weiler, A., Grossniklaus, M., Scholl, M.H.: Editorial: Survey and experimental analysis of event detection techniques for Twitter. J. C **15**(6), 753–773 (2011). https://doi.org/10.1093/comjnl/bxw056

16. Silpa, C., Niranjana, G., Ramani, K.: Securing Data from Active Attacks in IoT: An Extensive Study. In: Manogaran, G., Shanthini, A., Vadivu, G. (eds.) Proceedings of international conference on deep learning, computing and intelligence. Advances in intelligent systems

and computing, vol. 1396. Springer, Singapore (2022). https://doi.org/10.1007/978-981-16-5652-1_5

17. Silpa, C., Suneetha, I., Reddy Hemantha, G., Arava, R.P.R., Bhumika, Y.: Medication alarm: a proficient IoT-enabled medication alarm for age old people to the betterment of their medication practice. J. Pharm. Negative Res. **13**(4), 1041–1046 (2022)

18. Baseer, K.K., Jahir Pasha, M., William Albert, D., Sujatha, V.: Navigation and obstacle detection for visually impaired people. In: 2021 Fourth international conference on microelectronics, signals & systems (ICMSS), pp. 1–3 (2021). https://doi.org/10.1109/ICMSS53060.2021.9673618

19. Jyothsna, V., Sreedhar, A.N., Mukesh, D., Ragini, A.: A network intrusion detection system with hybrid dimensionality reduction and neural network based classifier. In: Tuba, M., Akashe, S., Joshi, A. (eds.) ICT systems and sustainability. Advances in intelligent systems and computing, vol. 1077. Springer, Singapore (2020). https://doi.org/10.1007/978-981-15-0936-0_19

20. Jyothsna, V., Prasad, M., GopiChand, G., Bhavani, D.D.: DLMHS: Flow-based intrusion detection system using deep learning neural network and meta-heuristic. Int. J. Commun. Syst.Commun. Syst. **35**(10), 5159 (2022)

Drowsiness Detection and Prevention Models for the Elderly People: A Promising Design

Shivendra Dubey[1]($\boxtimes$), Hameed Hassan Khalaf[2], Ausama A. Almulla[3],
Mustafa Asaad Hussein[4], and Israa Abed Jawad[5]

[1] Department of AIML, Manipal University, Jaipur, Rajasthan, India
shivendra.dubey5@gmail.com
[2] Department of Medical Laboratories Technology, Al-Manara College for Medical Sciences,
Maysan, Iraq
[3] Department of Medical Laboratories Technology, Al-Hadi University College,
Baghdad 10011, Iraq
[4] College of Nursing, National University of Science and Technology, Dhi Qar, Iraq
[5] Department of Medical Laboratories Technology, AL-Nisour University College, Baghdad,
Iraq

Abstract. Internet of Things (IoT) devices are becoming increasingly popular at an exponential rate in the modern day because of the ease with which they can be used to keep tabs on a person's everyday health status. By allowing devices to communicate with one another over the internet, IoT allows users greater freedom of choice in how they define their healthcare infrastructure. Because it utilises sensors and actuators to monitor people's health, it is essential that it is flexible enough to work with the limited infrastructure of the Internet of Things. In cases where an old person has fallen, it can alert emergency personnel using picture processing technology. Elderly people require a great deal of care, but in today's era of busy careers, working adults often lack the time to devote to that responsibility. Those who have virtual relationships with others now cannot imagine life without the Internet of Things. The application of these methods to the development of more effective scientific advances holds great promise for the resolution of many pressing issues. The Internet of Things not only facilitates communication, but also raises awareness of numerous pressing matters. This paper introduces the Elderly Fall Detection system, a monitoring system for detecting falls that uses Internet of Things (IoT) and Internet Protocol (IP) approaches to warn others before a loss occurs. There are many proposed algorithms for determining tiredness, however the majority of these methods focus on analysing the subject's facial expression or even skin tone variations.

Keywords: Elderly Fall Detection · Wearable Sensors · BAN · Cloud data

1 Introduction

As more people look to wearable electronics like smart glasses and wristwatches, wireless networking has become an even more critical concern. The wireless system collaborates with the BAND's wearable devices (BAN). The collection of a patient's health

© The Author(s), under exclusive license to Springer Nature Switzerland AG 2025
M. Gupta et al. (Eds.): MISS 2023, CCIS 1952, pp. 402–413, 2025.
https://doi.org/10.1007/978-3-031-69115-7_28

records requires the implantation of a significant number of sensors in order to continuously monitor critical health markers. These medical records or data must be sent wirelessly [1].

The rapid development of low-power integrated circuits made it possible for us to talk over a wireless network and physiological sensors. Additionally, it paves the way for future generations to utilise the Wireless Sensor Network (WSN) for keeping tabs on anything from crop yields and traffic to patient health and other forms of infrastructure maintenance.

Scientists claim that sensors can monitor a person's health by being placed within or outside the body (wearable sensors). These sensors can immediately relay the gathered data to a real person, alerting them to the imminent emergency [2].

Image processing (IP) has developed as a crucial method for detecting human motion in all BAN activities. This IP employs various techniques to glean relevant data from the gathered pictures. Data gathered by BAN's sensors and actuators can be sent to a mobile device or other central unit through Bluetooth or Wi-Fi. The data is then sent to a data centre through the Internet so that, in the event of an emergency, the medical centre can access the tracked record and provide better care [3]. The framework of BAN is depicted in Fig. 1. Elderly people require a great deal of care, but in today's era of busy careers, working adults often lack the time to devote to that responsibility. Those who have virtual relationships with others now cannot imagine life without the Internet of Things.

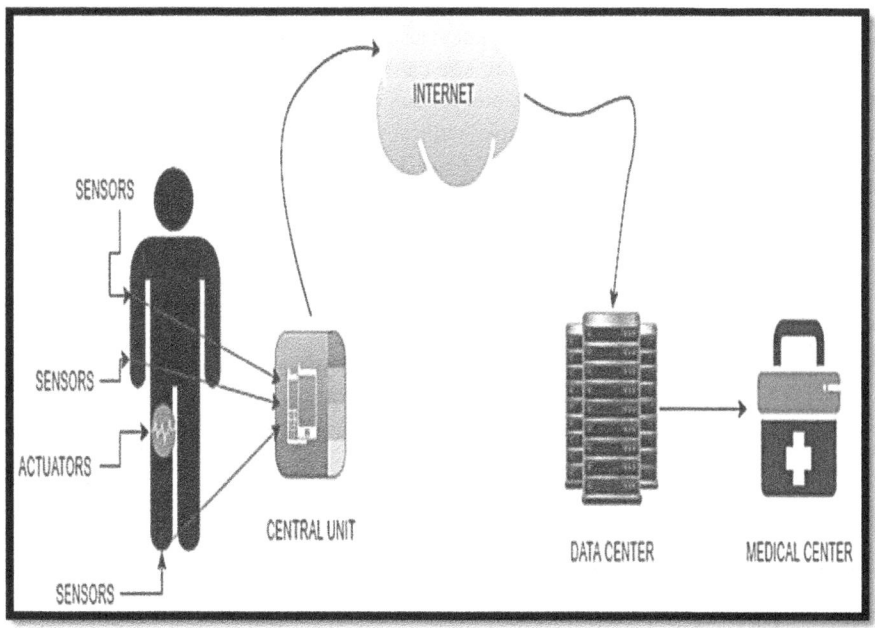

Fig. 1. BAN (Body Area Network) Architecture.

The application of these methods to the development of more effective scientific advances holds great promise for the resolution of many pressing issues. The Internet of Things not only facilitates communication, but also raises awareness of numerous pressing matters.

With the rise of the Internet of Things, security becomes a major issue for any type of communication, wireless or wired. Though safeguarding sensors and actuators is important, many industries, including those in medicine, physical training, the military, and others, place a higher priority on keeping sensitive data secure.

There are still a lot of big problems to be fixed with IoT, especially when it comes to things like security, useful applications, and ease of use. Since the elderly typically aren't up-to-date on the newest technological advancements, healthcare has risen to the top of the global agenda [4]. Although AI technologies play a crucial role in securing the data, a variety of Image Processing approaches are also used to detect and analyse falls.

Thanks to the widespread availability and accessibility of Internet of Things (IoT) devices, apps, and infrastructure, the IoT is ripe for surveillance and intrusion. Using the Internet of Things raises some security concerns and strains its capacity to process real-time data. As previous academics have noted, these factors increase the difficulty of the task of enhancing software security and precision.

The gathering of the data and its storage in the most secure area is the most vital component of the BAN security system. There is a wide variety of technology available that can assist in securing data or information that is susceptible to being attacked [5]. The BAN system is utilised so that users can carry out their routine tasks in an atmosphere that is comfortable to them. The parameter designed the key to keep the module size as small as possible or to prevent using any cables, to have a low power consumption in order to reduce the frequency of battery changes, and to give the system autonomy. This BAN has been intended to be usable while still being comfortable, unobtrusive, and discreet. A wide variety of sensors have been developed in order to enable the system connection, which in turn makes it possible to validate a variety of applications and signals. The following items are included in these particulars:

- Sampling Rate: The system that was created takes into account sensors that have a low sampling rate, and it is in the nature of the sensors to acquire the signals in order for them to continue.
- Connections for Interfaces: There are two different sorts of interfaces, and these interfaces define both forms of digital and analogue sensors.
- Monitoring Devices: These monitoring devices can take the form of electrodes, sensors, and a variety of other modalities. The ideas behind such a sensor are extremely all-encompassing. It is intended to check the connection that exists between the sensors, electrodes, and the monitoring equipment.

If BAN devices have been physically hacked, then the previous security context that was stored in the device will be accessible to the attackers that are targeting the device. Aside from this, the vast majority of techniques can either involve complicated key management or large computational overheads. Despite having a greater usability, it is not always worthwhile to secure the pairing device technique for the currently available alternatives that do not assume pre-shared secrets [6]. This is the case even though it is possible to do so. On the other hand, the existence has made an assumption for its unnecessary out-of-bond secure channel. It makes human-assisted verification easier, which is useful for processes that are not immediately obvious.

Wearable sensors enable users to communicate directly with one another, cutting out the need for a middleman in the process. It's possible that the BAN will play a significant role in the care given to elderly patients by medical technology. Falls among the elderly are the second most common cause of unintended and accidental injuries, and it is estimated that 6, 46,000 persons pass away each year as a direct result of injuries sustained in falls. Eighty percent of these people are residing in either a country with a low income or a country with a middle income on the global scale [7]. The primary objective of the research paper is to facilitate the prevention of falls among older adults through the use of a device that is compatible with BAN.

2   Existing Work Done

According to the findings of a study conducted by researchers, a variety of strategies have been utilized, ranging from specialised and more complex sensor-based strategies to the application of Machine Learning techniques in order to identify senior falls. The majority of approaches, which is a problem, fail to recognise a false positive.

Even the F1 scores and the other evaluation metrics only provide a limited amount of information regarding correctness in a wide variety of different circumstances. Within the framework of the Proposed Model, we will have a conversation about methods that are predicated on the positioning of sensors or other equipment that can detect falls. Wearable devices, ambience devices, motion devices, motion analysis, shape change analysis, and many other types of analysis are just some of the various ways in which a fall detection system can be integrated to monitor falls in the elderly. Accuracy, sensitivity, and specificity are three parameters that the authors believe should be considered when evaluating fall detection systems. Researchers used certain components of fall sensors to conduct descriptive tests [8]. A battery, microcontroller, acceleration sensor, and Bluetooth module were among these parts. The algorithm is straightforward, with an estimated sensitivity of 90–95% and a specificity of 91–93%.

The researcher looked into this, and found that sophisticated machine learning algorithms like DAGSVM and ellipse fitting techniques could be used to make sense of data from a wide variety of sensors like gyroscopes, accelerometers, and more, but that their use would be limited to the context of a smart home. In addition, it takes a significant amount of electricity to execute many mining algorithms on data each second [9]. Therefore, executing a complicated algorithm on a network with limited resources, such as a sensor network, is a difficult undertaking. In addition to detecting falls, it is able to recognise a variety of postures and actions carried out by a person within the context of a smart home.It does have an accuracy of 89–91%, a sensitivity range of 60–90%, and a specificity range of 98–100%. The large-scale deployment tests showed that it operated admirably. According to the findings of other researchers, one of the methods that is utilised is video feed to monitor home surroundings for signs of falls [10]. This method also utilises techniques such as elliptical approximation, projection histograms, and advanced foreground segmentation, among other techniques, in order to detect falls in a video feed. However, the question remains: who watches a video feed constantly enough to notice a fall? The workplace makes use of RGBD cameras to monitor for falls. The RGBD is a type of 3D camera that constantly analyses the video feed in order to identify different types of feeds and creates a 3D model of the surrounding environment. It employs innovative computer vision techniques on video data.

In the course of their investigation, the researchers made use of a gadget to gather data and a transmission medium to pass the data on for examination. The data has been organised into a variety of categories, but the primary issue with the system is that the algorithm only functions properly for data that has already been categorised in some way. Once any information is processed that does not belong to the defined classification, it will immediately begin giving false alarms [11]. Additionally, if another human is put in the position to wear the gadget, it will once again produce a false warning. In order to have a fully functional system, the entire system requires a lot of work.

The authors of this paper propose a method of feature extraction that consists of two segments. Following the completion of the feature extraction process with the help of the online tool, a machine learning strategy was utilised. According to the findings of the investigation, the results are mathematically accurate to a greater extent than 99.9%. The system must validate the information against the actual outcomes in real time. The primary goal of the researchers, which they successfully accomplished, was to reduce the amount of power that was consumed. The proposed model has one significant drawback, and that is that it will only be able to achieve success by working on the same frequency. This means that the human heart rate will need to remain unchanged for the model to be successful. If the heart rate increases, then the amount of power consumed could potentially increase as well depending on the heart rate [12].

Researchers carried out a study on this subject and came up with a model based on heat sensors to detect falls in elderly people's homes as a result of their findings. It is beneficial in the sense that one's privacy will be preserved by doing so. They were able to develop the system with the help of three algorithms, namely GRU, LSTM, and Bi-LSTM, and the results were accurate up to 93% of the time in the optimal setting. On the other hand, real-time observation presents a number of challenges, such as finding the ideal installation location, ensuring that there are no interruptions in the collection of data, and so on. Although it has potential, it has not yet been put into action or put to a real-world test [13].

In addition, the most reliable aspect of WSNs can be enlarged, which is especially useful in the healthcare industry. In the future, WSN-enabled smart environments will be able to monitor environmental factors and make other preventative measures dependent on the level of human effort applied to them [14].

The framework will eventually reach universality, at which point each individual will have an assessment module that is able to manage the other clever space that is accessible in the framework; this will even prevent the other medical problems. It is believed that the IP application that is used as a convention to interface the BSN in the healthcare setting is experimental [15].

It is essential to devote a significant amount of time and effort to researching new and improved technologies and ideas, such as senior consideration in the intelligent internet of things, because it is anticipated that the population of people aged 65 and older will experience a high rate of growth in the not too distant future (IoT) [16]. This is done in order to meet the needs of the elderly population.

Some large, medium, and small businesses have also become familiar with how to handle the requirements of the elderly, with specific objectives in investigating the primary IoT-based administrations and applications that have successfully launched for monitoring the wellbeing of elderly individuals in remote areas [17].

The need to screen the elderly arises from the desire to provide care for them through the application of unambiguous procedures [18]. This is done so that it can supply those benefits and simplicity to continue with their normal life by using these programmes in various regions. The reason for this is so that it can give those advantages and simplicity.

A system has been developed that provides an application with remote monitoring capabilities and that generates reports on a daily, weekly, and monthly basis. This framework has been made available to users. The specialist might be anyone, including clinical specialists, parental figures, or anyone who is affiliated with a crisis centre [19]. This framework is able to compile the feedback received from the specialist in order to provide the customer with personalisation and thereby enhance the customer's presentation. These frameworks give a few various issues, such as the genuineness of the information, the honesty of the information, and the confidentiality of the information.

The other available IP inside the smart sensor hubs is used to introduce the advantages in the environment of self-designed and information capture, but it also causes a communication bottleneck whenever a comparison is made to the current method [20].

In addition, it includes the typical behaviour that it expects such systems to exhibit at the present time. Many WSNs devices cooperate to accomplish the standard aim in a remote sensor network. Sensors implanted in humans will collect unique biological samples with the ultimate goal of checking patients' financial health regardless of their physical location.

When the data is transmitted wirelessly to the device's external Processing Unit, it assumes greater significance for the purposes of reaching the appropriate specialist or other recipient [21].

What's currently compelling is the amount of data provided by the framework and the necessity of resources suitable for powering the sensors. This study is just what the doctor ordered because it reveals the preliminary stages of research into the development of remote sensor organisations and concerned sensors for prosperity monitoring. This breakthrough is fascinating and has countless potential uses in fields as diverse as transportation, disaster management, medicine, the armed forces, entertainment, and more [22]. People have given these applications for the medical services regions a lot of thought. Researchers, academics, and businesses have all taken notice of the promising new developments in these charming wearables because they have made it possible for Bio-Medical Sensors to be used in evaluating people's well-being [23].

3 The Projected Work

The proposed study's fundamental dedication is in its peculiarity to spot elderly decline and even cross-approve it. It uses a reciprocal strategy to find the pit. At first, the sensors collect all the data associated with a fall, including, but not limited to, a Net speed increase of the body, Pitch and Roll, attributes that identify various exercises.

In addition, this study has made use of picture preparation to increase confidence in the model. In this case, the photos of various finger gestures made by a more mature adult at the time of autumn (Fig. 2).

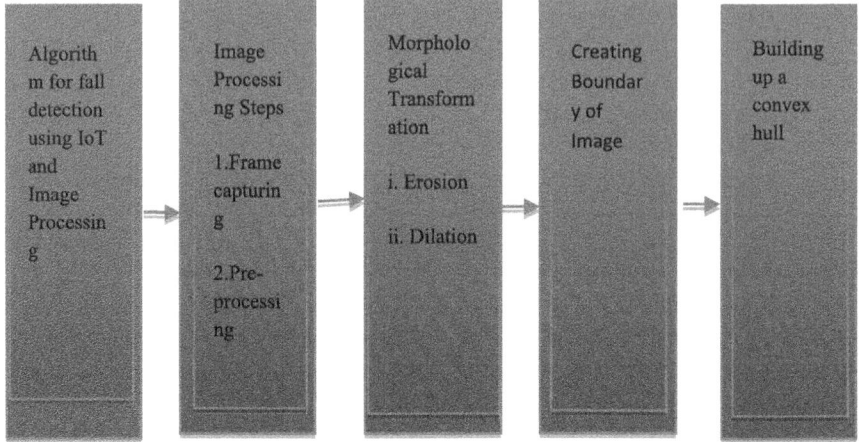

Fig. 2. The Projected Algorithm.

After extensive testing and calibration, this system activates fall detection and alert systems that are both reliable and timely. This study utilises a dataset collected while subjects sat, slept, walked, and performed other activities typical of daily life. As can be seen in Fig. 3, after a fall has started, that is, once the individual has made contact with the ground, there will be a momentary increase in the overall acceleration of the body.

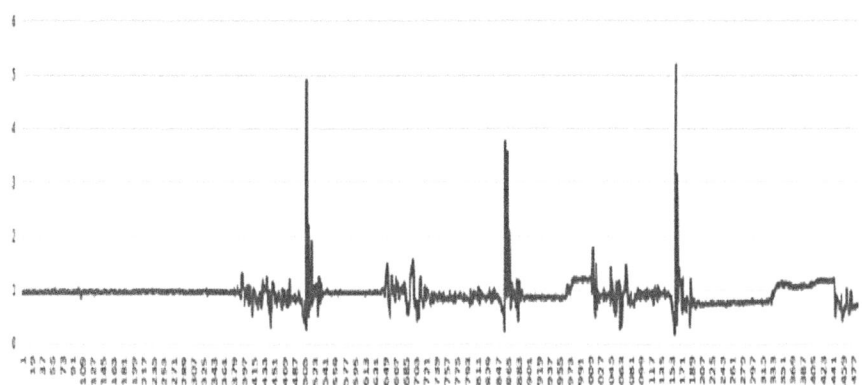

Fig. 3. Graph of net acceleration.

The orientation (angle) along the x-axis is referred to as the pitch. It is one of the most important aspects that will decide how far the descent will go. As depicted in Fig. 4, the algorithm predicts that there is a potential for a fall to occur if the angle of the pitch is more than sixty degrees.

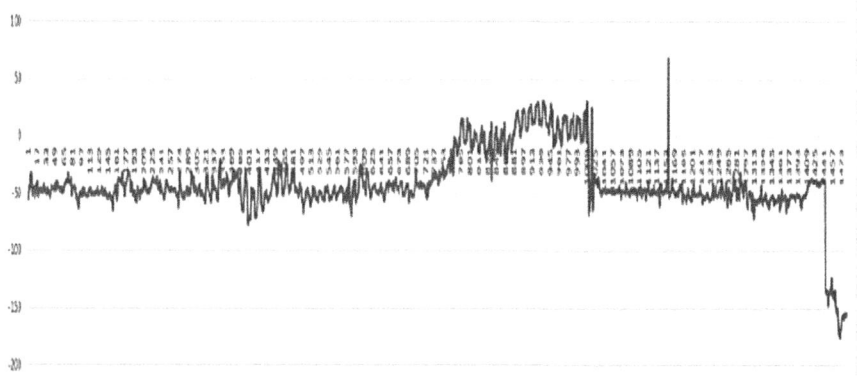

Fig. 4. Graph of pitch over.

Roll is the orientation (angle) along the Y-axis, and it is also a key component in the process of identifying and detecting the fall. According to our algorithm, the critical temperature is sixty degrees, as depicted in Fig. 5.

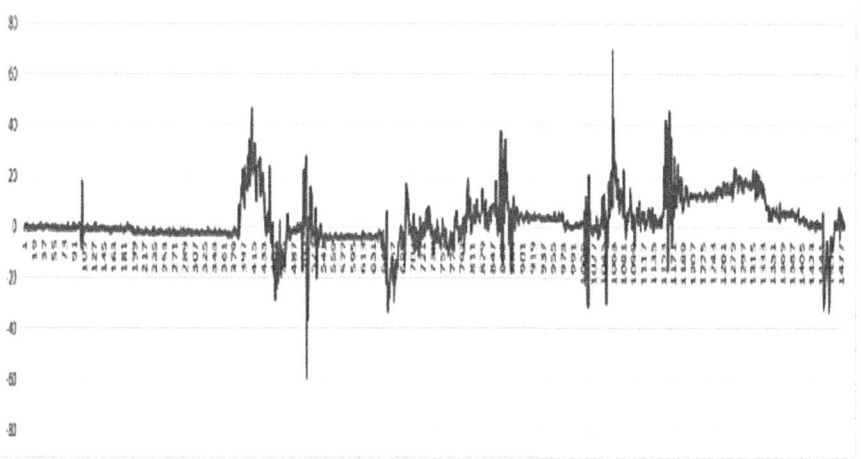

Fig. 5. Graph of roll over.

The system is able to say that a fall has been identified when all of the parameters concurrently exhibit abnormalities (there may be instances when one or two of the parameters are crossing the threshold because of ADLs), but otherwise it cannot. This manuscript has established a uniform parameter that is called the state, and it displays the state of fall in Fig. 6, as can be seen there. State 3 indicates that a fall is being detected,

State 4 shows that a fall has been confirmed, and State 5 is the SMS state. Condition 1 indicates that the patient is stable.

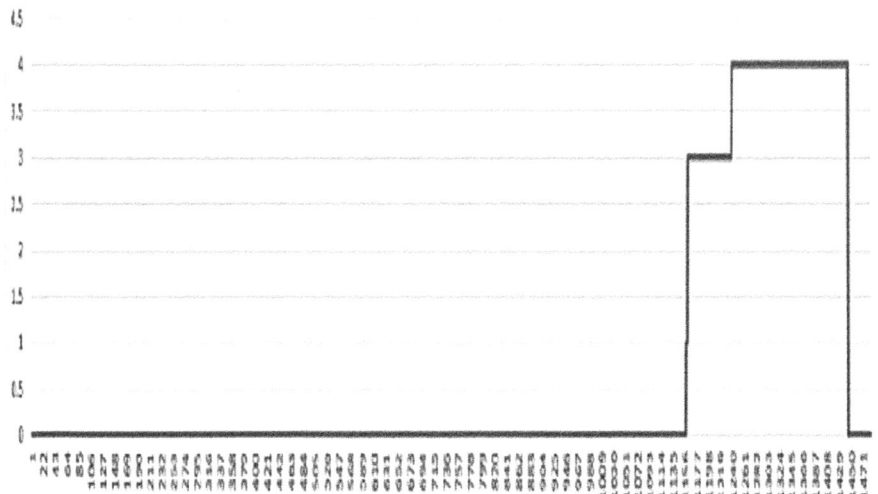

Fig. 6. Graph of the state.

The aforementioned sensors have been used to obtain all of these observations. Three algorithms, the Decision tree, the Support Vector Machine, and the Nearest Neighbour Classification, are trained using these features. All three machine learning algorithms were used, and Table 1 displays the resulting accuracy levels for the three characteristics (Roll, Acceleration, and Pitch). Among the various methods of accuracy measurement, SVM yields the best result, hence it is the one that will be used to implement ADA. The majority of SVM methods have an accuracy of 92.8% or above (Fig. 7).

Table 1. Analysis of Algorithm Accuracy.

Parameters	Pitch/roll (%)	Pitch (%)	Roll (%)	Acceleration (%)
Decision Tree	90.7	79.8	90.9	75.8
SVM	94.7	86.4	93.8	81.2
Nearest Neighbour Classifier	93.8	84.6	92.4	78.9

In contrast, the SVM model can only achieve an accuracy of 97.4%, 96.2%, and 98.3% when it is applied to walking, descending stairs, and ascending stairs, respectively. AFDI2, on the other hand, operates across all potential scenarios and delivers 99% accuracy most of the time in all possible conditions.

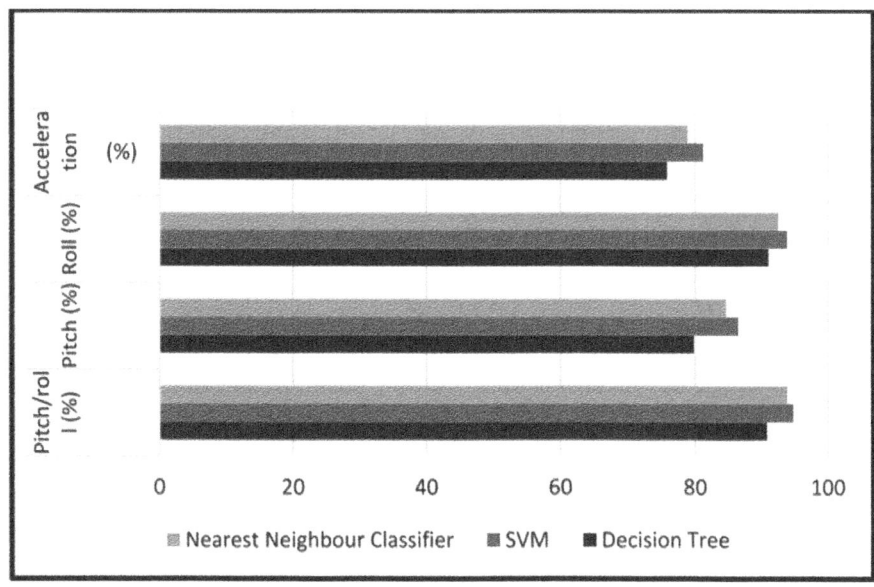

Fig. 7. Evaluation of Algorithm Accuracy.

4   Conclusion

Accidental or temporary memory loss might put older people at risk for fatal falls, if necessary, help is not provided to them as soon as possible. In this particular piece of research, we focus on the challenge of obtaining assistance precisely when it is required, as opposed to being constantly monitored. In this research project, we present a comprehensive strategy for developing a fall detection system for the elderly population. As a result, this method is dependable, it promotes an increased level of trust among members of the household, and it makes it easier to provide prompt responses. In order to improve evaluation metrics such as specificity, it is necessary to increase information flow while minimising the use of internet facilities. If a previously suggested model cannot be implemented in a place because it does not have internet connectivity, the model cannot be implemented there. In order to locate the hole, it makes use of a reciprocal method. The sensors begin by gathering all of the information that is associated with a fall. This information includes, but is not limited to, a Net speed increase of the body, Pitch, and Roll, which are distinguishable characteristics among the many activities. Algorithm for fall recognition using IoT and Image Processing (AFDI2), the work that we have done in this section, therefore works in all possible scenarios and provides 99% accuracy in all scenarios, even though the SVM model can only arrive at a precision of 97.4%, 96.2%, 98.3% while walking, walking on the ground floor, and walking on higher floors, respectively.

References

1. Kapoor, A., Bhat, S.I., Shidnal, S., Mehra, A.: Implementation of IoT (Internet of Things) and Image processing in smart agriculture. In: International conference on computation system and information technology for sustainable solutions (CSITSS). IEEE, pp. 21–26 (2016)
2. Sangwan, A., Pratim Bhattacharya, P.: Wireless body sensor networks: a review. Int. J. Hybrid Inf. Technol. **8**(9), 105–120 (2015)
3. Rupani, A., Whig, P., Sujediya, G., Vyas, P.: A robust technique for image processing based on interfacing of Raspberry-Pi and FPGA using IoT. In: International conference on computer, communications and electronics (Comptelix), pp. 350–353. IEEE (2017)
4. Nair, R., et al.: A survey on IoT (Internet of Things) emerging technologies and its application. IJEUCD **7**(2), 1–20 (2018)
5. Lakshmi Sirisha, B., Srinivas Kumar, S., Chandra Mohan, B.: Steganography based image sharing with reversibility. J. Discrete Math. Sci. Cryptograph. **19**(1), 67–80 (2016)
6. Patel, D., Tiwari, R., Pandey, S., Nikam, R.: Real-time fatigue detection system using computer vision. Int. J. Eng. Res. Technol. **9**(6), 1–5 (2020)
7. Zhang, H., Cheng, P., Shi, L., Chen, J.: Optimal denial-of-service attack scheduling with energy constraint. IEEE Trans. Autom. Control **66**(11), 3023–3028 (2015)
8. Zhang, H., Cheng, P., Shi, L., Chen, J.: Optimal DoS attack scheduling in wireless networked control system. IEEE Trans. Control Syst. Technol. **24**(3), 843–852 (2015)
9. Cai, L., Zeng, K., Chen, H., P., Mohapatra, H.: Good neighbor: adhoc pairing of nearby wireless devices by multiple antennas. In: Network and distributed system security, pp. 1–20 (2020)
10. Farooq, M.U., Waseem, M., Mazhar, S., Khairi, A., Kamal, T.: A review on internet of things (IoT). Int. J. Comput. Appl. **113**(1), 1–7 (2015)
11. Kerdjidj, O., Ramzan, N., Ghanem, K., Amira, A., Chouireb, F.: Fall detection and human activity classification using wearable sensors and compressed sensing. J. Ambient Intell. Hum. Comput. **11**(1), 349–361 (2020)
12. Khan, T., Singh, K., Son, L.H., Basset, M.A., VietLong, H., Singh, S.P., Manjul, M.: A novel and comprehensive trust estimation clustering-based approach for large scale wireless sensor networks. IEEE Access **7**, 58221–58240 (2019)
13. Nair, R., Bhagat, A., Shukla, P.K., Dutta, P.K.: Hybridizing deep neural network for genes expression classification using histone modification. IET Digit. Library **2020**, 252–256 (2021)
14. Kumar, V., Badal, N., Mishra, R.: Body sensor networks architecture and security issues in healthcare application. In: Presented in an international conference on computational research and data analytics organized by Suleyman Demirel University and Isparta Applied Sciences University, Turkey and College of Engineering Roorkee, Indiaon 24th October (2020)
15. Krawczyk, Z.: Application of wireless sensor networks to early warning and collision avoidance in road transport. Doctoral Dissertation, Radom University of Technology, (2010)
16. Li, Z, Huang, A., Xu, W., Hu, W., Linzhen: Fall perception for elderly care: a fall detection algorithm in smart wristlet mhealth system. In: 2014 IEEE International Conference on Communications (ICC), pp. 4270–4274. IEEE (2014)
17. Liang, X., Li, X., Shen, Q., Lu, R., Lin, X., Shen, X., Zhuang, W.: Exploiting prediction to enable secure and reliable routing in wireless body area networks. In: 2012 Proceedings IEEE INFOCOM, pp. 388–396. IEEE (2012)
18. Alsultanny, Y.: Systematic approach for image partitioning and shuffling. J. Discrete Math. Sci. Cryptograph. **10**(1), 55–71 (2013)
19. Lee, W.O., Lee, E., Park, K.R.: Blink detection robust to various facial poses. J. Neurosci. Methods **93**(2), 356–372 (2010)

20. Danisman, T., MariuosBilasco, I., Djeraba, C., NasaneIhaddadene: Drowsy driver detection system using eye blink patterns. In: International conference of Machine and Web Intelligence, pp. 230–233. IEEE (2010)
21. Carmien, S., Obach, M.: Back on track: lost and found on public transportation. In: International conference on universal access in human-computer interaction, pp. 575–584. Springer (2011)
22. Drutarovsky, T., Fogelton, A.: Eye blinks detection using variance of motion vectors. In: Computer vision - ECCV workshops (2014)
23. Fang, S.H., Liang, Y.C., Chiu, K.M.: Developing a mobile phone-based fall detection system on android platform. In: Computing, communications and applications conference, pp. 143–146. IEEE (2012)

An Empirical Evaluation of Pre-trained Convolutional Neural Network Models for Neural Style Transfer

Akash Sudan, Goutam Singh Chouhan, Dilip Singh Sisodia$^{(\boxtimes)}$, and Arti Anuragi

Department of Computer Science and Engineering, National Institute of Technology Raipur, Raipur, India

{dssisodia.cs,aanarayandas.phd2018.cs}@nitrr.ac.in

Abstract. Neural style transfer (NST) is a technique used to stylize an image with the style of another image. In the past, this required the skills of an artist, but later machine learning was used to tackle this problem. Various hand-crafted methods were employed to extract features from the image, but these local, global, or combined features still produced subpar results. The idea of image analogies was then utilized to translate one image's style to another by understanding their relationship. However, this method couldn't learn the content-based association between images. With the success of convolutional neural networks (CNNs) in image-related tasks like detection, CNNs were used in NST. This study investigates style transfer using four different CNN models such as AlexNet, MobileNet, VGG-16, and VGG-19. Six pairs of images, referred to as content and style images, were evaluated with these pre-trained models, and their performance was evaluated based on content loss, style loss, runtime, and memory consumption. Results showed that VGG-19 outperforms other pre-trained CNN models in terms of quality, as its content loss is lower than the style loss.

Keywords: Neural style transfer (NST) · deep CNN · Pre-Trained VGG-19 Models

1 Introduction

Painting is a beloved form of art that has captivated people for thousands of years, with iconic works such as van Gogh's "The Starry Night." Previously, creating a reinterpretation of an image in a specific style necessitated the expertise of a skilled artist and a significant amount of time [1]. The art theories behind visually appealing artworks have garnered the interest of both artists and computer science researchers since the mid-1990s. Numerous studies and techniques exist for converting images into synthetic artworks automatically. The field of computer graphics has seen significant advancements in non-photorealistic rendering (NPR) [2], which has become a well-established field. Despite this progress, many NPR stylization algorithms are limited to specific artistic styles and cannot be easily adapted to other styles. In the computer vision community,

© The Author(s), under exclusive license to Springer Nature Switzerland AG 2025
M. Gupta et al. (Eds.): MISS 2023, CCIS 1952, pp. 414–427, 2025.
https://doi.org/10.1007/978-3-031-69115-7_29

style transfer is generally studied as a generalized problem of texture synthesis involving the extraction and transfer of textures from a source to a target. The field of style transfer encompasses various image enhancement and generation techniques, which aim to extract semantic information from an image and modify it to achieve specific styles. The methods used can include colour correction, toning, image filtering, and texture synthesis and vary depending on the type of style desired. There has been extensive research into changing the style of an image based on a source image. Hertzmann et al. [3] introduced a framework called "image analogies" for generalized style transfer by learning the analogous transformation from example pairs of unstylized and stylized images. However, these methods have a common limitation in that they rely solely on low-level image features and may not effectively capture image structures.

Further inspired by the capabilities of Convolutional Neural Networks (CNNs), which have been already explored in many domains such as [4, 5]. Gatys et al. [6] recently explored using a CNN to imitate famous painting styles on natural images. They modelled the content of a photo as the feature responses from a pre-trained CNN and the style of artwork as the feature statistics. Their results showed that a CNN could extract both content information from any photo and style information from a well-known artwork. CNNs, as a type of deep learning network, are well-suited for applications that handle pixel data and image recognition. They are the preferred network architecture for object recognition among various deep-learning neural networks. This makes them ideal for computer vision applications that require object recognition. The architecture of CNNs comprises numerous layers of small computer units that process data in a sequential manner. The research of Gatys et al. established a new area called neural style transfer (NST), which involves using a convolutional neural network to apply various styles to a content image. Figure 1 illustrates an example of style transfer, where the style of a (b) style image (screaming image) is applied to a (a) content image (landscape image) to produce a (c) output image. The aim of NST is to empower people who don't have a background in art to create visually appealing images. The advancements and progress made in this field are significant.

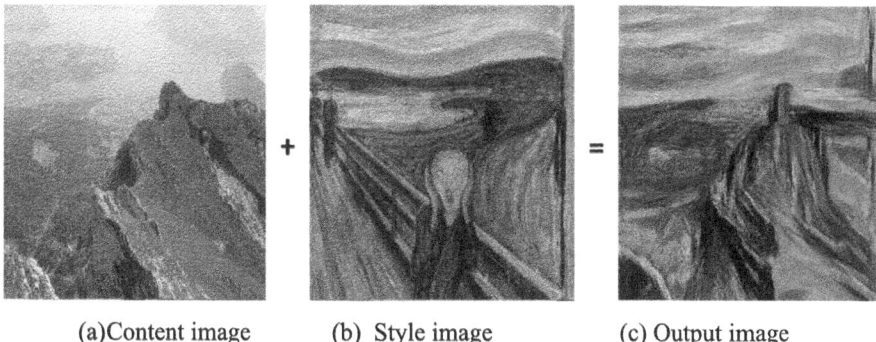

(a)Content image   (b) Style image   (c) Output image

Fig. 1. An example of an NST algorithm to transfer the style of (a) Content image onto a (b) Style image [1].

Wang et al. [7] proposed three quantifiable parameters to evaluate the effectiveness of style transfer. They also suggested two strategies, multi-objective networks, and cascade style transfer, to leverage these three attributes and improve upon previous work.

Park et al. [8] used a technique called semantic aware style transfer, which involves dividing the source and target images into various parts. By learning style from the source region and content from the target region, they were able to integrate both components to produce a stylistic result with the best semantic match between regions. This approach is an improvement over earlier methods.

Building upon current NST advancements, Ulyanov et al. [9] made improvements. Previous works by Gatys et al. [6] demonstrated that deep networks could generate stunning textures and stylized images from a single texture example. However, the optimization process for these methods is time-consuming and slow. Ulyanov et al. [9] proposed an alternative strategy that shifts the computational cost to the learning stage. Their approach trains compact feed-forward convolutional networks to generate multiple samples of the same texture at varying sizes from a single texture example and to transfer artistic style from one image to another. This method is much more efficient as it only requires evaluating the network once and eliminates the need for backpropagation.

Atarsaikhan et al. [10] used NST to produce digital fonts. With NST, they were able to effectively communicate the differences between the two font styles, reducing the time-consuming process of font creation.

NST is also applied to steganography, as noted by Aggarwal et al. [11]. The process involves hiding the content of a cover image using the style of another image, called the secret image, and then retrieving the original content, which is known as de-stylizing the image. This creates a way to hide sensitive information in the form of images through the use of steganography.

Research has been conducted in the areas of Visual Style Modeling and Image Reconstruction. There are two methods of Visual Texture Modeling (VTM), namely Parametric Texture Modeling (PTM) using Summary Statistics and Non-parametric Texture Modeling (NPTM) utilizing Markov Random Fields (MRFs). Image Reconstruction can be further divided into two categories: Image Optimization-Based Online Image Reconstruction (IOB-IR) and Model-Optimization Based Offline Image Reconstruction (MOB-IR) [12].

The paper will give a brief overview of neural networks and their architecture for neural style transfer. It will then explain the important loss functions and implementation of the transfer. The main focus of the paper will be to review the four major approaches for implementing neural style transfer, which has improved upon the original method. The paper will also compare the advantages and disadvantages of each approach [13].

Li et al. [14] introduced a solution to reduce artefacts and distortions in stylized images by incorporating Laplacian loss, called Lapstyle objective, into the optimization process. The Laplacian loss ensures the stylized image has a similar image Laplacian to the content image.

Future, this study aims to explore the use of NST to create new images by using pre-trained models such as AlexNet, MobileNet, ResNet, VGG-16, and VGG-19 with a loss function that transfers the style of one image to another. The focus is to evaluate the performance of these models and discuss the future scope of NST and its potential applications.

The paper is organized into four sections. Section 2 covers the methods and methodologies used to implement the NST model, including the datasets used to evaluate the model, the deep learning models studied, and the evaluation parameters used. Section 3 presents the experimental results achieved by using AlexNet, MobileNet, ResNet, VGG-16, and VGG-19 models. Finally, in Sect. 4, the authors provide a conclusion summarizing their findings.

2   Methods and Methodology

Figure 2 depicts the basic process of the proposed NST models, in which inputs, including content images and style images, were provided to pre-trained models such as AlexNet, VGG19, VGG16, and MobileNet. After calculating two losses, such as style loss and content loss, the total loss is calculated by adding them together. The final output image is then generated if the total loss has been minimized to the desired level or the maximum number of iterations has been reached.

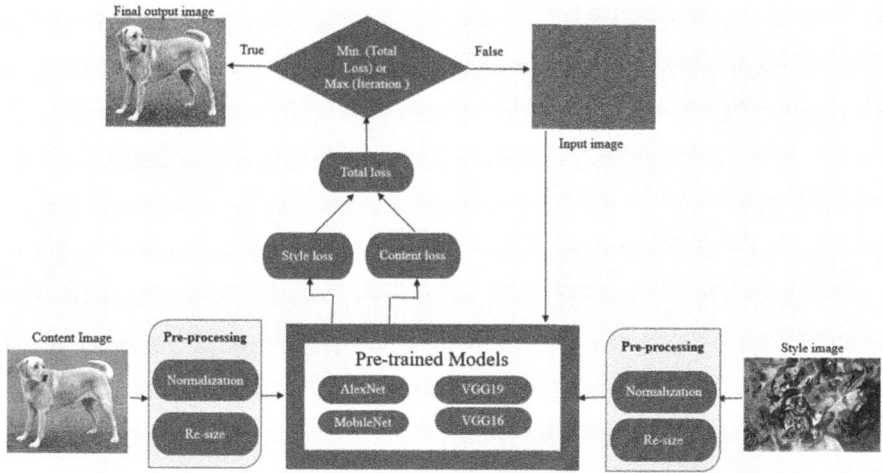

Fig. 2. The overview of our framework NST model using pre-trained models.

2.1   Dataset

This experiment used six diversified sets of style and content images. Each set consisted of a pair of content images and style images, which is also demonstrated in Table 1. The input images are utilized to create a style transfer image that features the texture of the chosen style image. The deep pre-trained models were run on a diverse range of image sets to assess their performance, and performance metrics, including the loss function, execution time, and memory utilization, were determined. An overview of the models is presented in the subsequent sections.

Table 1. Diversified content and style images were used in this study.

Different image sets	Content image	Style image
Set 1		
Set 2		
Set 3		
Set 4		

(continued)

Table 1. (*continued*)

Set 5

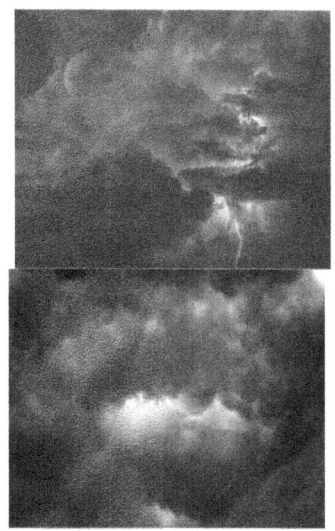

Set 6

2.2 NST Using Deep Pre-trained CNNs

Convolutional Neural Networks (CNNs) are a type of deep learning algorithm commonly used in image processing [15]. They are designed to process and analyze images by learning the relationships between pixels and features in an image. In a CNN, the input image is convolved with a set of filters, each of which is used to detect specific patterns or features in the image. These filters are then processed by multiple layers of neurons, which allow the network to learn increasingly complex representations of the image. The final layer of the network generates the output, which can be a classification of the image or a generated image. CNNs have been shown to be highly effective in a wide range of image-processing tasks, such as object detection, image classification, and style transfer. Figure 3 illustrates the standard architecture of a CNN, consisting of multiple layers, including convolution layers, pooling layers, and fully connected layers. A brief discussion of each layer in a CNN is provided below.

Convolution Layer
A convolution layer in a CNN is a key component that performs the convolution operation on the input image. The convolution operation involves taking the dot product of a small neighborhood of pixels in the input image (called a filter or kernel) with the corresponding pixels in the image. This operation extracts and amplifies certain features or patterns in the image, such as edges, textures, or shapes. The result of the convolution operation is a feature map that represents the processed image in a new feature space. Multiple filters can be applied to the same image to extract different features, and the result of multiple filters can be combined to form a single feature map. The Convolution Layer is typically followed by an activation function, such as ReLU, to introduce non-linearity

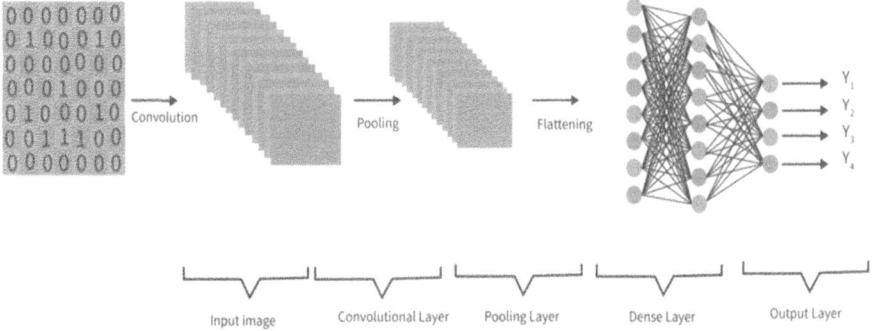

Fig. 3. Illustration of the various layers in a CNN.

into the network. Convolution layers are repeated multiple times in a CNN, with each subsequent layer learning increasingly complex representations of the image.

Pooling Layer
A pooling layer in a CNN is used to reduce the spatial size of the feature maps produced by the convolution layers. The main purpose of pooling is to reduce the computational complexity of the network, prevent overfitting, and increase the invariance of the features to small translations and deformations in the image. There are two main types of pooling: max pooling and average pooling. Max pooling selects the maximum value from each pooling window, while average pooling computes the average of the values in each pooling window. Pooling layers typically use small windows that slide over the feature maps and subsample the values in each window. The size of the window and the stride of the sliding operation are hyperparameters that can be adjusted to control the size of the feature maps. After multiple pooling layers, the feature maps are reduced to a smaller size but still retain important information about the image features.

Fully Connected Layer
A fully connected layer in a CNN performs a dot product between the input and a set of learned weights and applies an activation function. It is used as the final layer and produces the final output. The layer maps the flattened feature maps to the desired output through a supervised learning process. The number of neurons and the activation function used are adjustable hyperparameters.

Pre-trained models using CNNs are pre-trained models that have already learned image features on large datasets and can be fine-tuned for various computer vision tasks such as image classification, object detection, and segmentation. Popular pre-trained models include VGGNet, AlexNet, and MobileNet. They save time and resources and are widely used in industry and research. The following provides a brief overview of these models:

2.2.1 VGG-16

VGG-16 is a 16-layer CNN architecture introduced by the Visual Geometry Group at the University of Oxford (Simonyan and Zisserman) [16]. It was created to study the effect of accuracy in image recognition models and was presented at the 2014 ImageNet Challenge. With a top-5 test accuracy of 92.7% on the ImageNet dataset, VGG-16 has become a popular NST model for feature extraction.

2.2.2 VGG-19

VGG-19 is a deep CNN having 19 layers, including 16 convolutional layers and 3 FC layers [17]. It employs 3x3 filters with a stride of 1 and uses max pooling in the pooling layers. A ReLu activation function is applied to introduce non-linearity. The VGG-19 model was developed to improve performance by increasing the depth of the CNN and is based on innovative object recognition models.

2.2.3 AlexNet

AlexNet is a deep CNN introduced in 2012 by Krizhevsky et al. [18]. It was the first CNN to win the ImageNet Large Scale Visual Recognition Challenge (ILSVRC), surpassing traditional computer vision models. AlexNet consists of 8 layers, including five convolutional layers, two fully connected (FC) layers, and one output layer. The convolutional layers use rectified linear unit (ReLU) activation functions and max pooling. The FC layers also use ReLU activation. AlexNet demonstrated the potential of deep learning for computer vision and has inspired numerous other CNN architectures.

2.2.4 MobileNet

MobileNet is a lightweight CNN designed to efficiently deploy mobile and embedded devices with limited computational resources [19]. MobileNet uses depth-wise separable convolutions to reduce the number of parameters and computational costs, making it suitable for resource-constrained devices. The architecture consists of multiple depth-wise separable convolution blocks, each followed by pointwise convolutions and a ReLU activation function. MobileNet also employs a global average pooling layer to reduce the spatial dimensions of the feature maps before the final classification layer. MobileNet has achieved good accuracy on various image classification benchmarks while being significantly more computationally efficient than other deep CNNs.

2.3 Evaluation Parameters

This study employed the NST technique with four models: AlexNet, VGG-16, MobileNet, and VGG-19, to generate an artistic image by combining the content and style of two separate images. In Sect. 2.1, the images utilized for this experiment are described. The effectiveness of these models was assessed based on four performance parameter metrics: style loss, content loss, execution time, and memory consumption. Each parameter is described in detail below:

The content loss function aims to preserve the content of the original image in the generated image. CNNs are capable of capturing high-level content information, but they place more emphasis on specific pixel values at lower levels. To address this, the content loss function is defined using the activations of the top layer of the CNN. The purpose of the style loss function is to maintain the correlations of activations across all layers in both the style image and the generated image.

Let A^l be the activation of the Lth layer, j^{th} position and i^{th} feature map obtained using image I. Then the content loss is defined as defined in Eq. (1)

$$L_{content} = \frac{1}{2} \sum_{ij} (A_{ij}^l(g) - A_{ij}^l(c))^2 \tag{1}$$

Basically, $L_{content}$ captures the mean squared error between the activations generated by the content image and the generated image. The style loss is defined in Eq. (2):

$$L_{style} = \sum_l w^l L_{style}^l \tag{2}$$

where, L_{style}^l in Eq. (2) can be defined as follows:

$$L_{style}^l = \frac{1}{M^l} \sum_{ij} (G_{ij}^l(s) - G_{ij}^l g)^2 \tag{3}$$

where, $G_{ij}^l(s)$ in Eq. (3) can be mathematically expressed as follows:

$$G_{ij}^l(s) = \sum_k A_{ik}^l(s) A_{jk}^l(s) \tag{4}$$

Gram matrices are needed to compute the style loss.

$$G_{ij}^l = \sum_k F_{ik}^l F_{jk}^l \tag{5}$$

And then, the Final loss (L) is computed as defined in Eq. (6).

$$L = \alpha L_{content} + \beta L_{style} \tag{6}$$

where α and β are weighting factors for content and style reconstruction, respectively.

3 Experimental Result

This section exhibits the experimental setup. In this study, a comprehensive experiment was carried out to generate an artistic image through the NST technique using pre-trained models on six image sets. The process involved in the investigation is demonstrated in Fig. 1. It consists of the following steps: (1) acquiring the dataset listed in Table 1, (2) loading the model into the IDE, (3) pre-processing the input images by resizing, normalizing, and reshaping, and (4) passing the images through pre-trained models (VGG-16, AlexNet, MobileNet, and VGG-19) to produce the output image and evaluate performance metrics, such as style loss, content loss, memory consumption, and execution time over 500 iterations.

3.1 Experimental Setup

The experiment was conducted in a Jupyter Notebook and Visual Studio Code environment using Python 3.8.10. Key libraries utilized in the implementation included TensorFlow 2.12.0, NumPy 1.23.4, Matplotlib 3.6.2, Torch 1.13.0, and Memory Profiler 0.61.0.

After training the models, the achieved results of each model are shown below.

3.2   Achieved Experimental Results Using AlexNet.

The experiment involves using every pair of images from the dataset in Table 1 as input for AlexNet, with one image serving as the content image and the other as the style image. After training the models, the achieved results, such as style loss, content loss, execution time, and memory usage, are presented in Table 2 from each set. The results show that from all image sets, the average content loss (3.3042) is greater than the average style loss (0.09145), indicating that the generated image has more of the style's texture imposed onto the content.

Table 2. Performance of AlexNet model using different considered datasets.

Datasets	Style loss	Content loss	Loss	Time (Sec)	Memory (MiB)
Set-1	0.234	9.834	3326.69	51.8	211.14
Set-2	0.143	0.7724	219.75	60.8	209.59
Set-3	0.0454	2.94	697.75	41.1	315.68
Set-4	0.0267	2.6913	535.89	55.7	228.90
Set-5	0.0157	0.6016	217.65	36.1	205.21
Set-6	0.0839	2.9864	1137.46	40.1	221.62
Average	0.09145	3.3042	–	47.6	232.02

3.3   Achieved Experimental Results Using VGG-16

The experiment involves using every pair of images from the dataset in Table 1 as input for VGG-16, with one image serving as the content image and the other as the style image. After training the models, the achieved results, such as style loss, content loss, execution time, and memory usage, are presented in Table 3 from each set. The results show that from all image sets, the average content loss (16.696) is greater than the average style loss ($2.19E-05$), indicating that the generated image has more of the style's texture imposed onto the content.

3.4   Achieved Experimental Results Using MobileNet

The experiment involves using every pair of images from the dataset in Table 1 as input for MobileNet, with one image serving as the content image and the other as the style image. After training the models, the achieved results, such as style loss, content loss, execution time, and memory usage, are presented in Table 4 from each set. The results show that from all image sets, the average content loss (0.3382) is greater than the average style loss ($3.84E-09$), indicating that the generated image has more of the style's texture imposed onto the content.

Table 3. Performance of VGG-16 model using different considered datasets.

Datasets	Style loss	Content loss	Loss	Time (Sec)	Memory (MiB)
Set-1	1.99E−05	26.4532	28.45	796.2	625.51
Set-2	7.63E−06	4.5	5.27	601.7	561.82
Set-3	3.29E−05	23.288	26.76	614.9	577.47
Set-4	8.93E−06	24.3688	25.26	836.7	659.57
Set-5	4.56E−05	12.298	16.53	638.2	557.45
Set-6	1.63E−05	9.27	10.25	398.7	438.23
Average	2.19E−05	16.696	–	737.73	570

Table 4. Performance of MobileNet model using different considered datasets.

Datasets	Style loss	Content loss	Loss	Time (Sec)	Memory (MiB)
Set-1	3.24E−09	0.4225	0.457	960	180.75
Set-2	5.09E−09	0.3686	0.42	850	164.21
Set-3	5.57E−09	0.4556	0.49	586	190.45
Set-4	2.86E−09	0.2926	0.321	540	188.22
Set-5	3.19E−09	0.2131	0.245	570	189.84
Set-6	3.07E−09	0.2769	0.308	555	187.16
Average	3.84E−09	0.3382	–	676.83	183.4

3.5 Achieved Experimental Results Using VGG-19

The experiment involves using every pair of images from the dataset in Table 1 as input for VGG-19, with one image serving as the content image and the other as the style image. After training the models, the achieved results, such as style loss, content loss, execution time, and memory usage, are presented in Table 5 from each set. The results show that from all image sets, the average content loss (9.705) is smaller than the average style loss (1810160). VGG-19 is the only model in which average style loss is much higher values than average content loss, which indicates that the generated style transfer image has more features of content's texture and only has a slight texture of style image.

The results of using deep pre-trained models, such as AlexNet, VGG-16, MobileNet, and VGG-19, in generating NST output images are depicted in Fig. 4. Although the input content and style images are the same, the generated output images vary between the models. The VGG-19 model performed better than the other models, producing a higher-quality image with less content loss.

In addition to losses, execution time and memory usage are computed in this study. The bar graphs for execution time and memory consumption from all diversified image sets across all models are shown in Figs. 5 and 6, respectively. The study shows that

Table 5. Performance of VGG-19 model using different considered datasets.

Datasets	Style loss	Content loss	Loss	Time (Sec)	Memory (MiB)
Set-1	2453822	13.87	189.18	3060	403.3
Set-2	2046572	11.76	180.45	2750	458.23
Set-3	3327335	9.74	332.73	3180	835.80
Set-4	2118756	14.91	211.87	2880	795.21
Set-5	583822	2.32	58.32	5280	407.85
Set-6	330658	5.63	33.00	3240	406.24
Average	1810160	9.705	–	3398	551.105

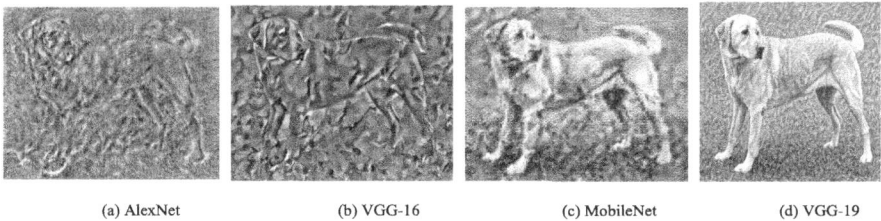

(a) AlexNet   (b) VGG-16   (c) MobileNet   (d) VGG-19

Fig. 4. Illustration of neural style transfer output images on different (a) AlexNet, (b) VGG-16, (c) MobileNet, and (d) VGG-19 models by providing Set-1 style and content images.

Fig. 5. Execution time bar chart for AlexNet, MobileNet, VGG-19, and VGG-16 models over all datasets.

MobileNet uses minimum memory among all four models, and AlexNet takes minimum time for execution.

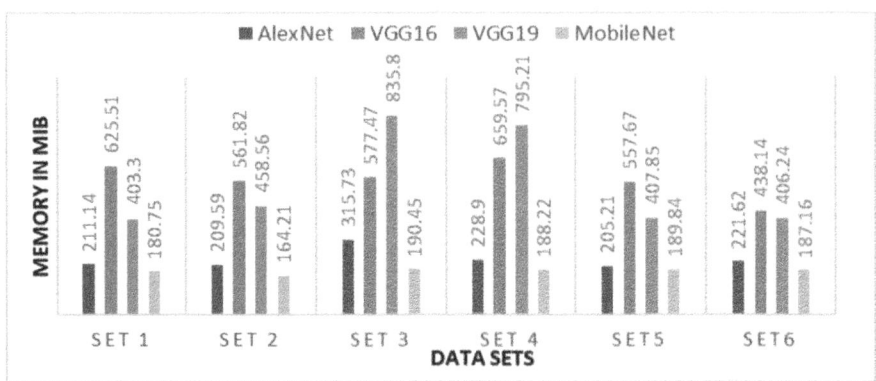

Fig. 6. Memory usage bar chart for AlexNet, MobileNet, VGG-16, and VGG-19 models over all datasets.

4 Conclusion

In this study, we evaluated the performance of four deep pre-trained models, AlexNet, VGG-16, MobileNet, and VGG-19, in blending the style and content of two images through NST, a method that utilizes CNNs. Using six sets of images, we calculated and compared the style loss, content loss, execution time, and memory consumption of these models. The results showed that the VGG-19 model excelled in terms of style loss and content loss, with a greater emphasis placed on the content than the style. This is reflected in the resulting image, where the VGG-19 model outperforms the other models by producing a better balance of content and style. Readers can apply these models to their own projects and choose the best model based on specific usage and evaluation criteria.

References

1. Chen, Y., Lai, Y.K., Liu, Y.J.: Transforming photos to comics using convolutional neural networks. In: Proceedings of International Conference on Image Process. ICIP, vol. 2017-Septe, pp. 2010–2014 (2018)
2. Gooch, A., Gooch, B.: Non-photorealistic rendering. CRC Press, New York (2001)
3. Hertzmann, A., Jacobs, C.E., Oliver, N., Curless, B., Salesin, D.H.: Image analogies. In: Proceedings of the 28th annual conference on computer graphics and interactive techniques, SIGGRAPH 2001, no. August, pp. 327–340 (2001)
4. Shrivastava, R., Sisodia, D.S., Nagwani, N.K.: Deep neural network-based multi-stakeholder recommendation system exploiting multi-criteria ratings for preference learning. Expert Syst. Appl. **213**, 119071 (2023)
5. Sisodia, D., Sisodia, D.S.: Feature space transformation of user-clicks and deep transfer learning framework for fraudulent publisher detection in online advertising. Appl. Soft Comput. **125**, 109142 (2022)
6. Gatys, L., Ecker, A., Bethge, M.: A neural algorithm of artistic style. J. Vis. **16**(12), 326 (2016). https://doi.org/10.1167/16.12.326
7. Wang, D., et al.: Evaluate and improve the quality of neural style transfer. Comput. Vis. Image Underst. **207**, 103203 (2020)

8. Park, H., Park, J.H., Shim, S.: Semantic-aware neural style transfer. Image Vis. Comput. **87**, 13–23 (2019)

9. Ulyanov, D., Lebedev, V., Vedaldi, A., Lempitsky, V.: Texture networks: feed-forward synthesis of textures and stylized images, vol. 3, pp. 2027–2041 (2016). *arXiv Prepr.* arXiv1603.03417

10. Atarsaikhan, G., Iwana, B.K., Uchida, S.: Neural style difference transfer and its application to font generation. In: Lect. Notes Comput. Sci. (including Subser. Lect. Notes Artif. Intell. Lect. Notes Bioinformatics), vol. 12116 LNCS, pp. 544–558 (2020)

11. Aggarwal, J.S., et al.: Neural style transfer for image within images and conditional GANs for destylization. J. Vis. Commun. Image Represent. **85**, 103483 (2022)

12. Li, J., Wang, Q., Chen, H., An, J., Li, S.: A review on neural style transfer. J. Phys. Conf. Ser. **1651**(1), 3365–3385 (2020). https://doi.org/10.1088/1742-6596/1651/1/012156

13. Jayaram, U., et al.: Different techniques in neural style transfer-a review. In: Proceedings of first international conference on computational electronics for wireless communications: ICCWC 2021, pp. 337–350 (2021)

14. Li, S., Xu, X., Nie, L., Chua, T.S.: Laplacian-steered neural style transfer. In: MM 2017 – Proceedings of 2017 ACM Multimedia Conference, pp. 1716–1724 (2017). https://doi.org/10.1145/3123266.3123425

15. T. Alshalali and D. Josyula, "Fine-tuning of pre-trained deep learning models with extreme learning machine," Proc. - 2018 Int. Conf. Comput. Sci. Comput. Intell. CSCI 2018, pp. 469–473, 2018

16. K. Simonyan and A. Zisserman, "Very deep convolutional networks for large-scale image recognition," 3rd Int. Conf. Learn. Represent. ICLR 2015 - Conf. Track Proc., pp. 1–14, 2015

17. Bhangale, K.B., Desai, P., Banne, S., Rajput, U.: "Neural Style Transfer: Reliving art through Artificial Intelligence", 2022 3rd Int. Conf. Emerg. Technol. INCET **2022**, 1–6 (2022)

18. S. Kavitha, B. Dhanapriya, G. N. Vignesh, and K. R. Baskaran, "Neural Style Transfer Using VGG19 and Alexnet," in 2021 International Conference on Advancements in Electrical, Electronics, Communication, Computing and Automation, ICAECA 2021, 2021, pp. 2–7

19. Howard, A.G., et al.: MobileNets: efficient convolutional neural networks for mobile vision applications. *arXiv Prepr.* arXiv1704.04861 (2017)

Offline Handwritten Signature Identification and Verification Using LBP Features

Chitvan Gupta[1]([✉]), Bhawna Singh[2], Vandana Bharti[3], Karishma Chauhan[2], and Shivam Tiwari[4]

[1] Department of Computer Science and Engineering, Noida Institute of Engineering and Technology, Gr. Noida, India
chitvangupta@gmail.com

[2] CSE Department, G.L.Bajaj Institute of Technology and Management, Greater Noida, India
bhawna.jadaun@gmail.com, karishmachauhan800@gmail.com

[3] Department of Computer Science, KIET Group of Institutions, Delhi NCR, Ghaziabad, India
vandnabharti233@gmail.com

[4] Department of Computer Science and Engineering, Sharda University, Greater Noida, India
shivam.tiwari.info@gmail.com

Abstract. Biometrics is the science and technology of automatically identifying people by their unique physical and behavioural traits. More and more people are being verified by biometric methods. In order to increase signature verification's classification accuracy, researchers are always looking into new feature representations.

As a robust feature extraction method, local binary pattern (LBP) has been proposed to capture the texture. LBP is a texture measure that works in any colour space, as it is based on a neighbourhood-level definition of texture. The LBP operator converts a 3×3 area's grayscale value to binary by using it as a threshold for a binary code. You may describe the texture using the histogram of these labels. To generate a binary code for each pixel in a greyscale image, the LBP approach begins with a basic processing step. Intensity differences between surrounding pixels and the current pixel are encoded by this code. Each subject is given a set of 100 signatures, of which 10 are authentic and 3 are forgeries, to use as training data. The testing process involves determining whether or not a submitted signature is a legitimate copy of the subject's original. In order to extract features, the LBP operator is used, which is a gray-scale invariant texture measure based on a generic definition of texture in the context of a particular neighbourhood. By thresholding a 33 area by its central grey value, the LBP operator generates a binary code. A texture description is then derived from the histograms of these labels. Tests are conducted on signature photos from the local dataset and the MCYT database. Signature verification systems have used KNN and SVM classifiers. KNN and SVM classifiers' overall recognition accuracy and FAR are provided.

Keywords: LBP · SVM · KNN · FD · FAR

© The Author(s), under exclusive license to Springer Nature Switzerland AG 2025
M. Gupta et al. (Eds.): MISS 2023, CCIS 1952, pp. 428–439, 2025.
https://doi.org/10.1007/978-3-031-69115-7_30

1 Introduction

Signature recognition refers to the process of verifying a person's identity by using their signature as part of the verification process. A signature in [1] not only establishes the identity of the party signing, but it also acts as proof that the signer offered their agreement after carefully considering the implications of their actions. [Case in point] A method for verifying the authenticity of signatures by using characteristics such as the number of inner contours and the number of components of vertical slopes.

The practise of using a person's fingerprints to confirm their identity extends back hundreds of years. The surface of the finger has ridges and furrows that form a core around which designs such as swirls, loops, and arches are coiled, making each fingerprint unique [2]. The fundamental purposes of a fingerprint scanner are to capture a photograph of your finger and to compare that picture to previously scanned photos in order to determine whether or not the ridge and valley patterns match. The scientific method of fingerprinting has developed alongside the passage of time.

It takes into account both the behavioural and the logical factors that play a role in the development of linguistic variety. Speech authentication makes use of speaker-specific traits like cadence, nasal tone, inflection, and fundamental frequency. Examples of these qualities include. Any kind of human speech can be translated into electrical signals and compared to a library of human sounds that have been recorded. Because something as simple as having a cold can cause a change in a person's voice, the accuracy of the system is limited, and in some instances, identification may be impossible.

In order to accomplish palm identification, a three-dimensional image of the hand must first be collected. Next, feature vectors must be created from the image and compared to those that have been saved in a database. These devices may be cumbersome, but they are capable of identifying their targets in a quick and accurate manner [3].

Researchers have determined that of all of the biological traits that were presented earlier, signature recognition is one of the tasks that is the most challenging. Because signatures are so important in the corporate world, the legal system, and the financial sector, it is essential to have a trustworthy mechanism for recognising and validating signatures. The authenticity of a handwritten signature obtained via a digital device can be divided primarily into two categories: static and dynamic. A visual comparison of two scanned signatures or a scanned signature and an ink signature written side-by-side is the most prevalent type of static. In dynamic digital handwritten signature authentication, the signer's signature is "read" from the signing device and then translated into coordinate values for use in the authentication process. Identifying who the signature belongs to and determining whether or not it is a fake are separate but interconnected processes in the process of authenticating a signature [4]. Methods of visual processing and feature extraction are utilised in the verification process of static signatures.

General procedures for recognising a signature include the ones below.

1. Obtaining a photograph of the suspect's signature.
2. a signature image analysis
3. using the retrieved features, check them against a known set of previously gathered signature photos in the database.
4. Decide if there is a match or not based on the results of the comparison.

During matching, the input signature is checked to see if it is a legitimate one or a fake (false signatures of an individual). Two stages make up this section: Verifying a person's identification using the first-signature part's database and other identifying information. Essentially, we check each subject's input signature image against every other sample in the database.

Input signature images are verified by being compared with samples of the same subject's signatures. The primary focus of this procedure is the confirmation of an individual's identity.

The method used to authenticate or verify a signature may depend on the writer or it may be able to stand on its own [5]. When a writer takes an autonomous method, they often stick to the steps outlined below. We calculate prior parameter distributions for means of 'genuine vs genuine' and 'forgery versus known classes' based on a distance measure. Posterior class probabilities for both genuine and forged signatures for a given author are calculated. Next, we calculate the probabilities for each group and select the group with the highest probability for a signature under suspicion. In the case of the author-dependent method, each user's classifier is created independently based on his enrolment data. In the verification phase, the classifier simply examines query signature samples. Having a large enough data set from which to train the classifiers is obviously crucial to the success of such systems.

2 Existing Work Done

The method presented by authors to recognise and verify signatures in less time with high accuracy and less effort. In [6], the Discrete Wavelet Transform (DWT) with Haar wavelets is primarily considered for feature extraction for both global features and grid features. A technique for offline verification of signatures based on a small number of simple geometric features was given. Area, Euler's number, eccentricity, standard deviation, centroid, skewness, kurtosis, and orientation are some of the features utilised in [7]. Offline signature verification using an AIRS and ANN has been proposed.

It is clear from the literature study that many different approaches have been proposed for signature verification, and that experiments are conducted on either local datasets or standard datasets like MCYT in [8] to report verification results. In other words, the methods described in [9] only work with signatures that are written in those languages, which limits their applicability. Recognizing that offline signature verification is a challenging endeavour that still requires investigation, we proposed the development of efficient verification systems that improve performance not only for locally-written signatures but also for signatures written in other languages defined in a standard database of signature images.

Learning-based signature recognition systems necessitate a sizable training set, ideally including examples from the vast majority of the intended users. For the database examples to be useful, they must account for all of the nuances that can arise in the course of a real-world application, including any personal nuances that may arise [10]. Additionally, some static signature datasets have been made public for study.

The purpose of pre-processing a picture is to get it ready to be fed into an algorithm for a specific purpose, such as target tracking, recognition, feature extraction, etc. Noise reduction, normalisation, illumination adjustment, blur and focus adjustment, thresholding, edge enhancements, morphology, segmentation, colour space conversion, scaling, and thinning are typical pre-processing processes [11]. In order to improve the feature description of the signature images we will be processing for verification, image pre-processing techniques are employed. Noise is a common by product of the signature collection and scanning process, and it is therefore not uncommon for an offline signature image to exhibit some degree of this. This necessitates pre-processing of the collected signature photos. Artifacts, false edges, invisible lines, corners, fuzzy objects, and disturbed backgrounds are all side effects of noise, which is unwanted information in digital photographs. Filters are proposed in the literature [12] to help mitigate these unintended consequences. Various types of error, such as broken pixel elements in cameras, insufficient storage space, poor digitization procedures, random bit errors in communication channels, and so on, can lead to salt and pepper noise, which corrupts digital images. Salt and pepper noise appears in scanned signatures due to the low quality of the paper and the scanning process. Scanned signature photos have had some of their pixel values altered [13].

Common nonlinear techniques for reducing salt and pepper noise include median filtering. The output pixel at the kernel's centre is chosen to be a pixel from the input image that falls within the window. Each original image pixel (m, n) serves as the centre of a kernel frame, and the median value of all pixels within the kernel frame is calculated. This median value is applied to the output image's pixel located at coordinates (m, n). In order to reduce random noise in an image while still keeping the image's edges intact, median filtering can be used [14]. The median filter can be used to remove the background noise from scanned signature images.

Signature verification systems classify a given signature as either authentic (made by the claimed individual) or forged, depending on the circumstances around the signature (produced by an impostor). This is a difficult task, especially for an offline (static) scenario that makes use of images of scanned signatures. The fact that a handwritten signature, which is already the biometric that is most frequently accepted for identification, is the primary benefit of signature verification in comparison to other forms of biometric technology, such as fingerprint verification or voice verification. Since the 1960s up until the present day, a large number of research activities on biometric person authentication have been carried out in [15]. A significant amount of progress has been suggested across the published literature.

The rotation of a signature is required in order to lessen the effect of angular changes that have been accumulated over time. It brings the axes of mass and inertia of each signature into alignment with one another along a horizontal axis. The edge of the signature is identified using an edge detector and then thinned (or skeletonized), after which the Radon Transform is applied, and the angle of rotation is measured in the anticlockwise direction [16]. This is one method for accomplishing the alignment. After that, the signature is turned anticlockwise to correct the skewness. The skew in the image can also be removed by utilising the bilinear interpolation technique, which is yet another method. The second method is the one that is utilised to rectify the skew in signature images.

The scanning and capturing technique can result in somewhat different dimensions for the signature. In addition, the width and height of a signature can differ from person to person and even across different versions of the same individual's signature. As a result, each of the photos containing signatures have had their dimensions adjusted so that they are all of the same size before being processed further. The size of the signature image has been adjusted to 40 × 60 pixels on average. The proportion of a signature's width to its height is preserved throughout the normalisation process. This is referred to as the aspect ratio [17].

This method makes use of the global, directional, and grid aspects of signatures. The Support Vector Machine, often known as SVM, was utilised in order to validate and categorise the signatures [18–20]. The investigators tested the performance of signature verification on two distinct tablet computers available for purchase. The use of a database containing three thousand different signatures allows for the reporting of authentication performance studies. Methods now considered state-of-the-art for signature verification have been analysed [21–23]. This method improves the way of turning a straightforward feature of a signature into a binary feature vector in order to improve the method's representational similarity with training signatures. They used a combination of spatial pyramid sampling grids and equimass sampling grids so that they could improve the representation of a signature that was based on gradient direction [24].

3 The Proposed Work

Researchers are still looking for more effective feature representations to aid with signature verification's classification challenges. It has been suggested that the powerful feature extraction method known as local binary pattern (LBP) can be used to accurately capture object texture. The Local Binary Pattern (LBP) is a texture metric that does not depend on colour or intensity. The LBP operator converts a 3 × 3 region's grayscale value into binary by using it as a threshold for a binary code. As a result, the texture may be described by looking at the histogram of these labels. In the most fundamental form of LBP, a greyscale image is processed to produce a binary code for each pixel. This number represents whether the nearby pixels' intensities are higher or lower than the current pixel's. According to the relative intensities of the surrounding pixels, a binary code of length 8 is created consisting of 0's and 1's, with the present pixel serving as the centre of a 3 × 3 neighbourhood. The frequency with which each binary code appears is then used to create a histogram, which can be used to characterise the distribution of frequent textural patterns. Multiple works have made use of LBP for its usefulness in offline signature verification. The reason is that LBP efficiently encodes pixel patterns that are close together. Many other forms of LBP have been proposed. The proposed works for LBP pattern selection are limited, though. The original LBP method's sparse histogram generation, for instance, of size 256 for a 3 × 3 region, is a major shortcoming. The vast majority of these motifs cannot be observed in a thumbnail image.

This study makes heavy use of the Local binary pattern (LBP) operator to extract statistically based texture features. The histogram of these local patterns is utilised as a feature descriptor, and the LBP operation explains them. In this paper, we provide a system for verifying signatures offline by extracting features using a Local Binary

Pattern (LBP) method. Both the KNN and SVM Classifiers use the retrieved features as input, ensuring their accuracy [25]. The MCYT dataset and a local database's worth of signature photos are used in the experiments. The proposed system has been shown to be effective through experimental testing (Fig. 1).

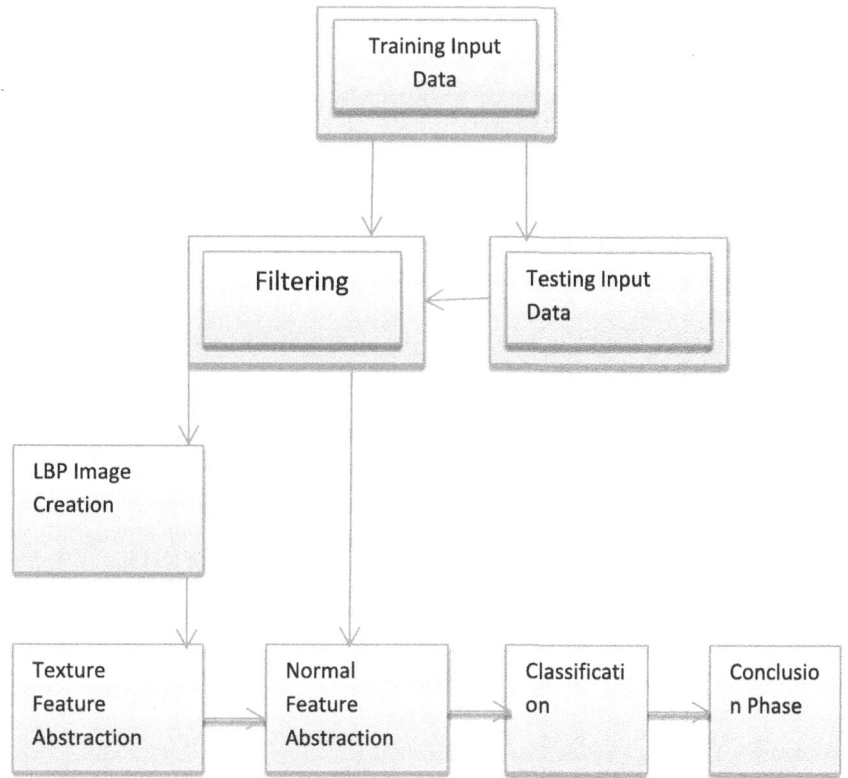

Fig. 1. The Proposed Block Diagram for LBP grounded Feature Abstraction.

- A median filter is applied to the signature photos after they have been converted to grayscale in order to remove the salt and pepper noise that is present in the images. After that, the thresholding method developed by Otsu is applied to the images in order to convert them from grayscale to binary.
- For the purpose of deriving statistically sound texture characteristics, this investigation makes extensive use of the Local binary pattern (LBP) operator. The LBP operation provides an explanation for these local patterns, and the histogram of these patterns is used as a feature descriptor in the process. In this research, we present a system for verifying signatures without an internet connection by extracting characteristics using a method known as Local Binary Pattern, or LBP. Through the use of several experiments, it has been demonstrated that the proposed system is efficient (Fig. 2).

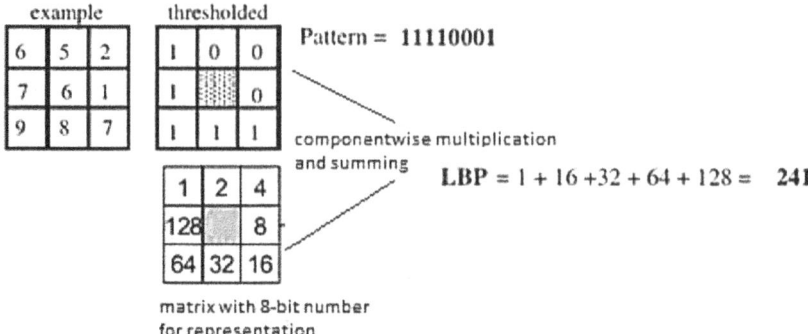

Fig. 2. Evaluating LBP Code.

Classifiers such as KNN and SVM are utilised in order to determine the type of script. The KNN is a slow classifier that assigns a label to a test image based on how similar its feature vector is to the full feature vectors of the images that were used during training. This similarity is measured using a Euclidean distance metric, which the KNN utilises. The SVM classifier employs supervised learning as a method for data classification. The technique uses the previously labelled training data to build an optimal hyperplane for categorising fresh samples. This hyperplane may then be used by the user.

The features of subjects' test signatures are extracted, and the features that are derived from this process are then fed into a classifier for the purposes of verification. The results of the verification are provided using the notations TAR, FAR, and FR (Figs. 3, 4, 5 and Tables 1 and 2).

Table 1. Consequences using KNN classification.

	Original	Fake	TAR	FAR	Accuracy	
Original	100	0	100	0	91%	K = 1
Fake	22	78	78	22		
Original	87	13	87	13	87%	K = 3
Fake	17	83	83	17		

A practical method for signature authentication/verification is proposed; employing characteristics extracted using the LBP operator. KNN and SVM classifiers are used in the experiments. KNN outperforms SVM when it comes to the MCYT database, but the two classifiers achieve the same average recognition rate when applied to photos of local signatures.

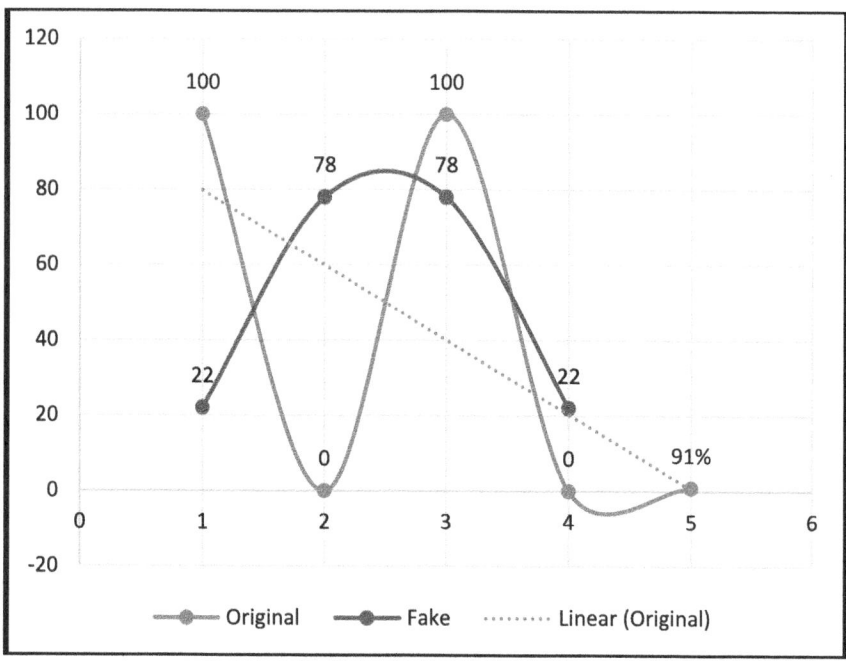

Fig. 3. KNN Classifier enactment (K = 1).

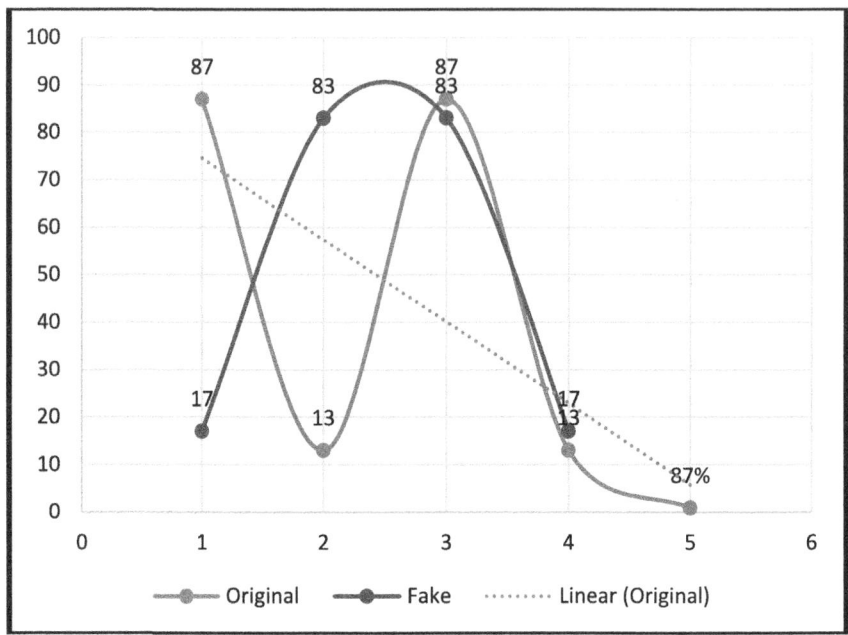

Fig. 4. KNN Classifier enactment (K = 3).

Table 2. Consequences using SVM classification.

	Original	Fake	TAR	FAR	Accuracy
Original	85	15	85	15	86%
Fake	14	86	86	14	

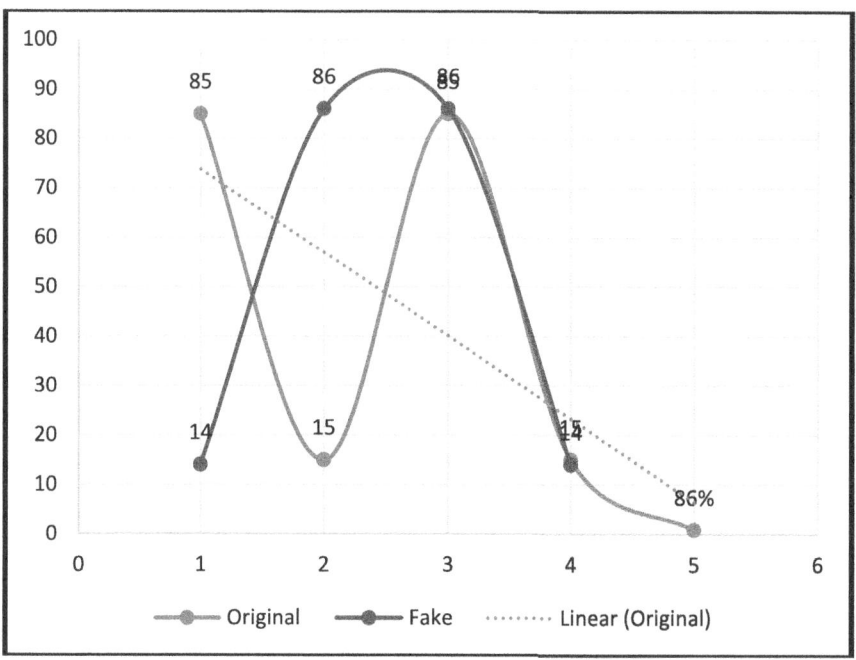

Fig. 5. SVM Classifier enactment.

4   Conclusion

Signatures have always been closely associated with police enforcement throughout history. When two or more parties use their signatures to establish their identities and enter into a contract or agreement, those signatures give the agreement the force of law. This can happen when two or more parties use their signatures. As a consequence of this, more people are interested in Signature Recognition than in other biometric technologies such as fingerprint scanning. Within the scope of this paper, we provide several powerful algorithms for the offline recognition and authentication of signatures. From the pool of genuine and forged signatures, ten genuine signatures and three forged signatures are chosen at random for each trainee from the set. During the testing process, it will be determined whether or not a signature that has been provided is an authentic copy of the subject's original signature. The LBP operator, which is used to extract features, is a texture measure that is not dependent on the colour space. The LBP operator will produce a binary code by thresholding a 3×3 area based on the central grey value of the area. The histograms of these labels are what have been looked at while attempting to describe the various textures. The MCYT database and the local dataset provide us with the signature photos that we use for our research. KNN and SVM are two examples of classifiers that have been utilised in signature verification systems. Utilizing the LBP operator to extract certain properties the experiments make use of the KNN and SVM classifiers respectively. When it comes to the MCYT database, KNN performs better than SVM, but when applied to photographs of local signatures, both classifiers reach the same average recognition rate.

References

1. Parviainen, P., Tihinen, M., Kääriäinen, J., Teppola, S.: Tackling the digitalization challenge: How to benefit from digitalization in practice. Int. J. Inf. Syst. Proj. Manag. **5**, 63–77 (2017)
2. Brennen, J.S., Kreiss, D.: Digitalization. In: The international encyclopedia of communication theory and philosophy, pp. 1–11. Wiley, Hoboken, NJ, USA (2016)
3. Krizhevsky, A., Sutskever, I., Hinton, G.E.: Image net classification with deep convolutional neural networks. Adv. Neural. Inf. Process. Syst. **25**, 84–90 (2012)
4. Kao, H.H., Wen, C.Y.: An offline signature verification and forgery detection method based on a single known sample and an explainable deep learning approach. Appl. Sci. **10**, 3716 (2020)
5. Cruz, S., Paulino, A., Duraes, J., Mendes, M.: Real-time quality control of heat sealed bottles using thermal images and artificial neural network. J. Imaging **7**, 24 (2021)
6. Zhou, Y., Zheng, J., Hu, H., Wang, Y.: Handwritten signature verification method based on improved combined features. Appl. Sci. **11**, 5867 (2021)
7. Saleem, M., Kovari, B.: Online signature verification based on signer dependent sampling frequency and dynamic time warping. In: Proceedings of the 2020 7th international conference on soft computing & machine intelligence (ISCMI), pp. 182–186, Stockholm, Sweden, November (2020)
8. Jia, Y., Huang, L., Chen, H.: A two-stage method for online signature verification using shape contexts and function features. Sensors **19**, 1808 (2019)
9. Poddara, J., Parikha, V., Bharti, S.K.: Offline signature recognition and forgery detection using deep learning. In: The 3rd international conference on emerging data and industry 4.0 (EDI40), April 6–9 (2020)
10. Tambade, V., Varma, P., Sonawale, A.: Bank cheque signature verification system. Int. J. Res. Eng. Sci. Manag. **1**(9), 265 (2018)
11. Nair, R., Bhagat, A.: A life cycle on processing large dataset-LCPL. Int. J. Comput. Appl. **179**(53), 27–34 (2018)
12. Tolosana, R., Vera-Rodriguez, R., Fierrez, J., Ortega-Garcia, J.: DeepSign: deep on-line signature verification. IEEE Trans. Biometrics Behav. Identity Sci. **3**(2), 229–239 (2021)
13. Ahmad, H.M., Hameed, S.R.: Eye diseases classification using hierarchical MultiLabel artificial neural network. In: Proceedings of the 2020 1st. information technology to enhance e-learning and other application (IT-ELA), pp. 93–98, Baghdad, Iraq, July (2020)
14. Saleem, M., Kovari, B.: Survey of preprocessing techniques and classification approaches in online signature verification. In: Image analysis and recognition. ICIAR 2020. Lecture notes in computer science. Springer, Cham, New York, NY, USA, vol. 12131 (2020)
15. Antal, M., Szabó, L.Z., Tordai, T.: Online signature verification on MOBISIG finger-drawn signature corpus. Mob. Inf. Syst. **2018**, 3127042 (2018)
16. Mersa, O., Etaati, F., Masoudnia, S., Araabi, B.N.: Learning representations from Persian handwriting for offline signature verification, a deep transfer learning approach. In: Proceedings of the 2019 4th international conference on pattern recognition and image analysis (IPRIA), pp. 268–273, Tehran, Iran, March (2019)
17. Lowe, D.G.: Distinctive image features from scale-invariant keypoints. Int. J. Comput. Vis. **60**(2), 91–110 (2010)
18. Justino, E., Bortolozzi, F., Sabourin, R.: A comparison of SVM and HMM classifiers in the off-line signature verification. Pattern Recogn. Lett. **26**, 1377–1385 (2005)

19. Zhang, B.: Off-line signature verification and identification by pyramid histogram of oriented gradients. Int. J. Intell. Comput. Cybernet. **3**, 611–630 (2010)
20. Ojala, T., Pietikäinen, M., Maenpaa, T.: Multiresolution gray-scale and rotation invariant texture classification with local binary patterns. IEEE Trans. Pattern Anal. Mach. Intell. **24**(7), 971–987 (2002)
21. Vargas, J.F., Ferrer, M.A., Travieso, C.M., Alonso, J.B.: Off-line signature verification based on grey level information using texture features. Pattern Recogn. **44**, 375–385 (2011)
22. Qi, X., Qiao, Y., Li, C.-G., Guo, J.: Multi-scale joint encoding of local binary patterns for texture and material classification. In: British machine vision conference (BMVC), (Sept 2013)
23. Zhang, L., Chu, R., Xiang, S., Liao, S., Li, S.: Face detection based on multiblock LBP representation. In: Lee, S.W., Li, S. (eds.) Advances in biometrics, Ser. Lecture notes in computer science, vol. 4642, pp. 11–18. Springer Berlin Heidelberg, Berlin Heidelberg (2007)
24. Sujatha, B., Kumar, V.V., Harini, P.: A new logical compact LBP co-occurrence matrix for texture analysis. Int. J. Sci. Eng. Res. **3**, 1 (2012)
25. Nair, R., Bhagat, A.: Feature selection method to improve the accuracy of classification algorithm. Int. J. Innov. Technol. Explor. Eng. **8**, 124–127 (2019)

Structured and Sparse Principle Component Analysis for Multi-modal Data Fusion Approach

Hameed Hassan Khalaf[1], Israa Abed Jawad[2], Ausama A. Almulla[3],
Mustafa Asaad Hussein[4], and Preeti Sharma Nair[5]($\boxtimes$)

[1] Department of Medical Laboratories Technology, Al-Manara College for Medical Sciences, Maysan, Iraq
[2] Department of Medical Laboratories Technology, AL-Nisour University College, Baghdad, Iraq
[3] Department of Medical Laboratories Technology, Al-Hadi University College, Baghdad 10011, Iraq
[4] College of Nursing, National University of Science and Technology, Dhi Qar, Iraq
[5] Bansal College of Engineering, Bhopal, India
preetirajitnair@gmail.com

Abstract. The field of image processing has paid a lot more attention to pixel-level picture fusion over the past two decades. Taking an average of each pixel in the source photos is the simplest method for fusing images at the pixel level. In, we see an example of this type of method where two medical images are averaged together to get a single coherent image. While the averaging approach is straightforward to put into practise, it has a number of downsides, including decreased contrast that can result in a significant defeat of data. The authors of this manuscript report deriving a novel fusion model in terms of an unsupervised classification model called principle component analysis, which takes into account the structure and sparse restrictions present in multi-modality images. Eigen value and Eigen vector correlation and covariance analyses make image fusion more adaptable and precise. CT images are ideal for studying rigid objects like bones and implants because of the minimal distortion they introduce, but CT scans are incapable of picking up physiological shifts. Soft tissue can be more clearly seen in an MR image. To measure the effectiveness of the proposed workbench, its performance is simulated in MATLAB for state-of-the-art methods. Key metrics include PSNR (peak signal-to-noise ratio), MSE (Mean square error), association, and precision. All metrics point to the suggested workbench's superior performance, and it also reduces computational load.

Keywords: CT · PCA · MR · SSC

1 Introduction

The development of sources such as wildlife photography, medical imaging, and video surveillance has led to an alarming increase in the rate at which digital photo libraries are expanding. When there is noise present in medical images while they are being

© The Author(s), under exclusive license to Springer Nature Switzerland AG 2025
M. Gupta et al. (Eds.): MISS 2023, CCIS 1952, pp. 440–451, 2025.
https://doi.org/10.1007/978-3-031-69115-7_31

captured or transmitted, the merging outcome is significantly diminished as a result. This is because the clamour signal might be easily mistaken as a beneficial quality of the input.

Because of this, the majority of traditional methods for picture fusion have difficulty dealing with the presence of noise. Therefore, investigating joint fusion and filtering for multimodality medical imaging is something that needs to be done, despite the fact that it is challenging [1]. The concept of medical image fusion has not yet reached its full potential in terms of design and development. This is done so that the deformation can be accurately described.

The pre-processing phase of picture fusion includes an extremely important step called image cataloguing. Image cataloguing, in a nutshell, is the process of superimposing multiple photographs of the same subject that were taken at different times, from different perspectives, and/or with different sensors in such a way that all of the individual pixels in each image relate to the same physical structure [2]. Our discussion of each of these topics will be as follows.

1.1 Identifying Distinctive Characteristics

Important landmarks like enclosed space are the starting point. Source image edges, contours, line intersections, and corner detection these characteristics should stand out, be evenly distributed across the image, and be easy to spot in any given source image.

1.2 Identical Features

The next stage is to establish connections between the features discovered. In this case, the origins of the photographs have been determined. To achieve this goal, we utilise spatial correlations among the features, in addition to a number of feature descriptors and similarity metrics [3]. Methods like random sample consensus (RANSAC) can be used to further perfect the matches between the retrieved feature points after an initial set of correspondences has been established.

1.3 The Art of Model Transformation and Estimation

Once the feature-correspondences have been established, the estimation of the model's parameters is required for the transformation. In this setting, there are three distinct varieties of transformations: similarity transformations, affine transformations, and perspective transformations. The similarity transform is the most elementary paradigm, requiring only translation, rotation, and scale to describe the world. The name "shape-preserving mapping" comes from the fact that angles are kept the same.

1.4 Transforming Inflexibly

Rotations, translations, and scales are all examples of stiff transformations. In order to define a rigid transformation, it is sometimes required that the transformation also maintains the handedness of figures in the Euclidean space, which rules out reflections.

With CBIR, biological information systems can gain access to robust services. CBIR methods have immediate utility in three major areas: education (both K-12 and higher), scientific inquiry, and healthcare [3].

1.5 Digital Libraries

Storage Facilities for Digital Collections Many online archives provide resources for businesses that rely on visual material. One such initiative is the digitization of Taiwan's butterfly collection, housed in the "digital museum of butterflies." There is a component in this electronic archive that is in charge of colour, texture, and pattern-based image retrieval.

1.6 Dissuading Criminal Activity

Among the police force's primary responsibilities is the apprehension of lawbreakers. However, in order to do so, the department of security investigation must quickly and accurately determine the identities of perpetrators. As the number of crimes committed each day rises, so does the amount of photographs of suspected criminals that must be processed by law enforcement [4].

1.7 Online Research

However, the Web is the most essential application because a sizable portion of it is devoted to photos and finding a specific image is a challenging operation. There are now a plethora of CBIR systems on the market, both commercial and experimental, and many web search engines feature CBIR capabilities.

1.8 Infrared Imaging and Spectroscopy

By collecting information from aircraft and satellites equipped with optical sensors, we may determine the characteristics of objects on Earth's surface. This process is known as remote sensing [5]. These systems, especially those mounted on satellites, offer a constant and reliable perspective of Earth, from which researchers can learn important details about both short- and long-term shifts and the effects of human actions.

In order to get an image that contains as much information of the same organ at the same time, medical image fusion is a crucial task that helps to decrease storage space by using only one image. Computed Tomography (CT) images, for example, provide the finest information on dense structures with minimal distortion, such as bones and implants, but they cannot identify physiological changes. Soft tissue is more clearly seen in an MR image. The determination of medical image synthesis is to yield a more comprehensive and precise description of the same object by combining data from many medical picture types [6]. As a result, radiologists are able to report CT and MR scans more efficiently and in less time.

As an increasingly important part of machine vision, medical image fusion has extended a lot of attention in current years. In the realm of medical instruments and measurement, this technology acting a vital part in both judgement and behaviour.

This necessitates either accepting longer execution times or increasing the amount of processing units and storage devices, both of which could result in costly solutions. When human operators are responsible for imaging systems, the sheer volume of images that need to be processed can be too much for any one person to handle, potentially resulting in a dramatic loss in efficiency. To address these issues, it is possible to substitute a composite representation that includes all relevant sensor data for the original collection of sensor information [7]. This multitude of methods is now commonly referred to as picture fusion, and it represents a potential topic of study in the context of image-based applications. Simply said, picture fusion is the act of combining numerous photographs into a single, more accurate representation of the scene than could be achieved by using the source images alone. Therefore, the combined image should be more suited for human review or automated analysis.

The purpose of multimodality medical image fusion and denoising is to transfer essential information from several source pictures to a single fused image without incurring any information loss [8]. This is done so that the deformation can be accurately described.

This research paper is organised as follows: the first part presents the topic that is being investigated, and the second part expands on prior work that was done in the same general area. In the third part, we are going to discuss the many techniques and results that have been proposed as potential solutions. The findings of the study are going to be summed up in the next section.

2 Existing Work Done

Orderly scene progression Filtering is a technique used to categorise unclear or noisy photographs. It makes an effort to mimic key aspects of human visual perception. The method relies on three distinct strategies [9].

Hierarchical classifiers take into account information from both the local and global processing chains. Combining GF (Gabor filtering) and PCA (principal component analysis) allows for the use of pictorial framework. A simultaneous pseudo-restoration technique is performed in conjunction with an affine invariant strategy to enhance the precision with which local prominent features are identified. Then, using a Monte Carlo method, we cluster the combined local prominent features and global essential signature [10]. Results from a large-scale experimental study are used to verify the validity of the suggested system.

An essential tool in many fields, Gabor Filters (GFs) improves and extracts Gabor features from photos. In order to improve the visibility of curvy edifices in otherwise strident images, in this model, the low-quality fingerprint photos' curved ridge and valley features are transformed using curved GFs [11]. To begin, we fuse together two distinct orientation-field estimating strategies to produce a more reliable estimate even for extremely noisy images. Following the appropriate local orientation, curved sections are then built. The local ridge frequency is then calculated using these curved areas as input [12].

Representing the target and understanding its look are two significant difficulties in solving this problem. For accurate target representation, this model uses an updated appearance model that estimates histograms of pixel intensities and local standard deviations for both the foreground and background [13]. Using covariance matching and a visual tracking application, as well as the more modern ALS (auto covariance least-squares) method, a solution is proposed. Although it has been shown that the ALS procedure is assured to converge lone in the scenario of universally BSSP (broad sense stationary process) and capacity clamours, it has often been shown to be quite effective even under the considerably stationary [14].

We use simulated stationary and non-stationary systems to demonstrate the enactment benefits of the ALS technique over the traditional covariance toning. The ALS-grounded approach outpaces covariance toning and classic histogram similarity algorithms when tested on real-world data [15].

Filtering images damaged by mixed Poisson-Gaussian noise falls within a broad category of transform-domain thresholding procedures. The LET description of the filtering process allows for optimization of the solely data-adaptive unbiased estimate of the MSE, calculated outside of the Bayesian framework [16]. This theoretical MSE estimate can be approximated in practise using threshold data fixation, allowing for the manageable optimization of freely chosen transform-domain thresholding. For un-decimated filter bank transforms, we use a point-wise estimator that is comprised [17].

Log Gabor Filter can be substituted for low pass and high pass filters in an improved contourlet transformation that uses multi scale decomposition and takes DFBs into account in order to produce an effective picture fusion [18].

To enhance the quality of the fused image, we employ the Log Gabor filter, which not only localises the image more precisely but also reduces the DC Component (image noise). An input image is broken down into pyramids using this technique [19].

A decomposed image is shown as a pair of pyramids: First, the average pixel values are stored in the Smoothing pyramid. Second, the edge differences between pixels are stored in the Difference pyramid. The input image is represented in this pyramidal shape using a multi-resolution edge representation. Each input image is subjected to a pyramid decomposition, the resulting decompositions are combined into a composite representation, and then the combined image is reassembled using an inverse pyramid transform. Many people look to the "Gabor Filter Bank" when trying to figure out which part of an image to focus on. By applying this method, we can separate a picture into its sine and cosine components [20].

Multi-modal image fusion is one of the most trustworthy and useful diagnostic technologies currently available. It finds application in diagnostic imaging equipment for the medical industry. This research [21] examined a medical picture fusion system based on the contourlet transform and the multi-level fuzzy reasoning technique. The research was conducted with the hope of influencing clinical practise for the better. The method takes two spatially registered medical images and creates a new, unified image by combining the useful information from each. This system allows for the development of more complex algorithms that account for both the time cost and the quality of the fused image [22], as the coefficients of high-frequency details in the contourlet are fused using a pixel-based fuzzy fusion rule and the low-frequency approximations are fused using a feature-based fuzzy fusion rule. By avoiding issues like fusion artefacts and the loss of visually vital information, the suggested fusion method increases the usefulness of the contourlet coefficients. The physical significance of the contourlet coefficients is taken into account to achieve this. Since experimental results show that the studied fusion system outperforms state-of-the-art fusion algorithms, it can be used to fuse medical images from a wide range of sensors, including those used to study the brain [23]. This is evident when comparing the outcomes of the investigated fusion system to the outcomes of the algorithms.

The authors have included a discussion of the literature and background information on image representation models, Image Noise filtering based on the different transformation, and the computation process [24]. Feature extraction and classification in image models have been briefly covered. Noise filtering employing methods like the Gabor filter, the kalman filter, and others helps to lower background noise.

3 The Proposed Work

Recent years have seen the emergence of multi-modal data fusion as an approach to comprehensive imaging analysis. This method often employs principle component analysis (PCA). However, the present fusion methods suffer from issues such as high-dimensionality, multi-collinearity, asymmetrical feature selection, unimodal feature selection, and the loss of spatial information when the imaging data are reshaped into vectors. In this study paper, a novel PCA method known as the Structured and Sparse PCA (SSPCA) technique is proposed as a potential solution to the challenges listed above. In order to evaluate the efficacy of the suggested algorithm, we analysed the performance of three different data fusion methods, namely the decision level fusion technique, the feature level fusion method, and the SSPCA method, and compared and contrasted their capacities to recognise multi-modal data associations.

This all-encompassing method of image analysis, known as multi-modal data fusion, typically employs Principal Component Analysis (PCA). The PCA is a mathematical tool that may be used to transform a large set of correlated values into a set of uncorrelated values known as principle components. Using an image fusion technique, many images are combined into a single coherent whole. This section develops an algorithm for un-stacking photos. With the use of principal component analysis, we may boil down a huge number of variables to a more manageable subset while retaining the majority of the information that was originally present. Principal component analysis is a statistical method that helps us identify and focus on the most important variables. It's considerably simpler to examine and make sense of a smaller dataset.

At the feature level, one may employ a CCA-based fusion strategy for similar effects. Recent research has proposed numerous CCA extensions based analysis, to fuse distinct modalities at the feature level (jICA).

To address these issues, this research suggests a new PCA method called Structured and Sparse PCA (SSPCA). We have analysed the efficiency of the suggested procedure by contrasting the enactment of three data fusion methods: the decision level fusion technique, the feature level fusion method, and the SSPCA method (Fig. 1).

The proposed algorithm is discussed in stepwise below:

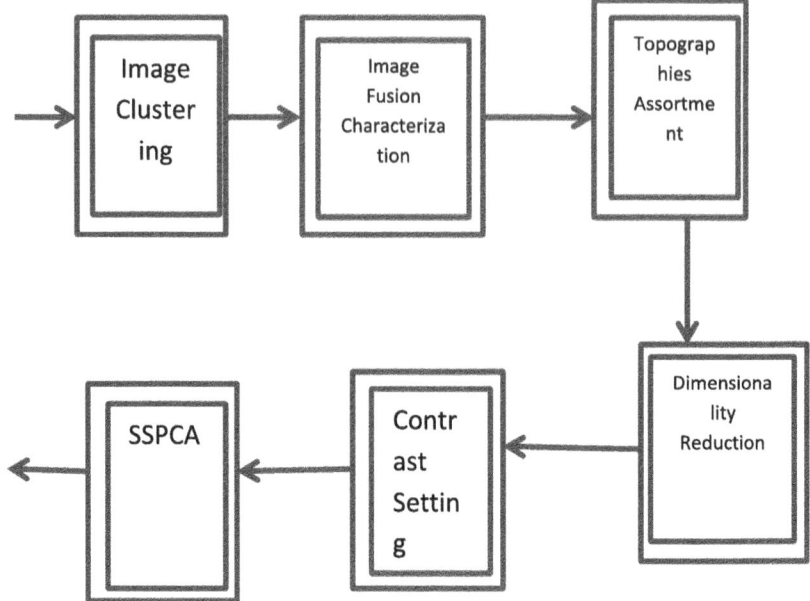

Fig. 1. The Proposed Block Diagram.

3.1 Image Clustering

When talking about an electronic picture, "to segment" it means to break it up into more manageable chunks. Image segmentation is typically used to search through photographs in search of items and borders that share comparable characteristics. When each pixel in the segment topographies corresponds, we will know that we have successfully segmented the image [25].

3.2 Image Fusion Characterization

By altering the imaging procedures and pulse sequence settings, magnetic resonance imaging (MRI) can produce a broad variety of image divergences, each of which has the potential to provide a treasure trove of facts regarding the structure and physiology of the brain. In the vocabulary of fusion, the term "modality" refers to a single contrast present in an image.

3.3 Topographies Assortment

The concept that underpins the feature selection approach is that a picture often contains a great number of characteristics that are either noisy or irrelevant, and that these features can be removed with only a marginal decrease in the image's overall quality. Common applications of feature selection methods include settings with a large number of features but a limited number of samples.

3.4   Dimensionality Reduction

When the input features of an algorithm are too extensive to analyse and there is a good chance that the process will be terminated, the features are converted into a simplified representation.

3.5   SSPCA

It is presumed that the weights of the most significant characteristics will be large, whereas the weights of the less significant and redundant characteristics will be minor or nil. It involves expansions due to the fact that the sign of the obtained correlation vectors is not restricted in any way. Because of this, you won't be able to use the negative voxels from the brain's reconstructed map as a substitute for weight.

For the purpose of fusing multimodal medical pictures, the performance of the SSPCA technique is evaluated and compared to that of the Hybrid technique. The findings are then analysed in terms of PSNR, MSE, and correlation (Figs. 2, 3 and 4 and Table 1).

Table 1. Evaluation Performance Comparison for SSPCA with DCT Algorithm.

	MSE		Correlation		PSNR	
	SSPCA	DCT	SSPCA	DCT	SSPCA	DCT
Image 1	31	34	0.96	0.9	33	31
Image 2	26	23	0.88	0.83	35	33
Image 3	29	25	0.93	0.92	35	32

When the findings of the SSPCA-based multimodal medical picture fusion are compared to those produced through the use of the DCT technique, the third image is the one that has the biggest standard error. The results suggest that the picture fusion-based approaches that our team has presented are superior to the methods that are already in use. The following figure illustrates the PSNR assessment repercussions that result from using the predicted SSPCA as opposed to the usual DCT technique. Using the method that was presented results in a PSNR that is noticeably better.

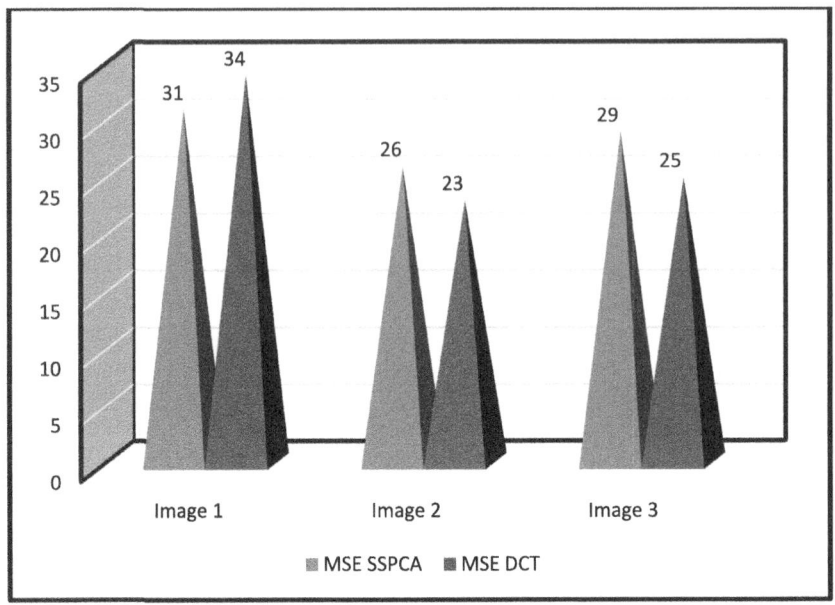

Fig. 2. MSE Comparison of proposed SSPCA with DCT Technique.

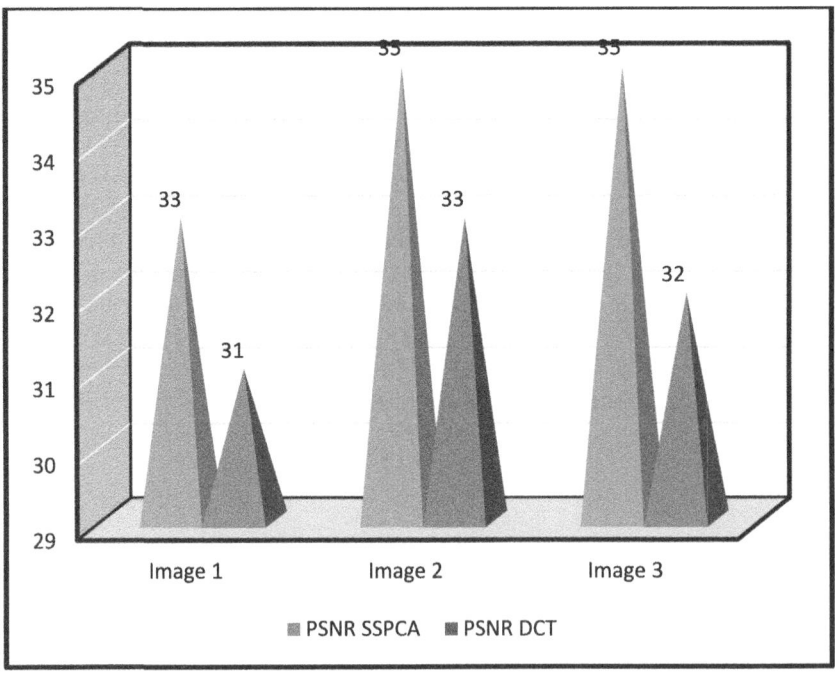

Fig. 3. PSNR Comparison of Proposed SSPCA with DCT Technique.

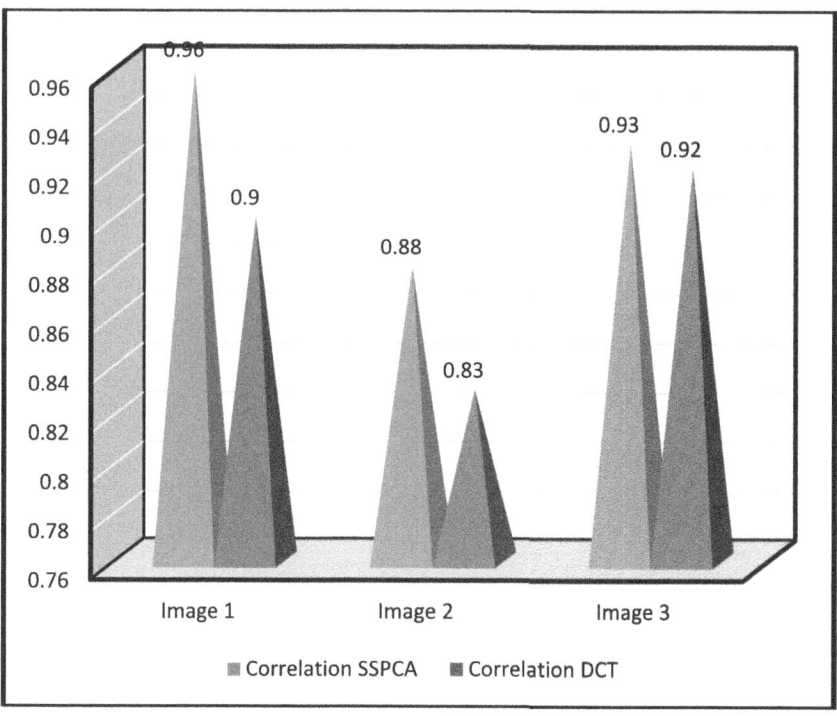

Fig. 4. Correlation Comparison of Proposed SSPCA with DCT Technique.

4 Conclusion

Recent years have seen the emergence of multi-modal data fusion as an approach to comprehensive imaging analysis. This method often employs principle component analysis (PCA). However, the present fusion methods suffer from issues such as high-dimensionality, multi-collinearity, asymmetrical feature selection, unimodal feature selection, and the loss of spatial information when the imaging data are reshaped into vectors. In this research paper, a novel PCA approach known as the Structured and Sparse PCA methodology (SSPCA) is proposed as a possible solution to the difficulties listed above. Over the course of the past two decades, the field of image processing has accorded pixel-level picture fusion a great deal more focus and consideration. The most straightforward approach of fusing images at the pixel level is to calculate the average value of each pixel in the source photographs. Our scalable methodology, on the other hand, makes use of an embedded topographies assortment technique that is both spatially learned and performs feature selection (which is typically performed in a PCA phase) in tandem with a model fitting that improves prediction accuracy. This technique is both spatially learned and performs these tasks spatially. As a consequence of this, the strategy that was suggested can successfully meet circumstances in which there is a small sample size but a big number of variables. Researchers are able to carry out research analysis and studies based on the data they collect if they use this approach.

References

1. Shen, H., Huang, J.Z.: Sparse principal component analysis via regularized low rank matrix approximation. J. Multivar. Anal. **99**(6), 1015–1034 (2008)
2. Zelnik-manor, L., Perona, P.: Self-tuning spectral clustering. In: Advances in Neural Information Processing Systems 17, pp. 1601–1608. MIT Press (2004)
3. Zou, H., Hastie, T.: Regularization and variable selection via the elastic net. J. Roy. Stat. Soc. B **67**, 301–320 (2005)
4. Dasarathy, B.V.: Information fusion in the realm of medical applications - A bibliographic glimpse at its growing appeal. Information Fusion **13**(1), 1–9 (2012)
5. Du, L., et al.: Structured sparse canonical correlation analysis for brain imaging genetics: an improved GraphNet method. Bioinformatics **32**(10), 1544–1551 (2016)
6. Yang, Y., Que, Y., Huang, S., Lin, P.: Multimodal sensor medical image fusion based on type-2 fuzzy logic in NSCT domain'. IEEE Sensors J. **16**(10), 3735–3745 (2016)
7. Bair, E., Paul, D., Tibshirani, R.: Prediction by supervised principal components. J. Am. Stat. Assoc. **101**, 119–137 (2006)
8. Tang, H., Xiao, B., Li, W., Wang, G.: Pixel convolutional neural network for multi-focus image fusion. Inf. Sci. **433**, 125–141 (2018)
9. Wang, K., Zheng, M., Wei, H., Qi, G., Li, Y.: Multi-modality medical image fusion using convolutional neural network and contrast pyramid. Sensors **20**, 2169 (2020)
10. Ma, J., Zhou, Z., Wang, B., Miao, L., Zong, H.: Multi-focus image fusion using boosted random walks-based algorithm with two-scale focus maps. Neurocomputing **335**, 9–20 (2019)
11. Chen, M., et al.: Improved faster R-CNN for fabric defect detection based on Gabor filter with Genetic Algorithm optimization. Comput. Ind. **134**, 103551 (2022)
12. Li, L., Si, Y., Wang, L., Jia, Z., Ma, H.: A novel approach for multi-focus image fusion based on SF-PAPCNN and ISML in NSST domain. Multimed. Tools Appl. **79**, 24303–24328 (2020)
13. Wei, B., Feng, X., Wang, K., Gao, B.: The multi-focus-image-fusion method based on convolutional neural network and sparse representation. Entropy **23**, 827 (2021)
14. Farid, M.S., Mahmood, A., Al-Maadeed, S.A.: Multi-focus image fusion using content adaptive blurring. Inf. Fusion **45**, 96–112 (2018)
15. Wan, T., Zhu, C., Qin, Z.: Multi focus image fusion based on robust principal component analysis. Pattern Recognit. Letter **34**, 1001–1008 (2013)
16. VonLuxburg, U.: A tutorial on spectral clustering. Statist. Comput. **17**(4), 395–416 (2014)
17. Barshan, E., Ghodsi, A., Azimifar, Z., Jahromi, M.Z.: Supervised principal component analysis: Visualization, classification and regression on subspaces and submanifolds. Pattern Recogn. **44**(7), 1357–1371 (2011)
18. Wang, K., et al.: Altered functional connectivity in early Alzheimer's disease: a resting-state fMRI study. Hum. Brain Mapp. **28**(10), 967–978 (2007)
19. Wang, Z., Ziou, D., Armenakis, C., et al.: A comparative analysis of image fusion methods'. IEEE Trans. Geosci. Remote Sens. **43**(6), 1391–1402 (2007)
20. Mehta, S., Marakarkandy, B.: Knowledge and research in electronics and communication engineering. Journal of Information **2**, 848 (2013)
21. Starck, J.L., Murtagh, F., Candes, E.J., Donoho, D.L.: Gray and color image contrast enhancement by the Curvelet transform. IEEE Trans. Image Process. **12**(6), 706–717 (2013)
22. Nair, R., Alhudhaif, A., Koundal, D., Doewes, R.I., Sharma, P.: Deep learning-based COVID-19 detection system using pulmonary CT scans. Turkish J. Electr. Eng. Comp. Sci. **29**(8) (2021)
23. Sarkar, S., Soundararajan, P.: Supervised learning of large perceptual organization: Graph spectral partitioning and learning automata'. IEEE Trans. Pattern Anal. Mach. Intell. **22**(5), 504–525 (2000)

24. Sanz-Arigita, E.J., et al.: Loss of 'Small-World' networks in Alzheimer's Disease: graph analysis of fMRI resting-state functional connectivity. PLoS ONE **5**(11), e13788 (2010)
25. Nair, R., Bhagat, A.: An introduction to clustering algorithms in big data. In: Encyclopedia of Information Science and Technology, 5th Edition, pp. 559–576. IGI Global (2021)

Squared Fault and Biased Entropy for Magnetic Resonance Imaging (MRI) and Computed Tomography (CT) Image Synthesis Using Firefly Algorithm

Nazia Abbas Abidi[1]([✉]), Hameed Hassan Khalaf[2], Ausama A. Almulla[3], Mustafa Asaad Hussein[4], and Israa Abed Jawad[5]

[1] Joint Step Consultants Pvt., Limited, Ranchi, Jharkhand 834010, India
jointstepconsultantspvtltd@gmail.com
[2] Department of Medical Laboratories Technology, Al-Manara College for Medical Sciences, Maysan, Iraq
hameedhassankhalaf42@uomanara.edu.iq
[3] Department of Medical Laboratories Technology, Al-Hadi University College, Baghdad 10011, Iraq
dentistrya@huc.edu.iq
[4] College of Nursing, National University of Science and Technology, Dhi Qar, Iraq
[5] Department of Medical Laboratories Technology, AL-Nisour University College, Baghdad, Iraq
israaabedjawad@nuc.edu.iq

Abstract. Multi-modal In recent years, medical image fusion has emerged as an all-encompassing analytic strategy that frequently makes use of a variety of various alteration methods. This trend began in the field of radiology. By integrating a large number of images obtained using a single or many imaging modalities, the purpose of multi-modal medical image fusion is to improve imaging quality while preserving certain properties of the images. Image processing, computer vision, pattern recognition, machine learning, and artificial intelligence are just a few of the many cutting-edge topics that can be categorised as part of the broader category of medical image fusion. In order to better understand the lesion, medical professionals have been making substantial use of a technique that combines a number of different medical imaging modalities. In this article, we present a framework for fusion that is based on a number of learning models that are known together as the firefly algorithm. The objective of this framework is to get rid of the squared error and optimise through weighted entropy. The performance of the suggested workbench is simulated in MATLAB and compared against approaches that are considered to be state-of-the-art in order to determine how effective it is. The peak signal-to-noise ratio, mean square error, correlation, and precision are some of the most important measures. The performance of the suggested workbench is superior according to all criteria, and it also decreases the amount of computational load.

Keywords: CWT · SSC · CT · EMD

© The Author(s), under exclusive license to Springer Nature Switzerland AG 2025
M. Gupta et al. (Eds.): MISS 2023, CCIS 1952, pp. 452–463, 2025.
https://doi.org/10.1007/978-3-031-69115-7_32

1 Introduction

The process of medical picture fusion is required in order to accomplish the goal of regaining an image that simultaneously provides as much information as possible on the same organ. Having access to a more in-depth explanation of the same object makes image-guided medical diagnosis and treatment a lot less complicated. Computed Tomography (CT) pictures, for instance, provide the most precise depiction of dense structures such as bones and implants and show them with the least degree of distortion; nevertheless, these images are unable to distinguish changes in the patient's physiological state. In an MR image, soft tissue can be viewed more clearly than in other imaging modalities [1]. The objective of medical image fusion is to provide a richer and more specific description of a target object by merging data from several types of medical pictures. This can be accomplished by combining the data from numerous types of medical pictures. It offers a straightforward user interface to facilitate the reporting of CT and MR scans by radiologists in the quickest and most accurate manner possible. Because of the significance it holds in the study of machine vision, the topic of medical image fusion has generated a significant amount of conversation in recent years. This technology plays an important part in the diagnosis as well as the treatment process, particularly when it comes to medical tools and measurement. During the course of the last few decades, there has been a great amount of progression made in the field of imaging sensors. Due to the increased durability and improved resolution of modern image sensors, as well as, most importantly, the reduced costs of production [2], the utilisation of a large number of sensors is commonplace in a variety of imaging applications [2]. As a direct consequence of this ground-breaking achievement, a vast quantity of information depicting the same image as captured by several sensors is now accessible. However, as the number of sensors increases, so does the raw amount of sensor data that needs to be saved and processed. Because of this, the subsequent processing of the information collected from the sensors can be slow and arduous.

This necessitates either settling for longer than necessary execution durations or increasing the amount of processing units and storage devices, both of which can lead to solutions that could be rather costly. Furthermore, when human operators are in charge of imaging systems, the sheer volume of images that need to be processed can be too much for a single observer, leading to a decline in overall performance. One possible solution to these issues is to abandon the original collection of sensor data in favour of a single composite representation that incorporates all pertinent sensor information. This multiplicity of methods is now collectively known as picture fusion, and it is a promising field of research in image-based applications [3]. The term "image fusion" refers to the act of combining disparate photographs into a single, more accurate representation of the scene than any of the original images alone could muster. Therefore, the combined picture ought to be more valuable for human review or subsequent computational analysis.

The development of a reliable algorithm for picture fusion requires the resolution of a number of issues. Choosing an appropriate set of characteristics to represent the photos is the first challenge. Images naturally include a wealth of data and attributes that can be mined for results. Still, we need to pick and choose the most crucial elements that contribute to a favourable fusion outcome in order to extract them. The second issue is the amount of processing power required to extract picture features and manage huge image databases. We require a system with high speed and low computing burden because massive image databases are used for feature extraction and image fusion [4].

Our study aims to present a new picture fusion system with good generalisation performance, high accuracy, and acceptable speed, as earlier designs may have suffered from insufficient generalisation performance and unsatisfactory accuracy [5].

The transfer of relevant information from numerous source images to a single fused image without causing information loss is the goal of multimodality medical image fusion and denoising, hence an efficient variation model is required for this task. In order to describe the precise deformation of a flat picture captured by a pinhole camera whose optical axis is not perpendicular to the scene, projective transformations can transfer a general quadrangle onto a square [6].

The organisation of the research paper is as follows: the manuscript is started with the introduction of the research area then in second section existing work done on the same area is elaborated. Third section is discussing the method which has been proposed to solve problem area and result is discussed in detail. Next section is concluding the findings of the research.

2   Related Work Done

For the most part, transform domain techniques are used by pixel-level image fusion methods to properly blend the input images. Although several transforms of this type have been suggested for use in picture fusion, the majority of transform domain methods rely on multi-scale decompositions. The reason for this is that most photos include a wide range of scales in their depiction of characteristics. Furthermore, it appears that the features of multi-scale transforms are strikingly comparable to those exhibited by the human visual system. In particular, there is substantial evidence that the human visual field is completely filled by neurons that are selective to a narrow range of orientations and spatial frequencies, and that can identify local characteristics like edges and lines [7]. This means that the response of the neuron is extremely similar to the fundamental operations of multi-scale transforms. In order to create a single composite multi-scale representation, multi-scale image fusion works by first applying a multi-scale transform to each source image. By inverting the composite representation, we may obtain the final fused image.

There are several examples of multi-scale picture fusion at the pixel level in the published literature. Here, we provide a brief overview of a few of these methods and then introduce a general multi-scale pixel-level architecture that can accommodate the vast majority of them. Better fusion outcomes can be achieved with redundant transforms as opposed to orthogonal transforms like the DWT [8]. The reason for this is that the over-complete collection of basis functions offered by redundant transformations is superior to orthogonal decompositions at capturing the intrinsic features of images. Orthogonal transformations' sampling typically degrades the quality of the resulting fused image by adding more severe blocking effects. In most cases, redundant transforms increase data bulk and complexity, which may restrict their application in particular circumstances. Keep in mind that redundancy and shift-invariance are frequently seen together.

2.1   Inverse Anisotropy

It is preferable to represent important information in the source images with as few coefficients as feasible because doing so makes the feature selection process more robust and reduces the introduction of distortions in the fused image. Discontinuity points, such as edges, positioned along the curves of a boundary of a physical item are generally used to display this kind of information in natural photographs. However, conventional transforms like the Discrete Wavelet Transform (DWT), the Unscented Wavelet Transform (UWT), and most pyramid-based transforms all have trouble accurately representing such structures since they use a dictionary of roughly isotropic basis components [9]. In other words, the smoothness throughout the curve cannot be "seen" by these transforms, and a large number of coefficients is needed to accurately describe it.

However, anisotropic basis components with an extended shape are used in transforms such as the CVT, ConT, and NSCT, which more accurately describe smooth curves. The result of representing a smooth curve with isotropic and anisotropic basic elements [10].

2.2   The Ability to Choose a Course of Action

Multi-scale transformations also have the advantage of providing 25-way directed decompositions at each scale. When it comes to directions, most pyramid-based transforms fall short, whereas wavelet transforms show three directional detail images that correspond to the horizontal, vertical, and diagonal directions. The CVT, ConT, and NSCT can be used to implement (nearly) any number of directions. Typically, a smaller representation of picture characteristics is the result of more directional selectivity. Commonly, anisotropy is closely linked to directional diversity [11]. This is because these fundamental elements are anisotropic, making multi-directionality feasible. The number of decomposition levels and the filter bank employed are also crucial factors that can affect the final fusion output.

2.3   Method of Choice for Image Joining

In most methods of image fusion, the choice of how to combine the source images is essential, as it determines how the fused, deconstructed image is built. In most cases, a set of fusion weights is generated and saved in a "decision map" as the result of the decision phase. There are several different types of decision-making systems, however they can be categorised as entirely selective, solely mathematical, or composite schemes. The most active coefficient, or the one with the most salience, would make sense to provide a fixed weight of one in selected systems [12]. Maximum selection rules, sometimes known as "choose max" selection rules, are among the simplest weighting methods. The maximum selection strategy, like other selective schemes, has low resilience to noise and random selections, which can lead to a "salt and pepper" pattern in the final decision maps.

One possible solution to the issues with selective and arithmetic decision schemes is to use a composite scheme [13]. To determine whether a strategy is selective or not, these decision-making approaches typically use a match measure. Therefore, the combination process picks the most prominent component at places when the source images are significantly dissimilar, and it takes an average of the source images at places where they are comparably diverse [14]. When the information in the source photos is similar, averaging can be used to minimise noise and offer stability, while selection can be used to preserve the most important details while minimising artefacts caused by contrasting colours [15]. Each coefficient in the previous cases is considered separately, without considering the others. Each coefficient refers to the same pixel in the source image and has a pair of corresponding coefficients in different directional bands and decomposition levels [16].

Although Otsu's approach to image fusion is computationally straightforward, unlike approaches that treat the entire image as a single region, it iteratively seeks for smaller sections of the image to segment. The iterative approach uses Otsu's threshold as a starting point, calculating the medians of the two groups defined by the threshold. The method uses the Otsu's threshold and the two mean values to divide the image into three categories, up from the usual two. The first two groups, foreground and background, are separated out and ignored [17]. The third category, TBD, is handled at the subsequent iteration. On the next iteration, we apply Otsu's approach to the TBD zone, calculating a new threshold and two class means, and we divide the TBD region once more into three classes: foreground, background, and a new TBD region, which is smaller than the previous TBD regions by definition. The newly discovered TBD area then undergoes the same treatment. When the difference in Otsu thresholds between two iterations falls below a certain threshold, the process terminates. The final fusion result is made by combining all the intermediate foreground and background regions. Simulations and experiments with real-world images demonstrated that the new iterative method outperformed the industry standard Otsu's method in a number of difficult scenarios [18]. This includes the detection of weak objects and the exposure of fine structures in complex objects, all at a negligible increase in computational cost.

Previously, Canonical Correlation Analysis was the go-to method for analysing neuro imaging data, but recently, multi-modal data fusion has arisen as a more all-encompassing option (CCA). However, issues with high dimensionality, multi-collinearity, unimodal feature selection, asymmetry, and loss of spatial information in transforming the imaging data into vectors plague the existing CCA-based fusion techniques. To address these issues, a structured and sparse CCA (ssCCA) strategy is being proposed as a CCA approach. As part of our research into the algorithm's efficacy, we have compared and contrasted three data fusion techniques: regularised CCA, ssCCA, and the more traditional canonical correlation analysis (CCA) to see how well they perform at detecting multi-modal data correlations [19, 20]. Non-negativity constraint, feature dimensionality, sample size, and noise power are all investigated as they may have an impact on the performance of these methods. As shown by the data, ssCCA performs better than both the baseline and regularised CCA-based fusion methods.

The authors have included a discussion of the literature and background information on image representation models, Image Noise filtering based on the different transformation, and the computation process [21, 22]. Feature extraction and classification in image models have been briefly covered. Noise filtering employing methods like the Gabor filter, the kalman filter, and others helps to lower background noise [23].

3   The Proposed Work

In this study, we propose a multimodality medical picture fusion system and conduct a quantitative analysis of its performance. Initial input consists of two image sets obtained via magnetic resonance imaging (MRI) and computed tomography (CT). To lessen the computational burden, a fusion formula is employed to combine these two images, and it is based on transparency labelling. We also use the firefly approach, which combines squared error and weighted entropy, to frame the issue as a discrete multi-label optimization. By minimising pixilation artefacts and excessive computational effort, the suggested medical picture fusion ensures almost global solutions. Quantitative matrices, such as the mean square error, are then used to examine the resulting data.

Although CT scans are commonly used to display skeletal structures, they offer little in the way of detail about the tissues surrounding them. In contrast, MRI scans reveal details about the soft tissues but are deficient when it comes to defining bony boundaries. To get around this, it is crucial to collect data from many clinical diagnostic modalities. In this context, fusion is a method for combining medical pictures from different modalities (e.g., CT, MRI).

The problem of image fusion is formulated as a discrete multilevel optimization problem, which the firefly algorithm can handle quickly and effectively. In Section A, we derive a multi-label objective model to address the fusion issue. We present a transparency-labelling formulation that drastically cuts down on computation time.

3.1   Squared Error Algorithm

1. To read the input image from a variety of sources.
2. Create an array of picture cells with a size of $M \times N = n$
3. A comprehensive statistical examination of all picture cells
4. Decide if image fusion was necessary for this particular image cell. If the value of ß is higher than the cut-off then it is now time to merge all the various image modalities. Optimize the stored image by computing the background node in the ith image cell.
5. The initial stage of picture fusion normalization. In this method, characteristics from picture cells I through n are fused into groups of k.
6. Node Computation
7. Image fusion for second iteration

3.2   Weighted Entropy Algorithm

Input: Density-biased sampling yields a collection of n data points, and you provide those along with the number of features you want to use (K)

Output: K modalities are fused.

1) Get the K fusion reactors started.
2) Repetition
3) We recomputed the fusion centre for each node based on the existing fusion memberships and weights, and finally, we assigned each data point to the fusion centre closest to it based on the membership function.

4) There must be no transfer of information to other fusion centres.

The weighted entropy algorithm's membership function is similar to the k-harmonic means algorithms. Introducing weight entropy speeds up the recompilation of the new centroids for the following iteration.

3.3 Firefly Algorithm

1) Randomly produce some fireflies (as explained in the section on how to depict the answer).
2) consider each firefly's potential for success in the population as a whole.
3) To start a new population, step three is to replace the update with the original version and wait for the new population to finish.
4) The algorithmic sum is updated to include the newly created population.
5) Iteration is to check if the condition being tested has been met, and if yes, return the top answer from the current population.
6) do it again, until you've accomplished your goal.
7) Get the best parameter values for the fusion strategy.

Haar wavelet technique is first used to split the input medical pictures like MRI and CT into four sub bands. Energy is compressed, the signal-to-noise ratio (SNR) is increased, fewer features are used, etc. are just a few of the many benefits of transform-based fusion over the coefficients of the transformation stand in for individual pixels in the image. Time-frequency localization and operations at several scales and resolutions are all made possible by wavelets (Fig. 1).

4   Result and Discussion

The following evaluation parameters are used to evaluate the performance of the proposed algorithm.

Accuracy: Accuracy is measured by comparing the number of correct identifications of an interruption to the number of incorrect identifications, or false positives and false negatives, for that same feature. It has been calculated as follows

$$Accuracy = \frac{TN + TP}{Total\ data\ Sample} \times 100 \tag{1}$$

Peak Signal to Noise Ratio: The visual quality suffers if the Peak Signal to Noise Ratio (PSNR) is low. A high peak signal-to-noise ratio (PSNR) is typically indicative of a well-reconstructed image.

$$Accuracy = \frac{TN + TP}{Total\ data\ Sample} \times 100 \tag{2}$$

Mean Square Error: Measure the overall quality of an image with the Mean Square Error (MSE). An MSE value above 0.5 indicates that the image quality is subpar. To define MSE, one may use the following criteria:

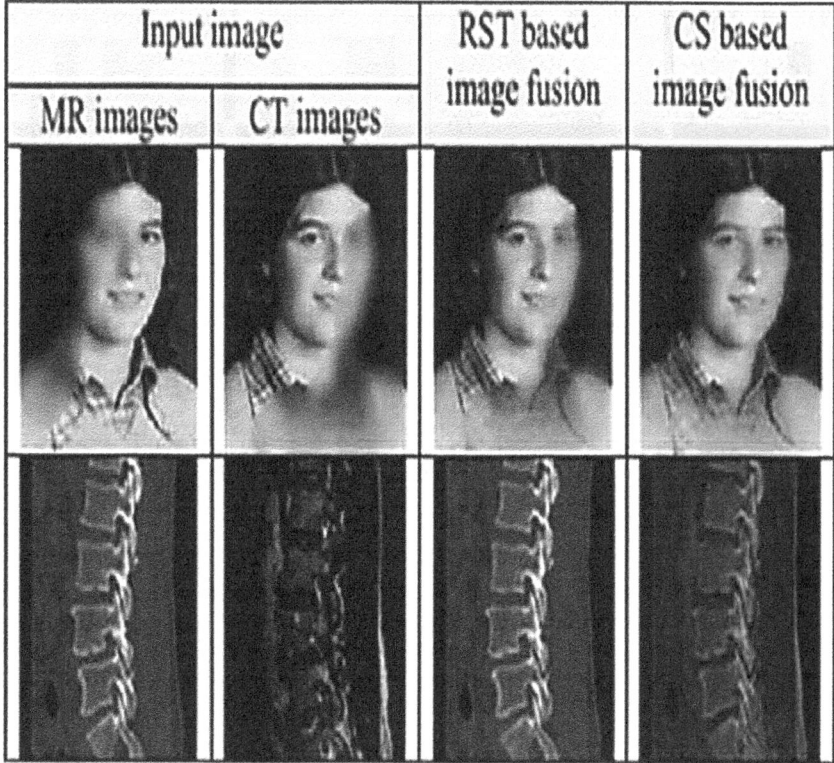

Input image		RST based image fusion	CS based image fusion
MR images	CT images		

Fig. 1. Enactment Consequences of Firefly System.

$$MSE = \frac{1}{XY} \sum_{i=0}^{X-1} \sum_{j=0}^{Y-1} \left[B(i,j) - b(i,j) \right]^2 \tag{3}$$

Correlation Coefficient: The correlation coefficient measures the degree of similarity between two pictures. The following equation is used to make an assumption about the correlation coefficient:

$$Corr\left(\frac{A}{B}\right) = \frac{\sum_{i=1}^{m} \sum_{j=1}^{n} \left(A_{i,j} - \hat{A}\right)\left(B_{i,j} - \hat{B}\right)}{\sqrt{\sum_{i=1}^{m} \sum_{j=1}^{n} \left(A_{i,j} - \hat{A}\right)^2 \sum_{i=1}^{m} \sum_{j=1}^{n} \left(B_{i,j} - \hat{B}\right)^2}} \tag{4}$$

The discrete wavelet transform is used to differentiate the computation of the proposed Firefly method. Several measures, including mean squared error, peak-to-average noise ratio (PSNR), and correlation, were used to compare the effectiveness of the existing and new methods.

A discrete wavelet transform is used to differentiate the computation of the proposed Firefly method. Variables such as mean squared error, peak-to-average noise ratio, and correlation were used to evaluate the efficacy of both the existing and new methods (Figs. 2 and 3 and Table 1).

Table 1. The Performance Evaluation of Algorithm proposed (FA) with existing DWT.

	MSE		Correlation		PSNR	
	FA	DWT	FA	DWT	FA	DWT
Image 1	33	34	0.95	0.9	33	35
Image 2	27	23	0.87	0.82	34	39
Image 3	29	26	0.92	0.93	35	40

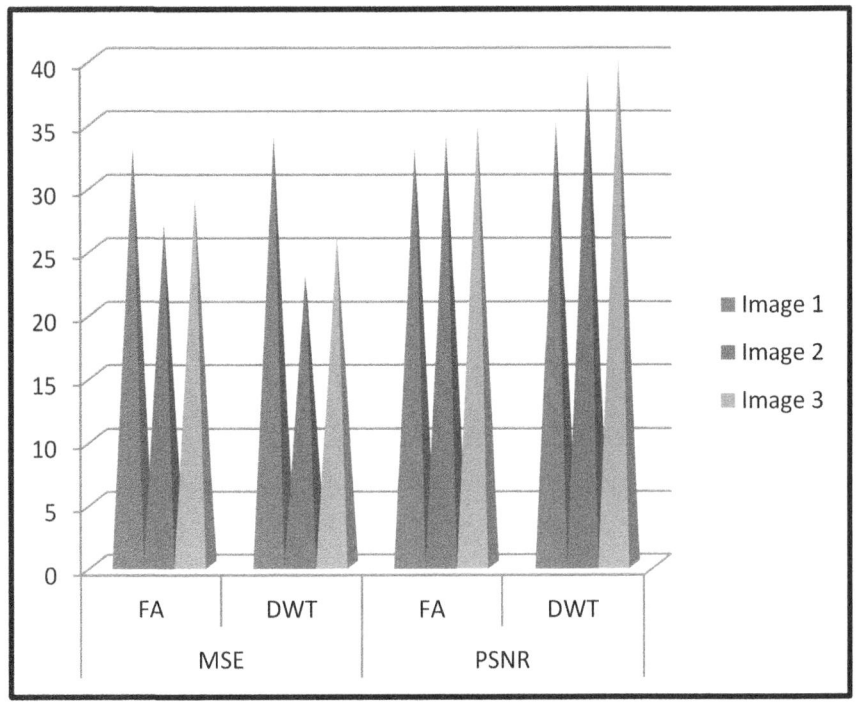

Fig. 2. PSNR and MSE comparison of FA algorithm with DWT algorithm.

The findings of the studies show that the correlation value for multi-modal picture fusion is significantly higher when using the proposed Firefly algorithm as opposed to the more traditional DWT approach. After being optimised using squared error and weighted entropy, the findings reveal that the suggested Firefly algorithm produces a high PSNR, which is indicative of a well reconstructed image. The firefly technique fuses the pixel data from a wide variety of modalities in an effective manner, and then applies DWT optimization to the pixels that are most relevant to the problem. Therefore, it provides the best PSNR value that is achievable for the brain images that have been specified, surpassing the value that the prior art had delivered. A comparison of the relative error, or measure of standard error (MSE), produced by the Firefly and DWT approaches.

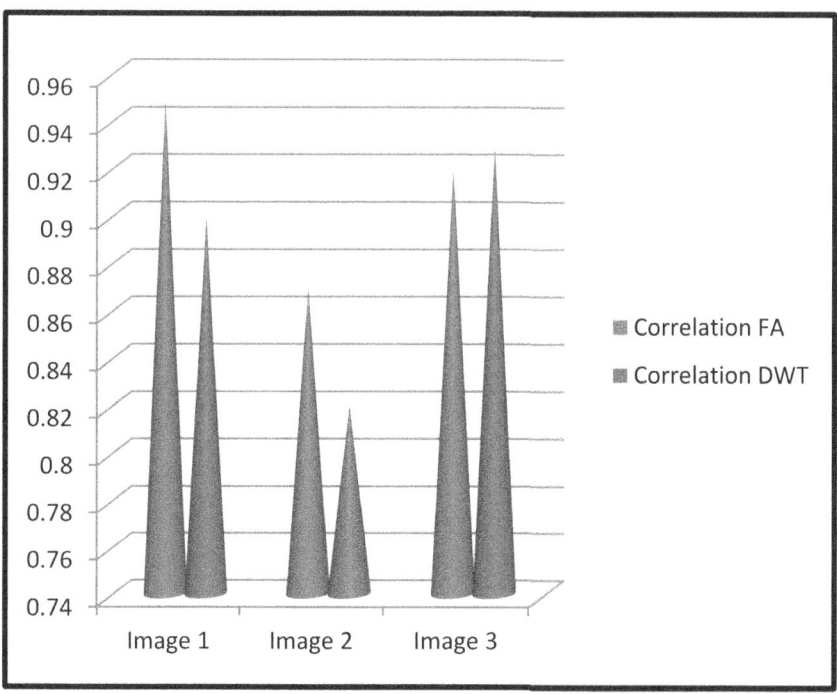

Fig. 3. Correlation Comparison of FA algorithm with DWT algorithm.

Within the framework of the proposed technique, the MSE is diminished. According to the findings, the proposed Firefly approach for combining many medical pictures into one provides superior reconstructions with significantly reduced mean squared errors.

5   Conclusion

The processing of digital images now has an impact on practically every other area of medical diagnosis. The human eye is the ultimate judge of whether or not a particular technique to the processing of images for virtual interpretation was successful. When it comes to fusing MRI and CT medical pictures, we have created and implemented a one-of-a-kind algorithm that is based on fireflies. As a result of our efforts, MR and CT scans have been effectively combined into a single image, providing clinicians with a novel and more all-encompassing diagnostic tool. At first, it was believed that it was possible to combine two distinct images obtained from different imaging modalities, such as magnetic resonance imaging (MRI) and computed tomography (CT). It is possible to combine these two photographs by making use of a fusion formula. This formula was established with the help of transparency labelling in order to cut down on the amount of processing time that was required. In addition to this, we characterised the problem as one that had several labels and distinct aspects. A combination of squared error and weighted entropy is used in the optimization process that the firefly technique employs.

The suggested method for fusing medical images ensures practically global solutions by reducing the number of pixilation artefacts and the amount of processing work required.

References

1. Andersen, K., Andersen, B.B., Pakkenberg, B.: Stereological quantification of the cerebellum in patients with Alzheimer's disease. Neurobiol. Aging **33**(1), 197.e11-197.e20 (2012)
2. Byun, Y., Choi, J., Han, Y.: IEEE Journal of Selected Topics in Applied Earth Observations and Remote Sensing **6**, 2212, 2013)
3. Faktor, A., Irani, M.: Co-segmentation by composition. IEEE International Conference on Computer Vision, pp. 1297–1304 (2013)
4. Guorong, G., Luping, X., Dongzhu, F.: IEEE Journals on Image Processing **7**(633) (2013)
5. Lin, D., Calhoun, V.D., Wang, Y.P.: Correspondence between fMRI and SNP data by group sparse canonical correlation analysis. Med. Image Anal. **18**(6), 891–902 (2014)
6. Basar, S., Waheed, A., Ali, M., Zahid, S., Zareei, M., Biswal, R.R.: An efficient defocus blur segmentation scheme based on Hybrid LTP and PCNN. Sensors **22**, 2724 (2022)
7. Zhang, X., Yang, Y., Li, Z., Ning, X., Qin, Y., Cai, W.: An improved encoder-decoder network based on strip pool method applied to segmentation of farmland vacancy field. Entropy **23**, 435 (2021)
8. Lopac, N., Hrži'c, F., Vuksanovi'c, I.P., Lerga, J.: Detection of non-stationary GW signals in high noise from cohen's class of time-frequency representations using deep learning. IEEE Access **10**, 2408–2428 (2021)
9. Wang, Y., Wang, Y.: Fusion of 3-D medical image gradient domain based on detail-driven and directional structure tensor. J. X-ray Sci. Technol. **28**, 1001–1016 (2020)
10. Luo, X., Zhang, Z., Zhang, C., Wu, X.: Multi-focus image fusion using HOSVD and edge intensity. J. Vis. Commun. Image Represent. **45**, 46–61 (2017)
11. Liu, Y., Wang, L., Cheng, J., Li, C., Chen, X.: Multi-focus image fusion: a survey of the state of the art. Inf. Fusion **64**, 71–91 (2020)
12. Wang, Z., Wang, S., Guo, L.: Novel multi-focus image fusion based on PCNN and random walks. Neural Comput. Appl. **29**, 1101–1114 (2018)
13. Wang, Z., Wang, S., Zhu, Y.: Multi-focus image fusion based on the improved PCNN and guided filter. Neural. Process. Lett. **45**, 75–94 (2017)
14. Zhou, B., et al.: Impaired functional connectivity of the thalamus in Alzheimer's Disease and mild cognitive impairment: a resting-state fMRI study. Curr. Alzheimer Res. **10**(7), 754–766 (2013)
15. Zou, H., Hastie, T.: Regularization and variable selection via the elastic net. J. R. Stat. Soc. Ser. B Statistical Methodol. **67**(2), 301–320 (2005)
16. Yao, H., et al.: Decreased functional connectivity of the amygdala in Alzheimer's disease revealed by resting-state fMRI. Eur. J. Radiol. **82**(9), 1531–1538 (2013)
17. Zhan, G.Q., Guo, B.L.: Fusion of multisensor images based on the Curvelet transform. Journal of Opto-electronics Laser **17**(9), 1123–1127 (2006)
18. Xu, M., Chen, H., Varshney, P.K.: IEEE Transaction on Geoscience and Remote Sensing **49**(5116) (2011)
19. Yang, Y., Han, C., Kang, X., Han, D.: An overview on pixel-Level image fusion in remote sensing. IEEE International Conference on Automation and Logistics, pp. 2339–2344 (2007)
20. Wang, Z., Ziou, D., Armenakis, C., et al.: A comparative analysis of image fusion methods. IEEE Trans. Geosci. Remote Sens. **43**(6), 1391–1402 (2005)
21. Wee, C.Y., et al.: Identification of MCI individuals using structural and functional connectivity networks. Neuroimage **59**(3), 2045–2056 (2012)

22. Wang, K., et al.: Altered functional connectivity in early Alzheimer's disease: A resting-state fMRI study. Hum. Brain Mapp. **28**(10), 967–978 (2007)

23. Von Luxburg, U.: A tutorial on spectral clustering. Statist. Comput. **17**(4), 395–416 (2014)

Author Index

© The Editor(s) (if applicable) and The Author(s), under exclusive license
to Springer Nature Switzerland AG 2025
M. Gupta et al. (Eds.): MISS 2023, CCIS 1952, pp. 465–467, 2025.
https://doi.org/10.1007/978-3-031-69115-7

The manufacturer's authorised representative in the EU is Springer
Nature Customer Service Centre GmbH, Europaplatz 3, 69115 Heidelberg,
Germany. If you have any concerns regarding our products, please
contact ProductSafety@springernature.com

Printed and bound by CPI Group (UK) Ltd, Croydon, CR0 4YY

24/04/2026

02096367-0018